AF717738

Propellers

The First and Final Explanation!

The Professional, Incisive, Understandable, Logic of Propellers.
The Explanation that has so very long, been so badly needed, by Aeronautical Engineers, Propeller Designers, Pilots -- the users.

That Should Have Been Available, at least a Half Century Ago!

The Longest, Most Fundamental Hole in Aerodynamics!

Jack Norris

Jack Norris Publisher

Propellers,
The First and Final Explanation

First Edition, Published July 2007
Northridge, CA
© Jack Norris

All Rights Reserved
No part of this Publication
may be reproduced in any form,
by print, copy machine, microfilm,
scanner, photography, or any other means,
without the written permission of the author.
Price purposely kept low to spread the insight!

ISBN13: 978-0-9620239-2-7

Library of Congress Control Number: 2007904697

Price: $14.95, a Double Book
A $2, $3 Shipping Charge is
encouraged to cover real cost.

propellersexplained.com

11613 Seminole Circle
Northridge, CA 91326
818 360 1105

A personal check, with your
name and address, telephone
number preferred, will serve
as an order. Return shipment
usually 1 day, to one week,
if away, by book rate mail.

Also Available, while supplies last:
Voyager, The World Flight, Jack Norris, Technical Director Mission Control.
ISBN 0-9620239-0-6 $15 + $2, $25 personally autographed by Dick Rutan.

The Book Cover - Key Insights

Several very important core insights are right on the cover, easy to grasp. **Seven** historic figures worked 83 years, from Rankine's first prop Analysis, 1865, to ultimately create **Betz's Classic Prop Logic,** by **1919**, and the **Goldstein-Theodorsen, essentially exact, rare genius math solution by 1948!**

We don't have to learn the math, in fact no one was ever able to see through the wildly complex math, and the heavily loaded, 3 D air flow, until my old college friend, Dr. Andy Bauer, and I took it on as an ultimate intellectual challenge, went through everything with gun and camera, played with it over a decade, until we were finally able to **see the core simplicities** no one ever had before, the **"easy enough"** **horse sense of what is happening physically**! The whole explanation is based on the core insights of **1. A Rotating Wing**, **2. Pulling in and Throwing Back Surprisingly Heavy Air, Newton**, and **3. An Airscrew**, which teaches **high Pitch Props are most efficient**!

See the perfect Helical Screw Airflow on the Cover, The Objective!

Simplicity can be a core of the best genius work, and Betz tells us to simply ***pull in*** **a perfect, Constant Pitch, Constant Slip, pure Helical Air Inflow**, ***throw back*** a **stretched, still pure, Helix, Archimedes Screws, 212 BC!** That **tells us how to set up the Air Inflow**, creates several fortuitous outcomes, one, a **constant ratio of Thrust to Torque, at every radius**, **if Drag Free!**

The Math!

Goldstein created **a rare genius solution** for the theoretical **3D airflow** by **1929**, **Theodore Theodorsen**, an **essentially exact**, real world, **rare genius, heavily loaded solution**, where the **speed increase at the blade** is **much greater than the average** of the **prop created stream tube** -- using **Goldstein**, by **1948!**

The Game!

The **core of the whole ball game** is simply to **counteract the aggressively wrong, excessive tip loading** of a rotating wing, **dumb as a Stump**, by **simply tapering the tips** much more than you usually see, **forcing the half tear drop radial loading** shown on the cover, right next to **the family of ideally shaped, and thus loaded, constant optimum angle of attack, coefficient of lift, C_L props**, simple, ***easy enough*** to grasp! **Low Pitch Props, the worst offenders, are more tapered! Hi Pitch, a Canoe Waterline is the Shape of the Perfect Prop!**

Make a List of All the Insights on this Page --- Start Nailing YOUR Insight!
Make That 10 Minute Effort and Begin to Launch Huge Insight

To my Lifelong Friend,
Teacher of Doctoral Level Aerodynamics,
and loyal, always ready, long time collaborator,
Doctor Andrew B. Bauer

As we both struggled through 83 years of the often Rare Genius Work of Rankine, Froude, Betz, Prandtl, Goldstein, Glauert, Theodorsen, it was Andy who saw that the Historic Masters had given us a simple Kx Blade Loading Chart, that, with Theodorsen's Equations, allowed us to Analyze, Program, Decode Propeller Art, Logic!

With Gus Raspet, number 8, who showed us all that overall Propulsion Efficiency, could be as bad as 58%, Andy well deserves to be number 9, the Professional, who made it possible to finally explain Propellers, 138 years after Rankine's First Prop Analysis, 1865, for the Wright's 100th Anniversary, Dec 17, 2003, when we first had it! It has been the Longest, Most Fundamental Gap in the Science of Flight, and Aerodynamics!

It was the ability to incisively collaborate, of two turned into more than two lifetimes of experience, that made the success finally possible, 138 years.

And to Milly, who Helps me in Everything!

Propellers

The First and Final Explanation

The Professional, Incisive, Understandable, Explanation that has long been very badly needed, Should Have Been Available, 50 Years Ago!

Though all Based on the Rare Genius Level Work of Betz Ideal Logic, Goldstein, Theodorsen Math, Perhaps the most Sophisticated Logic and Math Solution ever accomplished - Never Really Correctly Understood and Implemented in Ideal Form!

It is All Explained in the Understandable Logic of What is Happening Physically to a:

1. Rotating Wing
2. An Airscrew
3. Pulling in and Throwing Back Air, Much Heavier than you Thought

The kind of Insight and Logic we can all understand!

No Math, a few Simple Basics for Easiest Specific Clarity

All The Genius Level Math Works for us -- hidden in a Computer!

A Triple Ideal Prop Results - Actually 6 Ways Better, and QUIETER!

Jack Norris

Table of Contents

This Book is Different: It is not a Novel, not an Engineering Text Book. Its purpose is to teach the Logic, the understandable Physical Horse Sense Logic of how Propellers Work, previously the most complex, unintelligible Mind Swamp in Aerodynamics.

To help Pilots pick out, and tie together the many, many insights, the key logic insights are highlighted, ***superemphasized***, to make what you are trying to learn, **literally Jump off the Page at you!**

I've learned we **Professionals** need that more than I ever would have imagined, core issues, simple Aero 101 Insights in hindsight, that we **never saw in 138 years**. Leave all your hubris behind until you grasp the same logic that Novices need, then with the Superemphasis, the advanced reader can speed read, skim ahead!

This is a Different Kind of Book

It is Purposely Written a very Different Way - To Help You

It's Job is to: Explain How Airplanes, and Propellers Work

It probably has A Lot More Content, Specific Insights, Key Interrelationships -- Than you've ever dealt with before!

We've Superemphasized, all the Key Statements for You, Yellow Highlighted it for You --- Tied Everything Together!

It's apt to seem cumbersome, excessive, at first, but soon you'll see we make all the Insights Jump Off the Page at You

Hopefully, soon you'll realize you don't have to Pick Out the Key Statements, tie them all Together, we did it For You

You're in Training, to get Informed Enough, That You Can Grasp The Whole Logic of Props on Just 2 Pages, in an Hour ✓

Sharp Guys Can Grasp the Logic, Learn to Speed Read it!!!

Novices, who Never Tried Anything Like This can be Helped

Dive In - Go For It - A Lot of Content - but all Sorted for You

Cryptic in Places -- We Break a Few Rules -

If You can Grasp Props in an Hour -

We Broke the Bank in Monte Carlo!

Laugh at it -- We'll try to make it impossible for you to fail - So we can succeed together!

Come Prepared to Laugh, and Learn, a Ton

You can **make this a fun, challenging, *laughable* event! - It can be the most memorable learning experience in your lifetime, a real kick in hindsight!** We're taking the most complex, challenging subject in Aerodynamics, Propellers, never explained in 138 years, and **reducing it to all the understandable logic and *insights* of everything that is happening physically. No Math, each of the insights is logical, understandable, not a real brain strain, by itself. The big WOW, the big surprise, is how many things are happening, all at the same time, a wild and crazy subject!**

We Repeat, Tie it All Together - Give you Plenty of Time!!!

But, that need not be a big sweat for you, simply because we have it all layed out for you. That's what this explanation is. It's all layed out for you, all correctly tied together, correctly showing you how it all ties together, in sequence! You don't have to do that, we did all that for you. All you have to do is to first get aware, then, *at your own speed*, start catching on, first, to all the individual insights, and then, in your own time, how all the insights fit together, in hindsight.

If you're a pro, lose all your hubris right now. In 138 years we never saw, never caught on to all the "easy enough", Aero 101, and logic insights here, me too. It was pretty much an ultimate intellectual challenge to see through all the rare genius level math and complexity, to make it "easy enough" here. Forget all about criticizing how I wrote this. **I didn't write this, prop logic very specifically writes exactly what must be said. It's underlined, superemphasized, bold, yellow highlighted for you, because we pros need that as much as Pilots, Novices. It's what we all missed for 138 years, Aero 101 in hindsight.**

The first several pages, simply start getting you aboard, start *feeding you valuable initial insights*, **getting you practiced on superemphasis**, getting to where it can help you see, grasp the key statements, not seem too busy and unnecessary. *In an hour*, **all the key core insights on propeller logic are on two pages with pictures, the Superemphasis showing the exact words that can explain propeller logic to you. *Get that, You've won!***

Laugh at it -- We'll try to make it impossible for you to fail - So we can succeed together!

Making Propeller Logic Understandable

A **PROP** is *trying to load itself* ***INSIDE OUT*****, DEAD WRONG, DUMB,** ***compared to what a wing needs!*** A **Wing** creates **Max Lift at the Center** and simply falls off, pretty much like ***an Ellipse***, no matter what shape it is, because **Wing Lift** ***Must, Will Fall to*** **ZERO AT the TIP**, a ***TIP VORTEX*** **Swirling from the** ***Higher Pressure Bottom*****, to the** ***Lower Pressure Top***, (even tip Dams only partly effective). But here's our Dumb Prop with very ***high Velocity Tips*****, Huge Dynamic Pressure, q, Tips, (Proportional toV^2) trying to make max. Thrust and Drag AT the Tips, where Thrust will be LOST, must fall to Zero, creating an Extra Big Tip Vortex,** ***thus*** **Extra Induced Loss,** ***extra*** **Unrewarded Profile Drag,** ***at max. radius*****, Max Lever arm, extra unrewarded Torque, bogging down the Engine RPM, Losing HP, Losing (4), four Ways, DUMB AS A STUMP!!! Creating NOISE too!!!** ✓ **It's our job to be smart enough to COUNTERACT that, Taper the Tips, Pull the Excess Loading Back Away From the Tips,** ***EASY KEY INSIGHT!!!*** ✓

HOW ABOUT AN EQUIVALENT *TIP LOADED* WING

Ridiculous!!!

CHORD x C_L x q - (Shapes) Prop Loading!
Correct Thrust vs. Radius SHAPE --
is how we Load it right -- WIN ✓

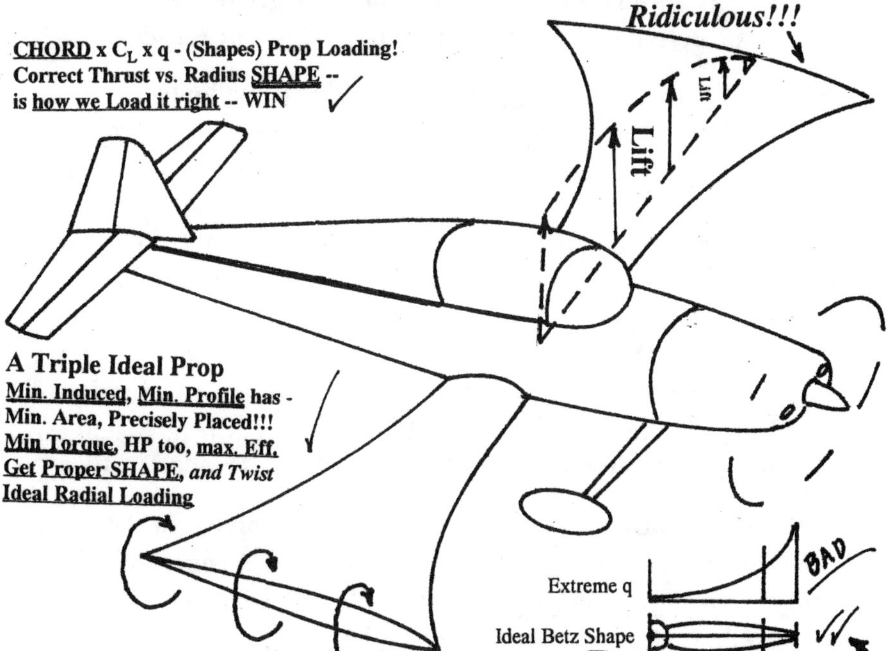

A Triple Ideal Prop
Min. Induced, Min. Profile has -
→ **Min. Area, Precisely Placed!!!**
Min Torque, HP too, max. Eff.
Get Proper SHAPE, ***and Twist***
Ideal Radial Loading ✓

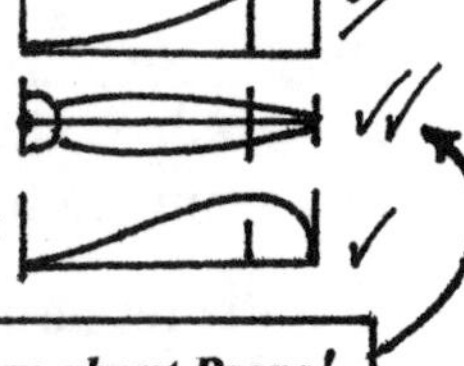

HUGE INSIGHT ON THIS PAGE!!!

> ***This is the most Important and easy Insight You'll Ever Learn about Props!***
> **It has been mostly missed, mostly not understood for the last 100 Years!**
> **You don't win** ***playing games*** **with a wider tip, make Lift, where we just pulled it back!!!**
> **(*Later we may be able to beat* the Betz 20th Century Ideal Loading and Shape!)**

Propellers Explained - for Pros, Novices

50 Years Late!

It took 138 Years, a Third of A Century after my Spacecraft Small Rocket Maneuvering Control Components, went to the Moon and Back, **to get a Correct, Understandable Explanation of Propellers**. Two Key Summary Pages later in this Explanation were written on the Flight Home from the **Wright's 100th Anniversary, December 17, 2003**. All work since then was simply refining, checking, making it better, ever **Easier to Understand**, the subject that was **Always the Impossibly Complex "Black Art", the Biggest Fundamental Hole in Aerodynamics.** I waited a Half Century, for someone else to do it, finally had to dive in with my old College Friend, **Dr. Andy Bauer, an Ultimate Intellectual Challenge, just because it finally needed to be done, a great, challenging, fun, productive Collaboration, two lifetimes of Experience!!**

It's **all explained in the Understandable Horse Sense of what is happening Physically, to a Rotating Wing, An Airscrew, Pulling in** and **Throwing back Air, heavier than you thought!**

Very Purposely No Math! We've Needed to grasp the Logic!

But, make no mistake, everything you learn here, is based on Classic Betz Logic, the **Rare Genius level Goldstein - Theodorsen Math, maybe the Most Elegant Logic and Math ever, in any of the Engineering Sciences, the final product of 83 years work, by 7 of the Brilliant, Historic individuals creating the Aerodynamic, Hydrodynamic, Math, Sciences, Rankine, Froude, Betz, Prandtl, Goldstein, Glauert, Theodorsen, Giants**

Propeller Math was essentially Exactly Solved in 1948 in Theodorsen's Historic Text, "Theory of Propellers", we simply never Understood what it Meant, Directed, Taught! Laughably, they gave us a simple Blade Loading Chart, no one seems to have correctly understood what was there!

We Professionals have needed to understand how to really optomize Propeller Design, finaly get it right, QUIET, less Loss, amazingly Props Trying to Operate Dumb as a Stump!

Some good work has been done Augmenting their's, better blade loading for very high Pitch, Advance. We'll tell you where! Modern work mostly missed, didn't understand!

There is A Substantial Human Problem Here to Solve!!!

Different Level Individuals have quite different **Needs, Desires, Expectations!**

Since this is **the first real explanation of Propeller Logic**, that **took 138 years**, it will be read by Lifetime Professionals, and an even larger number of just good interested individuals, Pilots, in many cases, most cases, with no technical background at all. **There will be very Different Personalities, very Different Expectations**, and of course, all very human, **we will all think that we are the ones who have the proper, correct expectations**, and **see best how the job should be done.**

Now, you may not think that you're starting with an opinion, but the truth is that we are all quite human, more predictable than we'd like to think, **really. We would all like to think of ourselves as the hero in Sinatra's song, "I did it my way"!!!**

As you'll soon learn, we have the core of Prop Logic on just two pages that can finally be very understandable, that took 138 years to find. It turns out that I did Not write those two pages, the Logic of Propellers wrote those two pages. I was simply the guy who found them for you. Those two pages incisively state all that you need to know to grasp the understandable Horse Sense of how Props Work Physically!

I've seen innumerable, **really laughable examples of just how diverse the expectations of we mortals are**, and of course, **we all unfailingly believe in our heart of hearts that we know better than anyone else how this job must be done.**

Props Set what you Must Learn -- I've seen What Can Work - what can't!

The truth is that neither you nor I have much luxury of opinion here! The job that needs to be done here is to get you ready to honestly understand just 2 Pages that are "easy enough" as soon as we can get you up to the point where you are prepared to learn what would have been impossible for you to grasp an hour before!

HELP, by simply Understanding, coming Aboard, Laugh a little, make it easy!

My job is: one part Human, one part Technical, one part Writing, and one Big part Explanation, that can work for EVERYONE! I'll be **teaching, leading, corralling** all we diverse human beings into **being able "to Get" 2 core pages!!**

Loosen up, drop any Hubris! We Pros missed the Aero 101 Core Issue for 138 Years!

If You're Particularly Sharp

I'll bet you'll think I'm being Way Too Repetitive!

Perhaps, but as I'm giving the slower guy lots of time and extra chances to get it, I'm often saying it a little differently, many times expanded insight, very purposely hoping that the sharper guys are getting more, and more facile in their grasp, ultimately being able to flip the logic around in their head, the place I'd love to get everybody. The sharper guys, more curious, more capable will go on to Advanced Chapters, and be that much more prepared to grasp all the Advanced Insight, end up with More Factual, Practical Grasp -- that has Never Been Available Before!

You Must Realize, I am NOT writing this Just for You, but for EVERYONE, from Novice to Pro! You'll see, I expect everyone to soon enough learn to read this at their Level --- Pros, Sharp Guys Learning to Skim, Cherry Pick, After you Grasp the Logic --- Your Contribution, Your Help and Participation to Help Me make this Work for All Levels.

The First part of this Introduction goes Slower to Help Novices Aboard! The Second part Skims ahead to a lot of great Advanced Insight, Great Stuff! Novices will get it too.

I've Purposely Put a great deal of Insight into this Special Introduction. Many Pilots Will have more insight than they ever expected in this Introduction, and be satisfied!

At the End here, I show how to Cherry Pick the rest of the Book. We're trying to set this up for anyone, everyone!

Your Contribution - Your Part of the Job - Learn how to read this book at your Level - Your Speed!!!

READ THIS!!!

How We're Going to Explain Propeller Logic To You!

Propellers are without any question the most ridiculously complex problem in Flight. Since we Aerodynamic Pros could never explain propellers -- it took 138 years, and over a Decade to finally get this one - It's going to take real insight to figure out how to correctly, understandably explain them to Pilots with no technical background! **If we tried to do it the normal way, there are so many things that we'd have to teach you, the story would go on and on, you'd get totally confused, lost, give up, and we both would have failed.** There would not be the chance of a snowball in hell that we could succeed that way.

Over about an hour, we're going to get you aboard, feed you **a lot of basic insight**, hopefully get you **a lot more informed**, far better prepared, ready to learn. **If you soak up what we're feeding you, about 20 pages in, you're going to be a lot more prepared to understand, learn, grasp what's happening!!!**

At that point, on **2 pages**, **two single pages**, I'm going to show you **the bottom line insight, on the understandable core of Prop Logic, How they Work, How we can Understand Props!**

Now, there are a lot of things happening, interacting, all at the same time, all interrelated. So how do we avoid confusing you there, solve that one? That could kill things!

We're going to start Superemphasizing the key points, make them Jump Off The Page at You!!! It will look Wild, and we'll encourage you to laugh at it!!! My hope is that with an hour's insight and practice, it will be "easy enough" for you to separate --- or integrate all the things happening!

Actually, we have pictures, little sketches in the margin, and it's easy to see the problem of a rotating wing, and the solution -- even before you read the text. The Pictures - Easy!

My bet: If you can fly an Airplane safely, you'll be able to noodle out those 2 pages, and we'll win! I'll help, Go for it!

You'll learn a Lot Here!!!

How Much Does a 100 foot Cube of Air Weigh at Sea Level?

No tricks -- Not Floating in the other Air, Separated, on a Scale, Truth!

76,000# Wow!!!

About the Weight of an 80,000# Highway Truck!!!

A 1,250,000# Airbus A 380 is Throwing Down Enough Air Fast Enough,

To React 1,250,000# - That's What Induced Loss is, The Core Induced!

The Tip Vortex Induced Energy Loss is Extra. Maybe + 25% Extra.
(At the Speeds we Fly all Induced may be 20, 25% of Total Loss)

NOT a Wing, Props Have Much More Induced Loss, lots More!!!

Nominally Twice the Loss of a Prop's Profile Drag, 200%, 2/3, Not ~25%!

Most Props Have Ridiculous Tip Loss - Noise!! We Never Understood!!

Props, Dumb as a Stump, Work Aggressively To Create Tip Vortex Loss.

And we Never "Got It". We Never Saw The Core Problem There!
(That's the Most Important Insight you'll Ever Learn About Props)

How Far is a Reno Unlimited Advancing Each Revolution? 37 1/2 ft!!!

How Much Air is That Unlimited Throwing Back? 4 1/4 Tons/Sec.!!!

Not Faked Out ---You Can Have Fun With All The Insight Here!!!

Planes, Props, We'll Deal With a Lot of Induced Thrown Air Here!!!

And We'll Get the Wild Induced Loss, Down to a Dull Roar - Quieter!

$(100\text{ ft.})^3 = 100 \times 100 \times 100 = 1{,}000{,}000\text{ ft}^3 \times .076\ \#/\text{ft}^3 = 76{,}000\#$ Wow!

Yes, a Lot More Air, a Lot More Weight, More Induced Energy Loss,
Than You Ever Realized!!!

Where Props Try to be Dumb As A Stump---Wasteful Tip Vortex Loss!
It's Our Job to be Smart Enough to Counteract The Tip loss!
We Just Taper the Blade, *Twist*, Pitch the Blade Correctly - EASY!
That's Been the Hidden Secret, a Big Reason We Needed an Explanation
For 138 Years!!! And it's All Here -- Understandable!
(It's All About How we Load the Blade - Blade SHAPE, with the correct *TWIST!!!*)

Laugh At It! We Did it Very Differently - On Purpose!

You'll Be Amazed At What You've Conquered, Learned - Proud, A Good Laugh!

The Game Here

To Make It Possible

For You To "Easily Enough" Grasp

The Most Complex Subject In Aerodynamics - Defused!!!

The Subject That Took 138 Years to Explain!!

That We Professionals Never Grasped!!!

If There Is Any Genius Here

It is in Converting The Ultimate Intellectual Swamp

Into The "Easy Enough Horse Sense"

Of What is Happening Physically

to

A Rotating Wing, An Airscrew, Pulling In, Throwing Back Air

It's Not For Speed Reading --- Grasp the Insights

The Game - Make It Possible For You to *Easily Enough See*

What You Could Not Possibly Normally See

The Core Insight is on Just 2 Marvelously Insightful Pages

My Bet is You'll End Up Smiling - Soon Laughing - Proud of Yourself!

We Try to Make the Key Insights - Superemphasized-

JUMP OFF THE PAGE At YOU!!!

Dive In, Laugh At It, You Can't Possibly Strain Your Brain

We Just Taper The Blades, Twist Them Right, That Easy!!!

You Can End Up Laughing, Plenty Proud Of What You Nailed!!!

You can Teach Your Friends The Easy Core - Laugh at it Together!

We'll Learn - Props - Fast Tips, Agressive Tip Loading - Really Dumb - Never Grasped!

The Impossible Subject, Propellers, Made Understandable!

How can we possibly take **the Black Art** subject that we pros could never understand and explain, now 141 years, and make Props understandable to Pilots, with no technical background. Well, we're certainly not going to do it with the rare genius level math that some of the most famous men in Science, Mathematics, Hydrodynamics, Aerodynamics took 83 years to create, **1865-1948**. The reason we've never had a propeller explanation: **no one could ever see through that rare genius level math, see what it Meant, Directed, Tried to Teach Us!**

The EASY Insight to Grasp: A Prop is a Rotating Wing, the British correctly call them **Airscrews.** Anyone standing behind a propeller, running up, or a helicopter landing, or lifting, realizes they're **Throwing Air,** and ***Newton*** teaches us his Equal and Opposite Reaction Force, and the ***simple, insightful, but a bit tricky way to understand what's happening there***.

Everything we teach you here is based on maybe the most **Elegant, Logic and Math** ever created in any branch of Engineering, over those often rare genius 83 years, but the game is to teach you that the **Understandable Horse Sense Logic** of **those three insights** into **what is happening Physically -- can exactly, correctly explain what the CORE of the genius level math is doing - never really nailed before!**

We've targeted this to help all the good, curious, most interested, most interesting guys who'll wade in and give it a good honest try - to try to make sure everyone can "Get It"

We had the guts to do this very Differently, Make it as Easy as Possible FOR YOU!

Since new guys can't pick out the gems of Insight from Bland Text, we **Superemphasize the key insights** to, as much as possible, make them **Jump Off The Page At You**. Since **most of us**, many pros included, **don't really catch on the first time** someone tries to teach us a lot of new insights, **we repeat, always tie together, as we broaden, deepen your insight.**

We **get you ready**, show you the **Bottom Line Conclusions in less than an Hour**, **FULLY AWARE early**, **much easier** that way. As we **repeat, expand your insight, tie it all together**, it **can all soak in**, and **you can grasp what we pros never did!!!**

We Pros Need to Grasp The Same Basic, Horse Sense Logic, Missed for 138 Years!

It can Look a little crazy, Laugh at it, make a Fun Challenge of it, Soon Smart!!!

IF YOU'RE A PRO - Don't Underestimate What's Here!!!

As a Pro, you'd recognize the names Rankine, Froude, Betz, Prandtl, Goldstein, Glauert, Theodore Theodorsen, T.T., NACA's marvelous Wartime Chief Physicist, Historic, often Rare Genius Level Names in our Science, Mathematics, Fluid Mechanics, Aerodynamics. Over 83 Years, from 1865 to 1948, they were the Historic, often Rare Genius Level Pros who created Classic Betz Prop Logic, essentially exact Goldstein - Theodorsen Math for us, brilliant, achieved an "essentially exact" Solution for Propeller Mathematics for us over a Half Century Ago, by TT's Historic 1948 Text!

Any fault was ours, we just never understood, what it all Meant, Implied, Directed, snowed, typically unable to Understand the Genius of their competence, we failed to grasp that the bottom line on their work was a simple Blade Loading Chart vs. Radius - for low to high Advance Ratios!

Prop Math, was essentially exactly solved over a Half Century ago!!! If you think that because you're smart, can do Math, have a Computer, know how to write code, you can start off and do better, smarter than those Historic men, likely very wrong! They were Brilliant! The trick is to finally Grasp what they created, far too long waiting for us!

What we teach is correct, based on maybe the Best Math Work ever!

Thinking it ridiculous, unacceptable, that we're into the Second Century of Flight, that my Spacecraft Controls went to the Moon, and back, a third of a Century ago, my old College friend, **Dr. Andy Bauer, and I decided it was time to Nail Props, explain them, not only to Pros, but Pilots, anyone interested!**

We went way past the Math to Find the *Easy Enough* Core Insights, Logic!

***The hardest thing of all is to make the most complex, daunting subjects "easy enough", understandable, but that's what Science really is, isn't it?* Nature, Science is always orderly, logical! Once someone solves it, sees through the complexity, we can all Understand, have Nature work for us!**

First - We pros Need to Grasp the Same Incisive Insight that Novices Need!

Over a decade, we went through all the Math, did all the Code, the Professional Studies, and as always, it all comes out the Understandable Aero 101 level Horse Sense, the understandable Logic of what is Happening Physically!

Drop the Hubris guys - You'll see we missed the Aero 101 Core Issue for 138 years!
Theodorsen is in the Appendices, the Math in a 19 Page Program!

The Task Here:

Take the Most Complex Problem in Aerodynamics

Propellers

Never Correctly Explained in 138 Years by Anyone

Find The Understandable Aero 101 Logic

Make it Understandable to Pilots, Professionals

Anyone Who'll Give it a Good Honest Try!!!

We Lead You In, Show You Some Basics

One Hour In, Getting You Aboard, Prepared

2 Very Special, Very Concise, Incisive Pages
Can Teach You The Whole Core of Prop Logic!!!

They Replace The Book That it Would Take -
To Teach Pilots a Whole Basic Aero Course!

By Essentially Starting With the Conclusions
It Becomes Much Easier, Faster, To Grasp, Learn!
You Can See the Logic, The Patterns Very Early!

The Normal Way, too Long, You'd get Confused, Lost, Drown, Give Up!

The Two Pages Look Pretty Wild at First, But
Since you Can See The Logic, *Bottom Line Answers*
It Can Quickly Start Soaking In, *gets Easy Enough!*

Laugh at it, Go For It, Make it a Fun Challenge, Smart Quick!

Pros, Propeller Designers, have badly needed this for a Half Century!!!

Designed Correctly, Props Get QUIET!!!

Anytime We Learn Something New ---

It **Tends to Seem Hard at First**, but then **Easy in Hindsight**, **once we catch on to the Logic, the Horse Sense of it**! That's what we'll find here, **Props** seemingly very, very complex at first, but **"Simple Enough, Easy Enough" in Hindsight!** The **Trick is to get you over the Hump, where it all makes sense -- show you that there are Aero 101 insights that nail Props**!

Pilots, Technical Novices, it would be dumb to expect you to pick technical gems out of bland text, especially when there's **a lot going on**, seemingly so complex, that professionals never found the understandable logic of Props, never an explanation in 138 years, **the biggest fundamental gap in Aerodynamics!**

Since there is **a lot going on, we need to be Creative, work out a way that we can sort out all the complexity for you, finally the bottom line decoding of the 138 year mystery!!!**

It will help a bunch if you're flexible, a little creative yourself, not an uptight, rigid person, someone who can laugh at something that looks very unusual at first, **that solves that core problem. The Task Here is to sort out the most complex problem in Aerodynamics, make it "easy enough" for you!**

The core math is at the rare genius level, we better not try that. There are a whole bunch of things happening, all at once, and interacting, so **the whole game is "can we be creative enough to show you that there are easy enough core simplicities", able to show -- but separate all the side issues -- or easily integrate them -- since they are useful, insightful, part of the final proper, correct grasp! Succeeding there is the key!!!**

We're going to use **Superemphasis, pretty much a hit you over the head with a 2 x 4 writing style**! We'll start off, feed you some key insights, give you time to grasp basics, get comfortable, building up. **About One Hour in we have the whole core logic of props on 2 Pages, wildly dense, but** ***simple enough*** **as it soaks in**, just **the Horse sense of what is Happening Physically. Laugh at it, go along with the gag, and see what we pros never did in 138 years! Wildly, We pros never saw the core issue!! -- and it's Aero 101 Simple!!**

OK, Here's Where We Start, and Where We're Going

Classic Prop Logic is based on simple, but Brilliant, Betz Insight and Logic - and Rare Genius Goldstein, Theodorsen Math. Impossibly complex, Math at the outer limits of what people could do, No one was ever able to see through it, what it Meant, Directed -WHY! That's where the Insight Hides! With a lot of very hard work, we have it down to the Aero 101 Horse Sense of What is Happening Physically, to a Rotating Wing, an Airscrew, Pulling in and Throwing Back *(Heavy)* Air to Make Thrust, (Newton's Equal and Opposite Reaction Force).

Most Important is to see What it - Directs - and WHY!

Easy Enough, there's a whole bunch of extra supporting things happening at the same time, with really important, and valuable insights. Great, but it becomes a problem of more things happening at the same time than Pilots, Technical Novices, have ever been faced with -- in **one big Gulp!**

So Here's Where We're Going! The only way that Pilots, or Engineers have a prayer to grasp it Quickly, easily enough, is to **put the Final, Marvelously Incisive, Bottom Line Insights of 138 years Work on 2 Wild, Superemphasized Pages**, that are **actually quite easy once it all soaks in, starts making sense**, but looks like a crazy head full the first time you see it!

But here's the clue. Rereading it a few times, vastly easier and faster than reading a whole Chapter, or a Whole Book for Novices, you can start seeing the core issues - separate from the supporting insights. Once you get to that point you can **see that you can read the core issues, with the *also Insights*, either skipped, separated, or then, fullly integrated, HUGE INSIGHT! In 10 Minutes you might grasp the really simple enough Insight that very smart Engineers never saw!**

If you don't Grasp it all, at first, You're Instantly Aware of the Final insight!

Don't fight the **superemphasis** that **separates *or* integrates the Total Insight**. Realize **the Logic, not me, writes the page. The whole trick is to be able to see, separate, then integrate the whole concise core**, the **final Insights of 138 years work, quick, easy. We give you plenty of time, help to get it! Long, normal writing, *the alternate*, way too long, no chance, you'd drown!**

Norris's First Law

Nature, Science, is Orderly, Logical, thus understandable, explainable, and can potentially be made useful for the good of man, part of our ever expanding grasp of knowledge!

In fact that's what Science really is: once someone is able to see through the Fog, the Complexity, gets an orderly, logical, proper explanation, anyone who tries can usually grasp it!

Norris's Second Law

However, if you think it's completely simple, you'd usually be very wrong, you'd fail to see, understand everything else that is happening in the background, the whole universe, and all its laws, a marvelously interrelated, web of Insight!

THERE'S MORE TO LEARN AFTER THE BASICS

Galileo, 1564 - 1642, figured out how fast dropped objects accelerate, but we're still trying to figure out what gravity really is, trying to master the logic of the universe!!! Watson and Crick defined the Double Helix of DNA, in 1962, so now we're deeply into working on decoding it.

Look on the Cover

Betz defined the Double Helix of Classic Prop Logic in 1919, but Theodorsen's essentially exact, genius level math took 29 years, 1948, 83 years after Rankine's 1865 start of prop analysis, but we never saw what it was trying to teach us!

A Double Helix? See there's one from Each Blade, 180° apart! ✓

So here, 2003 to 2006, 55 years after Theodorsen, you'll find the first real explanation of what it all really means, how to understand it all, use it fully intelligently. But, this is about a Half Century Late - Always before, too Complex to Solve!

Hard Work, a lot of Determination is always at the core of breakthroughs. So we finally have an explanation. It seems no one else was ever going to nail it, a half Century is long enough to wait! So the trick is to actually make it truly *easy enough*, not snow everyone -- truly easy enough to grasp.

You'll see we have the core down to Three Simple Insights, of What's Happening Physically, easy enough horse sense. We go on, help you grasp Props to any degree that you wish!
If you have trouble at first, we give a lot of **Time**, ***Repeat***, **Help** ✓

This Book is Written to cover THREE Needs

Since there has never been a penetrating, correct explanation of Propellers, this book will be read by **Pilots**, a major audience, with essentially no background, or experience, sorting out tough technical subjects, and **World Class Technical Professionals**, with a lifetime of technical insight, who, like me, have been waiting for the intelligent, incisive, correct explanation of Propellers, (for an extra Half Century). So, It is basically an impossible writing challenge to write in preferred style for such a diverse audience, and we both need to see that right up front!

Some People Like, Want, ***incisive summaries!*** Others like, need **a more conversational narrative, repeating, tieing together, to help them along as they figure it all out, absorb it all**, especially if it's a many faceted subject. Complete technical novices, who might wade in with no experience, or technical grasp, can be **helped with a super emphasized, hit you over the head with a 2 x 4 style** --- (We technical pros can benefit from that too, because there were very specific **Aero 101 core insights that we never saw in 138 years, the most basic gaps in the Aero Profession!!!**)

It's FAIR, Proper, to ask Pros, easier for them, to ADAPT!

It's all here, and all you have to do is realize it going in, **learn how to read it at your level**, but that's no detail, **that's your part of the job, your contribution** to make this book reach everyone that it needs to. Full busy, having to fit this book into an already busy lifetime, I simply don't have the time to write two books. Realize this is **not a Pablum Book, it nails Props!**

Pros can learn to Skim, Speed Read, BUT only AFTER they grasp the Core Basics.

With two lifetimes of experience applied, my old college friend, Doctor Andy Bauer, and I, have labored through all the Rare Genius Math that took 7 Historic technical figures 83 years to create, done all the lengthy studies to find out who had it right, wrong, or close but no cigar, and have it down to the **understandable** ***Technical Horse Sense*** of what is happening **Physically**, to a **1. Rotating Wing, 2. An Airscrew, 3. Pulling in and Throwing Back Air, Newton's Equal and Opposite Reaction. The Introduction first goes slow to help Novices! In the second half, nails a ton of Advanced Insight!!! We go on with a Primer, a Chapter 9 in Book I, then a whole Book II, a Summary, Total Insight!!!**

There's a ONE PAGE, 8 1/2" x 11" SUMMARY at the end. Go Peek!!! But, Don't Short Yourself, don't skip the Basics, or the Advanced Insights. A Day to Grasp it!

A Couple of Great Big Problems!

It's ridiculous really, unacceptable. We're into the Second Century of Flight, a third of a Century since we went to the Moon, and we never got a correct, understandable explanation of Propellers, their Logic, how we can correctly understand them, and their operation, proper design, the hidden horse sense, always the Black Art Mystery of Aerodynamics. I know, I've been looking for one, waiting for someone to do it, explain them, since I was a kid, building, designing competition models.

Worse, a propeller, a rotating wing, always has nominally, 2, 3 times as much efficiency loss as a comparable Aspect Ratio wing, even when designed by classic theory, and math, 4, 5 times more loss if designed unknowingly, and no one ever seems to questions why! In a giga byte world, how can that be?

All that extra loss goes into NOISE, Correct Props Noticeably Quieter!!!

Astoundingly, no one seems to have ever written, or said what the core problem is, when in hindsight it's really simple Aero 101, all of us blinded, faked out by the apparent Black Art complexity, when props are trying to operate aggressively wrong.

All that is even more amazing when we learn that 7 historic figures created classic propeller logic by 1919, exact theoretical math, but not real world math by 1929, a good approximate solution by 1934, and Theodore Theodorsen's essentially exact (~ 1%), real world solution by 1948 -- with the 1929, and 1948 work having a simple chart that showed us how to load the blades correctly vs. radius, for high to low Pitch, or Advance Ratio, as you'll learn. Nobody ever really caught on, nailed it.

It's time we Understand, get props Correctly Understood, Correctly Designed!

No Hubris here, my old college pal turned into Dr. Andy Bauer, one of Douglas's key, core Aero Pros, and collaborating, adding in my very full lifetime of experience, and a whole lot of honest hard work, we nailed it - the old fashioned way - we earned it! So it's our gift to you, and the ***Flight*** that gave us an E Ticket to life. We even have it down to the horse sense logic of what's happening physically. But you have to let me teach you in a different way, because done the wrong, long, complex, normal way you'd drown. Come prepared to laugh at it, learn incisively!

Aero 101 on One Page, with a Picture Page!!!

Pilots: We all **know how airplanes work**, but we probably ***don't specifically* understand** as well as we might. We know **Wing Area makes Lift**. At a **higher Angle of Attack**, -- (***a higher Coefficient of Lift***), we **can takeoff slower**. The **faster** we go, higher IAS, the **less angle of attack it takes**. You may have heard of **q, Dynamic Pressure**. It's simply the **Pressure our Pitot Tube feels**, **stopping the air** rammed into the tube, the same pressure we'd feel **sticking our hand out the window** - **EASY!**

Realize q is the pressure, energy motivator of Lift and Drag!

Simply, Lift, $\mathbf{L = Area \cdot C_L \cdot (\rho V^2/2)}$ $q = (\rho V^2/2)$

q, the Ram dynamic pressure, is just the mass density of the air, ρ, times V^2 divided by 2, so if we go **twice as fast** - the **q** pressure making lift, and drag **goes up by a factor of 4**, (**3x is 9x higher**, 4, 16) **Basic**.

Grasp that formula, it's the ~ same one for Drag, C_D, q, V^2, *Not Hard, Basic*!

We all learn it's the speedup of air over an Airfoil, dropping the pressure, that makes Lift, but we'll learn that **Airfoils *also* turn** and **throw down air**, causing **a reaction force, throwing air**, another way of looking at the **same lift**, ***not*** a duplicate lift. We'll learn that's what causes **Induced Loss**, the ***Energy Cost, the Loss of Throwing Air*** to make **Lift**, or **Thrust**, hugely important on **props, the biggest loss on Props**. **Profile** Drag, air **Friction drag**, the basic drag, may be only half induced.

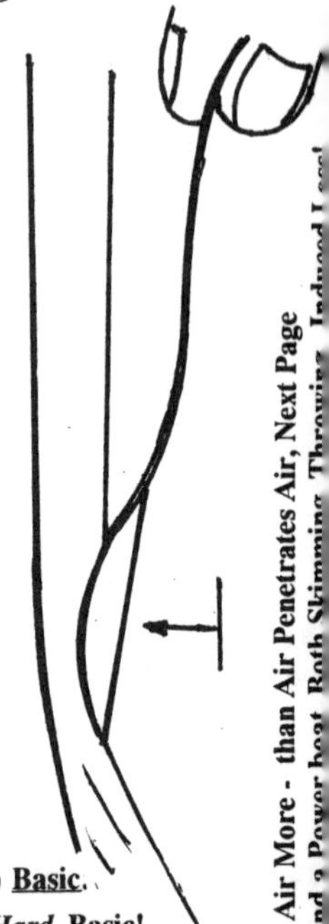

One Brief Page, a Ton of Insight, Grasp!

Try the Spoon, under the Kitchen Fawcet Trick, YOU CAN FEEL THE SPOON PULLED IN, SEE THE *BENT FLOW*!

ρ, rho, is Mass Density, More Later

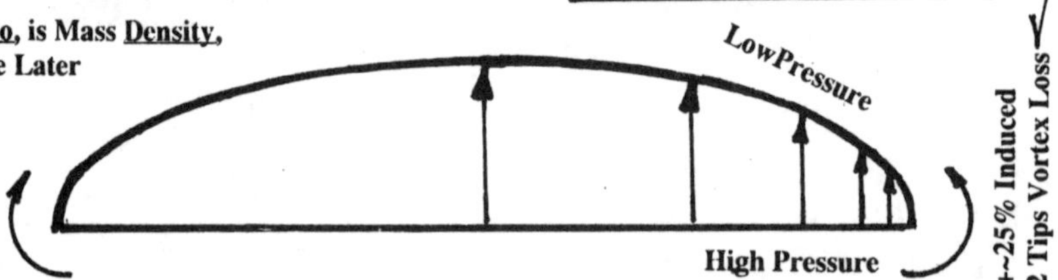

It's hugely important to grasp that **Wings tend to have Elliptical Lift Distribution,** fall off to a **Tip Vortex**, zero, maybe a +~25% Induced, **depending on Speed**, **Props Hugely Worse**! There's huge insight in the Lippich Smoke Tunnel Pictures, **Look!**

We Do NOT do the Scary Math in this Book, Rather, Insight!

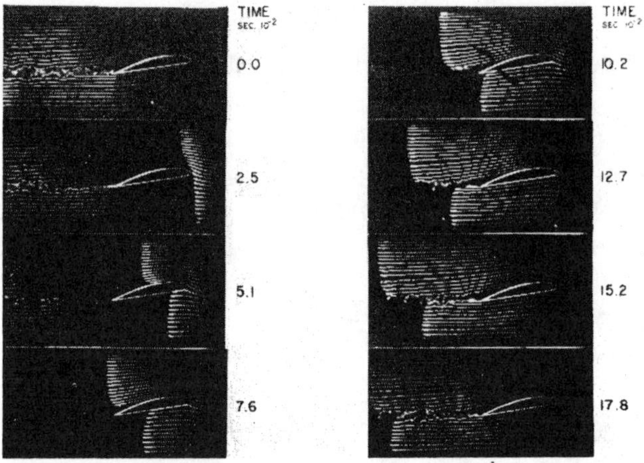

FIG. 13. Sequence photographs of time lines on a wing section with lift (reproduction from a high-speed film).

Notice the thickening upper Boundary Layer. Actual Separation can create an extra Pressure Drag.

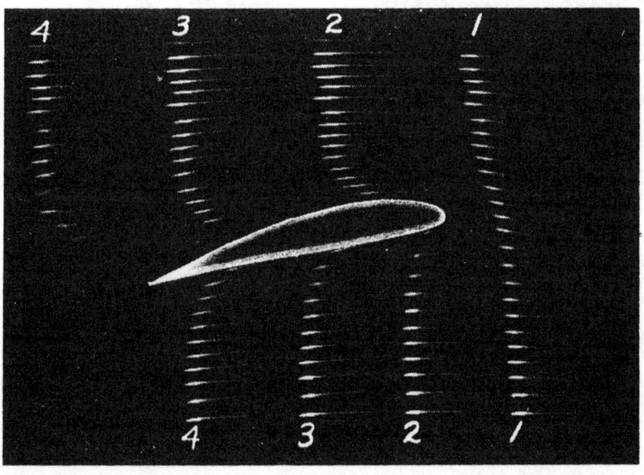

FIG. 15. Multiexposure of time line of a flow over a wing section. (It should be observed that the upper and lower branches of the same time line are largely offset. Equal numbers identify equal time line position.)

Marvelous Insight, from Lippich's Smoke Tunnel Work.

Fantastic Lippich Smoke Tunnel Visualization!

Lippich really understood Aerodynamics, really Smart, led Germany into Jets, before our team, **fantastic insight here!** The air over the top of an Airfoil does NOT meet the bottom's at the Trailing edge, **maybe ~133% faster**, as shown, **the mean camber line bending the Flow Downward, just a little, but vastly more Massive that you'd ever guess, or realize!!! Profile Drag, Friction retarding the Flow, a developing *thick* Boundry layer.**

Dumbed Down, or The Real Truth, *Easy Enough*

Everyone knows the drill: To sell a lot of books, make a lot of money, you "Dumb it down", write the Pablum Book. But we didn't do this to make money, but rather as **a gift, the real truth, a *Give Back*** to the **Flight that gave us an "E ticket" to Life!**

This Business gave me a leading edge Life most could only dream about!!!

Besides, if I write the Pablum book, that's done on the premise that I'm smug, you're somehow dumb, I have to dumb it down --- and that's just not me --- and my bet is, that's not you either!

It's ridiculous, really, that this correct insight is 50 years late!!!

You'll soon see that my bet is that, if you can fly an Airplane for several years without killing yourself, got through High School, maybe had at least a few teachers who tweaked your interests in smarter insight, took a courses, or so, with more challenge than Gym., you're not at all dumb, and **you can *soon grasp*** **Just Two Pages that Nail The Core Logic of Props!**

Besides, 10 of us worked 138 years, 141 years to this day in late 2006 when I'm writing these *last first pages* to get a wild subject "down to just 2 pages that I'm betting you can *get*"!

Most of us Pilots have some Adventure in our soul, ***confident***, think it's our right, love charging off into the **Wild Blue Yonder**. We're generally not techies, the nerd, geek type, don't want to be, and that's perfectly OK with me. I flew for years just for fun, a break, a get away from my work, flew at 18,500' in **the Sierra wave twice**, down the crest of the Sierras, in the upslope winds in a plane that wouldn't really fly at 15,000', through lots of Ohio Winters, Cold Fronts, wild winds, do ***power off landings from 10.5, 20 miles out*** - gliders can, we can too!

2 Miles High, a 12 to 1 L/D, 24 Miles Range, Gliding, 70 MPH, fast idle

My simple bet here is that there are actually a lot of you who love flying, have a fair bit of curiosity in your soul too, not a geek, but don't see yourself as a dummy, ***confident***, **not afraid!** So, simply, we've done everything humanly possible to get prop logic **down to 2 core pages, with simple direct words**. We've aimed it to give max. help to a guy who is *not* a geek, *may* sweat a little, **betting that you can and will get it! You might read those 2 pages twice, several times, finally see it really is *easy enough*, just *all the core, key insights* - collected!**

A Look Ahead, to p. 14 of the Primer

(Next, After this Introduction)!!!

Let me explain. There is nothing in the literature that tells us clearly, simply, that the objective of the math is to load the prop correctly vs. radius, to prevent excess tip loading, unnecessary, unrewarded induced and profile tip losses, that proper blade loading is accomplished by proper shaping of the blade with proper twist, that if you actually achieve a constant optimum angle of attack and C_L, you get min. profile drag too, min. area precisely placed, Min Torque, a Triple ideal optimum prop, not just min induced --- that you must accurately correct for nose velocity profile, slowdown, because we really need to predict inflow to all blade radii to .1 degree, yes one tenth degree, that a 1 degree, .1 C_L error will give us a 20 % error on a .5 C_L prop --- that the slowdown correction which results in less twist, lower blade pitch angles, especially inboard, can create a major Plus in fixed pitch prop operation slow, in take off and climb.

Betz is Min Induced - Norris's 6th. Law, Min Profile, Torque, Triple Optimum!

No Math! --- The Computer Does All the Hard Smart Work for Us! We just Learn the Understandable Logic, The Understandable Horse Sense Logic --- of What is Happening Physically --- Finally, *Easy Enough!*

The kind of things we'll learn is that Speed vs. RPM, with due consideration for needed inflow and angle of attack sets pitch, which sets max. efficiency attainable, that use of B-G-T, Betz-Goldstein-Theodorsen Math, will then deliver max. efficiency, that it will also deliver a **Precisely Loaded, Shaped, Twisted** prop, **a constant optimum angle of attack and C_L, minimum area, precisely placed, a Triple ideal, min. induced, min. drag, min. Torque prop, if we're only smart enough to ask for it.**

Remember, with 21st Century Insight, we can improve on the old Optimums

A lot of long, hard, dedicated work has been done here, simply because Andy Bauer and I were experienced enough to recognize that the explanation was badly needed, long overdue, and that with two lifetimes of experience we could sort it all out, do it. **There is a lot to teach you**, be you a novice or pro, **a long string of incisive insights. We'll simply build on and expand 4 basic insights here, ever deeper, broader, more sophisticated**, ***superemphasized so anyone who tries can get it!***

A Little Preliminary Overview

Our Whole Modern Life is Based on the Rapid Advances of Science and Engineering --- Made Understandable, Useable.

Do You Love Flight? Come Find out How it all Really Works, Propellers too, an *easy enough challenge*, if you'll go for it!

Think Engineers are Nerds, Geeks? Lose the Thought! There's more guts in Engineering than people ever realize! The Apollo, a 36 Story Tall Vehicle, 7 Million Pounds, Lifted with 7 1/2 Million Pounds of Thrust, was lifted, *leaned over, grabbed*, to fly it out of the Gantry, both too big to move clear of each other -- a bet that we could Stack 6 Separate Vehicles, make them all Work, Go Round-trip, the Whole World Watching -- *World Class Guts* --- *Audacity!!!*

Flight Controls, for Big Jets, and Spacecraft was my speciality.

Personally: We moved, Milly and I, 3 small Kids, on the Bet that I could Move us to L.A., Start a Company on the Run, Build a Team, beat all the Local, Established Competition, win the Lion's Share of the Small Rocket Maneuvering Systems Control Components --- 107 of ours used on the Mercury, Gemini, And Apollo, in the Milestones of Flight Gallery, the Central Hall of the Smithsonian, National Air and Space Museum, the day it opened --- July 4, 1976, our Country's 200th Anniversary. *Nerds* -- hardly -- savvy -- *guts football required.*

A Little Guts, a fair Try, We can get you huge insight - *Easy Enough!*

Before this, Props were the wildest Subject in Aerodynamics, never explained in 138 years, a combination of Black Art, Math, and guessing. **We explain them** with just the **Aero 101 Horse Sense of What's Happening Physically to a basic: 1. Rotating Wing, 2. Airscrew, 3. Pulling in and Throwing Back Air, *a lot of it*, a lot heavier than you thought, *No Math* - but all backed up, accomplished through a computer, with perhaps the most Elegant, rare Genius Math and Engineering ever accomplished, the work of 7 rare Geniuses over 83 Years - that you don't have to deal with!**

It's all Explained in the Specific Horse Sense Words of What's Happening Physically.

A Novice can't pick out the Gems of Insight, so we **Super Emphasize, Repeat, as we expand and go deeper - Thus, give you plenty of Time, and repetition to catch on - written to help you - your best chance ever to *really* end up understanding Flight, Props too! If you *scratch harder at first, learning*, you'll see it's *easy enough* once you grasp it.**

This Introduction, Pretty Complete, Gets Pilots Aboard - the Primer - then Ch 9, Book I, then Book II -- takes Anyone, Everyone as Deep as You Personally Choose, You Choose!

You'll soon see I had the Guts to do this Very Differently!

The Whole Core of the Story on TWO Very Dense Pages!

Normal Writing Style - Too Long a Story - People Get Lost!

A Little Guts, Desire on Your Part
Able to Laugh at it
Hang In
Let it Soak In

You'll Finally Realize it's Far Easier than you Thought!

Easy Enough!

You'll understand what we Pros Never Did! Props 101!!!

A bit More Insight,
I might have Tumbled to it A Half Century Ago
When I looked for it the first time, as a Kid.

Competing in Model Contests

I Wanted to Know --
What is the Right SHAPE?
What is the Right Twist?
How do we Get Pitch Right?
Stubby Blades, or Long Skinny Blades?

As a kid, my Questions were Spot on!

Blade SHAPE, *Twist*, Blade Loading is the Key!!!!!

Laugh at it. Read Exactly What the Words Say!

Not Faked Out -- You can get Very Smart -- Very Quick

A Needed Insight on the Real World, and us -- Mortals!

Never, in 138 years, have Propellers been adequately explained! Since there has never been a really correct, understandable explanation of Airplane Propellers, this will be read by every level of Individual from Novice Pilots, fully green technically, to Lifetime technical Pros -- with the full Spectrum of Knowledge, Experience, *maybe* the complete Lack of either - every Opinion, Prejudice, personal Bias that we mortals have.

I've been taught, some pretty enlightening, laughable insights!

There will be pros who think that if this isn't written in their preferred wording, that assumes a lifetime of technical insight and knowledge, their exact style, that it just isn't properly done.

Graduates of one course in English 101, effective writing can feel that they are the final arbiter of *the* one proper writing style.

The Game Here is to Help Novices Grasp what we Pros Never Did in 138 Years!

I'm saying this because specific experience has shown me that I'm dealing with the full spectrum of the Human Race here, every strongly felt personal opinion, prejudice, bias on how the task at hand should be done, and on page 1, I've been taught that I need to intercept that full spectrum of the Human Race here, and settle everyone down to the Real World of the Task at hand.

The task here is to **Explain Propellers to Everyone**, from **Novices to Lifetime Pros** - when in the Real World, Propellers have Never Been Successfully Explained, by anyone, to anyone, at any level, correctly, never once in 138 years, because the subject has always been impossibly complex! No one ever saw through them incisively before, the genius level math, the complex heavily loaded three dimensional flow, always previously beyond human grasp, never a proper explanation.

We have successfully reduced a correct understanding of propellers to the practical, understandable grasp of the Horse Sense of What is Happening Physically, just 3 *"easy enough"* core insights, but there are so many allied things happening, that we need a "hit you over the head with a 2 x 4 writing style"! You can see it quickly - in an Hour, Grasp it!

Dive in -- Go along -- Laugh at it -- Even Have Fun -- Get a lot Smarter!

The Bottom Line Core Insights, Conclusions - First, Up Front!

One Page --- Pictures Too

Induced Loss is simply the **Energy Cost**, **Loss of Throwing Air** - ***Down or Back*** - **to Make Lift, or Thrust**. **Profile Drag** is simply **Air Friction Drag**, a Loss. **Prandtl** teaches us the **Elliptically Shaped and Loaded, Minimum Induced, Min. Profile Loss (*Spitfire*) WING**, because any normally shaped wing loads itself very close to **Elliptical**, **Lift falling off to Zero at the tip** into a **Wasteful, Costly Tip Vortex**, maybe a 25% extra Tip Loss Induced Penalty, maybe, at the speeds we cruise, **Induced ~25% of total Loss**!

The **Core Issue in Propeller Design**, *Astoundingly, Never Clarified, Never Stated Before,* is that a **Propeller, a Rotating Wing**, Very Fast Tips, **Extremely High Tip Dynamic Pressure, q, Proportional to V^2**, is **Aggressively Tip Loading itself**, Dumb as a Stump, **Aggressively Wrong, Inside Out, Opposite to what any Wing Wants, Needs to be Efficient**, creating **2, 3, *done poorly*, 4, 5 times as much loss vs. a Wing**!!

You just Don't Tip Load Any Wing - Dumb - Extra Tip Vortex Induced Loss!

1. Losing the Lift, 2. Creating greatly excessive Tip Vortex Loss! 3. Excess, Unrewarded Profile Drag, at Max radius, Max Lever Arm, Max Torque, 4. Bogging Down available Engine Torque, H.P., LOSING FOUR WAYS, Propeller INDUCED LOSS is greatly Excessive, especially with an added Rotation Loss, Terrible, Never Identified before!!! A .55 C_L Prop, Induced 2/3 of the total loss, not 25% -- 200% of Profile - Twice the Profile!

Classic Betz Propeller Logic, Goldstein-Theodorsen Math, BGT, Rare Genius Level Work, available to us for over a Half Century, since 1948, **counteracts that** by controlling, directing classic **Half Teardrop, Radial Blade Loading**, given to us in a **Simple Blade Loading Chart, vs. Radius** for all Props from Low Pitch, Advance, to High, **amazingly never explained, never correctly understood, never correctly used before**, until the 2004 ***WhirlWind* 200 RV Prop**, we taught, a decade long professional effort to finally explain Propellers, **the lack of a valid explanation, the longest, most Fundamental Hole in Aerodynamics. A Half Century Late, into the Second Century of Flight without a Propeller Explanation, Ridiculous!**

A Family of ideally Loaded, SHAPED, *Twisted* Props results!
The Family of Shapes, The Loading, *with optimum Twist*, Shown on the Cover!!

The Key, Core Insight, ONE PAGE!!!

Low Pres. on the Top - High Pressure Bottom, Naturally, *a Tip Vortex!* Wasteful, Costly ------

A Wing

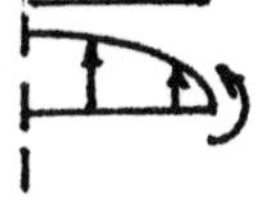

A Prop

The Problem

Get It - This is Easy to See Horse Sense!

Loading

Highly Tapered Tips!!!

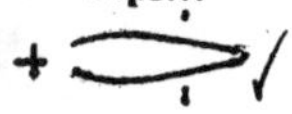

SHAPE The FIX

***Counteracts* q**

Right up Front - We Immediately - Pretty Brutally,

Dump you Right Into the Core Propeller Problem,

and the Betz - Goldstein - Theodorsen Solution!

Just Blade SHAPE, *Twist*, Controlling Blade *Radial* Loading!

(Always Tapered Tips to Counteract Excess Tip q, The #1 Core Issue!)

The Reason that's Best --- You're almost Instantly Aware!

That's Much Better than making it a long, Drawn out Story!

Get it Quick, If You Can -- *But*
***You Don't Have to Get It on the First Look* - Just Get Aware!**

We Repeat, as we Expand, Go Deeper, Always Tie together!

You'll Have Plenty of Time, and Opportunity to Catch On!

We Superemphasize, **(Novices can't pick out the Gems, at first.)**

Try to Make the Key Insights Jump Off the Page at You!!!

Give Novices all the Help Possible - Help get you Started!

You'll Find those Underlined Statements are Very Explicit!

The Explicit Words that Can Nail the Key Insights for You!

Laugh at it - Join in - Don't be Faked Out! -- Smart Quick!!

***If You Have Trouble* - There's a Cartoon on Page 6 - That can Help, a lot**

The Bottom Line Core Insights, Conclusions - First, Up Front!

No Math, we can completely **Explain Core Propeller Logic** by simply understanding, **1. A Rotating Wing, 2. An Airscrew, 3. Pulling in, Throwing Back Air**, *much Heavier than you thought,* **just understanding what's happening Physically, Logically!**

The Game is **Minimize Induced - Minimize Throwing of Air!** **3. Newton teaches $T = \dot{M}\,\Delta V$, Thrust equals Mass Flow Rate, $\dot{M}$, $\dot{M}$ dot, times the Delta V Speedup -- Get a Big $\dot{M}$, so we can Get a Small Delta V, ΔV, Speedup!!! EASY!**

Minimize the Throwing of Air!!!

So, we either need to Go Fast, or Have a Big Diameter Prop, Easy to Grasp -- Fast allows a Smaller Prop!!! Newton!!! Minimize the Throwing of Air - Minimize Induced, *Easy to Grasp*!

An Airscrew, the # 2 concept, brings it all together! We **Want HIGH PITCH!!! Simply, between two Props, the High Pitch Prop is the Faster one - that Minimizes Induced! High Pitch Minimizes Profile Drag Loss too,** so simple it's **Funny, a Shorter Spiral Path to Get to the Destination, a *low pitch spinning furiously*, way too far to get home, *wasteful*!**

***THUS* - It's SPEED *vs.* RPM that sets PITCH, thus sets Max EFFICIENCY Limit!**

Now, if you can **Grasp those Core Concepts, Quick, Early**, get **Quick, Facile in your Grasp**, You have this Book Licked!

By Simply **using BGT**, *choosing*, a **Constant Ideal Angle of Attack**, α^o, alpha degrees, and resulting **constant, ideal C_L, coefficient of Lift, at all radii**, we get a **Triple Ideal Prop**, actually **6 ways best**, **Minimum Induced**, **Min. Profile**, ***Min. Area Precisely, Optimally Placed***, Thus **Minimum Torque**, thus **Min. H.P., Max. Efficiency --- Norris's 6th Law!**

By **greatly reducing** the **Aggressively Violent Tip Vortex**, with **Optimum Loading, Optimum SHAPING**, ***Twisting***, we immediately **GREATLY REDUCE NOISE, excessive Noise, an immediate by-product of unaware, unknowing design!!!**

But There's a Lot Happening ---So!

To **Help Novices**, we **Superemphasize**, essentially Yellow hi-lite it for you, **make the key statements jump off the page at you - Repeat, as we Expand, Tie it all together for you!!!**

138 Years, 1865, to the Wright's 100th Anniversary, Dec 17, 2003!!!

Just TWO Pages to Grasp! Reread Get them!

Jack Norris 12/17/03

***A Reno Prop Throws 4 1/4 Tons of Air Per Second!!!**

Everything here depends on giving you an **Introduction -**

that pretty much **says it all** in a few pages - **an hour read**.

There's nothing else like it anywhere!

A Lifetime of Aero Insight - Propellers Nailed - One Hour!

You Won't get it All - You won't be able to repeat it Yet -

But you will be **Marvelously Aware, Quickly** - Easily Enough!

By "Easily Enough" we mean -

It will seem Like a Head Full at First - the First time.

But as it soaks in you'll see it is *Easy Enough - Makes Sense*

It's just the Horse Sense Grasp -
of What's Happening Physically
The Kind of Logic we can Understand!

No Math!

1. A Rotating Wing, That tries to work Very Wrong (Fix **Shape, *Twist***)
2. An Airscrew, Fast, Low RPM, High Pitch Best, min loss.
3. Pulling in, Throwing Back Air **($T = \dot{M}\,\Delta V$)**
*** Air a lot heavier than you thought!**
Get a Big $\dot{M}$ (Simply Go Fast, or Big Dia.), Thus Get a Small ΔV, Low Induced!
--- There's Nothing really very hard there! ---

The Game - In an Hour - You're Pretty Much Totally Aware

You can soak it up at your own Speed - Novice to Pro

(There's a 12 page Aero 101 Pamphlet to Help Novices who want it, get a good Start)

Getting an Introduction that could work determined the Publication Date.

So we just Skip all Formality - Just Start into it ---

Making Propeller Logic Understandable

Just Understand Aero 101 - What's Happening Physically - No Math!

Since there has never been a real, correct, understandable explanation of Props, this will be read by everyone from total Novices, to Lifetime Pros - an impossible writing task!

We want Pilots to be able to see through the simple fact that Props try to work Very Wrong, try to Tip Load themselves, Dumb as a Stump, Aerodynamically. We fix that, (*to the degree that it can be fixed*), by simply Tapering the Tips, Ideal Overall Blade SHAPES, with the *Proper Twist*, to hold Ideal Blade Angles, *most simply*, ideal Blade Shapes, varying vs. Prop Pitch, low to high - *accounting for* - Speed vs. RPM.

The Family of Ideal Shapes and Blade Loading - is on the cover!!!

We need Pros to see the same insight, because that Core Issue has been *missed for the Whole First Century of Flight*!

We Pros were so blinded by complexity, we missed the core of Aero 101!

To help Pilots, *Novices*, we've provided a 12 Page "Aero 101 Pamphlet" to get you familiar with the Definition of the Basic Terms, start you into the swing of understanding the core terms, how things basically work in flight, start you into the core *Technical Horse Sense* of how Airplanes work.

It's not Brain Surgery --- It's just Grasping the Physical Horse Sense!!!

Since this will be all new to Pilots, and you'll want all the help we can give you, we've adopted a "Hit you over the Head with a 2 x 4 writing style, *Superemphasis*, to *as much as possible* make the key statements jump off the page at you! Novice and Pro alike, We've Yellow Highlighted it for you!

Novices get HELP --- Pros can learn to skim, Speed Read it!!!

There are more Pilots than Pros, and the Pros, not sweating at all, it's logical and fair that we expect you to do the adapting to what is necessary, to help the new guys. Pros, don't feel you have to give me a writing lesson, I've been doing it in our Pro style all my life. The Challenge here is to help new guys grasp what we missed for 100 years!!!

Skip the Hubris guys. It's the open, adaptable mind that creates!

Since there is a lot going on, we very purposely repeat as we tie together - expand your insight, *repeat, expand, tie together*. The game is to get you *facile* in your basic thinking, grasp, so that it becomes second nature to you - *finally* able to grow!

The Definitions Novices Need are here: p. I-12-Bk II, more after Ch I Book II.

Making Propeller Logic Understandable

Just Grasp Aero 101 Logic - What's Happening Physically - No Math!

Over a lifetime, one can build ***a penetrating insight into flight***, but to **explain propellers**, pretty much ***the ultimate intellectual can of worms***, we took it on as **the final challenge, a final dare!**

The **First Challenge** was to **go deeper, farther, more explicitly** than anyone ever had, ***get past the 3D flow complexity***, the rare genius level math, ***grasp the hidden core logic***, the ***Horse Sense*** of the ***never explained - most complex Aerodynamic challenge!***

The Challenge was to find the Patterns of the Logic For You!

The **Final Challenge** was to **figure out** how to **make this most complex of all technical problems, never explained,** ***never really fully, correctly understood*** by engineers in 140 years --- **something** that **could "easily enough" be explained, actually be understood by *anyone* willing to go for it** - ***smart in 1 hour!***

NO MATH, the core Horse Sense of a 1. Rotating Wing, 2. Airscrew, 3. Pulling in and Throwing Back Air, *simply*, What's Happening Physically --- It's finally this simple:

The Computer, *behind the scenes, a pro's tool, out of sight,* does all the hard, smart work for us - Just grasp the Horse Sense!

***Let the Inflow, the Outflow, be a perfect helical Screw Surface,* just like the picture on the cover, *just* Constant Pitch Inflow!**

Let the Computer Precisely Load the Blades, *just Shape, Twist* them just like the Family of Shapes on the Cover, always with Precisely Tapered Tips, the WHOLE BLADE SHAPE actually, because, SURPRISE, Props, Dumb as a Stump are *aggressively* trying to load themselves *Inside Out, Max at Tip, a big mistake, compared to what any Wing needs to be efficient!*

Yes, Simply SHAPING, *Twisting BLADES Ideally,* *creates the bottom line answer, took 140 years - and the computer does it all!*

We purposely wrote it *a wild way,* *Blasts you into Awareness,* "a hit you over the head with a 2 x 4 style", superemphasis, because that's what works, leads you to all the smart insights. *Dive in, Be Brave, Don't quit. Go for it. A wild hour* - Smart!

Go Back - Reread those 2 Pages - if you need to! Let them Soak in!

Making Propeller Logic Understandable

I like and appreciate good writing, good writing style, and my intent here was to write an excellent, professional, great looking computer generated book, that explained how easy it is to understand flight, how to fly airplanes with more insight and intelligence --- and then the never explained logic of propellers.

As they say, the best laid plans of mice and men go astray! Propellers have always been the impossibly complex, unexplained "Black Art" of Aeronautical Engineering, the lack of an incisive, comprehensive, understandable explanation, the longest, most fundamental, gaping hole in the Science of Flight.

Actually, there has been magnificent, incisive, technical logic and Math work done on propellers, truly Elegant, and long ago, an essentially exact mathematical solution 57 years ago, 1948, and we have successfully made them vastly easier to understand, but there still is a problem. If we just start in writing, with professional style, the story goes on, and on, too long, too many interesting things to teach you, all tied together, and you're apt to just get lost, throw up your hands and quit.

The **first challenge** was to see through the ultimate technical swamp, unexplained for 140 years now, a whole century of flight, to the Moon, and back, and still no explanation - ***find the core logic for easy enough explanation***. The ***second challenge*** was to figure out how to take this subject, never adequately understood by Engineers, and **successfully explain it to Pilots**, who don't usually try to unwind wild subjects - and Engineers too.

Right there, good writing style lost out, because the central need here is to ***superemphasize***, make a complex basic subject, that engineers never adequately understood, easily enough understandable to technical novices -- and engineers too! It turns out the proven most successful method is to ***blast you quickly into broad awareness***, make the logic so clear, everyone can get it soon enough, not quit. The core is this easy: ***Dumb props try to load themselves inside out.*** We **counteract** ***that*** by ***forcing ideal radial loading***, simply with **SHAPE**, ***Twist***!

Twisted

Do You See? It's really Very Easy - A Family of Ideally Shaped Props!

Making Propeller Logic Understandable

Props have never been really correctly, understandably explained, never really correctly understood, because they **have always been a mind boggling black art swamp** of **wildly complex math**, and **unfathomable, heavily loaded 3 D flow**, where what is happening right at the prop is greatly different than what is happening, on average, over the Stream Tube pulled in, and thrown back to make thrust, more action *at* the blade.

The genius math of the 3D flow, is not the way *anyone* can learn it.

Actually, 7 rare genius historical figures worked 83 years from 1865 to 1948, and created **Betz Classic Logic**, and **Goldstein-Theodorsen Math**, that gave us **perhaps the most Elegant Logic and Math ever created** in any branch of Engineering, **Marvelous! But Pilot, or first time Pro, you Do Not want to tangle with the obscure inner workings of all that** -- ***we did!***

Happily, loaded into a modern Personal Computer, it will spit out **essentially exact Ideal Theodorsen Props**, as fast as we can type in the Specification Requirements, and push the Go Button!

So, we can skip the genius level Math, secure in the fact that **everything we teach you here, *all correct*, is fully backed up by some of the most Elegant Engineering work ever done!**

What we teach is much easier, correct! The Math far too obscure!

To **succeed here together, Pilot, or first time Pro**, we had to get the previous *Black Art* down to the **Understandable Horse Sense of What's Happening Physically.** **That's what we all can grasp!**

For the Pros who want to go on, once you understand the Horse Sense Logic, it is vastly easier to tackle the math, especially after we show you that Goldstein - Theodorsen gave us a simple Blade Loading Chart vs. Radius, that *simply* changes as we go from Low Pitch, or Advance to High!

Tibery and Wrench have refined loading for Very High Advance

Now, since there is **a lot happening**, and you'll want us to **sort it all out for you, make it as easy as possible, especially at** first, *embrace the style* - *emphasizing the meat, the core logic:*

1. ***Superemphasize,*** so the **key points jump off the page at you**.
2. **Dive right into the core Problems and the solutions.**
3. **Clarify Insight, repeat, tie together** as we **expand, go deeper.**

Making Propeller Logic Understandable

The **way we've been looking at props, is very, very simple** vs. the ***intellectual swamp that pros have had to face for decades***. If you look at the math in the Historic books it initially looks like a swamp of complexity that only a first class math pro could even begin to understand. The final 3 D Potential Flow Solution is something that few Engineers could even dream of accomplishing. **Theodorsen calls Goldstein's 3D solution the greatest, most important contribution in the History of Propellers. He then went on to use it, modify it for Heavily Loaded Flow**, ***where the ΔV pulled in, and thrown back at the prop, is much greater than the average of the Stream Tube*** the Prop Creates, a rare genius himself. His book is a Historic Work! But the bottom Line was, **nobody could see through that and the snakepit of the physical 3D flow well enough to nail the Practical Insights,** until **Dr. Andy Bauer** and I put in **a huge effort to finally nail it all -- a once in a lifetime task!**

But, fortunately ***there is a way*** to make the subject ***easy enough***:
1. People *easily enough*, see a prop is "*kind of*" a ***rotating wing***.
2. Many see a propeller is an **Airscrew**, the British description.
3. Pilots, *easily enough*, see a **helicopter** pulls in, **throws *down* air to make Lift**, **Propellers** pull in and **Throw *Back* Air to make Thrust** - too few realize wings are also throwing down air to make lift, a Newton's *Equal and Opposite Reaction force.* **Said simply**, the ***energy cost, the loss of throwing air*** is the ***Induced loss***, the other basic loss in flight, with skin friction, ***Profile Drag***.

GET THAT -- INDUCED -- IT'S BASIC

Fortunately, if we just teach you the comprehensive insight in those **3 very basic fundamentals**, *all backed up* by the **Elegant rare genius level math**, we can **skip the math**, because, the simple logic teaches us ***What*** the math is doing and ***Why***, better than the rare genius math, which hides what's happening physically, but ***the math precisely affirms the horse sense logic.***

All **three** of those basic insights are very important -- **teaching us specifically what is happening Physically, Logically**, the kind of **practical insight we can all understand, our brains working on understood logic** - complexities then easier to see.

Geniuses created the Logic, the Math, the Computers for us.
We just need to Grasp the Horse Sense of What's Happening Physically!

Making Propeller Logic Understandable

A **PROP** is ***trying to load itself INSIDE OUT*, DEAD WRONG, DUMB, *compared to what a wing needs!*** A **Wing** creates **Max Lift at the Center** and simply falls off, pretty much like ***an Ellipse***, no matter what shape it is, because **Wing Lift *Must*, *Will Fall to* ZERO AT the TIP**, a ***TIP VORTEX* Swirling from the *Higher Pressure Bottom*, to the *Lower Pressure Top***, (even tip Dams only partly effective). But here's our Dumb Prop with very ***high Velocity Tips*, Huge Dynamic Pressure, q, Tips, (Proportional toV^2) trying to make max. Thrust and Drag AT the Tips, where Thrust will be LOST, must fall to Zero, creating an Extra Big Tip Vortex, *thus* Extra Induced Loss, *extra* Unrewarded Profile Drag, *at max. radius*, Max Lever arm, extra unrewarded Torque, bogging down the Engine RPM, Losing HP, Losing (4), four Ways, DUMB AS A STUMP!!! Creating NOISE too!!! It's our job to be smart enough to COUNTERACT that, Taper the Tips, Pull the Excess Loading Back Away From the Tips, *EASY KEY INSIGHT!!!***

HOW ABOUT AN EQUIVALENT *TIP LOADED* WING

CHORD x C_L x q - (Shapes) Prop Loading!
Correct Thrust vs. Radius SHAPE --
is how we Load it right -- WIN

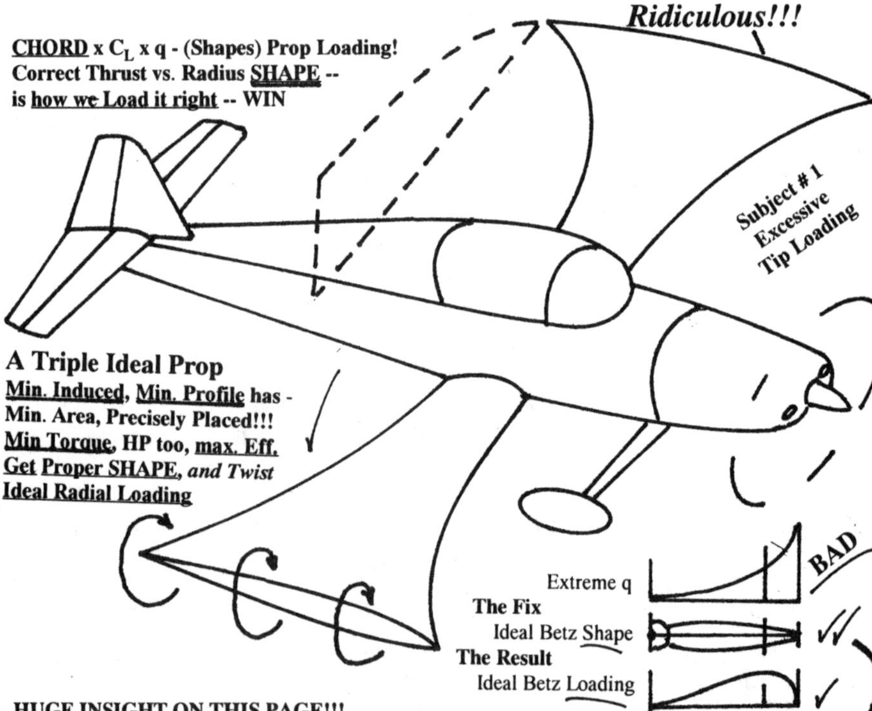

A Triple Ideal Prop
Min. Induced, Min. Profile has - Min. Area, Precisely Placed!!!
Min Torque, HP too, max. Eff.
Get Proper SHAPE, *and Twist*
Ideal Radial Loading

HUGE INSIGHT ON THIS PAGE!!!

This is the most Important and easy Insight You'll Ever Learn about Props!
It has been mostly missed, mostly not understood for the last 100 Years!
You don't win *playing games* with a wider tip, make Lift, where we just pulled it back!!!
(*Later we may be able to beat* the Betz 20th Century Ideal Loading and Shape!)

Jack Norris 12-17-03

PROPELLERS EXPLAINED

The easy core insight: Props naturally Tip load themselves, Bad! Just Tapering the Blade can *counteract* that problem! That Core Issue in Propeller Design, amazingly, seems to never have been specifically identified, stated, written anywhere that we have been able to find! If true, that is the greatest, fundamental oversight in the history of Aerodynamics, in 2006, over 140 years since prop Analysis started with Rankine in 1865.

(The Civil War "Monitor", others, had early Screw Propellers.)

Simply, **Propellers, a Rotating Wing, very fast tips, greatly amplified tip Dynamic Pressure, q, (Proportional to V^2), tries to Greatly Overload the Tips, *create Greatly Excess Tip Losses*, actually *"inside out, reversed loading"* vs., the Center Loading, any wing needs**, because the **Lift, Thrust of any wing surface MUST, will FALL to ZERO at the TIP**, the higher pressure below **swirling** into a **tip vortex induced loss** to the lower pressure above the airfoil surface ***wasting energy*** **- Aero 101 really!!!**

The Simple FIX is to *Counteract that very wrong q Loading* with tapered tips!

Simply, presuming we ***twist*** **the prop** to maintain an **optimum angle of attack, alpha, α°**, an **optimum coefficient of Lift, C_L**, we can **SIMPLY taper, SHAPE the Prop Blades**, into **a FAMILY of Triple Optimum Props, Min. Induced Loss, Min. Profile Drag, Min. Torque, Minimum Area Precisely Placed, Norris's 6th Law --- chord vs. radius, Shape varying vs. Advance Ratio, Pitch.** That's a **Betz Minimum Induced loss Prop, Goldstein-Theodorsen** rare genius level, essentially exact **Math**, available to us since **1948 --- with an optimum C_L to get the Minimum Area Precisely Placed - least Profile too!!!**

Try going for a higher C_L, less Area, and you get higher Induced, start losing.

Simply, since 1865, **Propellers have never been, correctly understandably explained**, never really properly understood, because a proper explanation has been missing, that long, the biggest, longest gap in Aerodynamics. I know because, a lifetime pro, with my products all over the world, I've been looking, waiting for that explanation since 1946, 1948 when I won the Senior National Model Championships, two of the three years I was there to compete my designs.

My college friend, and collaborator, **Dr. Andy Bauer, and I**, two Lifetimes of Knowledge, took it as - the *ultimate challenge!*

PROPELLERS EXPLAINED

Always, the never explained, most complex subject in Aero, I knew **we had to get it down to Aero 101 Logic,** ***Horse Sense***, so that **Pilots, technical pros too**, could finally grasp what was happening ***physically***, the **logic**, ***easy enough to see through it!***

It took years of professional work, of working all the math, all that everyone had done, **computer programming everyone's methods, running studies, cross-checking, finding who had it right, wrong,** *or close, but no cigar.* **Goldstein-Theodorsen had the math correct, essentially exact 57 years ago**, and all **we have to do is to be smart enough to honestly understand correctly, use it all,** ***their simple blade loading chart***, **let modern PC's spit out essentially exact answers for us** - ***easy!***

You, Pilot or pro, Do NOT have to deal with the genius math, ***we did***, **just understand the LOGIC, the Horse Sense**!

Specifically, because of the "Inside Out Loading" props always have 2, 3, (designed poorly, 4, 5) times the loss of a comparable Aspect ratio Wing!!!! If we simply use Betz Logic --- which demands the perfect Helical Screw Surface inflow, a stretched, still perfect helix thrown back, shown on the cover --- given to us by Goldstein-Theodorsen Math, we get the "Half Teardrop Loading" shown on the cover, the Family of Triple Optimum Shapes, shown, if we simply use an optimum angle of attack, α^o, and C_L - THAT EASY!!!.

We can understand all the Logic of What's Happening Physically, the core logic of Props by just learning the Logic of - 1. A Rotating Wing, 2. An Airscrew, 3. Newton's Laws, Pulling in and Throwing back Air, ***heavier than you thought!***

We've succeeded, made the logic easily understandable, but there is a lot happening, all tied together, ***interrelated***, **a whole book full, finally, depending on how far you want to go, a quite good grasp simply in this Introduction,** ***maybe some cherry picking where I'll tell you to look***, **matching anyone's level of interest.** --- ***But enough to confuse you if we did it wrong!***

Making Propeller Logic Understandable

It's been **140+ Years**, into the Second Century of Flight, the **Astounding fact** is that, **no one ever stated this Central Core Problem in Propeller Design***! Actually, B-G-T did the **Marvelous, Rare Genius Technical Work**, maybe in all, the **Most Elegant Logic and Math ever, in any Branch of Engineering** - but it was such an impenetrable Swamp of 3D Flow, Genius Level Math, it Faked Everyone Out. **No one ever** dug enough, **dug deep enough to grasp what it was trying to direct and tell us**. We Never Got the **Triple Optimum Prop**, ***Min. Induced Loss, Min. Profile Drag, Min Torque***, Norris's 6th Law, the **true Optimum**, that delivers the **Needed Thrust, for the Least Torque, H.P., turns easiest for the Thrust, Max Efficiency!**

* Actually we had this core long before 2003, Wright's 100th, continuing our work.

We don't even have to deal with the Math - Just let a Computer design an Optimum, Exact Betz Logic, Goldstein - Theodorsen Math, Minimum Induced Loss Prop, but ***Twisted*** **to achieve an Optimum Angle of Attack, Coefficient of Lift**, for ***Minimum Area, Precisely Placed***, Norris's 6th Law, **all on a Computer, Easy!**

That previously Secret Key, Startlingly Simple, is to **just grasp the True Aero 101 Logic of a Prop:** That Fast Tips, often Extremely Excess Tip Dynamic Pressure, q, relative to the prop inner radii, works to Violently Overload the Prop Tips, *Losing Paid For Thrust*, creating greatly excess Tip Vortex Loss, NOISE, Excess Profile Drag, at Max Lever Arm, Excess Torque, Bogging Down Engine Torque, Losing RPM, Losing H.P., Dumb as a Stump, Easy, once you see it. Dr. Andy Bauer, and I, two old pros, nailed that and many other key Aerodynamic Insights, had the experience to solve it . We were willing to do the hard work, **as a Gift, an ultimate challenge, a "Give Back" for a fantastic life in Flight** ***in this Greatest Time and Country ever.***

We did it the old fashioned way, we Earned it. Honest Work Pays off!

The big Funny: we launched the 57 year old Math. It tapered the Prop Blade Tips, ***counteracted*** the **excess tip Loading**, got a **Family of Theoretically, essentially Exact Optimum Props!**

Looking for answers, Hard Truth as a kid in 48 - winning the National Championships, *T T's genius waited a half Century for Andy and I - wild!*

Propellers - *Just Understand What's Happening Physically!*
(It's all backed up, *Proven*, by genius level Math -- *that You can Skip!!!*)

Pilots,

If we simply look at Airplane Propellers as **1. Rotating Wings**, **2. Airscrews**, **3. Pulling in and Throwing Back Air**, that's a **lot heavier than you thought***, we can understand all the **core logic** of **how they work**, and **should be designed**! We thus get **understanding them** down to the **horse sense** of what's happening physically, and **you**, ***and engineers too***, **can understand most easily!**

It turns out this is a big ***breakthrough***, because, if *pros* dive into it as an Engineering, and Math problem, it quickly becomes a morass, that has defied understandable explanation for 140 years

WE HAD TO SEE THROUGH IT --- YOU DON'T!!!

As Pilots, this hidden logic can seem harder than it really is. ***Unexpected***, **we engineers, need time too. To help *everyone***, we purposely do things that **look pretty wild at first**, but **succeed!** **1.** We ***Superemphasize*** the Horse Sense Logic, ***make the key words***, **the logic** jump at you. **2.** We're purposely going to dump you into the whole core of the Logic in a pretty brutal, dense way, to ***Blast you into Huge Awareness, Quickly, not draw it out!***

You don't have to grasp it all, at first, we repeat, *just get aware, just start to get it!*

Experience has taught us there's very good purpose in both approaches. If we just start in explaining everything in a more gentle manner, the explanation goes on, and on, gets deeper, too drawn out, and pretty soon novices and engineers both feel confused, maybe overwhelmed, might give up. If you can live through a few wild looking pages, without going into shock, **blasted into Awareness**, you can start seeing that there is some pretty powerful logic, very quickly. Then, it quits looking crazy

It's not for Speed Reading. We want you to see all the phrases, how they fit together!

You will surely be shocked to some degree, but if you just hang in, you'll find we wrote it to help pilots get aboard on what is a pretty wild subject. We keep repeating, drawing you in, leading you. *I'm sure you'll feel out of control*, but ***hopefully***, it will start sinking in, start making sense to you. You don't have to figure it out, *we lead you to all the right insights*. Have guts. Dive in, hang in. Treat it like a game, a challenge, *a worthy Mystery novel*. You'll ***get aware, very fast***, maybe even laugh about it!

*** A Reno Racer Prop *Flows* 4 1/4 Tons of Air, 8500 #, per Second, at 480 MPH!!!**

A Special Early Page for Pros

As a pro, I've been really surprised to see that **a smart novice can sometimes "get it all", right now, quicker than a long experienced pro!** I've seen some pros tend to have a head full of ***barrier ideas*** they're very sure of, *don't really get it*, read, all that is said here, don't see all the implications, going off on their own thoughts before they get it. I've had pros immediately start telling me how to write it, (for them), **clearly before they understood what is here**, **also, not a clue** it has to be **written very differently for Novices**. I've seen we pros can need the Superemphasis, *at first*, as much as Novices!!! ***Sobering!***

One can see why it can be *the adaptable guy* who gets the new idea!!!

How did we pros never nail the core issue here, the bad loading? As pros we seem to maybe accept the high q prop tips, assume we'll deal with that correctly, maybe misunderstand that's the natural place to get high lift, not recognize that's **a very bad situation for any Wing** --- that **any wing is incapable of efficiently holding significant loads, *relative to the root*, anywhere near the tip** - lose the thrust we're paying for, a prop sure to create greatly excess tip Vortex and Profile Drag losses, at max. lever arm, bogging down the available Engine Torque, losing H.P., ***losing 4 ways, a prop always 2 to 3 times more loss than a comparable wing, 4, 5 times done poorly!!!***

How come no one ever asks why props lose 2, 3, 4, 5 times more efficiency?

As a kid, before I had the formal education, had a clue about circulation, horseshoe vortices, I had to figure out, understand the horse sense of Aero, and that can help, even be better. I first understood **High Aspect Ratio** as a very long sailplane wing where we ***wanted a lot of lift, a long way from the tip***, the long narrow tip itself with less loss --- a **60/1 L/D Sailplane**, a 1/60 loss, .0166, 1.66% loss, **98.33% efficiency for the whole plane. That helped me to tumble to the terrible prop loading!**

Aspect Ratio teaches - Don't tip load ANY wing - the Truth!

If you're a pro, learn how to best read this book, (Written for Novices, not you)! Don't bitch, think broader, understand. Smart enough, once you grasp the insight, you can **speed read skim the superemphasis! It's Yellow High Lighted for you!**

You want a **Long Span** for High Mass Flow rate, **min ΔV**, **min Induced - Tip Far Away too!**

Now, We'll Move On, Learn a Lot more

Some Aero 101 Basic Definitions --- and Insights

Very Important, Enlightening Basic Insights

Profile Drag - Is Simply **Skin Friction Drag - by the Air**, a **Basic Loss!**

Induced Loss - We're taught **Wings produce Lift**, because of a lower pressure over the top of an Airfoil - but all less than infinite span wings **also throw air down**, *another way of looking at the same lift*, just like Helicopter Rotors, or Props throwing air ***back*** to make Thrust - Also a ***Tip Vortex Loss!*** Both, are an **Induced Loss**, the **Energy cost of Throwing Air!!!** **The core problem of Propellers** is a **Greatly Excessive Tip Vortex**, from **excess Tip Speed, V**, **magnified Tip q, Dynamic Pressure**, and a **Rotation Loss too!** **Induced is only a fraction, of Plane Profile where we fly Planes, ~+25%**, but on a **.55 C_L Prop**, **Induced is ~TWICE Profile, A HUGE LOSS!!**

q, Ram Dynamic Pressure - Is simply the **Pressure Rammed, into our Pitot Tubes, Stopped**, used to **measure** Indicated Airspeed, **IAS**, but also measures **Energy at that V**. Thus, it is the **Pressure and Energy** used to create **Lift, Profile Drag, Induced Loss**, often thought of, used as a **Drag**. **q is Proportional to V^2**, V squared, so it's 4 times as big at 200 MPH, vs. **100 MPH, 25 times as big at 500 MPH** (a Huge Problem on a 100 MPH Plane with 500 MPH Tips)! **Excess TIP q on Prop Tips, causing Excess Tip Vortex Loss is the Central, Core Problem, and Source of Excess Loss on Props!**

Yes, Our Airspeed Indicator -- Extracts the Square Root of q!!!

Helical Pitch - Is Simply **how Far Forward a Prop Will Screw Through a soft solid, in one Revolution, No Slip**, acting as a ***Perfect Screw*** at **all Radii**, like the **Perfect Betz Helical Airflow on the Cover**. (A Prop won't end up with Pure Helical Pitch, after we correct it for Slowdown, the Slowed air in the Plane of the Prop, from the Embedded Body, decreasing, the steeper inner Prop angles.) But that can be very Good, Perfect at Design Speed, **Better Slow, at Takeoff and Climb!!!** **So, Props have *Less than Theoretical Twist, Better!*** **Pitch Measured at 3/4 r!**

Pure Helical Screw Pitch

Model Builder's Layout of Pitch

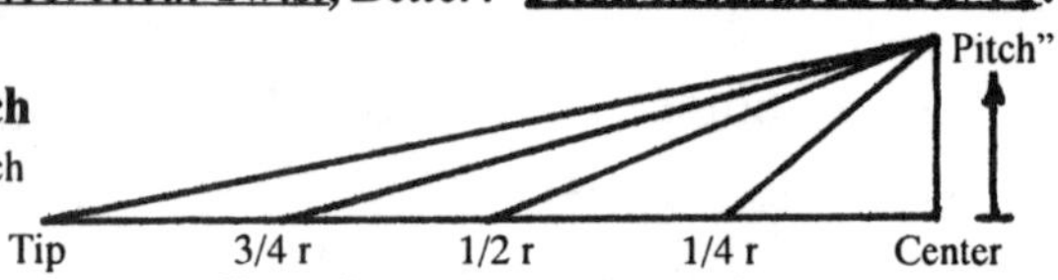

Circumference layout gives angles

Slip - Advance - **Propellers are *Overpitched* to Account for Air Inflow** Thus, **they SLIP**, actually **ADVANCE less than Pitch! Simple, *makes Sense!***

A Key Insight: **Betz Logic Demands Constant Slip**, at each radius, **sets up the Flow Geometry Perfectly**, to make the **Pure Helical Inflow, Pure Helical Stretched Outflow** of the **air Thrown Back, faster, *Simple Genius*.**

Advance is very **important**, fundamental to **a Graph of Max possible Efficiency**, since **only High Pitch, High Advance Props can have Max possible Efficiency**. Advance is **often called Lambda, V/πnD***, very **Simple once you see it, just Forward Velocity / Rotational Velocity of the Tips. Think** and you can see that **defines the *actual Angle of Advance*** of the **Tips!**

***All measurements in ft., #, sec. V in ft./sec. n = rev./sec. λ = πnD V**

Some Aero 101 Basic Definitions --- and Insights

Very Important, Enlightening Basic Insights

Propeller Efficiency, Eta, η, is Simply: **Energy, or H.P. out / input of related Energy, or H.P.,** *easy enough*, **a ratio, shows "useful percent".** **Nominal Efficiencies** on **High Pitch Props** range from **90%**, a little better on fast Optimum Cases, to **85% max. on** *slower*, **Low Pitch Classics at 100 MPH**. On old, Aerodynamically poorer planes, with wider Tip Props, not correctly designed, it's possible **to** ***lose 10%***, a **75% Prop,** ***not*** a proper **85%**, and **often another 10%,** on extra **scrubbing Drag**, and **poorer flow**, maybe some flow **separation,** η x .9, ***only 67% overall Propulsion Efficiency,*** η_P. In the 60's, Gus Raspet, found only **58% overall on a Bellanca Cruisair, charging Engine Cooling to the Engine, not the Airframe,** ***an eye opener!***

1/.58 ***requires*** **172.4% more HP, Powered, vs. Gliding, WOW!!!**

The Lift, or Drag Formula **Area* x C_L x q = Lift** **Drag = Area x C_D x q.** Naturally, **Wing Area is used for Lift**, but **also for Drag,** (which makes less sense, because the whole Airplane is involved, but that's convention). **For Props**, of course, **Blade Area is used. q, Dynamic Pressure,** as above, **is the pressure available to be put to work,** (interestingly, as above, it's also a measure of Energy, per cubic foot of air). **C_L maybe ~ .5 on a Prop,** to ~ 1.2, or more at Stall. Let's Call it a **Fudge Factor** that **defines how hard the Area x q, is working**, maybe 2,5 with Flaps!

*** Using Chords for a Prop, local q,** **gives local Thrust, for integrating vs. Span!**

Angle of Attack, α^o **Simply** the **Angle between the Wind Line, and the straight Line connecting the nose of the Airfoil and its Trailing Edge.** Significantly, with ***a cambered airfoil,*** **Zero Lift C_L may be at minus 4°,** thus, **to get a .5 C_L** we may only need a **~1 degree positive angle of Attack,** since ***typically*** **1 degree angle of attack nominally changes C_L ~ .1** ***Handy.***

LATER YOU'LL LEARN INFLOW CHANGES *EFFECTIVE* ANGLE OF ATTACK!

Coefficient of Lift, C_L It's really just a dimensionless **coefficient that simply ratios, how hard the airfoil is working, thus the "Lift per sq. ft.", little at a low angle of attack, a lot at high α^o, simply ~ +.1 C_L per 1α^o, until that nice easy relationship quits being linear near stall! A Small, or Slow, Low Reynolds number chord,** ***may*** **stall at a C_L of 1.2,** a **fast, huge wing at 1.6,** with max. **flaps ~ 2.5.** Notice anything over 1 could look like it's getting **more than 100 percent of q,** ***not true,*** **huge Flow Rates of air,** ***of q, energy!***

Aspect Ratio. Very Big **Span/Chord**, or **Span²/ Area, Long Wings, narrow chords, is what drastically reduces Induced Loss on a Sailplane wing.** Physically, **a lot of efficient Lift** is made **far away from the Tips**, far **less loss rolling into a Tip Vortex --- More Important, on a Prop !**

A LONG SPAN WORKS ON A BIG $\dot{M}$, THUS LESS ΔV, THUS LOWERS INDUCED

Blade Shape As we've seen here intelligent **Shaping of a Prop Blade, (with proper *Twist* and Pitch, to fly all Airfoil stations with proper, α^o, and C_L)** is far more important on a prop than a Wing with Hugely exaggerated Tip q.

***Further*, Complete Definitions at the end of Book II Chapter 1**

Remember, the Book Was NOT Written Just For You, but for Others Too!

Prop Loading vs. **Radius** - **Pictures Worth a Thousand Words**

A Dense Summary - The Core of the Whole PropLogic Story

In 1919 Betz Conceived the **Ideal Minimum Induced Loss Propeller**, *initially* done on a *Profile Drag Free Basis*, a **key basic core** math optimum. It creates **pure helical inflow, throws back a stretched pure helical backflow**, in a "**Streamtube**", thus creating **a reaction force Thrust**, per **Newton's Laws**. It's overpitched a little to *account for the inflow*, thus **slips a little, advances a little less than the pitch**, creates the pure helical flow *ideal* by **designing the inflow for constant Slip**. The **excess tip velocity, greatly excess V^2 tip Dynamic Pressure, q**, of a **Rotating Wing, greatly overloads the tips**. The **Excess Thrust is Lost rolling off the tip into an excess tip Vortex Loss**, extra, unrewarded **Profile Drag Loss** at a max. lever arm, **bogging down the Engine**, losing H.P., **losing 4 ways**!!!! By **precisely loading the Blade vs. radius**, by precise **SHAPING** of the blade with **proper matching** ***Twist***, Betz ***counteracts*** the **Excess Tip Loading** Losses, **prevents excess losses**, **minimizes losses, actually achieves the Ideal Propeller, even a constant ratio of Thrust vs. Torque, or H.P., constant efficiency at all radii** -- all, if first considered Profile drag free -- ***Amazing Insight!!!*** We learn **only** (**Fast**, *or geared*) **High Pitch, High Advance per Rev Props can have Max. Efficiency, Min. Induced, min. Profile Drag Energy loss** (a short Spiral path home). **The Ultimate Prop is a Graceful Canoe Shape**, (low Pitch props more Tapered), **by Theodorsen Math.**

Betz Helical Inflow, ***pulled in***
Stretched Outflow, ***shown*** -
***Downwash*, Backwash.**
Inflow helix + ideal α° = Twist β°'s

Helical just means a perfect Screw, all radii.

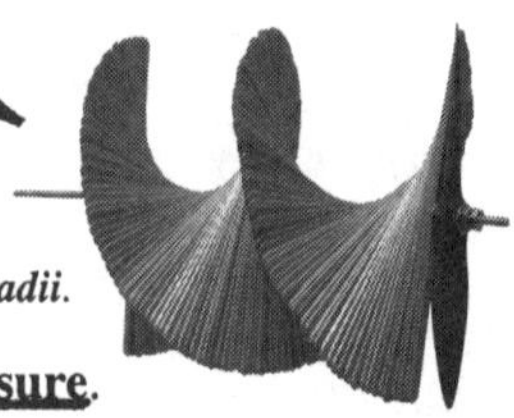

Tapering Tips
Pulling Back Loading
Is the Crucial Key Step!

Exaggerated Tip Dynamic Pressure.
"q" proportional to V^2!!!

*q magnification on **this ultimate prop** (1) is **only 3 1/3**, but is **25 on a Luscombe!**

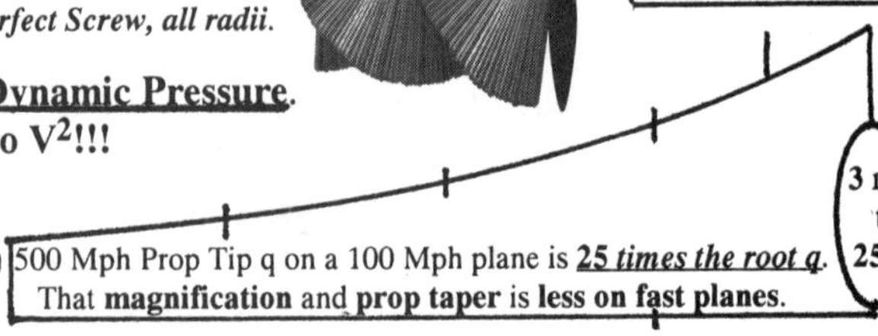

500 Mph Prop Tip q on a 100 Mph plane is ***25 times the root q***. That **magnification** and **prop taper** is **less on fast planes**.

Ideal Blade Shape Counteracts Excess Tip q. *Ultimate Hi Pitch, Max Eff. Shape*

Hi q magnification props, ***are far more tapered***, narrow tips, almost triangular, at *very* lo pitch, ***moves*** the *max. chord inward*.

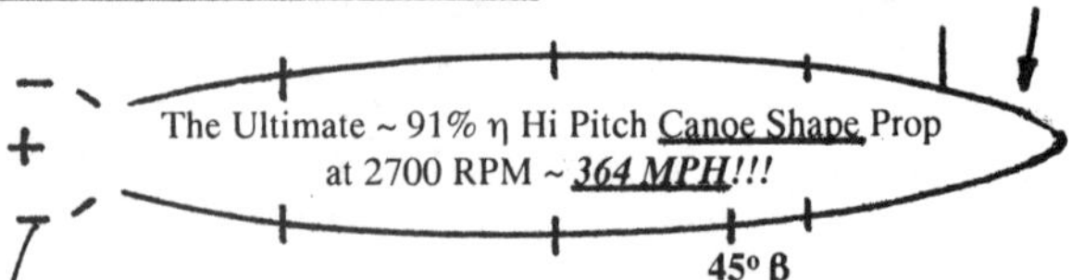

Betz Optimum Blade Loading. (Going for More Tip Load, steeper drop-off just loses!)

The Right Blade Shape, Loading - all the Magic Happens!

See page 147 for all the complete insight on Shapes, Loading.

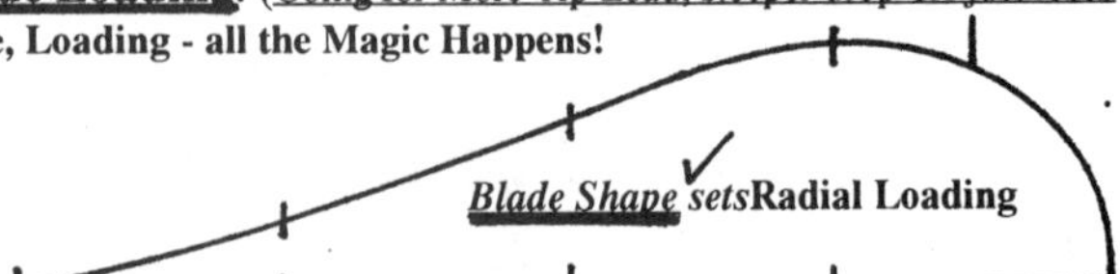

Ideal Prop Shape Changes a Lot vs. **Pitch, Advance Ratio, λ. Loading is fairly Similar**

Beginning to Expand our Insight

An Interesting Overview of Propellers

The British never went to **broad tip Props**, not in WW II when our guys did, (to absorb the high H.P. of BIG engines, also to avoid excess Mach tips, *limiting diameter*, with High Activity Factor, tip loaded props), not after*, because it's **a Bad Idea**, fails to understand the basic engineering logic of propellers, the essential horse sense of how to make props work well, or poorly. ***Brits*** call them ***Airscrews***, and that's a proper term, but we get really key insight by learning what's bad about a **Rotating Wing**. They went to more blades, kept blades closer to the correct shape, **the four bladed Spitfire, a smart WWII example**.

* Some British Planes had American Props. I saw such a Spitfire On TV.

It turns out **only High Pitch, High Advance** per revolution props can be **most efficient**, **Faster vs. RPM**, they have the **least Induced loss, throwing of air to make a Reaction Force Thrust, and the least Profile Drag Energy Loss**, simply a shortest path to get home, so easy its funny! **Fast**, **low RPM**, thus hi pitch, is the easy way to get a high efficiency prop, a higher cap on the possible efficiency. A fast prop can be smaller in diameter, than a slow prop, because **Newton** teaches us the **game is to get a big Mass Flow Rate**, **$\dot{M}$**, so you can get a **small Delta V**, **ΔV**, **thrown air speedup** for any given thrust, because ***throwing air is what causes Induced loss, lost Energy.***

SLOW PROPS MUST HAVE BIG DIAMETER, WANT LOW RPM, HIGH PITCH!!!

It turns out that props want to be **a steep, effectively 45° Helical Screw**, and the **final efficiency ball game** is all about **loading them ideally vs. radius**, and presuming we use proper twist to hold a constant ideal angle of attack, and C_L, that simply means **SHAPING** them optimally, since ***chord places Area***, **Load Distribution**. We've learned **dumb props** are trying to work **inside out, *opposite* vs. a proper wing loading**, our job is to be smart enough to correctly ***counteract the inside out!!!***

Where you let the MAX LOAD happen, the Distribution, *is the game!!!*

Despite magnificent technical work that took 83 years, but got us a Classic Prop Logic and essentially exact math by 1948, we ***never got a comprehensive, understandable explanation*** of the *logic*, **the horse sense of it all**. **That's what this book is all about!!!**

Beginning to Expand our Insight

It is absolutely essential that we first show you how to grasp the Basic Horse Sense Logic of Props **Quickly**, ***superemphasized***, because, explained the wrong, long, complex way, it's just too easy to get lost. **We nail it, repeat, expand!** We very purposely use ***this Introduction to get you to understand the big picture***, then slowly take you into a quite sophisticated grasp in bite size steps, repeating, helping, summing up, ***so you can go as deep as you wish!***

Here's the way we do it, ***superemphasized to help novices get it:***

1. This **Introduction** to get you started with ***great overall*** **Insight**.
   The 1 page X-Ray, Incisive, Magic Basic Core Insight. page **I - 21**.
   The slightly more sophisticated **2 page, very concise Grasp.** p. **I-22, 23.**
2. The ***Helpful*** **~1 hour Book II Primer** to get you well into it all.
3. **Chapter 9 of Book I**, the Airplane Book, targeted for the "Thinking Man Pilot", *p.147, ultimate Blade Loading Family of Shapes, Twist.*
4. The ***Smooth Read*** **Chapter I, Book II**, really the **Advanced Introduction to Book II, Propellers**. ***Advanced a ~1 day read!***
5. The **Pro's Insight, all of Book II**, no math, but more incisive insight than has ever been available anywhere before.

We use a few key Graphs, PICTURES, of how things work --

Despite some of the most Elegant Logic and rare genius level mathematics work ever done, there has never before been a comprehensive, incisive, understandable explanation of props, their logic, and **what all the math means, instructs, and implies**, in a way that even pros could fully grasp.

The great misfortune is that despite some absolutely magnificent technical work by 1948, the lack of an incisive explanation has resulted in even Aerodynamacists ***not*** correctly understanding the real core engineering logic, the understandable horse sense of prop logic, almost no one, even among propeller designers, correctly understanding the ultimate way to go.

It was only with the advent of **early smaller computers in the 70's** that (only) the best companies got started on Betz caliber design, but complex, **the accurate, simple insight here was pretty much missed. We never got an ultimate Triple Optimum Prop, our 2004 *WhirlWind* 200 RV 8 *a first!***

You can spot a Triple Optimum Prop, Min Area Precisely Placed, by its Shape!!!

It's all simply proper loading vs. radius, and that means Shape, with correct Twist.

The **Objective** here is to ***get Propellers correctly understood for the second Century of Flight, accurately designed***, no more seat of the pants guesses, **precise designs easy enough!!!**

(We may now even beat the Betz Ideal - simply a modified SHAPE, *correct* twist!!!)

Beginning to Expand our Insight

Getting Twist Right - *The* Betz *Minimum Induced Loss* Prop

Getting a Major Fundamental Nailed -- Easily!

In Gottingen Germany, at the University, Prandtl concluded that (at a constant angle of attack and C_L), the **Elliptically Shaped and Loaded Wing was Ideal**, because that lift naturally fell to zero at the tip, gave Minimum Radial Flow, **Minimum Tip Vortex**. His friend, **Betz**, concluded the much more complex Propeller had the **Minimum Induced**, if it ***pulled in*** a ***Perfect Helical Pitched Air Inflow***, and ***threw back*** a ***stretched still Perfect Helically Pitched Outflow***, downwash, backwash, perfect **Archimedes Screws**, (212 BC), **four**, a pair from each blade, **two separate rotating wings, 180° apart** in their **Stream Tube**, ***nominally half the speedup in front, half behind!*** **See the cover picture**.

To **Advance Forward**, the prop should logically be ***Overpitched a little to account for the Inflow***, (not the outflow, it turns out), and if we make the resulting **SLIP the Same at Every Radius**, it would ***act like* Even Inflow Velocity**.

Logically, we might expect that there is an **Optimum angle of Attack** and C_L, (and there is, ***nominally*** **.5 C_L, ~1.1° angle of Attack, α°**). Do you see the hint on where we're heading? If we use an **optimum α°, and C_L**, we should be able to get **min. Profile air friction Drag**, **if we distribute the minimum area optimally!**

Now we've thrown quite a bit at you, ***perhaps too quickly***, but we're right up to **an insight of Major Importance**. If we simply **add a constant optimum α° to our perfect helical screw inflow, we have accurately, and pretty easily defined our propeller *blade twist* and Specific Pitch Angles, β°, at any radius!!!!!**

Plane Speed, + Inflow, + α°, *vs* Rotation Speed, sets the Pitch Ramp Angles!!!

So **our prop blade does not have perfect helical Pitch, just a tad different**, but actually, **simple enough, once you catch on! The prop is Pitched a little extra to account for pulling in air to throw back, so it Slips, *Uniformly*, Advances a little less,** but you have **Major Insight in one page, *Twist Understood!!!***

Much later, we'll correct angles for *Slowdown*, the *nose Velocity Profile!!!*
This is pretty typical of Prop Logic - tricky details, but OK, soon Understandable!!

Making Propeller Logic Understandable

A **PROP** is *trying to load itself INSIDE OUT*, **DEAD WRONG**, **DUMB**, *compared to what a wing needs!* A **Wing** creates **Max Lift at the Center** and simply falls off, pretty much like *an Ellipse*, no matter what shape it is, because **Wing Lift *Must*, *Will Fall to* ZERO AT the TIP**, a ***TIP VORTEX* Swirling from the *Higher Pressure Bottom*, to the *Lower Pressure Top***, (even tip Dams only partly effective). But here's our Dumb Prop with very ***high Velocity Tips*, Huge Dynamic Pressure, q, Tips, (Proportional toV^2) trying to make max. Thrust and Drag AT the Tips, where Thrust will be LOST, must fall to Zero, creating an Extra Big Tip Vortex, *thus* Extra Induced Loss, *extra* Unrewarded Profile Drag, *at max. radius*, Max Lever arm, extra unrewarded Torque, bogging down the Engine RPM, Losing HP, Losing (4), four Ways, DUMB AS A STUMP!!! Creating NOISE too!!! It's our job to be smart enough to COUNTERACT that, Taper the Tips, Pull the Excess Loading Back Away From the Tips**, *EASY KEY INSIGHT!!!*

HOW ABOUT AN EQUIVALENT *TIP LOADED* WING

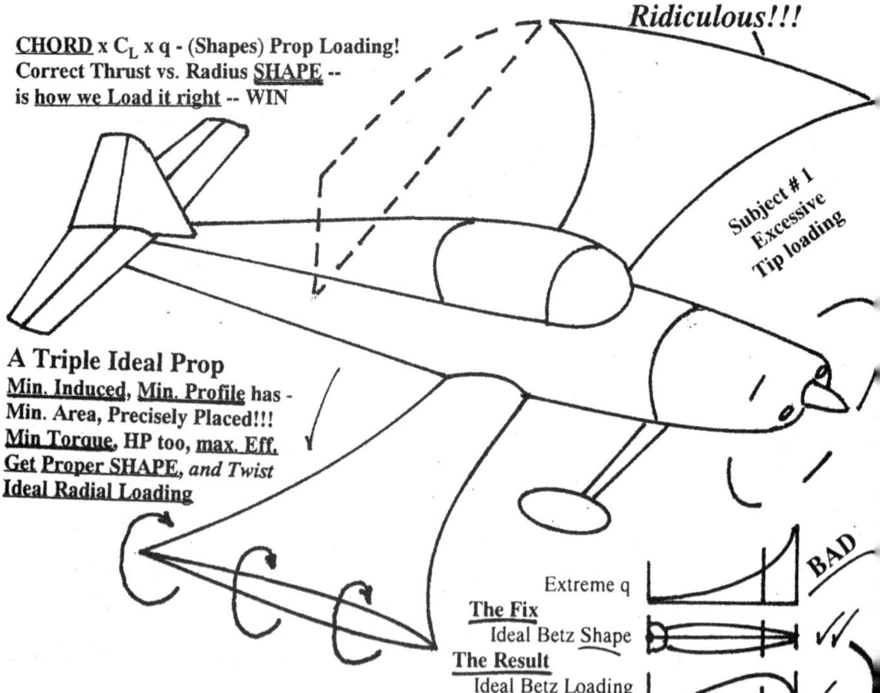

HUGE INSIGHT ON THIS PAGE!!!

> ***This is the most Important and easy Insight You'll Ever Learn about Props!***
> **It has been mostly missed, mostly not understood for the last 100 Years!**
> **You don't win *playing games* with a wider tip, make Lift, where we just pulled it back!!!**
> **(*Later we may be able to beat* the Betz 20th Century Ideal Loading and Shape!)**

Moving on to the Advanced Introduction

We'll do a Little Mid-Stream Review

To Solidify Your Grasp of the Core Logic

(The More Facile Your Grasp, the Easier, Better Going On!)

Then, We'll Bring in Other Expanding Insights

Give you a Very Good Grasp in Just this Introduction!

We'll also Show you How to Cherry Pick -
Many of the Other Interesting Insights in the Book!

Some Pilots may be Satisfied with just the Introduction

So we'll Help, Show You a Way to Maximize That!

If you tried to Learn 25 insights, on one Pass, You Wouldn't Remember!

But the More We can Build This into an Integrated Story,
Repeat the Basics, *continually Tie Together*, as we expand Your Insight,
Make a well Integrated, Logical Story, You Might win a Great Grasp!
So That's the Way We're Building This!

(If we could Discuss Props for a Day, we could get you a Pro Grasp)
So there's a Day's Worth of Meat, Discussion Here!

Using the Classically Perfect Prop to Understand Prop Logic

Because, with ~ Perfect Math we can teach you Thoroughly, Accurately!

Later, we'll show you how to do even better!!!

Rankine did the first prop analysis in 1865. Betz defined the Ideal Minimum Induced Loss Propeller in 1919, and by 1948, Goldstein and Theodorsen had created rare genius level math for an essentially exact, Elegant Math Solution. But we never got a comprehensive, understandable explanation that went way past the math to fully understand **What** the math was doing, and **Why**, fully understanding the Logic, the Engineering Horse Sense of the Engineering Swamp involved. That's what this book is all about, **showing you the understandable horse sense that has been hiding in the Genius level Math since 48.**

We learn that a prop **Pulls in and Throws back (heavy) Air** to make a Reaction Force Thrust that ***causes induced loss***, that it's an **Airscrew**, and ***only High Pitch, High Advance per rev props can be most efficient***, that as a **Rotating Wing**, a prop is being "Dumb as a Board," trying to ***greatly overload the tip*** area, causing greatly excess and unnecessary Induced and Profile Drag losses, turning those *losses* into objectionable **Noise**, ***drag***, **losing engine RPM**, thus **Power**. And we learn that we can stop ***both*** unnecessary losses by loading the prop Ideally vs. radius, **by just Shaping it, Twisting it Ideally vs. radius, an Ultimate Triple Optimum, Min Induced, Profile, Torque Prop, precisely defined by what is maybe the most Elegant Logic and Math you'll ever find.** The math and a lightning fast computer, working in the background, allows us to give you **precisely correct, Horse Sense clear, X-ray Insight**!!!

But there's a lot to teach you, all interrelated, so we repeat, expand, tie together, *Superemphasize it*, make it super clear. We need to get it Super clear to see all the possibilities.

The *core* that the logic and math is doing is pulling excess loading back away from the tip to prevent *both* excess losses, *~optimally!!!* But that mathematically optimum blade loading is max. at the 70 to 80% radius, **still very high at the 90% radius**!!! As we use the logic and math that has been available for a half century, but never fully understood well enough to get Triple Ideal Props, ***teach you***, we'll see that even more tip unloading may beat the theoretical ideal with a lower, more elliptical outer blade loading.

First we'll learn the classic Betz logic Prop - Then Look Beyond - A New Century.

3 Physical Insights can Help us Understand Propeller Logic

This is the *1 Page Magic, Specific Core Insight*

Propellers **1. Pull In** *and* **Throw Back Air** to **Make a Reaction Force Thrust**. They are **2.** an **Airscrew**, **3.** a **Rotating Wing**.

- **Magnificent Math** *can precisely* **Computer Design Betz props**!

3 Core Basics Teach us the Underlying Physical Logic of Props!!!.

- **Betz taught us the core Logic of the Ideal Min. Induced Loss Theoretical Prop in 1919**, first done Profile Drag Free, *basically* **creating pure helical Screw Inflow,** *stretched* **Outflow**.

1. A **Fast Prop** has an **amazingly high Mass Flow Rate of Air** going through its Disk area, and **thus it can make its required thrust with a Lower Delta V, ΔV speedup**, and since **throwing air to make lift or thrust is the root cause of Induced Loss, we want to go Fast to minimize Induced Loss**.

GRASP THAT THROWING AIR -- IS THE BASIC CAUSE OF INDUCED LOSS!!!.

2. A propeller must be understood as an **Airscrew**. Since a **FAST**, low enough RPM, **High Pitch Prop**, has ***both*** a **minimum ΔV thrown air Induced loss**, and a **Shortest Spiral Path to the Destination, minimum Profile Drag Energy Loss. Only High Pitch, High Advance per Revolution Props, can have max. Efficiency --- pretty simple once you see it!!!**

SPEED and RPM -- SET A CAP on MAX EFFICIENCY!!!

3. A Rotating Wing has a serious, basic flaw: The Prop ***tip*** is going so fast, its V^2 Dynamic Pressure is so much greater than the root, that it trys to **Greatly Overload the Prop Tip**, where the **Thrust must, and will fall to Zero** in an extra big **Wasteful Tip Vortex, Losing Thrust, much higher Profile, air friction Drag**, at **max. lever arm**, *(high torque)*, **bogging down the Engine RPM**, **losing H.P.**, **losing 4 ways**. If we **simply taper the tip**, pull the excess loading back away from the tip, we can ***counteract***, **prevent the excess loss** - ***grasp that!!!***

Precise Radial Loading, A family of Ideal Shapes, vs. Twist, sets the Ideal!!!

- **Goldstein,** *then* **Theodorsen**, by **1948** had created an *essentially* ***exact heavily loaded 3D flow solution, magnificent!***

We learn a **Modern Computer, Elegant Math** can give us a **Mathematically Triple Optimum,** *min. induced, min. profile*, min. torque, **ultimately Minimum Area Precisely Placed, *fantastic insight*!**

We can **Learn the Basic Horse Sense Logic** with 5 Overview Insights, 3 Core Logic Insights, expand, broaden, deepen that insight, **let the Computer create ~perfect Designs for us!**

The Propeller - *Understandably Explained* - after 138 Years

The 2 Page Incisive Summary

A Propeller is **1.** a **Rotating Wing**, **2.** an **Airscrew**, **3.** it **Pulls In** and **Throws Back Air** to make a **Reaction Force Thrust**. That's all accurately constructed by some of the most **Elegant, Rare Genius Level, 3D Optimizing Math** ever created. Happily, it is those **three simple insights that can teach us the Logic**, **the understandable horse sense insight of WHAT the Elegant Math is doing, and WHY it is doing it!** Thus, the Elegant math and the lightening fast computer can simply work in the background to **give us a precisely calculated, Triple Optimum, Minimum Induced, Min. Profile, Min. Torque Propeller**, a task any computer buff with the program can do!

See the Cover Picture of Betz's Perfect Helical Screw Inflow, Stretched Outflow

The computer will **precisely Shape, *Twist*** the prop blades to create **Betz's perfect screw air inflow and stretched outflow of the thrown air**. If requested, a constant angle of attack, α^o, a constant chosen optimum Coefficient of Lift, C_L, can be used, **correctly Pitched**, marvelously accounting for the **required extra inflow** for **Heavy Loading**, where what is happening right at the propeller is significantly different, and ***greater than the average speedup*** over the **stream tube the outflow creates!**

The core key to making all this Optimum Magic actually happen is in Optimally Loading the Blade vs. Radius, by simply Shaping the Blade vs. Radius, along with the Matching Precise Twist, Pitched for a Constant Slip vs. Plane Speed, which acts like perfectly uniform axial inflow, actually created by highly varied axial inflow and rotation, perfectly, geometrically combined for the Constant Slip!!!

See the Family of Perfectly Shaped Props on the Cover.

Actually, it is the Engineering Horse Sense of the Logic that teaches us that the Dumb Prop is trying to work Inside Out, OPPOSITE to the logic of a proper wing which wants max. load at the center, because the lift or thrust will fall to zero at the tip - and here the very high speed tip, extremely high *Velocity² Tip dynamic pressure*, q, is trying to create extreme overloading of the tip, a greatly excess, wasteful tip vortex loss, *noise*, losing the thrust, which **cannot be held**, also

extra profile drag at max. lever arm, bogging down the engine torque, the **RPM, and thus H.P. -- losing four ways!!!**

See the Ideal, Half Teardrop Loading, next to the Ideal Blade Shape.

It is actually the precise, systematic, accentuated tapering of the Outer Prop Shape, the whole shape, that precisely, **optimally counteracts** this **wasteful natural tendency of the rotating wing propeller.** The whole, **simple, hidden secret of the Ideal Propeller,** is a **Family of Ideal Shapes, varying vs. Pitch, or Advance Ratio,** the low Pitch, Low Advance Propellers, being more highly tapered - **The Ideal, Fast,** ***sufficiently*** **Low RPM,** quite **High pitch, High Advance prop** ***wanting*** **a perfect Canoe Shape, max. Chord at Half Radius.**

A 2700 RPM ***sleek craft*** **at ~ 364 MPH seeks this Optimum Shape!!!** **See p. 147**

They all are, in fact, a **fully integrated family of Optimum Designs, mathematically essentially perfect, all done by the computer and the genius level math, sight unseen to those who wish to ignore the math, but quite understandable to the professional trained to appreciate the** ***calculus*** **Elegance!**

The Fast, High Pitch, Short Spiral Path prop can be Most Efficient!!!

The designs are finally ideally corrected for the actual average velocity profile at each radius at the nose of the embedded body. That ***final key step*** **makes the prop better** ***slow, T.O. and Climb.***

Reduced angles at slowed inboard radii can avoid stalling - slow!!!

It's perhaps the most Elegant total solution in all of Engineering, but it's ***easy enough for an informed amateur to grasp and use.***

The key is to Superemphasize the magic Insights, make it easier!!!

It is as important for an Amateur and Pro to grasp the Logic, the Understandable Horse sense Insight, as it is for a pro to grasp the Elegant Math, because it is the Logic, the Horse sense, that tells Why it is all needed, How it works out!

Two separate rotating wings, 180° apart, there's a root vortex too!

The Simple Bottom Line: It gets the **SHAPE, TWIST, BLADE ANGLES, PITCH, DIAMETER, the ASPECT RATIO,** the blade width distribution, all **Precisely Correct** for the Specification Requirements, exactly the **Understandable SPECIFICS** the professional or amateur wants to know!!!

It is an Elegant Masterpiece of Logic and Math took 138 years!
It is as comprehensive, incisive a Science Insight as you are ever apt to find -

Rankine, 1865 to Dec. 17 2003, 138 Years!
The 100th. Anniversary of Flight.

Jack Norris 12-17-03

A Complete Outline of *Classic* Betz *Min Induced* Prop Logic

Betz Core Objective: Getting Ideal Minimum Induced Loss!

(Basically Achieved by Pure Helical Inflow, *stretched* helical Outflow)

- **It's purposely First Done on a Drag Free basis, the simple case first ---**
- **Drag can then be easily added, Min Induced, *then* Actual Drag** *(separated)*.
- **It's an Archimedes Screw into, out of each Prop blade** (see Cover Photo).
- **Achieved by designing for Constant Slip, *holding constant Pitch Inflow*.**
- **Slip is the difference between inflow speed and Airplane Speed, *constant*.**
- **(Mathematically that is done by holding r tan ϕ constant. (ϕ inflow angle).**
- **The Geometry teaches how *highly variable* inflow and rotation combine.**
- **Amazingly, a Constant dT/dQ results, a constant ratio of Thrust vs. Torque.**
- **That means a constant efficiency at every radius, *steep root angles too!***
- ***But remember, it's first done Drag Free, then Drag Added,* separated.**
- **Bottom Line: that results in "Ideal Betz Loading", Thrust vs. Radius.**
- **Betz Loading is like a half tear Drop**, weak root, stronger, hi q outboard.
- **A Family of Ideal Blade Shapes *and* Matching Twist** *results, if constant α, C_L*.
- ***Most Significantly*, Tapered Tips *vs. Advance Ratio*, Pulls Back Excess Tip Loading.**
- ***Not Allowing* Greatly Excess Tip Induced and Profile Drag at hi q Tips.**

That final Hidden Imperative is the SECRET, Never Adequately Seen!.

--THEN--

We teach the Logic of Props *by just looking at* 3 Physical Insights.
But, it's all Backed Up by Elegant, *rare genius* Goldstein Theodorsen Math.

Pulling in, Throwing back *Heavy* Air - a Reaction Force Thrust

(We teach Props without Math, but two little formulas here, give us the easy key insight.)

- **Induced Loss *is just the* Energy Cost, Loss of Throwing Air.** *It drops to calm.*
- **Newton's 3rd Law teaches us how to understand Reaction Forces.**
- **Thrust $T = \dot{M}\,\Delta V$, M dot, *just like it looks*, Mass flow rate, *x speedup* Delta V.**
- **Axial Efficiency loss is half Delta V, $\Delta V/2$ vs. Speed, V, -- *thus*, $\Delta V/2V$.**
- **There is actually 3D Induced Loss, Rotational and Radial, (Tip Vortex) too.**
- **Rotational, Radial are substantial, but it's Axial that teaches us the logic.**
- **(Airplane Logic teaches us Induced is a $1/V^2$ loss,** at twice the speed, 1/4 loss).
- **(*Props the same*. If V doubles, ΔV can drop to half,** thus loss 1/2, divided by 2, 1/4.
- **Thus the Objective is: go Fast, *have a* big $\dot{M}$, a Doubly reduced Axial Loss!!**
- **Slow, to be efficient, requires a Big Diameter - Fast** *allows a Small Diameter.*
- **(Radial Induced, can be as big as axial, too big! A clue, it can, should be less)**

After a simple enough start, there is finally a lot of insight that we can teach you. Reading onward you might get into overload, get confused. A *complete* list like this can keep you straight, *cut off confusion*. Read on, *use it for that*. The final objective is to *have the whole list make sense to you*, and this helps.

Understanding a Prop as an Airscrew

• A Prop wants to be a FAST, *Low Enough* RPM, Hi Pitch, Hi Advance *Airscrew* pulling in and throwing back a Betz pure Helical Screw Inflow, a stretched pure helical Outflow, *for a* Big $\dot{M}$, *min.* ΔV, *min.* ΔV/2V Induced axial loss, *and* total 3D circumferential, and Radial Induced Loss too, but also a shortest helical path, for least Profile Drag Energy Loss, as well.

• A Prop wants to be *effectively* a 45° Airscrew, nominally a 45 degree Blade at ~ 2/3 + radius, because that steep screw is the most efficient, yields the Optimum Blade Planform, a Canoe Shape, Max Chord at ~50% r!

• There is a constant mechanical screw ratio of Thrust to Torque, dT/dQ, at low or high blade angles, constant efficiency, at every radius, if first considered Drag Free, but profile Drag loses efficiency at both shallow and Steep β angles, *a classic rounded crown corner* hat shaped curve.

• Airplane Speed, Engine RPM are the Primary Controllers of Propeller Efficiency, η, both setting, limiting Pitch, Advance, the Max η Cap possible.

Understanding, Correcting, Optimizing a Rotating Wing

• Excess tip V^2 Dynamic Pressure, q, loads a rotating wing inside out, 1. loses thrust, 2. creates excess tip vortex Induced, 3. Excess *unrewarded* Profile Drag loss at max. lever arm, 4. excess torque limits RPM, HP. Pulling back tip loading by tapering tips, is *the core of intelligent design*.

• The Classic Betz Min Ind., half Teardrop loading gives key Insight!!! *Forcing Basic* Wing Elliptical Loading offers higher η for 21st Century!

• The Betz Pure Helical *Inflow* + Constant Optimum α° *easily* sets basic *blade Twist* and β angles -- later *accurately corrected* for body nose Velocity Profile.

• Using Optimum α° and C_L on a Betz Prop, we get *a Triple Optimum Prop, Min.Induced, Min.Profile, Min.Area Precisely Placed, Min Torque!*

• The Family of Double Optimum, Betz- Goldstein- Theodorsen Props shown on p.147 of Ch. I, which varies from *almost triangular* at low pitch, low λ Advance, to Canoe Shape at High Pitch, High λ Advance, is the proper bottom line objective of Classic Prop Logic. Using *High Drag wide tips,* is a negative!

• Loading a Prop Blade correctly vs. Radius, By Correct SHAPE, *Twist*, also optimum α°, C_L, is the heart of intelligent Propeller Design!!!

Goldstein- Theodorsen Math

• Goldstein solved the 3-D flow Math in 1929, *but* for evenly loaded flow.

• Theodorsen, using Goldstein, solved 3D Heavily Loaded Flow *in 48* by considering the needed greater full ΔV, his w, Far Back, with a slightly smaller D. His Kappa relates his $\overline{w}$, to the Stream Tube average ΔV.

• *Their* Kx factor, (vs. Radius and Advance Ratio) *sets* Radial Prop Loading!

• Simple Charts of Kx, Computer *interpolated, greatly simplifies the Math!*

• Appendix T, shows Theodorsen's Math to design Props, using Kx, *etc.*

The Ideal 45 Degree Blade Angle, and **where it's located**

When I was a late teenager competing and winning Nationally, I had very good plane designs, but was really frustrated trying to find correct prop design specifics. What was the **really *correct shape*, *twist*, *pitch***? **It seemed that those high angle prop root radii, pulling sideways against engine torque, more than forward, could not possibly be good.** ***But***, we've learned here, that **drag free, a Betz prop can have a constant Thrust vs. Torque Ratio, every radius equally efficient, productive,** (*if first considered drag free*). We'll look more at high inboard blade angles shortly, see that they're perfectly OK, working as a screw on very high pitch props. We'll also see a **Blade Angle of ~ 45° is optimum**, thus it's insightful to see ***where*** the **45° blade angles are vs. Advance.**

Also --- Where is the Max Chord vs. 45°?

The study below looked at **J/π**, (Lambda's, **λ**'s), of **.2 to .88**, **(of low to very high pitch props)**. **(J/π, λ Lambda, is the advance ratio of Prop Tips.)** The chart below and the chart and Shape Plot vs. load distribution on the next page give great insights into the characteristics of ideal propellers and are worthy of your study.

	Reasonable Advance		Extremely High Advance				
J/π, λ	.2	.3	.4	.5	.6	.7	.8
Overall η	.8607	.8940	.9068	.9125	**.9149**	.9144	.9112
45° β,% r	23.25%	33.60%	44.15%	54.82%	**65.74%**	76.70%	87.82
Local η	.8781	.9018	.9104	.9141	**.9159**	.9158	.9135
Max Chord.	25% r	38% r	43% r	47% r	**49%** r	53% r	56% r

Amazingly, a Reno Racer can have a .88 λ, almost a 45 degree Tip!!!

So we see the most efficient **45°** blade can fall way inboard to way outboard - but is **most efficient at the ~2/3+ Radius at .63 λ**. Now if the ~**45°** Blade Angle is most efficient, it might be that the splendid Math might put the **Max Chord** there. Well, if you look close at the data above, the interesting insight is that's trying to happen, at the lower **J/π, λ's** where fixed pitch props operate --- but are pulled outward a little at the low **J/π, λ's** away from the inner root vortex **--- much stronger inward at high J/π, λ's away from the stronger tip vortex!!! Now, isn't that *smart*! It fits right into the logic, not overloading tips!!**

Interestingly, the max Chord is at ~50% radius on the max η ~.63 λ Prop!!!

The Change in Blade Shape and Twist --- Low Pitch vs. Steep Pitch.

We know it's the **Slow planes, Low Pitch, <u>Low Advance</u>**, that have **the *huge magnification of Dynamic Pressure, q,* <u>tip vs. root</u>**, thus **need the <u>most highly tapered blades</u>**. ***<u>Conversely</u>***, the **<u>Steep Advance</u> props will be <u>less severely tapered</u>**. Below, see the blade outlines for this blade study, λ from .2, low, to a steep **.88λ**, (a Reno Racer class prop, though this study of all 6' props for comparative consistency is a miniature for Reno). Note how the **<u>steep λ props</u> *<u>taper in greatly inboard</u>***. (**<u>Slowdown</u> *lowers blade angles, especially inboard, widens blades.*)**

MAKES THEM MUCH BETTER AT SLOW SPEEDS, TAKEOFF, etc.

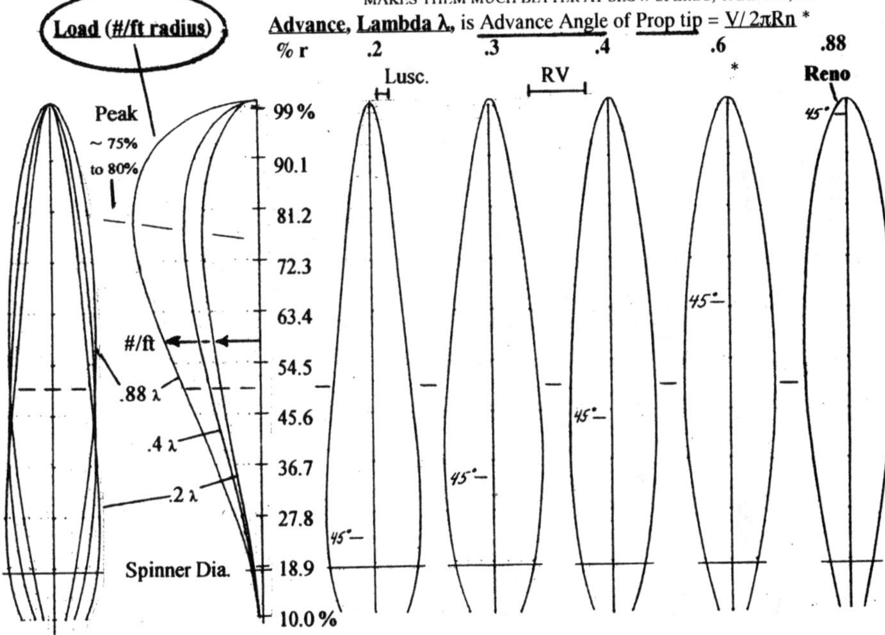

- - RV 8 Cowl Width -- Locations of 45° β °. Ideal ~.63 λ * ~364 MPH.

To keep all props on the same apples vs. apples comparison, starting with a 6 foot prop we end up with low thrusts and H.P.'s for the fast planes, like a miniature, slick Reno Racer finally, but this shows shapes and twists well, the intention here, so just imagine and understand.

*THEODORSEN ACTUALLY HAS A MORE COMPLEX DEFINITION OF λ THAT INCLUDES $\bar{w}$

***<u>Prop Blade Study basis</u>*: 6'Dia, α° 1.1°, .5 C_L, AR 15, <u>2700 RPM</u>, 8000', Area 1.944 ft²**

Ref. Data pages 99.56 -1 to 99.56 - 8.1 — BLADE ANGLES ARE *<u>BEFORE</u>* SLOWDOWN CORRECTION

λ	Vmph	HP8000'	Thrust#	Eff. η	β°, 18.9% rad.	β°, 99% rad.	ΔTwist°
.2	115.6	56.5	155.3	.8607	50.365°	13.600°	36.765°
.3	173.5	94.4	179.7	.8940	60.583	19.046	41.536
.4	231.3	140.5	203.6	.9068	67.053	24.263	**<u>42.790</u>**
.5	289.2	195.9	228.5	.9125	71.382	29.142	42.240
.6	347.0	260.5	253.8	**<u>.9149</u>**	74.440	33.634	40.806
.7	404.8	334.5	279.1	.9144	76.715	37.760	38.955
.8	462.7	418.4	304.3	.9112	78.482	41.556	36.926
.88	508.9	494.2	324.6	.9058	<u>79.653</u>	<u>44.411</u>	35.242

This Page is really the Bottom Line on 138 Years Work --- 1865 to 2003! ✓

A Look at Elliptically Loading a Prop Blade

At a constant angle of attack, constant C_L, **Prandtl's** theoretical Ideal Wing Shape, and Ideal Wing Loading is ***Elliptical***, max. at the center, falling to zero at the tip, ***minimum radial flow, min. tip vortex, no extra unwanted tip induced loss***. Of course, very much unlike a prop, all radii are going the same speed, **same q!**

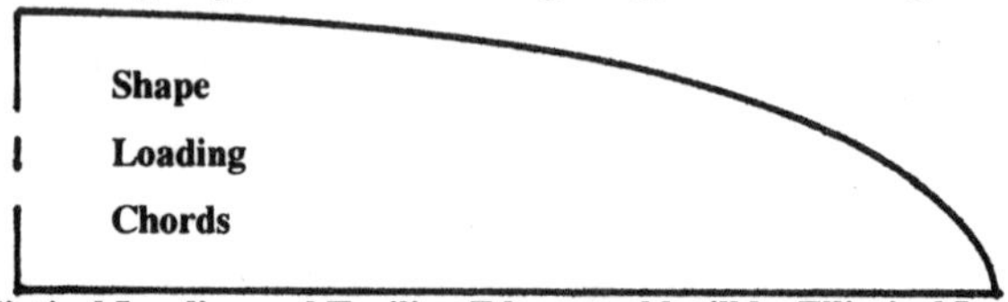

Elliptical Leading and Trailing Edges would still be Elliptical Loading

Now, if we **try loading** a constant angle of attack, α^o, and C_L, **prop blade, elliptically** vs. radius, like a wing, with far higher dynamic pressure, q, outboard, we get very narrow chords outboard, **a wild shape curving to excessive chords inboard**, due to rapidly falling q -- can really only do it to 40, or 50% radius, must then taper the blade to a practical root, with lift falling dramatically. If we tried to hold blade forward thrust elliptical, not blade lift, it would be worse, (at steeper inward blade angles).

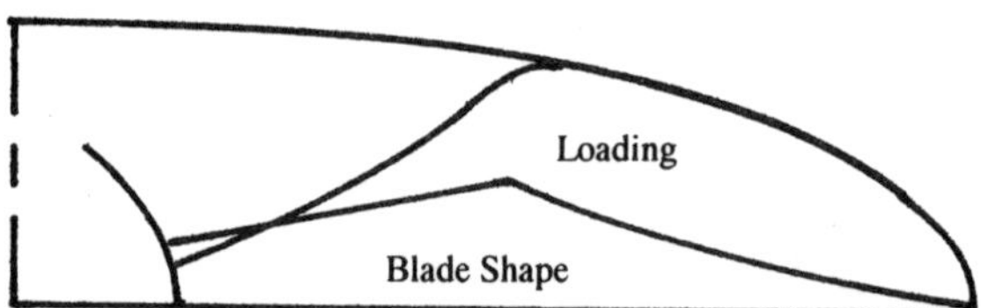

A Propeller Blade Shaped to hold Elliptical Blade *Lift* Distribution, (not Thrust)

But if we superimpose this elliptical lift distribution on the half teardrop radial lift distribution of an Ideal B-G-T prop, we can see that in moving the blade loading more inward, we would **logically lose less off the tip**, ***better than the B-G-T Ideal!!!***

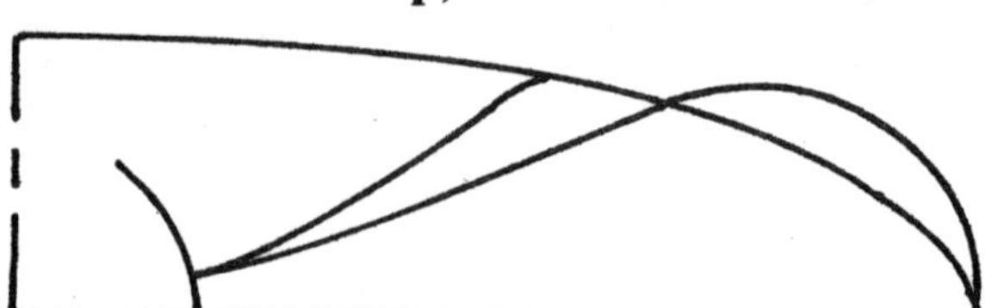

The Betz and Elliptical Blade Loading Directly Compared

With a wild looking prop, we can beat the Classic Betz Ideal!

A Look at Elliptically Loading a Prop Blade

There is a lot of potentially important insight to be gained here. We'll learn that Ideal B-G-T propellers tend to have potential efficiency caps of nominally 85% for a 100 MPH Luscombe, to 90%, or a tad more, for a Fast, High Pitch, High Advance RV, but if it were an efficient elliptical wing, a 20:1 L/D, only 5%, or less, loss would be easy to achieve, actually better than that. When we look deeply into Theodorsen's book, we find Radial Flow, Tip Vortex Loss, an auxiliary, add on loss, can actually be bigger than the axial loss we must have to make Thrust, certainly not a good situation, we'd like to have to live with.

So, the creative new thought, after a Century of Flight, of making a prop act more like a wing, has real merit that was never seen by anyone in Aerodynamics before, so let's look closer.

A high Aspect Ratio Sailplane Wing has very low Induced loss, because there is a lot of Center Wing maintaining max lift, undiminished, far away from the tips, can have small narrow tips, an Ideal Situation! Now, that prods us into a very important insight. Looking more closely at the prop teardrop, or outboard elliptical lift distribution, we tumble to the important insight that, either way, **a prop has almost NO Center Wing working**, it's **effectively all tip**, **almost only two outer halves working**, **much more vulnerable to excess tip vortex loss**, **a major insight**!

So, simple bottom line, even if we switch and start forcing a prop to act more like a wing, we're not going to get down to very small percentages of loss, but, with a creative, but wild looking prop, **we can potentially do even better than a classic optimum prop** - and do it by just doing what the horse sense tells us to do, **get the load pulled back away from the tips**!!!

Now, we can get over 90% efficiency, η, maybe 92, 94, a task for Century 21.

We have to be careful, because a metal prop like this, might fail much more easily in vibration, potentially kill you in a crash, perhaps giving more advantage to a composite prop. Requiring new math, this creative new idea by Paul Lipps of Arroyo Grande California is the new idea for the new Century of Flight!

The Simplest, Best Summary Insight on Propeller Logic

For the smaller, slower, non jet, personally owned airplane, the slow, or very long range military reconnaissance planes, the **propeller should always be the most efficient, lowest cost, lowest fuel usage method of propulsion**. With duct losses, no form of Fan Jet, Ducted Fan, or other method of propulsion *should* ever be able to match the maximum 90% plus attainable propeller efficiency. With amazingly high compression ratios, over 40:1 on the GE 100,000# Thrust engine for the Boeing 777, the advancing Fan Jet is amazing, but propellers, especially with optimum radial loading, especially if it is possible to optimally gear and pitch them, should, it seems, stay in the lead.

The Fatal Flaw in Propeller Logic is that with greatly faster tip velocities, even more greatly magnified V^2 dynamic pressure q, at the tips, vs. the root, the dumb prop really is trying to work inside out, just exactly ***opposite*** from the way a wing is going to work, max. lift at the center, where lift must, and will, fall to zero at the tip - in a losing radial flow tip vortex. Automatically magnifying that, the prop case is the worst possible situation, and the best we can do is to **counteract** that unfortunate tendency by **significantly tapering the tips to significantly pull back the loading radially inward, to a more manageable lift or thrust distribution**, that can better be retained, not lost.

This is the HUGE Fundamental that was Never Adequately Recognized!!! .

The **Betz, Goldstein, Theodorsen Classic Logic and Math** clearly is a marvelous integration of creative, multifaceted, **Elegant Logic and rare genius level Math**, aimed at minimum Induced Loss, initially done drag free. Actually, profile Drag, even correct for applicable low Reynolds Numbers can be added, the prop subsequently corrected for the embedded body nose average velocity profile at each radius -- or analyzed for off design point, are beautifully possible -- **Elegant, a tribute to amazing human intelligence, and the modern personal computer.**

Unfortunately, we can see that the Half Teardrop radial thrust distribution of the Classic B-G-T Minimum Induced, Optimum Prop, is still biased heavily outboard, max. at 70 to 80% radius, nominally 75%, still high at 90%, that it is logically farther out than is best. Other than bringing in the Modern PC, creative, genius level work on the propeller stopped in 1948. **We never even got a comprehensive, understandable explanation** until now, a half Century later. That did hurt, propeller designers never did really understand the true final logic, **we never did get the Triple Optimum Prop, Constant Optimum α^o, C_L, Min. Induced, Profile, Min. Area Precisely Placed, Min. Torque!**

The 2003 WhirlWind 200, RV 8 Prop, is the First Triple Ideal Prop ever!

Worst, maybe never actually realizing all the ultimate Engineering horse sense, bottom line grasp, the fact that the B-G-T radial thrust distribution is still heavily biased far outboard, no one in the first 100 years of flight backed off, thought outside the box, as they say, and saw that if we simply pulled in the distribution more, went for the ideal elliptical thrust distribution of a wing, that a practical improvement of a few more percent, certainly desirable, if safely attainable, is available. **Paul Lipps, of Arroyo Grande, California, a highly intelligent Radar Pro, and Private Plane refiner, clearly deserves to become the 11th. benchmark achiever in the history of Propellers, for creatively providing us this path to ultimate efficiency, for the Second Century of Flight.**

Now, that takes care of **completely explaining and correcting the harmful effects of a rotating wing, excessively loading the tips, creating excess losses.** All the other key insight that we can learn from the basic logic and understandable engineering horse sense of propellers is equally applicable to any prop design

----

Pulling in and Throwing back a heavy, high Mass flow rate, $\dot{M}$, allows us to minimize the ΔV speedup required to generate the required Thrust, minimize the $\Delta V/2V$ axial induced loss, actually complete 3D Induced loss, thus FAST, is the easy way to have an efficient prop, Low RPM, as well,

gives High Pitch, a High Advance per Revolution, **both Min Induced Fast, and Min Profile Drag Energy Loss, with a short, steep, least energy path to the destination!!!**

A Slow Plane, a Geared Down RPM, High Pitch, Advance -- Works too!!!

Thus, the **1. Steep, Hi Pitch, Hi Advance AIRSCREW, 2. Maximizing 3D Efficiency of Pulling in, Throwing Back Air to make thrust,** ***(decoded by Newton's 3rd Law),*** **3. Tapering the ROTATING WING Prop Shape, crucial outboard, to pull excess loading back away from the tip, where it would be excessively lost,** ***NOISE,*** completely decodes what can look like a frighteningly, massively complex 3D flow situation, all backed up by the Elegantly integrated Betz Logic, and Goldstein-Theodorsen Math, which does completely deal with, and essentially exactly solve, the heavily loaded, 3D flow complexity, with us completely free of the details, *if we wish.*

Regarding Betz: It should be said that his core objective is to **pull in a pure helical inflow**, and **throw back a stretched, twice speeded up pure helical outflow**, two separate **Archimedes Screws**, (212 BC), one to each blade, blended at the center, finally stretched. This is done by over pitching the inflow and prop to account for the speeded up inflow, (not outflow), to **hold a constant pitch of that inflow at all radii**, a **Constant Slip**, **vs. Airplane Speed**, which acts like constant axial inflow velocity, (even though you'll ultimately learn that a close look at the inflow and rotation diagram of the actual airflow is quite variable, but magnificently combines in constant pitch, constant slip to act like uniform inflow)!!!

Equally **magnificently**, simple High School math, Algebra, Geometry, Trigonometry, can prove **a constant ratio of Thrust to Torque, dT/dQ, exists at every radius**, (*if initially drag free*)!!! If an **optimum constant angle of attack and C_L is used**, for ***Triple Optimum, Min. Induced, Min. Profile***, **Min Blade Area Precisely Placed, Min. Torque,** ***Min HP, Max*** η, **the blade twist is simply the perfect helical inflow, plus the constant angle of attack** --- ***Voila, a 140 year mystery put to rest!!!***

All up, **the entire Betz Logic, Goldstein solution of the partial Differential equation based math for 3D flow, Theodorsen extending Goldstein's Math for real world heavy loading**, where the full ΔV, right at the prop blades, much greater than the average of the resulting Stream Tube, is **perhaps one of the most Elegant integrated Logic and rare genius level Math Solutions in all of Science and Engineering, a tribute to the towering intellect of man at his best.** ***But alas, human Hubris is disallowed, there is a flaw!!!***

The high humor in the situation is that we never got past the rare genius level math, to see the basic horse sense insight, apparently never saw the family of ideal Blade Shapes, the half tear drop loading of it all, that the teardrop loading was still too heavily biased toward the tip. Not in the half Century after Theodorsen did a professional decode and go past the marvelous work, completely understand it, get it down to specific drawings of the hard actual results, ***a picture one could grasp***, **the understandable horse sense of it all**. **Math can clarify, but not carried out, studied, really understood,** ***complexity can obscure, prevent the understanding that is there!***

The Computer, of course, was the Final, and NECESSARY Tool!!!

Betz and Prandtl, ***might*** have seen that the true proper objective of propeller design was to force it to act like Prandtl's optimum, elliptically loaded wing, with a wild shape, not let it act as a propeller with extreme tip overload, but the math solution of Betz's great insight was a decade, 1929, and 29 years in the future, 1948, before the true picture could be seen. Who's guilty? No one, the Aero Profession maybe. No one ever finished the job, got it down to real understanding before.

A Half Century later, Andy and I just decided to finally get Props Understood!

A huge credit goes to ***Paul Lipps*** creatively thinking outside the box, not encumbered by knowing, buried by the Classic logic. Not in the profession, not encumbered by all the complexity and detail, but very smart, creative, knowing that Prandtl sought **elliptical lift distribution**, he **went for that simple ideal**, ***great work***, *a big belly laugh after 100 extra years*, **Good going Paul!**

It's Repeating, Expanding, Saying it Different ways,
That Makes you *Facile in your Grasp* -allows you to go *Easily, the Goal!*

Props are **1.** a **Rotating Wing, 2.** an **Airscrew**, they **3. Pull in and Throw back Air** to make a **Reaction Force Thrust** per **Newton's 2nd and 3rd Laws**, but that's what causes **Induced Loss**, the ***Energy Cost Loss of Throwing Air*** to Make Thrust. In Props there's **a *3 Axis Induced Loss***, not just axial, the air finally just settles back to static again, finally the energy lost to friction.

The Game is to make a Triple Optimum Prop, Min Induced Loss, Min Profile Drag, Min torque, HP, Max η!

Slow, Demands a Big Diameter Prop, Fast, Allows a Small Diameter, because the game is to work on a **Big $\dot{M}$, Mass Flow Rate**, so the Delta V, **ΔV speedup** can be ***small***, **Low Induced! High Aspect Ratio**, just like a long sailplane wing is ***thus*** **Best.**

It's Finally exactly correct Twist, and SHAPE, Tapered Tips, the Key, Min Area, Precisely Placed!!!

Very High Pitch, High Advance per revolution is **Most Efficient**, a **~ 45° Blade Path** best, a **45° Blade Angle**, way out near the max. blade loading point, **~ 2/3 to 3/4, radius**, farther out the higher the **Pitch, Advance**. Surprise, it's thus **RPM vs. Speed** that **sets your efficiency cap**, ***Speed most Important***, but **RPM** more important than you realize, ***since it's*** **fundamental to Pitch!**

A Short Spiral Path Home, Min. Profile Energy Loss, -- Fast, Min. Induced too, SIMPLE!!!

Betz showed us the **Sophisticated Logic** of the **Minimum Induced Loss Prop** way back in 1919, and he thus established **Classic Propeller Theory**. **Goldstein** gave us the *Basic*, Partial Differential Equation, 3D Potential Flow Math Solution in 1929, but **Theodorsen** showed us how to get the **Real World Heavy Loading, High Advance ratio essentially exact, 1948**, all **Rare Genius Level work**, correct answers, a gift to us all!!!

We make it Much Easier, but there's a lot to teach you. We ***Super Emphasize*** to ***help!***.

Always before, an unexplainable dismal swamp of complexity, **never comprehensively, understandably explained before**, even for pros, the **breakthrough here** is that we can show you the **Engineering Horse Sense** of the **4 subjects, 1.** a **Rotating Wing, 2.** an **Airscrew, 3. Throwing Air, 4. Betz**, the **Basic Logic of Props** down to **Understandable Engineering horse sense Logic, a few key Pictures!** It's this **easy**: you **load it optimally** vs. **Radius**, by simply ***twisting it, Shaping it correctly!***

It's looking at it as a Rotating Wing that teaches us to taper the tips .
The stupid prop is trying to work inside out, trying to load itself Inside Out, max. load at the very fast, extremely high dynamic pressure, q, tip, where thrust must fall to zero!!! Dumb as a board, our job is to be smart enough to counteract that by tapering the tips!!!

So, How are we going to Teach you?

There's **<u>more</u> to learn in the Classic Betz Theory**, but you can **thus learn a lot more going through it**, and with **essentially exact math** developing our data and incisive studies with clear conclusions, ***we can both be sure what we teach you is correct!***

Tibery and Wrench 1964 have later, better Blade Loading, at High λ.

Paul Lipps used his own math to design his props. Full credit to him, he went about it quite professionally, but depended on the ability of modern P.C.'s to iterate calculations where such things as induced angle of attack are not known, the computer looping to finally, supposedly, eliminate errors from assumed inputs. Such basic approaches have no way to cope with 3D flow and induced loss, whereas, radial, tip vortex loss is the core issue we're trying to refine and evaluate. No put down, because his math and computer seems to give excellent design predictions with results that agree with the computer outputs.

On the other hand, his initial work was not corrected for slowdown, a major correction, but we can show him how to do that easily, professionally, accurately, and plan to as we write this, as we're checking the validity of doing it with Glauert's quite good approximate method (~ 1 1/2 % pitch error, etc.), as we also look at possibly adapting Goldstein, Theodorsen math.

All that aside, **the obvious right approach is to teach you the B-G-T logic and math**, *quite valuable in itself*, because it's all logical, understandable, with **essentially exact math, accepted, respected by the very best pros, thus completely creditable** answers, so you can be **sure you're learning it right**, hugely *valuable for us both*, no Mickey Mouse, good incisive learning.

All we really do for the (partial, outboard only,) ***elliptical lift distribution is shape the narrow tips even more severely than the Betz prop***, pull the load in even more, just what we did with B-G-T, just do it a little more severely, **the same basic game**!!!

In the Betz Chapters, next, we'll still call B-G-T *Elegant, call solutions Optimum*, because they are -- until, and *if*, Testing, we finally go for the new, potentially next 21st Century way, going for more elliptical Radial Loading. *If you're interested in props,* **BGT is the way to learn it all,** *correctly!!!*

How to Cherry Pick!

As I built the story here, I purposely repeated, **said the basic logic several ways vs. the other factors,** *because* my objective was to get you as close as possible to being able to flip the logic around in your head, be **facile in you ability to grasp it, use it**.

Many Pilots may have satisfied themselves by now, and I purposely created a pretty incisive Introduction to fit that - and if you wish it, **here's how to Cherry Pick the rest of the Book**:

The Family of Ideal Props - p. 146, 147 of Chapter 9, Book I, p. I 26, 27 here, really is **the Bottom Line on 138 years work!**

Inflow Geometry - p. 128 Ch. 9 Bk. I, then p. 82, 86 Bk. II, shows the first, then "all up" insight into the key inflow geometry! The creation of a **Constant Slip Geometry is basic to Betz.**

Chapter 1, Book II might be considered the more normally written, **"easy read" "Summary Introduction to Book II"**, but with less content, less Superemphasis. It has the very fundamental, important **Advance Ratio vs. Efficiency Graph**. p. 9, (also more extensively covered on p. 48 II), **an RV and a Luscombe Prop** design and characteristics, (a bit of extra Thrust on each), and, at the end, a pretty complete **Definition of Terms List**.

Chapter 2, Book II

Super Magic Chart, p. 45 II, **Sea Level Thrust, Efficiency, ΔV, $\dot{M}$, α^o, C_L -- Zero Speed -- to Zero Thrust Speed** - where Prop runs out of sufficient Pitch to make Thrust, **at actual RPM vs. available H.P.**, for a **.55 C Prop designed at 12,500'**

Magic Chart, p. 50 II, **Sea Level Efficiency vs. Zero Thrust Speed**, without a Slowdown Correction of **Various P/D Props**.

Baldeing Efficiency, Profile Drag Efficiency vs. β^o, p. 60 II, Profile Drag effect, **low at normal angles, worse low and high**

The Five 800# Gorillas of Prop Design, p. 19 II

Chapter 2 Conclusions, p. 62,65 II

Chapter 3, Book II

More on an RV and Luscombe Prop, a little extra Thrust, p. 90,92 II, the page 92 data with considerable detailed insight!

Slowdown, p. 106, 109, **A Source Sink Analysis**, getting the **Velocity Profile**, in the **Plane of the Prop**, at the nose of an **Embedded Body**, a **method by Rankine**, taught by Prandtl.

Overall Propulsion Efficiency, η_P, Gus Raspet's Discovery, p. 110, 115, the **extra losses over and Above Prop η losses.**

Design Study, C_L vs. Diameter, p. 120, 126 II. An Important, Interesting Study that shows the **effect of Changing Cl**, it's **huge effect on required Diameter**, and **Induced vs. Profile** Drag Loss, but most significantly finding **.5 Cl to be most Optimum across the Performance Spectrum**, an Alt Cruise Design, at T.O., Climb at S.L. and Altitude, V_{Max}, .55 Close!

Theodorsen's Special Factors, p. 147, **w, Kappa, κ, Epsilon, ε.**

Airfoils, Low Reynolds Number Drag, p. 148, 149

Of Course There's a Lot More Insight than this provided in the Chapters, well worth seeing all the insight provided.

A Word of Caution: being a good engineer, I'm always careful to keep my statements sound, not get involved in optimistic claims. Having been the guy who founded and ran the Spacecraft Controls Company, one works on Hard Fact, that meets the payroll, not an engineers fond Hope.

Everything we've seen so far says Goldstein - Theodorsen Math is Spot On, Essentially Exact, the Props doing just what the Computer Program predicts, but use Tibery and Wrench Loading for very High Pitch. **When off spec Performance is Calculated, however, of necessity, we are dealing with normal Engineering Trigonometry, and Math Calculations, less Sophisticated, and exact than the BGT rare genius core solution.** As such, my appraisal is that **studies, etc.**, are valuable, insightful, helpful, but will **simply not be of the same caliber as the core, heavily loaded math,** ***can't be.***

Off Spec. -- Better than you might think

Of course, if you design a prop for a specific condition, **altitude cruise**, for example, **a very good choice**, because that's where you should spend your time, fuel, and money, fly there every time you can, it's not going to be perfect at other conditions. But, of course, **if you have it essentially perfect at its design point, loaded, shaped, twisted, ideally, you have a basically excellent prop**, a bunch better than someone's amateur try, and the rest of the story, fortuitously, is surprisingly favorable, with a properly designed prop, so, **we're heading for a good place**.

The first surprise is that later, **at low altitude, Vmax**, if you just must play macho with your buddies, it turns out that it's apt to be **even more efficient**, going faster, at a lower angle of attack, **a bigger mass flow rate, $\dot{M}$, less ΔV speedup** of the air in proportion, a dandy surprise the computer teaches us.

The really big piece of good fortune is that when we correct a perfect prop for the nose velocity slowdown, far bigger inboard where the embedded body is pushing the air ahead, and outward, the **slowdown correction**, which we can do quite accurately, **results in significantly less twist, and the prop is significantly better flying slower in climb, and takeoff**, because slow, the uncorrected, grossly steep inner blades would be stalling, so either fixed pitch, or constant speed, variable pitch props are much better slow, than they would have been.

The **correction at the 19% radius on an RV 8** with an 18% radius, or diameter spinner, **is a wild 17.65 degrees**, and remember, we're trying to hit blade angles to ***<u>.1 degree accuracy</u>***, **2%**, a **1°, degree, .1 C_L, being a 20 % error on a .5 C_L prop!!**

The correction is still a hefty **7.56 degrees at the 35% radius, still a .953 degrees at the 75% radius**, the difference between a climb and a cruise prop, **.383 degrees at the tip**. Not only do you get a bunch of good help doing it with the master's math, you can see how much better off you are using the master's math, compared to some hubris type thinking that, because he is smart and has a computer, he can start cold and get it right! Our **forebears** were **very smart men, and did 83 years work!** √

Everyone Needs a Day to Assimilate Prop Logic

We can show you the **core of Prop Logic in 10 Minutes**, but I've found *everyone*, pros included, ***needs time "to really get it"***, really grasp it, soak it up, make it their own. The three core insights, **1. A Rotating Wing, 2. An Airscrew, 3. Pulling in, Throwing back air**, are simple enough, but there is **a long enough list of insights, and implications**, that it comes down to **a whole new way of thinking about props.** ***You need a day***

Since we can't talk. I give you a day's conversation, a day's reading.

Trust me, if you start out cold, trying to understand all the math, the heavily loaded 3D flow, with no real insight on the logic, it's a world class technical swamp, a morass, a nightmare! It's really a snap by comparison, the way we explain it, but my point here is quite the opposite to that. **Anytime we learn something new, it's a challenge**, especially if it has a lot of complexity, as props do, even simplified, especially if you're a greenie, a technical novice, and this is your first try, at logic!!!

No one ever got this simplified before, no one ever took the very long time it took to see through it and get it simplified. Andy and I took it on as ***the never solved Intellectual Challenge, did it over years!*** If you're a novice, relax, pros need time to get their head lined up with the new way of thinking about props, seeing, believing, that **the simple horse sense teaches us what no one ever learned from the genius level math**, just too hard to see what all the hidden implications are in the math. **Making hard things simple, is the *final*, hardest task of all!**

A retirement hobby intellectual challenge, over years, the big funny is that *very gradually*, my *subconscious would sort out - the pearls of insight, that I could say, must say to nail it, make the complexity clear, easy*, as I woke up in the morning!

Knowing it will take you time to get it, soak it up, make it your own, **I purposely repeat, expand, add insights, give you time, lead you to what you'd take years to see through. Just *stay going*,** *as far as you wish.* **and *you'll end up with a pro's insight***

Pro's: Understand, This book is to help Pilots! Learn how to Cherry Pick it!!! That's your contribution. I expect you to be smart, wise enough to see that!

Teaching You Classic Propeller Theory and Logic

Rankine, Froude, Betz, Prandtl, Goldstein, Glauert, Theodorsen were the 7 Historic, Scientists, Mathematicians, and Engineers, who created the Benchmark, Milestone steps in the Development of Betz Classic Propeller Theory, ***Elegant, with essentially exact*** **Goldstein-Theodorsen Math -- from 1865 to 1948 - 83 years of often rare Genius level work.**

We teach it to you, not only because it is the Classic Theory, but because it is **orderly, and logical, understandable, simply explainable, with clear Engineering Logic**, that we can present to you as just **good understandable Horse Sense**, ***clearly, logically correct***, but **backed up with essentially exact math**, maybe **some of the best, most sophisticated math work ever done -- unquestionably accepted as the Gold Standard by top Aerodynamic Leaders.**

With the use of an **Optimum angle of attack, α^o, and C_L**, it all ends up in an **orderly family of Triple Optimum Props, Min. Induced Loss, Min. Profile, Min. Torque, Minimum Area, Precisely Placed, Norris's 6th Law, now all explained, all with the natural Elegance and beauty of form and function, the "Good Art" that we so often find with Nature explained.** ***What could be more beautiful than a Sea Gull riding the bow wave energy of a breaking wave coming in to shore, the finger feathers of a soaring Eagle converting a rectangular wing tip Vortex into thrust, an equal and opposite solution to the Gull's pointed tip, our human work only a fond hope?***

The **"Introduction" Blasted You *Quickly* into Sophisticated early Awareness, but then repeated, expanded, *justified*, hoped to cajole you into really grasping the new insight, really making it your own! Now the "Primer" will carry that further, begin to add more, deeper insights to help make your knowledge ultimately whole!!!** ***You need time, more and more insight to make it your own!!! Onward!!!***

PROPELLER PRIMER

THE PROPELLER

PROPELLERS FOR A PILOT'S *FIRST INTRODUCTION* .

We can look at Props in several ways: **1.** as simply **pulling in** and **throwing back air - to make thrust - a reaction force**, per **Newton's Laws**, **2.** as a **Rotating Wing**, **3.** as an **Airscrew**, the British term, **4.** as **Betz's Minimum Induced Drag Theory** and **Logic**, pulling in and throwing back a **perfect helical screw shaped downwash, backwash**, by **controlling the loading vs. radius**, set up by very carefully, and accurately controlling the **Shape**, thus loading, with matching **Twist** and **Pitch**. We're going to want to look at and understand propellers from each of those standpoints, because we can learn very much from each view, and learn how it all fits together into a quite sophisticated, but understandable overall grasp, that has never been satisfactorily explained in the 138 years since prop analysis started in 1865.

Have some fun with your friends. Ask them **how much a 100 foot cube of air weighs at sea level**? Not a trick question. Presume you can cut out that block of air, put it on a scale, not floating in the other air, **real weight, no trick**. Ask yourself, pilots, Engineers, a sophisticated group at a cocktail party. Have a good laugh together. The real answer is **76,000 pounds**. Air only weighs .076 pounds per cubic foot, 3/4 of one tenth of a pound, but there are **a million cubic feet in a 100 foot cube**. Try it, a **1** followed by **six zeros**, x **.076**. A long airplane wing, even a propeller disk is processing, ***moving*** far more volume and weight than you'd ever guess. It shocks experienced engineers as much as a cocktail set. A 747, 700 to ~900,000#, almost a million pounds, is throwing down *enough* air, *fast enough* to hold itself up, just as ***prop thrust is throwing back air***.

Sailplane pilots know the game is to have a long, skinny, high aspect ratio wing, **low span loading**, **involve a lot of air**, as much as possible, because Mr. Newton teaches us the game is to involve a lot of air, a **High Mass Flow Rate**, in his terms, because then you can hold up the plane with **a *surprisingly small speed up of air*, throwing the air down in an amazingly low angle downwash.** That will hold up a sailplane with just a gentle whoosh of air as it flys over, and a 747 with a lot more whoosh, but still a quite small angle of downwash.

The 747 has a second trick, it's **going fast horizontally**, in addition to a big wing span, also moving a vast flow rate reaching up and down vertically. Think of a vast symmetrical water melon encasing the wing, (see how the big chord gets involved too), not taking off until ~**170 IAS**$_{MPH}$, with **flaps**, flying ~ **270 IAS**, ***540 Alt TAS***, the numbers can all work out, enough mass flow *rate*, enough speed up. You not only don't want to get behind a 747 in a light plane, you don't want to in another big jet. There can be **massive tonnage still swirling minutes later**. Fish Salmon got upside down in a Bonanza on final at Burbank years ago, but as Lockheed's chief test pilot, got it right side up and lived.

That old story may be wrong. It may have been Tony Levere, with Sammy Mason.

Throwing air, using extra energy to hold the plane up, or make prop thrust, is why **Induced Loss** happens, air thrown down to make lift, or back to make thrust, or into those great big swirling tip vortices that Aero Engineers account for with a fudge factor, called the Oswald efficiency factor increasing the theoretical induced drag calculation, that basically accounts for throwing the air down to hold the plane up - the tip vortex extra.

Now we make a big point out of this because the way we end up designing a prop, the **induced losses can nominally be twice the skin friction profile loss from the prop blades.** Now realize **we can reduce induced loss on a prop by having a big diameter, high aspect ratio blade, but also by just going fast**. ***Just going fast is a key to high prop efficiency***.

1. A Prop Throwing Back Air - Grasping Newton's Insight

Learning from the ***Axial*** **Induced Loss**

Pilots, Civilians are usually faked out if they see even a simple little math formula, but let me show you simple **math can be** the **EASIEST way to see the Horse Sense of this subject**. Newton teaches us Thrust, **T = $\dot{M}$ ΔV**, Mass flow rate -- times the Delta V speedup, dead simple, if you don't let yourself get faked out. If we can get a **big Mass Flow Rate** through the prop, by either having a **Big Prop**, or even better, **Going Fast**, we can get by with a **small Delta V, ΔV speedup --- Dead Simple, really.**

$\dot{M}$ is called M dot, just like it looks. Delta, Δ, just means difference to engineers.

Now that basic, **simple insight is of huge importance**, because the **efficiency game is to minimize the axial ΔV wind you make, DON'T MAKE EXCESS WIND**, because the efficiency loss is **half the ΔV wind speed increase --- compared to the plane speed**, or big wind coming toward the prop. Can you see that **going fast gets you a good double whammy, favorable result on efficiency**? **Fast** gets you **a high Mass Flow rate, thus a small ΔV**, and ***half that*** **divided by a big V_1** gets you **a twice lower *axial induced* efficiency loss. That's why GOING FAST is the EASY WAY to GET GOOD PROP EFFICIENCY.**

3 induced losses happen, Axial, Rotational, & Radial - feeding the Tip Vortex

Now, if you can grasp that little drill in math, and Logic **you have learned the most *basic* Horse Sense Logic of Propeller Efficiency.** Yes you have to be gutsy enough to not be faked out by a little math formula, **but the reward is you can easily grasp the most *basic* Horse Sense of Prop Efficiency Logic.**

Of course, we cover Profile Drag Efficiency Loss shortly, also

Our basic objective in this Primer is to write it without math, just the basic technical horse sense logic, but that one simple little math formula is actually the easiest way to show you the *specific* little ***axial*** efficiency logic that underlies everything in propeller logic, in the fewest possible words. We actually show you the ***axial induced efficiency loss formula**** in the next section, which looks more scary, but isn't really, because again it's easy and basic, for those who aren't faked out. **Next, fast is Good, Slow can cost.**

***Total 3D Induced losses simply have a higher percentage loss.**

Gearing - Used Very Slow or Very Fast

Very **Slow** and very **Fast** planes are each **special cases** that **need to be geared for quite opposite reasons**. To be efficient a prop ***wants*** a high mass flow rate passing through it, so that it can have a low Delta V, ΔV speedup to make its required thrust, since ***axial induced*** **efficiency loss** is ***half*** of ΔV ***compared to, ratioed to***, that is divided by the speed of the air coming at it, **ΔV / 2 V_1**.

The full 3D Induced losses simply have a higher percentage loss. Don't be confused.

Slow: A very slow plane like a Wright Flyer, a man powered plane, or an ultralight, being slow have a low Mass Flow Rate passing through their **D**iameter, ***Disk Area***, so they tend to need a relatively high ΔV, and when that is compared to a low, slow V_1 a *double whammy loss, both unfavorable*, it comes out to be low efficiency. If the normally high RPM of an Engine is **geared way down**, like in the Wright's chain drive, we get **High Pitch**, which we'll learn is desirable, a tell tale of good Efficiency, and a **Big Diameter**, a Big Mass Flow Rate, finally a **low ΔV**, good efficiency even compared to the **still low V_1**, when run through Prop Math. That can fix the basic problem if the gearing is sufficiently low. The Wrights ended up with two **Big 8'6" props** (12 HP), the **Daedalus** man powered plane had a **giant 11.3' prop** (.2+ HP). Ultralights start with a terrible situation, slow speed and maybe 5000 RPM, very flat pitch which we learn is bad, the sign of an inefficient situation. Ultralights are often only partially fixed, insufficiently geared to limit prop size.

Slow, *you need Geared*, Lo RPM, Hi Torque, Big Diameter, High Pitch.

Mach, Fast: Prop tips must be kept below .9 the Speed of Sound, nominally 1000 Ft./Sec. at standard temperature, slower cold, at altitude. If you gear down, slow the RPM, ***but*** don't let Diameter expand as much as the efficiency suggests, **use more blades**, the vector combination of rotational and forward speed can be kept below .9 Mach at Prop driven Fighter Plane speeds, at some efficiency penalty, right up to the stretched limit at Reno max speeds. **Slow and fast are the special cases. Normally, we want the light weight, simplicity, low cost of an ungeared engine**.

2. What's Very Wrong about a Rotating Wing Hi q Tip?

A Prop Tip Shape, its Loading, is Far More Important Than a Wing!

How is a Wing Tip Made Right, Much Less Loss than a Prop?

As a Pilot, you probably already realize that the **Lift of a Wing *must* fall to Zero at the Tip, a big swirling vortex created as the higher pressure air below the wing swirls around to the lower pressure top**, no dam there to stop the swirl -- the reason we tend to taper wings, a smaller tip vortex -- the reason **Prandtl** taught the ***Ideal*** **Elliptically Shaped** (Spitfire) **Wing**, that is thus elliptically loaded, the **lift naturally falling to Zero in an equivalent half ellipse, max. at the center span, zero at zero tip chord**, the math optimum **least Induced, *and Profile* Loss**.

Constant α°, C_L, *Elliptical, has a constant Downwash Angle*, tip loss a min. extra.

The Seagull and the Sea Birds, the Land Soaring Birds, Buzzards, Eagles, Condors have found two very opposite, even more sophisticated ideal solutions, **both varying the wing tip twist as conditions demand** --- the **sea birds then using a pointed tip**, **the land birds an almost rectangular wing** with **twisted finger feathers that convert the swirling tip vortex loss to thrust**, *the genius of* natures's perfect adaptive evolution, a natural wonder, our **tip sails a poor man's copy**. (We don't get the soaring bird's full effect, ***no added thrust***, ***if*** there's no twist!)

The **birds** are even more sophisticated, **twisting for least muscle load**, man copying these days for **least structural load and weight**, a Jet with typically a huge root chord and thickness, which also conveniently houses the landing gear, a tapered, very narrow tip chord to minimize wing and root bending loads, while still achieving high aspect ratio, big span, minimum induced loss, our best, most sophisticated shot at emulating what the birds do naturally, hardly a thought, a reflex.

A Boeing 787 Dreamliner Ad shows a long, Swept, Seagull like, Pointed tip!

When you see a Jet flying low overhead, start noticing the often gross root chord, the sometimes very small tip chords, notice that some tips are narrower than others, nothing like an ellipse these days, and you can start building insight into the Structures Group and Aero Group noodling out the Weight-Aero Optimum.

(The tips stall quicker, if too narrow - untwisting, **less α°**, the usual fix.)

So Here's Our Propeller, Dumb as a Board, ***Inside out, dead wrong*****!**

With **very High Tip Speeds**, 5 times the root velocity on a Luscombe, the V^2 Dynamic Pressure that controls Lift and Drag is 25 times greater at the tip, compared to the root, **trying to make max. Thrust and Drag at the tip -- where Thrust *must* fall to Zero**. Begin to appreciate the wrong headed, inside out, dead wrong way the prop is trying to operate, max. thrust and drag at the tip, minimal at the root, at low dynamic pressure, low q, ***exactly inside out, opposite, compared to a proper wing***.

With huge tip q, vs. the root, Prop Tip Shape is far more crucial than a wing tip!!!

The dumb prop is **going to lose most of the extra thrust it is trying to produce out there**, **rolling off the tip** into an **extra Big Tip Vortex Loss, extra Drag way out there at max. lever arm, trying to bog down the engine torque, thus RPM, lower the available H.P. You not only pay for extra thrust in extra induced and profile drag, you lose much of the thrust, while reducing the available engine H.P.,** ***losing 4 ways!***

Realize props try to act about as inside out, opposite, as you can do it!!!

Now fortunately there is an easy, simple, **good technical horse sense fix for that problem, simply tapering the tips, but significantly more than the elliptical wing!** The chords represent wing or prop blade ***area*** as you integrate out along the wing half span, or prop radius, so we can control the prop blade loading vs. radius, in pounds per foot, or inch of radius, by just tapering the shape as we choose. **This is the easiest and most important insight you will ever learn about propeller logic**

Do you see the simple enough horse sense of that? We can *counteract* the excess tip loading that exaggerated Dynamic Pressure, q, tries to produce at the prop tips, pull the max. loading inward away from the tips to an Optimal Radial Thrust Distribution that we can control and choose.

P. 147 shows **Optimum Blade Loading and Shape**, vs. **Advance Ratio**, (or Pitch)!

In a few more pages Betz will show us how to do this Ideally, Optimally, for Minimum Induced Drag Loss, genius level insight, that gives us ***Ideal Prop Shape, thus Loading!***

3. The Prop as an Airscrew - Screwing Home Efficiently.

The English had it right, a prop **truly is an Airscrew**, and must be, its job much more than just a rotating wing, its objective to **Screw you to your destination as efficiently as possible**.

There are ***two remarkably simple insights*** that teach us that a **High Pitch, High Advance per revolution prop is the most efficient Screw Propeller**. Simply, a High Pitch, High Advance prop screws the **shortest, least energy consuming path to the destination, a low pitch prop simply spinning an excess distance to get home**, simple as can be, the **least profile drag energy cost with steep pitch**. **(45° is best!!!)**

It's almost as easy to see that a **High Pitch, High Advance prop has the Least Induced Drag Loss,** from what we've already learned. A high pitch prop is going to be going relatively faster for any given case, **Fast, a bigger $\dot{M}$, Mass flow rate, thus a lower ΔV, a lower $\Delta V/2$ divided by a bigger, faster V_1, a lower efficiency loss.** I hope you can grasp this, see that this is really pretty easy once you tumble to it all. If you can grasp this OK you're getting really great insight, the real meat, pretty quickly, that demystifies the most complex basic subject in all of Aerodynamics, the **Basic subject that never got coherently, Incisively explained before**

A Tricky Little Logic Conquered!!!

Let's Review, and Sum up to this Point:

We've learned that a **prop naturally tries to greatly overload both the Thrust and Drag at the Tip**, **(where thrust must fall to zero, the excess lost)**, simply because the Dynamic pressure can get way excessive at the tip, vs. the root -- that **we can fix that flaw by simply tapering the shape, pulling the radial tip loading inward** - that we **want high pitch, *if* Speed vs. RPM permit** to get a **big $\dot{M}$, Mass flow rate**, a **low ΔV speedup, low vs. the plane speed V_1**, to **get a low *axial induced efficiency loss*, a low $\Delta V / 2 V_1$**, if math symbols don't scare you, **a Short, Low Profile Drag Path too**.' It's ***simple enough***, **the core logic**.

7 **It's SIMPLE: Taper the Tips --- Hope Speed vs. RPM allows High Pitch!!!**

GET THIS.

4. Albert Betz's 1919 Genius Level Insight

The beautiful thing about **Science**, the **explanation**, the **logical application of** the principals of Physics, Chemistry, Biology, **all of Nature**, is that (although there can be mind boggling complexity), there is almost always **finally beautiful simplicity, orderly logic**, like God, ***smiling***, gave us the challenge, but the Brain to sort it all out.

Albert Betz was a contemporary of Ludwig Prandtl in post WW I Germany, just after the age of the Spads and Fokkers. Prandtl saw the logic of the ideal elliptical wing, and Betz, amazingly, saw through the ideal propeller, a far more difficult problem.

He must have seen that **controlling the radial loading of a prop is the key to Ideal Operation and Efficiency**. In marvelous insight he saw that the ideal solution is simply to **pull in a perfect helical screw surface *of air***, perfect **constant pitch at each radius**, and **throw back a theoretical stretched, still perfect Archimedes screw downwash, backwash from each blade, just like the picture on the front of the Book!!!**

It's the Air that's a perfect Helical Screw Surface, *not* the prop.

He first considers this **drag free**, because the simple case yields the **Minimum Induced Drag Loss**. He seems to ignore the Tip Vortex, which in the real world swallows up this Ideal, Perfect Math Model. **He doesn't ignore this, however, and the subsequent, 3D Flow Math Approximate Solution by Goldstein in 1929 and Theodorsen's essentially perfect solution in 1948 allows us to design an essentially perfect Ideal, minimum Induced Loss Prop on a Computer, as fast as you can type in the parameters and hit the Go Button.**

The 3D Radial Tip Vortex Loss *is* accounted for, rotational flow too.

Concurrently, the proper Low Reynolds number drag coefficients can be added, for X-ray insight on a ~ perfect Prop. There are magnificent parallel insights from Betz, shown on page 129, as incisively as you'll ever find it, **constant Slip the hidden insight** that optimally balances radial flow, Thrust, and Loading. **Page 147 shows optimum Shapes, Loading, vs. Advance.**

5. Seeing the Whole Propeller Logic Picture

Of course there's more to learn, a lot more, but if you understand this **Primer**, you've grasped **the core of Prop logic**, **easily worth reading these 11 pages over (if you need to), because this is the heart of the subject, and if you got this Ch. 9 is licked!!**

A **computer** uses an **amazingly simple Goldstein-Theodorsen Kx blade loading factor *chart* vs. Radius, and Advance Ratio**, to interpolate, ideally load a ~ perfect theoretical prop, give two or three pages of Data, for ~ exact definition of an ideal Betz, drag free prop. But Low Reynolds number drag can be added to the program, if you're a pro, as my partner Dr. Andy Bauer has.

SHAPE, thus CONTROL of LOADING is the KEY!!!

There is a highly varied velocity profile at the nose of a plane, the imbedded body pushing a bubble of air ahead, especially if there is a bulbous cowl, that acts like a Slowdown, that is **a very important correction, adaptation** of the basic theoretical prop. We really **want to predict inflow angles to +/- .1 degree, yes, 1/10°**, a normal shop tolerance, for prop product consistency, so anything but the real math is fooling yourself if you want real answers that match real results.

Home Grown Computer Programs are *Anti* Productive!!!

A separate program analyzes an Ideal Prop design off of its normal altitude Cruise design point, for example, Sea Level and Altitude Climb, Vmax on the Deck and at 8000' Altitude, for those who want to fly at max continuos power, 75% of Rated.

We set this book so you can read each subject at any level you wish, this very insightful 11 page primer on props, Chapter 9, more, progressively deeper, a good bit more, after its core 20 pages, Book II the whole Magilla, pro insight for anyone who cares to go after it all, more and more insight, the first comprehensive, understandable explanation ever. It takes **an hour or two for Ch 9, a day or so to truly get it all. We purposely Superemphasize, repeat, lead, tie together, sum up -- *all* so Novices can get a pro's insight.**

Summary Insights and Conclusions, Insights to Realize:

1. **Definitions** of **Propeller terms** found **after Chapter 1**, Book II

2. **Pitch: Simply, Perfect Helical Prop Pitch** is the **inches** or feet a **Propeller would Screw through a soft solid,** *per rev.*, with **no Slip.** (If the twist is not pure helical, especially after correction for slowdown, it is taken as the pitch of the 3/4, **75% radius.**) Realize that since a propeller is **pulling in and throwing back air** to make thrust it is ***typically overpitched*** a small amount, to pull in the air, **there is some Slip** thus the actual **Advance** per Rev. is **less than the Prop Pitch**.

Ideally, *Air Inflow Slip* is held constant, Betz's constant pitch *Air Inflow V_2 minus V_1*.

3. **Propulsion Efficiency:** Realize **Propeller Efficiency, eta, η, is Traveling Efficiency,** and thus **at static Runup,** burning fuel, straining, **going nowhere, propeller efficiency must be Zero! As the plane speeds up,** more slowly throwing back air, going faster, thrust is logically dropping, but **efficiency is improving,** still **reduced during climb,** only reaching cruise design efficiency at cruise, typically highest at Vmax. If there is an **extra interference (in)efficiency,** scrubbing drag, faster air blowing across the fuselage, etc., any extra separation, any extra pressure drag, propulsion efficiency, η_P drops farther, prop efficiency, η, and Interference efficiency, η_i multiplied together! In a **shallow dive,** faster than the pitch capability, ***thrust*** drops to ***zero, efficiency again drops to zero,*** ***graph on next page.***

4. **Newton's Laws Fundamentals: 3rd, Thrust,** an **Equal and Opposite Reaction Force to air Thrown Back, 2nd, Thrust, Equal to Mass Flow rate $\dot{M}$, times** the **Delta V, ΔV *Speedup, Axial Induced Efficiency Loss****, Half the **ΔV, ΔV/2, vs.,** that is **divided by, Inflow Velocity V_1, --- (ΔV / $2V_1$** in math terms, if that does not scare you), is a real fundamental of propeller logic, can be seen operating all through Propulsion Efficiency, above, and if you can grasp the horse sense of it you have made a giant stride. The actual 3D airflow used for actual propeller design is much more complex, but grasp this simple, basic logic and you've won.

 ***3D Loss, Axial, Rotation, Radial, (TipVortex), is Simply a higher % Loss.**

5. A Rotating Wing, has a **Bad Basic Flaw.** With very High Speed Tips, **Much Higher Dynamic Pressure, q, at the Tip vs. the low q root,** a prop is **trying to make excess thrust and drag at the tip,** where **Thrust Must Fall to Zero,** physically **rolling off the tip, making extra induced loss,** a bigger Tip Vortex, unproductive **extra Profile Drag,** *at max lever arm,* max radius. The **propeller begs for the good technical Horse Sense to highly taper the outer prop Profile Shape** to **pull the max load point inward, allow a natural drop off profile** of **thrust vs. radius** to **prevent excess, unproductive tip losses, better efficiency,** just good sense, once you see it. See p. 147. Ch. 9.

6. A Prop is an Airscrew. Thinking of a Prop as an Airscrew is very Important, Fundamental. **High Pitch, a shorter Spiral Path** to the destination is the key to both **Low Profile Drag loss** and **Low Induced, high pitch** ***going with*** **high V, big $\dot{M}$, low ΔV.**

NEXT, SEE BETZ'S PURE HELICAL AIRFLOW PHOTO ON COVER

7. Betz, p.129, teaches us how to get a **Minimum Induced Drag Prop,** (first considered profile drag free), **by Ideally Shaping,** thus loading, a blade, **constant pitch** **air inflow, pure helical inflow, constant slip,** vs. plane speed, ***stretched pure helical backwash.*** More p. 129

8. Theodorsen, has a ~perfect math solution for a **Betz** prop!

9. Advance Ratio, Pitch/Diameter, P/D Ratio. A look ahead. Low Pitch, low P/D, props have **low efficiency, high Pitch, hi P/D, high η, each fixed pitch prop has a limited speed range!**

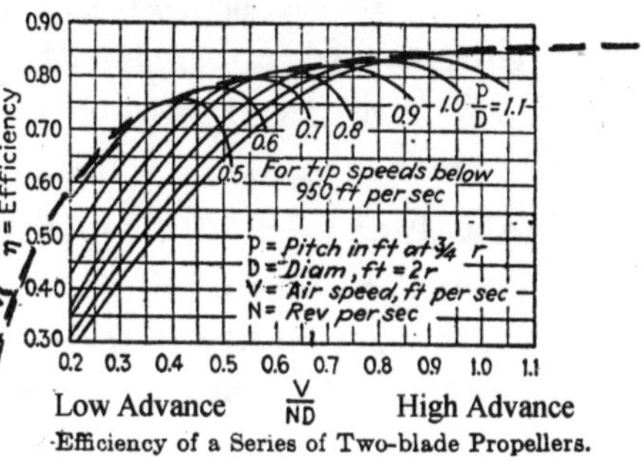

Efficiency of a Series of Two-blade Propellers.

This old graph from a 1941 Marks, Mechanical Engineer's Handbook*, though inaccurate, ~5% low on Max Efficiency, shows how a series of various P/D Props have a limited speed range, starting at zero efficiency, at zero speed, but nest together to form what we will learn is the **Advance Ratio Curve of Max Efficiency of *ANY* Prop ----** Valuable Insight. By permission of *McGraw-Hill Book Co. Inc.

Advance Ratio, J, V/nD, like J/π, V/πnD, ratios forward to rotational velocity, an insightful picture. This is an old, outdated Graph with poor plots, the efficiency ~5% low, but marvelous insight if you can grasp it this early.

Wow!!!!

Of course I realize that's a whole new world of logic to lay on a civilian, a pilot, not experienced in technical thinking. In fact that's never been explained to engineers before. Pro's have never had a starting explanation of propellers that is that **coherent**, that **incisive**. We don't have correctly designed props with properly tapered blades because that introduction to propeller logic was never correctly put together before for pro's or civilians. **Few pro's have really understood that correct loading of the blade, correct SHAPE, with matching proper twist is the final basis of properly efficient propeller design**.

Understanding this Primer on Props is Pivotal, a challenge, it's your entry ticket!

So now understand what's going on here. Airplanes are relatively easy to explain, but propellers have been so mystically complex they've been impossible to correctly understand or explain. The pros never had a proper explanation before, 138 years, 1865 to Dec. 17 2003. But now, we finally have one, and **if you have a little guts**, don't get faked out, **you can understand all the things that never got nailed before!!!**

Props are the special case, the previously impossible challenge.

No, I'm not some weird techie who's too dumb to understand that you'd like the simple, dumbed down, easy explanation of everything, written in good simple normal writing style. This book started out to just explain planes, and matching engines to them intelligently, easy enough, but then it became possible to teach you what has always been impossible, understanding props

If you can understand the horse sense of this Primer, you can learn Props!

So here's the deal. **Recognize your awareness just went up by a huge factor!!!** It's really not possible to strain your brain, and maybe you only spent an hour or so to get through this Primer. **If you simply understand that I lead you through the morass of propellers, superemphasizing, I can lead you to all the nifty insights and you can understand what would have been impossible for you to know before. Have guts, have faith, hang in, and you'll be amazed at what you learn!**

Realize we keep leading you to the Truths. You don't have to figure it out, we lead you!

A Real Engineering Breakthrough --- Understanding Props

For the first time in 138, years we can understand and explain both the big picture and the incisive details of prop logic, design, efficiency, and operation --- both as the engineering horse sense involved and the rare genius level math, and what the math really means, and what it dictates. And happily, we find that the horse sense supports and better explains the logic of the math, making it more clear and understandable to both the novice and the professional engineer -- as the math completely supports and verifies the horse sense logic, the mutual agreement supporting that we do, in fact, finally have it all correctly understood!

Before computers there was no really adequate analytical methods for using Theodorsen's marvelous math efficiently in a cost effective way for non government, low cost private flight. Talking to the old timers, in one key company, it was an old pro from NACA with real feel replacing analysis. In hindsight my 1947 model propellers were much better than the prop on my 47 Luscombe. I got a specific, accurate speed answer every time on a speed model, better test data than pros.

In the very best companies there is always someone sharp enough to get to the technical truths, and as soon as earlier computers became available that started, but some got it less correct than the best, and it was proprietary, secret, and vibration often took precedence. The best did very good work.

The brilliant few who created the logic and the math, Betz, Goldstein, Theodorsen certainly had to understand, but there is an interesting insight there. People at that level of brilliance, locked on to the genius level math, are almost unable to come down two notches and say and teach all the practical insights on what all the math really means and dictates. The technical horse sense that we have here has been the missing link, can teach novices and non genius level pros what the resulting logic really is, what is dictated that has not been clear enough before.

Let me explain. There is nothing in the literature that tells us clearly, simply, that the objective of the math is to load the prop correctly vs. radius, to prevent excess tip loading, unnecessary, unrewarded induced and profile tip losses, that proper blade loading is accomplished by proper shaping of the blade with proper twist, that if you actually achieve a constant optimum angle of attack and C_L, you get min. profile drag too, min. area precisely placed, Min. Torque, a Triple ideal optimum prop, not just min. induced --- that you must accurately correct for nose velocity profile, slowdown, because we really need to predict inflow to all blade radii to .1 degree, yes one tenth degree, that a 1 degree, .1 C_L error will give us a 20 % error on a .5 C_L prop --- that the slowdown correction which results in less twist, lower blade pitch angles, especially inboard, can create a major Plus in fixed pitch prop operation slow, in take off and climb.

There is no criticism whatsoever of the truly magnificent, rare genius level math and technical work done by our marvelous forbears, who gave us superhuman solutions, ready for easy use.

The problem is that with no explanation, what it meant was not understood! ✓

The kind of things we'll learn is that Speed vs. RPM, with due consideration for needed inflow and angle of attack sets pitch, which sets max. efficiency attainable, that use of B-G-T, Betz-Goldstein-Theodorsen Math, will then deliver max. efficiency, that it will also deliver a **precisely loaded, shaped and twisted** prop, **a constant optimum angle of attack and C_L, minimum area, precisely placed, a Triple ideal, min. induced, min. drag, min. Torque prop, if we're only smart enough to ask for it.**

Remember, with 21st Century Insight, we can improve on the old Optimums

A lot of long, hard, dedicated work has been done here, simply because Andy Bauer and I were experienced enough to recognize that the explanation was badly needed, long overdue, and that with two lifetimes of experience we could sort it all out, do it. **There is a lot to teach you**, be you a novice or pro, **a long string of incisive insights. We'll simply build on and expand 4 basic insights here, ever deeper, broader, more sophisticated**, ***superemphasized so anyone who tries can get it!***

I'll take this page as the bottom line on a lifetime of experience + insight — Jack Norris 8-16-06

 We had to go years past the math to get you the incisive clarity here.

An Important Advanced Insight

***If* I can do a little calculator math without scaring you ---**

Airplane Math Hides Some Pretty Bad Prop Design!!!

Going faster, plane drag goes up as the Velocity Squared, **$\sim V^2$**, but **power Cubed, $\sim V^3$** -- because in going faster you have to put in the higher Thrust, and Energy, **(pounds x feet traveled, # ft.)** for that higher speed, *at a faster rate*, **ft. # / sec.**, because that's what Power is, rate of putting energy in, **ft. # energy per second.** (It's actually a little less than V^3, because faster, induced loss is dropping, but let's keep it simple, to look.)

To go 1% faster, you need **$(1.01)^3$**, 1.030301, **3+% more power**, to go 4% faster, you need **$(1.04)^3$**, 1.124864, ***~12.5 %*** more power!

Now let's say a **220+ MPH Triple Ideal RV 8 Prop**, min. Induced, Profile, Torque, **can be 90% efficient**, because it can be that efficient --- but we **really Klutz up the design and get 50% unnecessary, extra loss, lose 15% not 10 %,** we have an 85% efficient prop, not the 90% that we can and should. But that only cuts our available HP in the ratio of 85/90, .944444, or 1.0588, inverted, and the good prop only goes $(1.0588)^{.333333}$ faster, **1.019**23555, using all the decimal places, less **than 2% faster**, maybe a typical difference, fair vs. perfect. **With few valid tests, *poor props have hid for decades*,** the perfect prop gypped.

***If* you can cope with a good calculator, play with this, get a lot smarter!!!**

Now, let's say we design it really poorly, and get **twice the min. loss, 20% loss, not the ideal 10%**, 80% efficiency, not the 90% we deserve, a 80/90, or 90/80 power ratio. We only go $1.125^{.3333}$, **1.04**0042, **4% faster** vs. a really poor prop. **WOW!!**

The numbers hide quite poor props, but there is no reason to not get them right.

Guess What? On Dick Van Grunsven's, Van's Aircraft, recent tests of available commercial props for the RV 8, the best prop, **Hartzell's latest,** ***basically*** **designed to Theodorsen Math**, with some extra area at the tip, went 208.9 MPH at 2500 RPM cruise, the worst European prop only went 200.7 MPH, a **1.04**086 speed ratio, **the klutzy prop had double the ideal loss, bad!!!**

It's time we get props correctly understood, correctly designed!!!

We're after **2 objectives** here: **1. *Accurately, Understandably, explain prop Logic*, 2. Get props correctly designed, easy with *Betz-Goldstein-Theodorsen math*, available for 50 years.**

Throwing Air - Really understanding What Induced Loss is

There are two basic types of loss, **Profile, Skin Friction Drag,** just based on ***surface area skin friction***, just like it sounds, and **Induced Drag**, or **Loss**, just the **energy cost loss** from ***throwing air, to make Lift, or Thrust***, as a **reaction force**. Mathematically, it acts like another drag, but I very much **prefer the term loss,** not drag, which ties in to a much more simple, smarter, incisive way of looking at it, because you **simply throw air to get a reaction force lift, or thrust, spend energy to do that throwing**, the **energy simply lost** to **friction, turbulence**, as the air settles back to static, ***but a fundamental need***. -- Yes, you can get separation, interference, bad high drag designs, but we're talking basics here.

You never hear it referred to that way, but, you see **it really is the core essence of the subject.** The day that first dawned on me, it was like someone just turned on the sunlight, and I really understood for the first time. **I hope that, maybe, it can be an equivalent awakening for you!!!** Anyone can make these things complex, **the magic is to see the simple, basic insights**.

That's very important for propellers, because if we design a prop for **a .55 C_L the basic losses are nominally 2/3 Induced, 1/3 profile, *Induced the bigee!*** You'll learn that if you design for a very low C_L, you can interchange those losses, and have a prop almost as good at the design point, ***but*** **much bigger, heavier, less good overall**, not what we do, **.5 to .55 C_L best.**

Ideal Props are all about minimizing both Induced and Profile, KILLING TIP LOSS.

Profile increases as V^2, twice the speed 4 times the drag, 3-9, 4-16 etc. But we learned in the Airplane Book I that **Induced is a $1/V^2$ Phenomenon** that **drops with speed**, 1/4, 1/9. 1/16 etc. **Propellers disguise that, but work exactly the same way**, the Induced, the reason ***fast***, *high pitch* props are best.

The tricky **$T = \dot{M}\ \Delta V$, and $\Delta V/2V_1$ logic is easier to see through here!!!**

See Through $T = \dot{M}\ \Delta V$. Big speed, **big V, big $\dot{M}$**, only need a **smaller ΔV**, a **twice lowered $\Delta V/2V_1$**, a **big V lowering the $\Delta V/2$, dividing it by a bigger V, *a $1/V^2$ phenomenon disguised.***

**Try very hard to get this if you're a novice, if this is tough for you.
It can be a huge fundamental breakthrough, that can clarify the tricky basic logic!!!**

Book II

Advanced Propeller Insight

Following Propeller INTRODUCTION, and PRIMER,
Read Chapter 9, Propellers, of Book I, The Logic of Flight.
Then return to this start of Book II, Advanced Propeller Insight.

Propellers -<u>Starting a Professional Understanding</u>
A Lifetime of Insight in an Hour, a Day, ***as much as you choose!***

There has **<u>never been an understandable, comprehensive, incisive explanation of propellers and their logic</u>, not for the 138 years since Rankine started water Propeller Analysis in 1865**. It's been t**he ultimate dismal Swamp** of the Engineering profession. Quite simply, **no one ever got their head around the whole subject well enough to make sense of it all, <u>explain it in a way that either novices or pros could</u> <u>see the horse sense of it all</u> -- as well as <u>the key technical specifics</u>**.

With a lot of long hard work we've conquered the 138 year challenge!!!

Actually, 7 historic, benchmark technical masters achieved **an *<u>Elegant</u>*, essentially exact solution for the Ideal Betz, Minimum Induced Loss Propeller by Theodorsen's 1948 book**. But it mostly scared even pros away, not realizing that his book showed **a very practical, easy enough way to solve the Ideal Prop**. I saw that yawning gap in Aeronautical Engineering as a kid model competitor in the later 40's, and it still existed at the Millennium. **Dr.Andy Bauer** and I took it on as **a dare, a retirement <u>final intellectual challenge</u> a decade ago**. It was **a worthy challenge, a final gift** to the **time and profession that gave us an E Ticket to life**. We could **fly**, help **design and build Jets, even manned Spacecraft to the Moon and back**, ***<u>what a fantastic, fun, satisfying life we've had</u>!!***

We found **<u>Rare Genius Level Work</u>, Rankine, Froude, Betz, Prandtl, Goldstein, Glauert, Theodorsen, the seven masters**. Our good intent was rewarded by finding the opportunity to finish and explain **some of the greatest, most challenging, Elegant engineering work ever done**, the chance to make it understandable **to pros --- and novices too -- <u>Put It To Work</u>!!**

There was a lot more to finish, <u>Adapting an Ideal Prop</u> to the <u>Highly Variable Velocity Profile of an embedded body</u> nose, adding <u>low Reynolds Number Drag</u>, <u>analyzing it off spec Speed, RPM, Altitude ρ</u> --- *<u>Making it understandable</u>!!!*

There is so much here to be taught that, done wrong, we could drown, confuse, lose even good pros, no progress there. **Betz, Goldstein, Theodorsen** were **not just brilliant**, they were the **once in a century, rare genius level wonders who came along in the right sequence, spread out from 1919 to 1948**. They could see through and solve what normal pros, mere mortals, could not. We had to dig through all their work, **understand what it all really meant, do analytical studies, get way past the math to finally see all the hidden, obscure technical subtitles -- finally see the real horse sense of it all.**

The marvelous thing about Nature, Science, is that amidst unfathomable complexity, there is finally orderly logic, Betz's Double Helix, one from each blade, no less a discovery challenge than Watson and Crick's later Double Helix DNA.

You and I, mere mortals **understand and remember what makes sense to us**, so our job here was to dig and dig, not quit until we had been through everything, **not quit until we saw *ALL* the horse sense of it all**, checked everything twice, went around, consulted with, taught, **checked with every prop expert in the country, wrestled it down to the irreducible technical horse sense of it all, nail *all* the technical specifics.**

Novice or pro, there's enough here that it takes everyone a little time to actually absorb it all, all the new thinking. **To succeed we had to get it down to technical horse sense, the 4 element Horse Sense of the Primer, 1. Newton logic, 2. a Rotating Wing's Flaw, 3. the need to understand a prop as an Airscrew, 4. the Elegant, *multifaceted* total logic of a B-G-T Propeller.**

We purposely set up the Book for each reader to go as far as he chooses.

We ***teach*** the logic in the **Primer**, ***expand*** **it in the Primer Chapter 9**, ***overview*** it to **start this Book 2**, a nice smooth read, ***many pilots satisfied, and hopefully honestly understanding in 3 shots*. There's a ton more, a really insightful education, a pro's grasp for those who choose to go all the way through Book 2, *a day*.**

We Build on the Same Basic Pattern, Go deeper and Deeper, For Novices and Pros.

A STARTING POINT TO A TRULY ELEGANT LOGIC

Let us presume that you have read, and generally understand the **11 page Propeller Primer for Pilots**, and **Chapter 9 of Book I, Propellers for a Pilot's First Grasp,** Book I having covered Airplane Logic. ***Maybe***, you have **first read** the **Overview here**, were **able to start the Primer with the very helpful benefit of knowing where the whole subject is going**, insight that can make a complex subject like propellers far easier to learn and understand. **We want you to get the horse sense of Props so we build it all on the same base, lead you ever deeper!**

You can start with 12 pages here, or the Primer, but read them both and Ch.9.

Propellers have always been the **unexplained, unexplainable subject**, just too many things happening, all interrelated, affecting each other, **pretty much the ultimate technical swamp**, the subject that even experienced engineers couldn't understand, or explain to each other. It's taken two of us, two professional lifetimes of experience to **tame it**. We took it on as a **retirement intellectual mystery challenge, taken on just for the challenge, like Everest, because the dare was there, the job that never got solved, and wasn't going to be solved!**

Often engineers can resolve usual problems to a math equation, maybe a couple of equations, and when you can do that, see what's happening there, engineers can see the logic of it all. Props were always too much of a morass, too many things happening, interrelating, affecting each other, everything else. It took a long time to learn everything that has been learned, to figure out the logic, to figure out how it could all be explained. **To succeed here, my challenge was creating an easy enough start so people can grasp the technical horse sense here, fast.**

Compared to the usual morass this overview explanation is pretty tame.

It's all here. We show you the logic of it all. ***Props still try to be complex, tricky, so we superemphasize***, repeat, bring it all together, **do everything possible to help you get it**. If you're **interested enough to try, care, you can get it**. **It's up to you now!!!**

CHAPTER 1 BOOK II

THE PROPELLER

AN OVERVIEW INTRODUCTION - TO A PROFESSIONAL GRASP

The **brilliant work** of the **7 Historic Giants of the field**, over **83 years, Rankine, Froude, Betz, Prandtl, Goldstein, Glauert, ultimately the *rare genius* of Betz, Goldstein, Theodore Theodoreson,** we have **a truly Elegant, Classic Propeller Logic, essentially Exact Math for our Computers!**

Theodorsen's 1948 book brought together the very practical, and **essentially exact mathematical solution** for the ***Heavily Loaded, High Advance Ratio, Ideal Minimum Induced Loss Betz Logic Propeller.*** It's *first* done on a **Profile Drag Free basis.** Theodorsen credits **Goldstein's 1929 basic solution for the Mathematics of the (3D) Potential Flow Problem** as, "unquestionably the greatest single step in the evolution of Propeller theory". ***Theodorsen finished it, for heavy loading!***

It is possible to show with **simple technical horse sense** that there is a **harmful basic flaw** in the **operation of a propeller as a rotating wing!!!** There is **a fundamental need to significantly taper the tips of a propeller!!!** The **Shaping, twisting, therefore *Radial Loading* of a propeller, logically, is significantly more important than the shaping of a wing!!!**

Simply, the **tip velocity of a propeller, so much faster than the root**, at an **even more magnified tip Dynamic Pressure, q,** proportional to V^2 --- **tries to greatly overload the thrust and drag at the tip, where the thrust must fall to zero!!!** ***Excess Thrust swirls into the tip Vortex --- and is lost,*** **causing extra unrewarded wasteful tip vortex induced loss,** ***noise*** **- extra tip profile drag at max. lever arm -** ***excess torque, bad design!!!****

 ***Now we can *nail*, Min Induced, Min Profile, Min Area Precisely Placed, a Triple Ideal Prop. We'll ultimately learn how to beat that 20th Century Ultmate!**

Fortunately, there is a **simple, good technical horse sense fix** for that very fundamental basic flaw, ***simply taper the tips***, **thus the radial blade loading,** ***pull the excessive tip loading back to a more sustainable, mathematically ideal radial shaping of the blade loading***. **Betz-Goldstein-Theodorsen, Math** simply optimizes the tip loading, indeed the **loading of the entire blade into a mathematical optimum, Minimum Induced Drag loss, a unified propeller theory with several advantageous, ancillary, insights and features**, one of the most elegant, all encompassing logics that you will find in all of engineering -- constant pitch pure helical inflow, constant slip, stretched pure helical outflow, constant dT/dQ, a constant ratio of thrust vs. required Torque at every radius, every radius equally efficient - *if* first considered free of Profile Drag. ***Elegant***

PERFECT SHAPE VS. TWIST GIVES IDEAL RADIAL BLADE LOADING

Profile Drag can be added concurrently in a subroutine program, in the same basic overall propeller design program. Since propeller blades tend to have the speed of Jets, but the chords of model planes, **low Reynolds numbers**, well below to above 1,000,000, it's **proper to use drag coefficients that have proper higher drag coefficients vs. Rn's, also earlier Stall**.

Norris's 6th Law, teaches the useful insight that if you simply design an ideal BGT prop, that **actually achieves a constant, optimum angle of attack, α°, and C_L**, you will have a **Triple optimum prop for the design condition**, because you'll have **ideal minimum area, precisely placed**, ***Ideal Shape***, ***Loading!***

Min. Area, Precisely Placed, gives Min. Induced, Profile, Torque for that C_L

The whole, **very simple, very understandable Bottom Line** on **Ideal Propeller Design** is simply that the game is to **SHAPE the blade, (with matching, proper TWIST), thus ideal LOADING** for the design conditions, which, **with ideal constant α° and C_L produces a Triple ideal, Minimum Induced, Min. Profile Drag Loss, Min. Torque Prop Design!**

It is subsequently **necessary to adapt a theoretically ideal BGT prop, for the slowed velocity profile at the body nose!!**

 For Century 21, we can try to force Elliptical Loading, do a few percent better!

We initially learn prop logic by looking at them in 4 ways.

1. As **pulling in and throwing back air**, to get **a reaction force** per **Newton's Laws** - Which **teaches us basic prop logic**.

2. As a **Rotating Wing** which **exposes the need for tapered tips**

3. As an **Airscrew** which teaches us that a propeller is **more than a rotating wing - must be considered as an Airscrew**.

4. As a ***Triple*** **Ideal Minimum *Induced* Drag Betz Propeller** designed for **constant optimum α^0 and C_L - min. *Profile, Torque* too.**

Pitch of a perfect helical screw prop, is simply **the inches or feet that a prop will screw forward in one revolution** through a soft solid with **no Slip**. Since a prop blade is not apt to have perfect helical pitch, especially after it is retwisted to match the slowed velocity profile in the prop plane at the nose of an embedded body, the **pitch at the 3/4, 75% radius** is used.

Advance. Since a prop must be **overpitched a little bit to pull in and throw back air to make thrust**, and have **an angle of attack too**, a prop will **advance a little less each revolution than its pitch**. The **Difference is the SLIP**. We will learn that a core characteristic of an Ideal Betz prop is to have **Constant Slip at every radius --- a basic of Minimum Induced Loss.**

With Constant Pitch Air, Constant Slip, it has Constant dT/dQ, if drag free.

We will learn that **only High Pitch, High P/D, High Advance Ratio Props** can be **Most Efficient**, because the technical **horse sense** will teach us **all three go with Minimum Induced, Min. Profile Drag Loss, *Min Torque too***. We will find three similar, but different ways to define **Advance Ratio**, the most useful, most easily understood one being **V / πnD. Look**, and you can see that **V** is simply **forward Speed**, or **Velocity** (in ft./sec.) **vs. the circumferential velocity of the prop *Tip*** in ft./sec., **n** simply being the **rotation** in revolutions per second, **D** the diameter in feet, ***the Tip Advance Angle***, which **acts much like P/D Ratio, both tied to max. efficiency possible.** Realizes there's a few % **Slip** there.

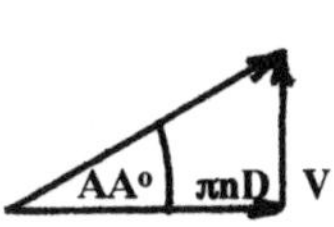

High Pitch, High P/D, High Advance Equals High Efficiency

In the Propeller Primer and Ch.9, Propellers, in the Airplane, Book I, we learn that Isaac Newton teaches us the way to **minimize Induced efficiency loss** is to **minimize throwing air, minimize** ΔV, and.Δ**V/2V$_1$** the ***Axial* Induced Efficiency Loss.** There is also a ***Rotational*** and a ***Radial*** flow Induced Efficiency loss that makes the total Induced efficiency loss **significantly bigger than just the *axial*** Δ**V/2V$_1$** loss, but just looking at the axial loss, at first, **most easily teaches us the tricky basic logic.**

It's Really a ~1/V^2 Induced, just like aWing, *a double effect of V, just disguised!*

Newton teaches us **the way to minimize, (all three) Induced Losses** is to **have a BIG Mass Flow Rate** flowing through the **Prop Disk**, since Thrust equals Mass Flow Rate times the Delta V, ΔV speedup of the air, simply, **T = $\dot{M}$** Δ**V**, for a flowing fluid. Simply, a **big $\dot{M}$, M dot**, allows a **small** Δ**V** for any given required thrust. Now logically the way to have a big mass flow rate is ***either*** have a **Big Diameter Prop, or Simply Go Fast! Simply going Fast** is the **easy way to get high prop efficiency**.

Fast lowers the Optimum Diameter, both Profile and Induced!!!

Now, that sets us up to **very easily see** that **only High Pitch Props can be Efficient**, that **high blade angles** are the pro's math based, and horse sense **clue to high efficiency**. It's easy, a cinch to see that a **high pitch** prop has the **shorter Spiral Path** to the destination, **less profile loss cost**, low pitch wasteful. Since **High Pitch correlates with Fast**, at usual RPM's, it correlates with **Big $\dot{M}$, Low** Δ**V, low** Δ**V/2V$_1$, low axial Induced loss** too - tricky yes, but pretty easy once you catch on, especially once you realize that's the easy enough answer to decoding the logic swamp that befuddles smart guys, before it's explained.

Now, we'll be teaching you the **classic Advance Ratio Graph** of **Max Potential Efficiency** for ***any prop*** shortly, done **including Profile Drag, full sophisticated math**, but, you see, **High Pitch, High P/D, High Advance, High Blade Angles,** all correlate with **High Efficiency**, **just *a little slip insight hiding in there*** ---

Gearing Slow Planes for Low RPM, High Pitch , Big Diameter

Gearing for Lo RPM, increases the ***Torque***, can turn Hi Pitch, Big Diameter.

We've purposely been teaching you the much more simple **basic technical horse sense of propeller logic, (wonderful to finally have what has been a 138 year morass down to horse sense),** because horse sense is what technical novices can understand, and this book is for **Pilots who properly want insight that makes sense**, not intellectual elitism. But pros should understand that the only way we were able to make it this simple, was going through and past all the marvelous Historic BGT Math, two old pro's with **two lifetimes of experience**, enough to **finally clarify the logic**.

The Sense, the Logic, is what pros grasp when they really understand.

Either the **sophisticated math** or the horse sense can show us that High Pitch, **High Blade Angles**, High Advance is the **fundamental tell tale of High Efficiency**, ~ **45° the optimum**, you'll see, but that tells us that we can gear down the RPM on a too Slow, too High RPM case and ***always*** be able to get a max ~ 91% prop efficiency - and that's true, the math proves we can.

Aspect Ratio can nudge the ~ 91% up or down.

But, as always seems to be the case with props, there is some trickiness hidden in the shadows. If you gear down, slow down the prop for **lower RPM, Higher Pitch**, it's going to take a **bigger prop** to make the required thrust at now lower q. **Good**, that will get the **mass flow rate up**, the $\Delta V/2V_1$ **induced loss down. But,** when the diameter goes up, for the same speed the **tip angle**, tip **advance angle goes down**, ***counterproductive***. Notice two pages back, at the bottom that **V / πnD**, that best defines Advance ratio, is really just the **prop tip Advance Angle**.

Now, what happens is, gearing slow planes for efficiency works, but **Diameters get a lot bigger, props heavier, RPM's a lot slower**, a process that takes ***a Lot of change* to get *a very Low ΔV vs. a V_1 that starts low* --- to actually get high efficiency**. The man powered ~ **.2+ H.P. Deadalus used an 11.3' prop**, the **Wrights** were **smart** enough to know they needed **(2) 8.5'** props for ***only* 12 H.P. I'm impressed** they were that smart in **1903!**

 Paul MacCready used a 12' Prop on Gossamer Albatross.

Propeller Efficiency, eta, η, Overall Propulsion Efficiency, η_P

If you're either lucky or smart, you may have realized that **basic propeller efficiency** doesn't depend on the propeller, but rather on **Airplane Speed, V, and Engine RPM**, because **that's what controls Pitch!!!** You'll see **High Aspect Ratio helps**, a **bigger Diameter**, for a given blade area, **more $\dot{M}$, Mass Flow Rate,** and, of course**, BGT Logic and Math** to get **perfection, the last few percent.** But it's **Speed vs. RPM** that determines what ball park you're in, **85% max for a Luscombe, ~89% max for an RV 6**, is the **best possible, the BGT max**.

The Advance Ratio Graph Teaches us that.

Efficiency is really Traveling Efficiency.

It really helps people to teach them early that **Prop Efficiency, eta**, η, **is really Traveling Efficiency, Zero at runup**, roaring, shaking, **going no where**, **hurt in climb**, throwing back extra air, making extra thrust to **hoist the plane too**, going slower, **finally up to design efficiency at cruise**, if designed there, actually **better at Vmax, actually at a lower angle of attack, less ΔV, divided by a now bigger V_1.** Grasp it that way and you have **a pro's insight, just 6 pages** from the start of Book II.

Super Magic Graph, Ch. 2 II, will show how everything varies vs. Speed.

Overall Propulsion Efficiency

I never met him, but **I just love and respect Gus Raspet**, a **creative maverick**, I'm led to believe. He did propellerless glide tests, from a towed altitude, found normal drags, but terrible overall propulsive efficiency on light planes when compared to required H.P., only **58%** overall on a Bellanca Cruisair, **172% more H.P., vs. the Drag x V gliding H.P.!!!**

Gus had sealed the cooling ducts, taped leaks for min Aero Drag.

Zero Thrust Glide Testing my Luscombe we found only **67%** η_P **overall propulsive efficiency** with **cooling charged to the Airframe**, ***(~10%),*** at efficient altitude cruise, 85 IAS, 100 TAS, equivalent to the Bellanca. Presuming bad ***75%** η **prop**, that's an **89.33%** η_i interference efficiency to equal .67η_p. **More to come!**

Realize a .67 η_p *inverted*, demands 150% of Gliding Drag H.P.

 An RV 8 Prop Test, showed a poor prop could be 10% worse than a proper Prop!

Implications of Blade Angle Accuracy Requirements

McCauley uses **a shop tolerance of +/- .1** degree, yes **One Tenth Degree**, to ship **a consistent Prop Product**. I certainly would not expect that their old props, designed decades ago, before computers, were designed anywhere near that accuracy, but let me show you with **our usual technical horse sense** that **we need to predict the inflow to all our prop radii that close** to have props that agree with our computer calculations.

Nominally a 1 degree angle of attack change, changes the coefficient of lift .1, 1/10. Now, simple horse sense can tell us that **if we design a prop to .5 C_L, a 1° error, .1 C_L error** we have a **20% error in our calculation!!!!!** Thus, with a **+/- .1° shop tolerance McCauley would be shipping a +/- 2 % variation, quite good**, when you realize they've been **hand grinding and twist bending,** ***accurately tweaking*** **2025 Alloy fatigue resistant Aluminum forgings for over a half century.**

The **Goldstein-Theodorsen Kx Blade Loading Chart**, **(vs. radius and Advance Ratio)** printed to three decimal places, i.e., .xxx, is said to **be good to 1%**, actually therefore **twice as good as McCauley's shop tolerance**, just what you'd like to have from an analytical method. Actually, Ribner and Foster, University of Toronto, did **a modern computer analysis confirmation** in the 90's, generally found the basic loading chart was about **1%, excellent, when you fully understand the complexity here.**

The Wild Flow We're Analyzing Here ***Grasp This One*** ----

We're pulling in and throwing back a big Mass Flow Rate of air to make thrust, with **minimum ΔV**. It's **rotating too, much more at the steep inner blades.** The ΔV is not even close to uniform out along the radius, but with rotation, you'll see we form it into a **perfect helical** constant pitch inflow, constant Slip, which acts like uniform inflow, you'll learn, **stretched perfectly helical downwash, backwash,** the picture you can see on the cover. There's **radial flow feeding the tip vortex, out on the bottom of the blade, in on the top,** opposite, but there is **a root vortex too, radial flow opposite to the tip. With Heavy Loading, the ΔV at the prop is much bigger than avg.**

 ~1% accuracy with that complexity! You tell me how Smart BGT were.

Elegant -- Theory, Logic, Mathematics -- *Rare Genius Work*

Top Scientists use the word **Elegant** when they grasp the ***magnificent order and logic of Nature*** which is so often **simple**, while being **complex and sophisticated, all at once**. As you come to fully understand, let me suggest that you will see that, in total, **propeller theory is as elegantly simple and beautiful in its complexity as any logic you will ever find.**

The Genius of Betz, in 1919 taught us to **simply guide inflow into a *perfect screw surface*, a pure helical Archimedes Screw inflow of air to each blade**, then **throw back a still perfect stretched helix, downwash, backwash emanating from the blade trailing edge, as the second $\Delta V/2$ is added**, a good art form if you add **the vari-hued beauty of varnished wood** in a simple stack of ice cream sucker sticks, the model on the cover.

Archimedes died 212 B.C., and his screw is still pumping Irrigation Water in Egypt

Hiding in that **sophisticated constant pitch simplicity**, is a **"Constant Slip" at every Radius**, when **compared to Airplane Speed,** which acts like constant inflow speed at each radius, when axial inflow is ***not*** constant, a perfect blending of axial and rotational inflow, you'll learn, for **balanced inflow out along the radius, no added radial flow loss, the least induced loss!!!**

To create perfect, Constant Pitch, Helical Inflow, the Blade Shape, Twist must be Perfect

Appendix T-Q, proves a **constant dT/dQ**, a **constant ratio of Thrust vs. required TorQue, or H.P. at every radius, equal efficiency at every radius, no penalty for prop twist**, if first considered profile drag free, all provable with simple high school Algebra, Geometry, Trig. **Elegant Simplicity from gross Complexity**

It's all Topped off with an Elegant, *Rare Genius level* Math Solution!!!

It's all accomplished by **perfectly Shaping and Twisting, *thus* precisely controlling the radial distribution of Loading** to **deliver precise, pure helical inflow, stretched outflow** --- the **magnificent Goldstein-Theodorsen "Potential Flow" Math Solution *directing* each air molecule in the precise 3D direction to solve it all - even for heavy loading** where the ΔV **is much greater at the prop**, vs. the Stream Tube. ***Elegant!!!***

Elegant Simplicity - Pure Helical Inflow = Optimum Prop

Propellers were always impossible to explain, too many complex things happening, interplaying, a too complex technical morass, too much for engineers to get their head around, ridiculous for Civilians and Pilots. **The solution is to show people we have an essentially exact, *canned*, Elegant 3D math solution for Betz's Elegant perfect helical inflow and stretched outflow, easy, indeed beautiful to visualize.** Then **with people *not* snowed**, we went off to **teach everyone the 4 different ways of looking at props**, all the little individual things that are happening, **their logics, how they fit together, *superemphasize* so novices can really get all the hidden points, and pretty soon it can all start coming together for people who want to try, who never tried anything like this before. Go for it ---**

Let's take a simple, quick look at the famous **Advance Ratio Graph**, that shows low pitch, Low Advance props have poor efficiency, High Pitch, High Blade Angle, High Advance, High Efficiency. Simply, The **horizontal Axis is V / πnD**, as before, simply the **Actual Advance Angle of the Prop Tip at D**. The **vertical Axis plots Efficiency, (vs. tip Advance), angles shown.**

(The full coverage of Advance Ratio is on p. 48II, but more complete, is more complex.)

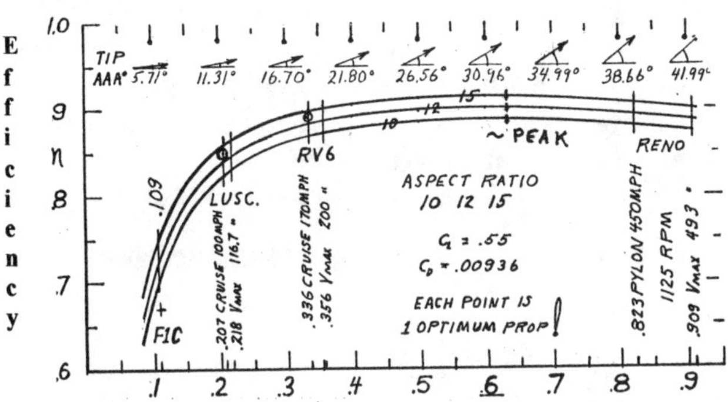

Tip Advance Angle vs. Efficiency η (optimum ~32°)

Actual Advance Ratio **J/π = V / πnd** (often called Lambda, λ)*

Next, let's look at an **actual Ideal Luscombe, and RV 6 Prop,** their **Ideal Shape and Loading, then on to a pro's explanation of props.**

 * Think of Tip AAA, Actual Advance Angle -- Shown on Graph, 32° optimum!

Really Understanding --- The Elegant Betz Propeller Logic

The **Betz Propeller is truly Elegant** in the **sophisticated, yet simple** way that it **ties together several subtle but hugely important features.** It's **hugely important, most important of all, that you *finally* understand this core of all the logic. This page will help you see where it's all going!!!!**

The **pure helical Inflow** is the **cornerstone of the logic**, simply because it is **Constant Pitch Inflow**, which results in **Constant Slip at every radius**, simply vs. the **Airplane Speed**. It's the **Constant Slip that acts like uniform inflow speed**. *Imagine **loading each radius uniformly, a balanced radial loading***, not causing unbalanced extra radial flow, **avoiding extra induced radial flow loss, (right at the heart of min. induced loss logic).**

We set this **Constant Pitch, Constant Slip inflow** up in the geometry we **teach you in Chapter 3-II**, (p.94-II, 98-II), that you **first saw** in **Ch.9, p.128.** It results in the **blade shape**, which, with the constant C_L twist, results in the **ideal blade radial loading** you see on the **facing page**, produced by the **Elegant 3D potential flow BGT Math** solution. Now, **(in reverse order)** if you learn that the **Characteristic Betz SHAPE (tapered tip, root, *twist*)**, and thus **Ideal LOADING**, that pulls back the excess tip load, **(we learned a rotating wing needs)**, we **produce an Ideal Prop**. Constant pitch, constant Slip also produced **constant dT/dQ, a constant ratio of thrust vs. required Torque, equal efficiency at every radius** if first considered drag free. **(proven in Appendix T-Q), that cuts like a knife through all the complexity!!!**

Realize Radial Flow is Tip Vortex Flow, not shown in the Picture

When you finally grasp it all, and then finally grasp its elegance, you see **a pure helical Archimedes screw inflow to each blade, a constant Slip, (elegantly produced by the Trigonometric combining of axial and rotational inflow), acting just like constant axial inflow, minimizing radial tip vortex flow, minimizing induced loss,** by preventing excess *radiial* induced loss - ideal **shape, ideal blade loading (excess tip loading pulled back to a sustainable loading profile). (In addition to the facing page, see the ideal load distribution and shapes on p.147, the Betz page, p.129, the flow geometry taught on p. 94-II through 98-II.**

Look ahead to the grasp of the integrated logic you can end up with!

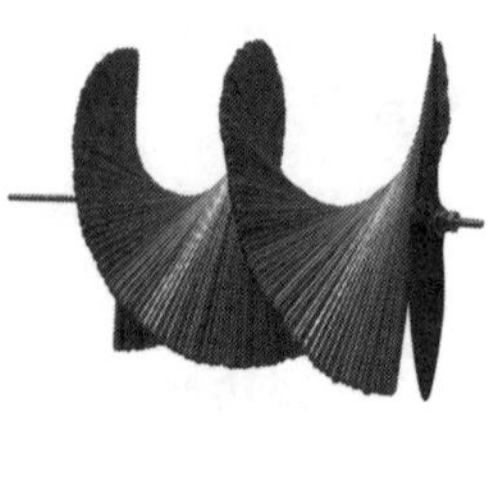

Betz's Rigid Vortex Sheet

<u>Shapes</u> and <u>Resulting Loading</u> of Constant C_L Betz Props

(These examples are *<u>not</u>* modified for Slowdown) ✓

Luscombe 1400# GW	**RV 6 1600# GW**
73.8" Diamater vs. 71" normally	70" Diameter, (or 72" is good)
151.5# Thrust (140# T at new η_2 OK)	**164# Thrust,** (156# T at η_2 OK)
.55 C_L 1.612 degrees alpha	**.55 C_L** 1.612 degrees alpha
2280 RPM 48.15 H.P. Alt Cruise	**2400 RPM** 82.9 H.P. Alt. Cruise
100 MPH TAS 85 IAS	**170 MPH TAS** 140 IAS
10,500 ft. Density Altitude	**12,500 ft. Density Altitude**
Ref Analysis 99.36 - 21-23 & 99.75 - 67,68	Ref Analysis 99.36 - 24-26

Realize Betz SHAPE, thus LOADING vs. radius are shown here. ✓

Thrust vs. Radius #/ft.	**Blade Planform** Chords"	**Thrust vs. Radius** #/ft.	**Blade Planform** Chords"

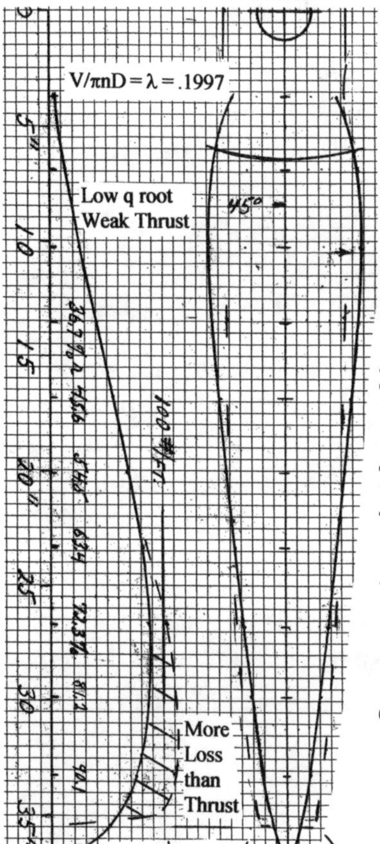

<u>Optimum, High Pitch Propellers</u>, a V / π n D = <u>λ = ~.63</u>
Have Symmetrical, ~Canoe Blade Shape, Max Chord 50% r.,
not excessively pointed tips, a pleasant, ***Nature's design***. ✓

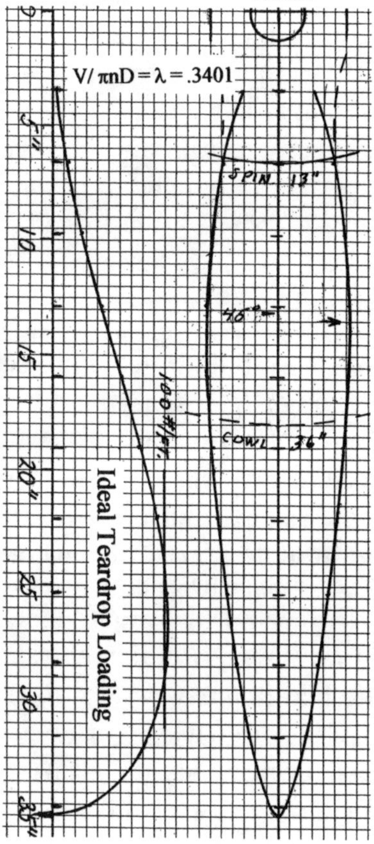

Luscombe — Square Tip Loading, Thrust won't happen, Losses will! ✓

High Tip q, but Zero T# result *wants* Outer Chords Narrow for efficiency!!!

Ultimately Realize the Good and Bad of *half* Teardrop Radial Loading -- ✓

It's <u>Precise Betz SHAPE</u> and twist that gives <u>Ideal Loading vs. r.</u>, min Induced Drag Loss

Is a Betz Prop Really Better*? What Happens Off Speed?

Models and private planes prove **poorer props that load the engine fairly correctly will work, just not as good as they should**. Knowing that we want +/- .1° inflow accuracy control, and most props have only a wild guess for slowdown, flow field correction, shapes far from ideal, we have a world full of less than great props, ***but* *hidden by Arirplane V = Power³ Math.***

I'm not pushing fancy theory, or math. I'm after the Facts here, TRUTH!!!

Looking at the just better engineering horse sense of a properly shaped and loaded tapered tip, and a professionally designed prop with an intelligent correction for the flowfield, slowdown, I'd expect a proper prop to be 5, 7, maybe 10% better than what we have, not huge, but **now we can both understand it all correctly**, and **know we have the Optimum Classic Prop**. If a perfect Luscombe prop can be 85% efficient, a bad one only 75%,* 15%/25% loss, we've cut our prop loss to 60% of what it was, **166.66% better**, *a lot, but only* **.85/.75, 1.1333% more H.P. available,** ***and*** **Speed is** ***only*** **the cube root of Power --- 1.0426!!!**

Many props miss Chord and Blade Angle distribution badly, 5, 7, *10% off!!!

What we'll do is get rid of the poorly designed props, wildly varying C_L's, senseless chord and load distribution, flow field corrections far off, wildly varying vs. the +/- .1° inflow accuracy we should shoot for. **We can now understand the logic exactly and what we should shoot for, and we have the math to do it credibly, accurately**. ***We finally understand!!!***

Getting rid of Poor Props, finally knowing what we're doing is my objective.

Off Speed, fixed pitch, understand the angle of attack goes up as we slow down, finally way up at static runup, the **good high pitched props worse, *stalled a lot, most inboard, but* helped by BIG slowdown correction**, *lower inboard angles, less twist*. Realize **fast, even at higher RPM** ***the angle of attack continues to decrease***, α° **goes to zero, finally zero average C_L's in a shallow dive, no thrust, back to zero efficiency**, But we ***do better starting from a known, a Classic Ideal, accurately designed prop, especially with big slowdown angle corrections!***

 Surprise - We're finding BGT Props are QUIET - Proof Positive of Losses Eliminated!

* A good RV 8 Prop Test showed a European Prop had ***Twice the Loss***, ~80% η, not 90!!!

An Early Incisive Overall Evaluation of the Math Method

We're **purposely** trying to give you a **good, quick, insightful grasp of overall BGT Logic to start Book II** Propellers, so you **grasp early where it's all going**. We're purposely writing the text to quickly teach people that we **finally have a correct Logic to understand, and *essentially exact* math,** because all checks tell us the **Betz logic is correct**, indeed ***truly Elegant!***

We do also have ***essentially exact math***, a proper claim because it does address and solve the very, very complex **(3D), Heavily Loaded, High Advance Ratio Potential Flow Problem,** but we must now properly qualify that, not overclaim, because that is the fundamental no no in engineering **where credibility is everything, *absolutely* must be maintained.**

I've always thought of Theodorsen's basic core accuracy as close to 1%.

Theodorsen notes that **Goldstein solved the Potential Flow Problem**, but **strictly speaking** only for the **multiblade, thus lightly loaded, low advance ratio part of the problem. Theodorsen,** I believe, accurately **solves heavy loading** by treating the math **far back**, for a **higher Advance Ratio,** to account for the greater ΔV effect right at the blade, **and a slightly reduced effective Diameter, D_0.** Then Theodorsen notes that his (Kx Blade Loading Chart *(multi blade*) work **depends on his (pre computer) very accurately done analog, voltage field Potential Flow simulation.**

Theodorsen's method makes Goldstein's basic chart directly, accurately usable!!!

With Computers now available **Ribner and Foster**, of the University of Toronto, checked Theodorsen's work, publishing in 1991, a **Cray X-MP/24. Their work was extensive**, adding **3 bladed props**, up to **12 blades**, checking not just the **basic Kx Circulation Blade Loading vs. Radius and Advance Ratios** but the **lesser efficiency and energy loss plots of associated variables** used by Goldstein -- and Theodorsen's extended work.

It would be as big a mistake to underrate the answers, as to think exact!!!

Ribner and Foster made **a few changes in the math details**, that they felt were appropriate and presented **many pages of carefully plotted data** in their **64 page report**, only providing 2 numerical chart checks of Goldstein's basic work, generally much **closer than 1%,** 3 local exceptions of 1.2 1.7 and 1.8%, along with a few charts of associated variables showing a few % errors in some less central numbers. The **graphs show excellent agreement and support of the core numbers**, a **digital comparison not possible** from the plots, **a hint of a bit heavier loading proper outboard** and **at very high Advance**. In all, **as close to *coveted* 1% basic core answers on such a complex problem as we could ever hope to have**. With modern Computer Power, **future checks can clearly refine the answers even more.**

Later work by Tibery and Wrench at David Taylor Model Basin confirms loading a t Lambdas, λ of ~.5 and below, but higher above, Applied Math Lab Rpt.1534, 1964.

 Now, we'll sum up, start going deeper, discussing more, in Book II

A Summary Overview

After you understand the basic logic of Propellers there is a **wealth of more incisive Insight Available**, going back in more depth, and **that is what we'll do in the rest of Book II.**

But first, a comprehensive ***Summary Statement*** can ***nail a lot***. The trick is to get **a grasp, "a feel" for how props really work.**

By now you should realize a prop is really two separate rotating wings 180° apart, in a stream tube that it activates, moving the whole tube backwards, pulling in a pure helical inflow sheet, a slightly reduced pressure in front, throwing back a stretched faster, pure helical downwash, backwash sheet, two Archimedes screws, one feeding and stretched from each blade, nominally half the speedup as inflow, half as outflow, two almost equal $\Delta V/2$'s. (Theodorsen's Math can divide that precisely.)

As two separate rotating wings, they each have both a root vortex, and a tip vortex, of course the root weak, the tip much stronger, relatively weak q, dynamic pressure at the root compared to an exceptionally stronger tip q, proportional to V^2, with the much higher rotational velocity at the tip, making a diagonal spiral path with the forward speed. **Intelligently managing the spiral velocity effects is the central key to intelligent prop design, *nailed below.***

But first, realize that a low pitch prop rotates an excessively long distance getting to the destination, very wasteful of profile, skin friction drag *energy consumption*, so easy to see, once you see it, high pitch props far more efficient vs. profile drag *cost*. But, also from what we've learned from Newton, a fast, High Pitch prop naturally has a big $\dot{M}$, and that allows a small ΔV for any required Thrust, minimizing the thrown air velocity, minimizing Induced loss too, high Pitch, hi P/D, high Advance per revolution, high Tip Advance Angle the key to efficiency!!!

But now, just like a Wing, where Induced loss drops as $1/V^2$, simply going fast is the easy way to get high Induced Efficiency, but on props $\Delta V/2V_1$ is the specific math, low $\Delta V/2$ allowed by a big $\dot{M}$, divided by a big fast V_1, two good effects!

It's still is a $1 / V^2$ effect, dropping fast, but acting indirectly through $\dot{M}$ and ΔV

But with Hi Pitch, Hi Advance, the key to both Profile and Induced loss reduction, Hi Efficiency, **Speed, V** becomes the most important factor in prop efficiency, but surprise, **low RPM** which allows High Pitch, Hi Advance, is almost as important!!!

Speed, V, and RPM set the max limit of your efficiency!

Now, having established that basic insight, it sets us up to show now that the use of Betz-Goldstein-Theodorsen Logic and Math is the Sophsticated Professional Classic "Master Stroke" to intelligent Propeller design, because with sophisticated radial Loading, through sophisticated SHAPING and Twisting, we make the Classic Ideal radial distribution of loading, don't create an extra radial flow over the mathematical minimum, minimize both root and tip Vortices, get the Classic mathematical minimum Induced loss from throwing air back to make thrust!!

The whole game there is simply recognizing that the Dumb prop trys to work **Inside Out, Dead Wrong, trys to make max thrust and drag** at the undammed, **ridiculously high q tip**, ***where thrust must fall to zero***, **manufacturing excess tip vortex loss, NOISE, WASTED ENERGY, high drag at max lever arm, bogging down the engine Torque, RPM and available H.P.** and **losing thrust it tried to make** there, *dumb as a stump.* In addition to proven performance tests, BGT props are **QUIET**, proof positive **the significently excess tip loss is gone!!!**

The final Coupe de Matre, Master stroke, is Theodorsen's Math accounts for **Heavy Loading**, the fact that Betz's stretched "Rigid, Vortex Sheet" is moving back through the stream tube faster than the stream tube average, actually nominally 2, 3, 8 times faster on a Luscombe, RV, a Reno Racer. **Theodorsen is the *only way* to get the Pitch, all the calcs precisely correct*!**

 Now, we're trying a More Elliptical Tip Loading, for a few % better!

But we've learned, and are going to learn much more -----

On a Sea Level Buzz Job, a 200 MPH RV 6 can make 280# of Thrust with only a 10 MPH full ΔV, only a 5% full ΔV, half that, 2 1/2%, for ΔV/2, the axial efficiency loss. But surprisingly on a high pitch prop Theodorsen's magnificient math shows us the radial vortex loss can be a tad bigger, another 2 1/2%, 5% together, maybe 2% or so rotational loss, 7% total. That makes sense because an RV 6 Prop can be 89% efficient, an 11% loss. That would leave 4% for Profile Drag loss, and that makes sense because if a ,55 C_L is used the loss is 2/3 Induced, 1/3 Profile Drag, a .5 C_L prop roughly as shown here, Great Insight.

On a bigger .5 C_L prop a ~4% profile is more than 1/3, logical.

Long narrow, High Aspect Ratio Blades reach out farther, a bigger stream tube, a bigger $\dot{M}$, more efficient, less Induced loss, just like a long Sailplane Wing, but a long prop blade can be more vibration prone, a lowered natural frequency. We must temper our quest for efficiency, because a fatigue failed blade can rip the engine out, your plane uncontrollable, kill you!

But, **a Luscombe Prop needs a 14:1 AR to *lower* a critical RPM below Cruise RPM!**

Page 146 and 147, may be the two most valuable, practical pages you'll ever see, because they show the **ideal teardrop loading**, unfixably weak at the root, low q, **max at 75% r to 80% r**, **a fast drop off to the tip**, ***not smart to try to get more tip loading, once you see that plot***, the ***perfect BGT SHAPES that generate that optimum loading***, highly tapered tips, tapered roots too on the very hi pitch, hi advance optimum props with those naturally graceful ~ Canoe shapes, the low pitch, advance props surprisingly triangular, the max chord at mid span on the optimum .63 λ, Lambda, prop, optimum 45° blades from root to tip on low to high pitch props, Reno racers near 45° tips, WOW!

The basic Sophisticated Game is just **be smart enough** to realize you need to ***taper the tips***, perfection proven by math, understandable by horse sense. ***Now let me show you how to sort out, make sense, understand the logic of all the variables:***

Understanding all the Variables in Prop Design for T# reqd:
* **Speed**, **RPM**, **Altitude Density**, ρ are usually set by the application, **3 items set by the Specification, 3 Spec Items**.
* **Aspect Ratio**, **Coefficient of Lift**, C_L, **Coef. of Drag**, C_D, are **3 Choice Items**. Physics really sets C_D, ***but it takes judgment vs. Rn.***

I purposely down play the Airfoil, bugged leading edges killing super foils.

Now, we really want to find **Diameter**, **Area** and **Shape**, but with the **extremely varied q vs. diameter**, those 3 very basic variables are ***interdependent, can't be separated***, but once **Aspect Ratio** is ***chosen***, **BGT math just scales them all up and down to perfectly meet the Thrust required,** ***SIMPLE!!***

Of course what **BGT math** does is get **SHAPE** and **Twist**, thus **LOADING, *precisely correct* -- *min. area precisely placed*** for a **Triple Ideal Prop**, if you're smart enough to use an ***Ideal*** C_L.

We learn how to set up the key Inflow Geometry in Chapter 3, Book II

We rough out prop design with an NACA 4412 airfoil, 12% thick, 4% mean camber chord line - ***then vary t %, but hold camber,*** **the easy way to vary thickness with minimal effect!**

How's that for making complexity easily understandable!?!

The core concept to understand, remember, **visualize** in your mind is the **great layout** of **p.131**, the **extreme magnification of tip q**, a **tapered blade that** ***counteracts* excess tip loading**, the **half teardrop blade loading** - a **mistake to try to load the tip more** because you'll **just waste energy**, **attempted thrust that will just roll off the tip, manufacturing extra Induced Loss, making NOISE, unrewarded drag at max. lever arm bogging down engine torque, RPM, H.P.,** not smart.

It's *simple* **TAPER the TIPS,** ***hope*** **Speed vs. RPM = Hi Pitch**

And now for those who wish, we'll go deeper in the rest of Book II, get some really professional insights. Think of it as a Mystery Novel, not a difficult chore. Even if you miss a lot, you'll learn a ton of insight, and you can't strain your brain!!!!!

It's all available because two young WW II modelers, got a lifetime of pro experience!!!

So You Can't Possibly Fail to Understand - The Tapered Tip

It's almost funny, after going through all the complex math and gross complexities of propellers, decades where people argued prop shape was not important, to find that **<u>SHAPE, with correct *Twist*, is the Central Issue of Ideal Radial Loading of Ideal Props</u>, missed for a half Century**. In hindsight, it's as simple as seeing broad tips flunk Aero 101 horse sense, that we missed, were **blind to - the easiest key insight in Prop Logic!**

You simply <u>Can't Maintain High Loading at the Tip</u>, *<u>you create Extra Losses</u>!!!*

The **<u>Elliptical like drop off</u> of <u>Wing Lift</u>, at <u>constant q</u>**, fairly independent of wing shape, **DECREES --- *<u>tip lift fades to zero</u>!***

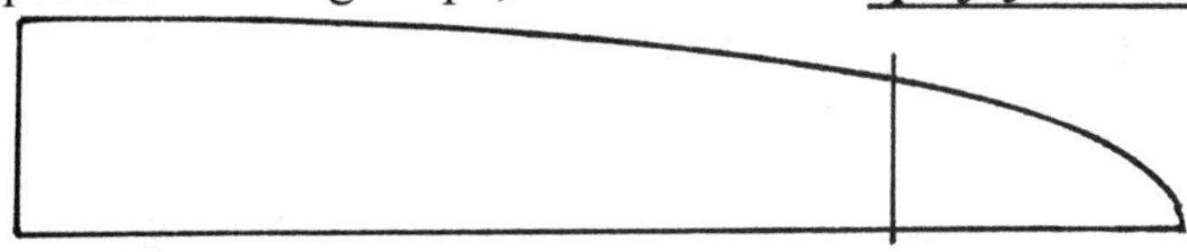

Even a Hershey Bar Rectangular Wing Drops off pretty much like this, no stopping it!

<u>The Tip Thrust Drop-off of Perfect Prop Thrust Distribution</u> vs. radius is <u>already Quite Severe</u>, due to a very, very excessive tip dynamic pressure. To not strongly counteract **q** with a fairly severely tapered tip, **to leave a tip wide, to try to get more tip Thrust in any way, is to <u>just manufacture unrewarded extra losses</u>, <u>completely miss the basic engineering horse sense</u> of it all, flunk basic Aero 101!!!!!**

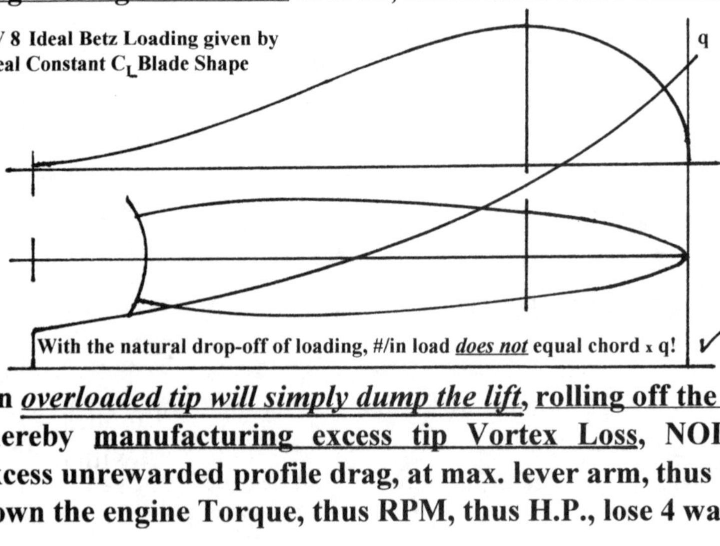

With the natural drop-off of loading, #/in load *<u>does not</u>* equal chord x q!

An *<u>overloaded tip will simply dump the lift</u>*, <u>rolling off the tip</u>, thereby <u>manufacturing excess tip Vortex Loss</u>, NOISE, excess unrewarded profile drag, at max. lever arm, thus bog down the engine Torque, thus RPM, thus H.P., lose 4 ways!

 This is the Easiest, most fundamentally Important Insight in Prop Logic!!!

The Five, 800 pound Gorillas of Prop Logic

As you begin to get a feel for Prop Logic, you'll come to realize there are -- FIVE 800 Pound Gorillas -- that **control the results!**

1. **Dynamic Pressure, q, proportional to V^2**. The fact that the **Prop Tip** is going **so much faster than the root**, *especially on slow, high RPM planes*, and that **q is : V^2**, is the ***key fundamental*** that makes props, their theoretical ***constant C_L Shape***, and ***Ideal Lift-Thrust Distribution vs. Radius, different than wings***

This Key,Tapered Tips for Ultimate Props has gone Unrecognized for 50 Years!

2. **Diameter**: is obviously ***hugely important***, because it controls the **size of the stream tube** and the **mass flow rate, $\dot{M}$**, (*also considering Speed*), the **lever arm of the blade drag**, especially when you learn it's **D^4** in the thrust coefficient, **D^5** in the power coefficient, but surprisingly **C_L**, the next Gorilla **can trump it.**

Stream Disk Area needs D^2, bigger D and V needs a second D^2 for Thrust & Drag

3. **Lift Coefficient, C_L**. Surprisingly, a **.6 C_L** shrinks the needed RV 6 prop **D**iameter to **69.1"**, a **.3 C_L** balloons the **needed D"** to **83.7"**, proving to be **an even more powerful basic than D!!!** A real surprise is that the **profile drag** on a **.55 C_L** prop grows from ~**1/3** of the total loss to **62%** at **.3 C_L, ~78% heavier: ~$(D_2/D_1)^3$!**

C_L actually controls the ratio of Profile Drag loss vs. Induced Loss!!!

4. **Speed**, is Hugely Important because it **raises the $\dot{M}$, Mass Flow Rate**, reduces the ΔV **(required to make thrust)**, ***twice*** **reduces the $\Delta V/2V_1$,** ***twice*** **reduces Iinduced Loss, rasies the Pitch** and **Advance Ratio**, thus lowers the corkscrew path length, the **profile energyy loss**, all **raising the Cap on Efficiency, eta, η. Speed** also **directly reduces the optimum Diameter.**

Speed and RPM Controls the Cap on Efficiency, BGT *gets that* Ultimate!!

5. **RPM is much more important than realized**, because with Speed, it **directly Controls Pitch, Advance Ratio**, thus directly controls the **Cap on Efficiency**, **(with Speed)**, and of course, the **Induced loss** and **Profile Drag corkscrew path**. Engines want high RPM for max H.P. vs. Weight, **a direct conflict** with **props that want High Pitch for Efficiency!!!** ***FIVE GORILLAS!!!***

 This is a Very Important, Valuable Page to Grasp Early!!!

Symbols, Definitions, Clarifications

Realize Engineering calculations *in the English system* are *always* in consistent units - foot, pounds, seconds, - ft. # sec. (never in MPH, inches, RPM, HP) for the *very good reason* that the *units then all cancel out* to give *the correct units of the answer*, say rps, or #, and the correct numerical answer, if no wrong constants are used. Engineers check long calculation, to see if the answer comes out in the units they intend, ***to check that they set it up correctly*, a hugely important double check. You'll see it's the dynamics calculations with Mass and g that get weird, (but as simple as possible below*), *or* skip it for now.**
Power is expressed in ft #/sec, 1 HP = 550 ft #/sec.
MPH x 22/15 = Ft/sec   Ft/sec x 15/22 = MPH.   Simple.   60 MPH = 88 Ft/sec.
~ The "similar sign" used to show where things are *not* exactly equal, say, a detail effect.

α^o alpha   Greek a -- *Angle of attack* to airstream.
β^o beta,   Greek b -- Propeller *blade angle(s).*
κ kappa   Greek k -- Cuts Theodorsen's w to avg. ΔV of stream tube.
γ^o gamma,   Greek 3d letter -- Airfoil *drag to lift angle.*
Δ Delta,   Greek Capital D -- Meaning *difference*, ie. ΔV.
ε epsilon   Greek e -- Theodorsen's 3D losses for each axis.
ϕ^o phi   Greek f -- Wind *inflow angle* vs. prop plane.
η eta   Greek h -- Propeller *efficiency*, regarding energy, power.
ρ rho   Greek r -- Mass density of air, .002377 slugs/ft^3 at S.L. (see M)
Σ Sigma   Greek Capital S -- Meaning a summation, "summing up".
Ω Omega   Greek w, Capital and lower case ω -- Rotational velocity in Radians per sec. $\Omega r = 2\pi rn$ -- both are ft./sec.
a   Glauert's *ratio* of the *inflow velocity*, half the total ΔV, ΔV/2, to the airplane speed, or the *slowdown speed V_1*, thus $\Delta V/2V_1$.
a'   Glauert's *ratio* of *half the prop caused stream tube rotation Speed at any radius*, $\Delta\pi$nd, to the *rotational velocity* πnd, thus $\Delta\pi nd/2\pi nd$
AR Aspect ratio, span/average chord, or span^2/area, for a wing, but here we use 90% of D, the hub not effective, later 80%, 19% to 99% D.
C_L The coefficient of lift, based on the the air inflow angle of attack α^o.
ΔV The *total axial velocity difference caused by the prop*, ie, ΔV/2 in, out
D, Propeller Diameter in *feet*.   d = partial diameter.   (~ Radius R, r).
g   *Acceleration of gravity, 32.174 ft per sec, per sec*, ie., ft/sec^2.
J   *Advance ratio* of propeller = V/ nD. (Forward velocity vs. rotation).
J_π J/π, often called Lambda, λ. With V vs. πnd, the more easily grasped rotation *velocity*, actual *prop tip* advance angle, or ratio, AAA°, AAR
Easy: V is the forward velocity, πnD the circumferential velocity of the tip, Tan AAA = V/πnD.
Kinematic Viscosity, ν, nu, is .0001576 @ S.L. .0002015 at 10,000 ft.
Lift, L = C_L Area ($\rho V^2/2$) pounds. Similar, Drag, D = C_D Area ($\rho V^2/2$)

M Mass, in slugs, which is weight / g, # / g, # / ft per sec^2, = #sec^2 / ft, !!! *Newton's Dynamics calculations need weird units, slugs to get correct numerical answers --- and come out in pounds, a smaller number of more massive units.* SEE NEWTON BELOW!

$\dot{M}$ Mass flow *rate* in slugs / sec, which is # *per sec* divided by g, the acceleration of gravity 32.174 ft/ sec^2 which comes out # sec/ft, which seems weird, except that makes $M\Delta V$ come out in pounds, T#! Dynamics calcs using Mass, Mass flow rate $\dot{M}$ and g are weird until you catch on. *Ignore all that engineering complexity if you wish.*

$\dot{m}$ The natural mass flow *rate* through the prop's disk area from the plane's speed, no speedup from the prop -- no prop there.

n Revolutions per second, rps. = RPM/60.

\# The pound symbol used for both force and weight.

P Propeller *Pitch, (no slip), in feet* for calculations, P/D OK in inches.

q Dynamic Presseue, $\rho V^2/2$, #/ft^2, 25.565 psf, 100 MPH, S.L. ~ .1776 psi.

Rn Reynolds Number = Vc/ν = Velocity x Chord / Kinematic Viscosity.

V Velocity in ft./sec. V_0 is plane speed = V_1, *if no slow down.*

V_1 Plane air speed V_0, but slowed by the body pushing air ahead.

V_2 Velocity *at the prop disk*, after the *first* prop $\Delta V/2$ speed up.

V_3 Velocity far back, after the *second* prop $\Delta V/2$ speed up.

V_4 Final Velocity farther back, from the initial slowdown pressure, p_1.

ΔV The *full speedup* caused by the prop, = 2 times $\Delta V/2$

W Wind *inflow velocity* to prop blade, (at angle ϕ to the plane of rotation of the propeller. See the prop geometry sketch).

w Total Displacement Velocity: Theodorsen's heavily loaded flow total ΔV, much greater right at the prop. w/2 sets the constant pitch inflow angles ϕ, that all radii of a prop blade must maintain due to the inflow, rotation, the pitch angle geometry, that sets ideal, minimum energy loss props for Betz's loading. See geometry sketch, Ch. 2. $\overline{w} = w/V_1$ See κ

*Newton's 2d Law *for solids*, F = Ma, is the equivalent of F = Wt. (a/g) That is amazingly simple once you see it, Force = Wt. times *the ratio* of the acceleration, to the acceleration of gravity. *The force is a simple proportion to the two accelerations*, but it complicates the units by using W/g = M, makes a simply thing seem hard, confusing units to all, at first.

$T = \dot{M}\Delta V$ the *fluid flow* version of Newton's Second Law. Simply, the *"per second"* is removed from the "a", in the acceleration of Newton's F = Ma for the *reaction force* for *throwing solids*, and applied to the M, making $\dot{M}$, a flow rate per sec. the formula now $\dot{M}\ \Delta V$. (a becomes ΔV) Velocity V and ΔV are ft./sec. Acceleration, a, is change in velocity per sec, ft./sec.^{2}

Downwash - Backwash. In addition to creating a pressure difference lift, wings make equivalent lift from the reaction force to throwing air down, the downwash, comparable to props throwing air back, backwash, their full ΔV to make Thrust.

Getting Fully Comfortable, <u>Deeper Into Newton Basics</u>

A Special Lead in to Chapter 2 II -- Nailing Newton Insights.

Pros, ***as below, Cherry Pick this,*** **set up to <u>help Novices Really Get Newton</u>!**

Novices seem to have the most trouble with Newton, so we go slow, show it all!

PROP LOGIC is BASICALLY THIS EASY!!!

<u>Prop Horse Sense Logic</u> - is really pretty easy once you catch on. You simply ***<u>need adequately tapered tips</u>*** to **<u>stop excess q</u>, dynamic pressure, from forcing excess tip lift, dumping that <u>excess lift off the tip</u>, causing <u>excess tip Vortex induced loss</u>, <u>noise</u>, unrewarded profile drag out there <u>at max. lever arm</u>,** ***<u>torque dragging down the engine RPM</u>,*** **thus H.P. - actually just properly SHAPED, TWISTED, PITCHED vs. radius,** ***<u>thus IDEALLY LOADED</u> (vs. r), <u>a Triple Ideal Prop</u>,*** **min. Induced, min. Profile drag loss,** ***<u>Min. Area Precisely Placed</u>, min Torque.***

Next, Look at it as an Airscrew

You want **<u>High enough Speed</u>, <u>Low enough RPM</u>** to get **High Pitch, High Advance, because that's what <u>sets your Limit on Max Possible Efficiency</u>, <u>Caps It</u>. Then the computer, Betz, Goldstein, Theodorsen can give you an essentially perfect <u>Triple Ideal Detail Design</u> -- <u>right up to that Cap on eta</u>, η!**

You must **think of a Prop as an AIRSCREW,** wanting **High Pitch, High Advance. Obviously, <u>a steep, high pitch prop travels a much shorter, less wasteful profile drag energy consuming path to the destination</u>. But a little surprisingly, it's Newton's flat Actuator Disk concept that teaches us we want a fast, Big V, (or big Diameter) Big $\dot{M}$,** ***<u>thus</u>*** **Low ΔV, twice lower $\Delta V/2V_1$,** ***(if fast)*****. That shows us why a** ***<u>FAST</u>*****, High Pitch Airscrew prop has <u>min. axial *(and total 3D)* Induced loss too</u>!!!**

Now we Look at it as Newton's Laws

As we begin to go deeper, you want to clearly, easily grasp Newton, $T = \dot{M}\Delta V$, and $\Delta V/2V_1$, hopefully <u>be *facile* in your grasp of all the basics</u>. But since <u>those look like formulas, math, they scare many beginners</u>! -- We're going to do a deeper lead in to Chapter 2 II, help beginners get <u>better insight on the Stream Tube, *not* be afraid of Newton's math</u>.

This Page is GOOD -- GET IT!!!

<u>YOU</u> <u>may</u> want to <u>Read</u> or <u>Skip</u> this Special Lead in to Ch. II-2.

It was originally intended to Help Novices Having Trouble with Newton's little Formulas: $T = \dot{M}\,\Delta V$ and $\Delta V / 2\,V_1$ --

We <u>say it different ways, repeat, go deeper</u> to <u>help Novices</u>.

The Formulas are <u>not hard once you catch on</u>, but can look hard if you tend to be afraid of Math, even a little formula, never tried one before, and need more time and some extra help.

But going deeper, <u>it can look TRICKY, Pros may want to read it</u>

<u>Novices may want to Skip it</u> --- <u>You Choose</u>!!!!

<u>If you grasp Newton well enough, just go to 38 II and 39 II</u>.

BUT:

Everyone should read p. 30 II, and 31 II,

Really Important Insights on Stream Tube Specifics:

My biggest challenge, it seems, is to ***<u>not</u>*** have $T = \dot{M}\,\Delta V$ fake out, stop, and drive away technical beginners, too many Pilots, scared stiff of even a simple math formula --- so I go as light as possible at first. There is more to be learned about $\mathbf{T = \dot{M}\,\Delta V}$ and $\Delta V/2V_1$ so after going as light as possible at first, now we'll ***go a little deeper***, get a pro's insight with minimum brain strain.

We *<u>learn more insight about the stream tube, looking closer</u>*.

Getting Props down to our <u>4 Insight Logic</u>, is **good for pro engineers too** - but too often they think this must be light weight if I'm not immediately into explaining Theodorsen's Math. The fact is, ***<u>we have the key logic insight here</u>, that was <u>not</u> grasped for a half century***, that **<u>a novice or pro BOTH NEED</u>**. *<u>The broader thinking pro can understand</u>* that if he just accepts **a format that helps more novices get it**, he can learn all the great insights, many of which have been missed for a half century, <u>***only one*** precisely triple ideal prop out there ***now***</u>.

Smart Guys, *<u>be willing</u>* to <u>help the novices trying to get all this</u> ---

Novices get scared of *<u>EASY</u>* math, like $\mathbf{T = \dot{M}\,\Delta V}$ and $\Delta V/2V_1$ don't see it's **pretty easy**, once you catch on, but they certainly **don't see it that way, especially at first!!!** So, **relax,** here in this lead in. I'm **purposely going to be repetitive** enough to **give new guys, novices, *<u>extra time and help to catch up, catch on</u>!* *<u>Realize</u>*, we're <u>dealing with a very diverse audience here</u>!!!!**

Amateur or Pro, if you just go through this, **let us lead you**, in an hour or so, **in a day**, you can know important insights about designing an essentially exactly correct, and slowdown corrected **Betz-Goldstein-Theodorsen** propeller, insight not really correctly, fully understood in the past. We have yet to have a perfectly shaped and loaded propeller from the propeller industry that really, accurately and completely understands **B-G-T logic and math**, and gets **a Triple Ideal, Minimum Induced, Min. Profile, Min. Torque ~Perfect Propeller.**

The **Whirlwind 200 RV prop** for Van's RV 8, designed with **Jim Rust**, **the founder of Whirlwind**, is that **first** somewhat historic, benchmark **Triple Ideal Prop**, that understood **all the math, all its implications**, *the full logic of everything*. No, we're not bragging, don't have a distorted ego problem, we just put in the long hard work, after **a 55 year gap** to finally nail prop logic and math. We did it the "old fashioned way", we earned it with long, hard, comprehensive work, **the hero, Dr. Andy Bauer**, who **comprehensively finished the problem after 138 years**.

Now, to help struggling Novices, we'll use the same format, repeat, but **go deeper, and then broader**, **keep on summing up**, **bringing together**, so if you miss something you can still end up knowing how things come out, a quite sophisticated final grasp. Pros, intelligently skim, cherry pick, as you must be doing. From **p.147 I**, you all know what the perfect props look like, their loading, the core keys. ***So Go for the whole grasp!***

Since I've learned that it does take at least a day for most of us mortals to grasp a whole new set of prop logics, I do very purposely repeat, say it different ways, always tie together, sum up, because I found the smarter guys can get more facile, the less gifted guys can actually get it if we talk about it for a day!!!! A day is not too long to solve a 140 year void. I'd like to see you end up with an easy grasp of these basics. Then you'll more easily gain a grasp of all the pro insights.

Newton, and the "**Rankine-Froude Actuator Disk**".

Looking Deeper, we want Novices and Pros to have a solid grasp of basic prop logic. Newton holds the incisive insights.

Novices - we lead you deeper - we repeat - it's Important. You *Can* Get It.

The **Basic Logic of Props, can be Tricky**, but it's actually quite **helpful and insightful**, and **Newton makes it easy enough**!!! This is **so basic and helpful, important** to understanding the **basic horse sense of props**, that I want to **say it concisely**, but **helpfully enough** that you **really get it --- make it your own!!!**

Yes, I repeat, but I'm taking Novices Deeper, so you can finally get it all.

A prop is **a *Rotating wing Airscrew***, is computer calculated using amazing math, but it can also be seen as **simply pulling in and throwing back air that is far heavier than you'd think**, and the **simple little formula for Newton's 2nd Law makes it a real opportunity to grasp the basic horse sense logic of props**.

The Key Insight is Very Simple!!! The Detail Insights get Trickier!!!

If you try to throw a Medicine Ball, **a heavy blob** of sand in an oversize basketball, you find **you can't throw it very fast**, and it **pushes back on you hard**, just as hard as you try to throw it, **Newton's famous 3d Law --- Equal and Opposite Reaction**.

Of course we want a BIG $\dot{M}$, a SMALL ΔV ---

The **Fluid Flow version of Newton's 2nd Law**, **$T = \dot{M}\Delta V$**, is **Thrust in Pounds = Mass flow rate, $\dot{M}$, M dot**, *times* ΔV, **delta V, in ft/sec.**, (***not*** MPH). Now, in case you're not a math guy, don't let that simple little formula fake you out. ***It's this simple:* The thrust simply depends on how big the Mass flow rate is, and how much you speed it up**, *easy, simple multiplication!*

M and $\dot{M}$, are ODD UNITS, Slugs and Slugs/sec. See DEFINITIONS p.21-II

The reason that is **Big Time Important**, is that the **mass flow rate is far heavier than you'd ever guess**, and the ΔV, **delta V speedup needed is much lower than you'd ever guess** --- and **thus you can quickly get a whole different feel for prop thrust than you'd ever figure out for yourself**. You'll soon see an **RV6 at Vmax can make a 280# Thrust**, only throwing back a **10 MPH ΔV average wind**, because it has **614.5#/sec. of air flowing through its disk,** *like* ***1 1/4 cubic feet of steel/sec!!***

A **Reno Racer at 480 MPH** has ***4 1/4 Tons of air per second***, 8500 #/sec, may ***only*** make a **7 MPH average ΔV wind**, numbers *no one would guess anywhere near correctly*, **big time insight!**

WOW!!! Even Engineers are Blown Away by Those Amazing Numbers!!!

That's known as the **Rankine-Froude Actuator Disk**. Rankine the famous Scottish Scientist-Engineer who first figured out **the correct Steam Engine Thermodynamic Cycle was doing the first rudimentary water prop calculations by 1865**, Froude a famous Hydrodynamicist, (the Froude Number) had it named by 1889.

$\dot{M}$ is Explained More at the top of p. 21 II

The **actuator disk concept can teach us several important insights, below**, so it's **absolutely basic to understanding the horse sense logic of props**. (There are *oversimplification*s that make it **inadequate for real prop design calculations**. It **assumes a constant axial ΔV**, as the term Actuator Disk implies) You'll learn in the Flow Geometry of Ch. 3 II that the axial ΔV varies a lot from blade root to tip, and with heavy loading the helical vortex sheet actually moves back 2, 3, 8 times faster than the average ΔV of the stream tube, but **the simple concept can teach us a ton of insight, so we want to grab that insight**.

With a small pressure reduction in front of the prop, and a small pressure increase behind the prop, nominally **half the ΔV, ΔV/2 happens as inflow**, *half as outflow*. **Now the prop is logically designed and pitched to account for *only the ΔV/2 inflow!!!***

Next, it **pulls in air**, *speeds it up*, **from a larger diameter** than its disk --- and that **increases the *natural* $\dot{m}$, mass flow a little bit** to **$\dot{M}$**. Per Bernoulli, ***the stream tube shrinks again*** as it speeds up, the second ΔV/2 behind the disk, but that ***does not*** **increase the mass flow rate again, *because* it's still the same tube flow rate that came from the larger diameter in front of the prop!** At cruise that's a ***small*** increase, *more in climb*, big at runup.

Natural $\dot{m}$ is the Heavy Flow Rate thru the Prop Disk, *no prop there*, from Plane Speed.

Now what you're learning **here, and next**, is **Hugely Important, fundamental**, so I want you to **read this all very carefully, take time and really get it, because props are licked *if you do***.

A Little Fair Warning - I Drag you Screaming to Success

Novice or Lifetime Pro, I've learned it **takes a day of talking, or reading** to **really catch on to the real technical horse sense** of props. Surprisingly, pros sometimes catch on slower, a head full of **preconceived** *(wrong)* **ideas, fighting** the "easy enough" **real truth**. But **a day is a hell of a bargain**, compared to a lifetime! The missing insight is: it's **Shape** *vs. Twist*, that **loads it correctly**.

I give you BOTH this Narative, and Very Dense, Concise Summations!!!

Novice or pro, people just don't grasp what they don't expect, the first time they see or hear it, so my approach is to be constantly **repeating, building and expanding the basic story, so by the time I'm done, even the guys who didn't get it at first, have caught on and are working on the more insightful expansion.**

What's happening is that we have a **fantastic modern computer**, and **rare genius level math** that took the smartest, **rare geniuses** that ever came along, **83 years to finally get it right**, a gift from heaven far **beyond the capability of all mere mortal engineers**, that **you *don't* have to learn. We simply show you the engineering horse sense**, that is essentially **directing the math**, and in turn is directed by the math, the **insight that's been missing for 138 years**, the product of 10 years of lifetime experienced digging, checking, decoding, ***a gift!***

You earn your right to use that gift in this **unusual introduction** to this Chapter 2 II. I must warn you that it's going to **seem Ridiculous, repetitive to most of the audience**, for sure the pros, and the less smart guys, who care less than the eager guys, **but there is good reason, slower, eager guys get it!**

Dense Summations - Longer Narative - Read this the way it works for You!!!

Were going to dig deeper, all the way, to see all that $\mathbf{T = \dot{M}\,\Delta V}$ and $\mathbf{\Delta V/2\ V_1}$ can teach us, **professional depth insight on those**, but we're going to be doing it **in a way to help**, and drag along, where necessary, new guys who are struggling, ***repeating, going deeper, broader***. Your job, **don't bitch, just get it, *pro insight!***

Now there is **this simple little fraction**, that **looks complex, but is not hard at all,** **$\Delta V/2$ vs. V_1, $\Delta V/2V_1$ in math form.** That and $T = \dot{M}\,\Delta V$ are the ***only*** **two little formulas we'll ask you to grasp** to **understand *all* of propeller logic, *not a sweat!***

That V_1 is just plane speed, (or thinking of a case with a ***slowdown profile***, we could think of it as the slowed speeds coming at the prop), so that little fraction that might look complex to you is simply the **ratio of half the full ΔV, $\Delta V/2$ vs. V_1 the air speeds. EASY!!!** Now **in our RV example above,** it's simply 10 MPH/2 vs. 200, **2 1/2%, .025**

A prop also pulls in air from a larger diameter, increasing Mass flow Rate, $\dot{m}$ to $\dot{M}$.

Now that is of Huge Importance, because we can use it twice:

- It's the **Fraction, Decimal or Percent that the *natural mass flow rate* (thru the prop disk), $\dot{m}$, (*mostly from the plane's speed*), gets increased to get $\dot{M}$, from a bigger diameter, (increased just a little, 2 1/2%)!**
- **It's also the nominal *axial* energy or efficiency loss, 2 1/2 %** which is the **axial, *in and out* component of induced loss.**

We learn the logic from *axial*, BGT Math nails radial, rotational too, all 3d Losses.

The Insightful Logic that comes out of $T = \dot{M}\,\Delta V$ math is:

You want a BIG $\dot{m}$ and $\dot{M}$, a SMALL ΔV --- for Efficiency ---

- **Fast, high pitch props are efficient, slow, low pitch props can't be efficient, (they create high Induced Loss, *and* Profile too*).**

*** A Long Shallow Helical Path Wastes Profile Drag Energy Traveling Too Far.**

- **Plane Speed,** and **Engine RPM, set and cap the limit on a prop's efficiency** - as shown on the classic **Advance Ratio Curve** p.48 II!

Realize High RPM Lowers Pitch --- BAD!!!

- **The Prop Blade is *Not* in Control of its Own Efficiency:**

--- *Only* able to help itself with a high Aspect Ratio Blade.

--- *Only* able to match the Max Ideal Efficiency Cap with Betz-Theodorsen Ideal design!! The eta, η, on the Advance Ratio Graph!!

- **Max Efficiency, Min. Induced Loss, achieved by Hi Speed, Big $\dot{M}$, Low ΔV Wind, Low $\Delta V/2V_1$,** per **Newton's 2nd Law.**

We gain Huge Insight from grasping **$T = \dot{M}\,\Delta V$**, and **$\Delta V/2\,V_1$.**

$\dot{M}$ is slugs/sec, ***simply $\dot{W}$ #/sec divided by the Acceleration of Gravity*, g, 32.174 ft./sec. per second, $\dot{M}$ a smaller number of bigger units of Mass.**

$\dot{M}$ Slugs/sec. *Flow Rate* Gives answer in pounds, and not 32.174 x too high!!

IMPORTANT!!! If you have to read the last page 5 TIMES to REALLY GET IT, I want you to do it - ***because*** **Newton's whole basic horse sense logic of props is there as concisely and completely as I know how to say it. If you get that page, you have Newton's basic prop logic licked!** ***Read on.***

In Ch. 2II you'll learn a ton more about props, The **Super Magic Graph**, that shows *how everything changes As Speed Changes*, The **Advance Ratio Graph**, *the classic basis of prop efficiency*, The **Magic Graph** that shows *efficiency vs. P/D ratio vs. Speed*, ***All*** **in Chapter 2 II.** The fancy **ideal helical flow is in Ch. 3 II, design Studies, much more**, but ***if you really get the last key insight page, you have the Basic Logic of Props Conquered!***

Now, Going at the Speed of Explanation for New Guys

We very Purposely led into this Nitty Gritty Chapter for very good Reasons: We don't want to lose Novices or Pros.
1. Moving concisely, covering **a lot of ground fast** can quickly **give everyone,** (**including non technical Pilots**), **a quick overview!**
2. Moving too slow would badly frustrate pros, engineers, more advanced amateurs seeking the real facts, Prop Truths.

We certainly don't want to loose the advanced audience, because this is **the first time there has been a comprehensive, yet understandable explanation of prop Logic.** We want it to be **possible for everyone to quickly get a true grasp of how props work**, ***even great for pros who can then tackle the Math.***

But this Book is First and Foremost for Pilots, generally **new to taking a serious look at the engineering secrets**, especially a wild case like props, with **a TON of content for a civilian**!

In the rest of this Introduction and in the Chapters we're going to purposely go slow enough, repeat enough, say the same things in different ways to give the new guys who will be struggling more time to catch up, catch on. To help them, the central audience, ***I expect the pros to astutely cherry pick!!***

A NOMINAL RV 6 PROPELLER STREAM TUBE

Simple case, no slow down of the air from the body pushing a bubble of air ahead.

Propellers make thrust by pulling in and throwing back a stream tube of air passing <u>through the prop disk area</u>. **T = $\dot{M}\,\Delta V$ #**

There is a **<u>heavy mass flow rate</u>**, $\dot{m}$, ***<u>fed to the prop disk area</u>***, ***<u>just from the plane's speed</u>***, without the prop even being there.

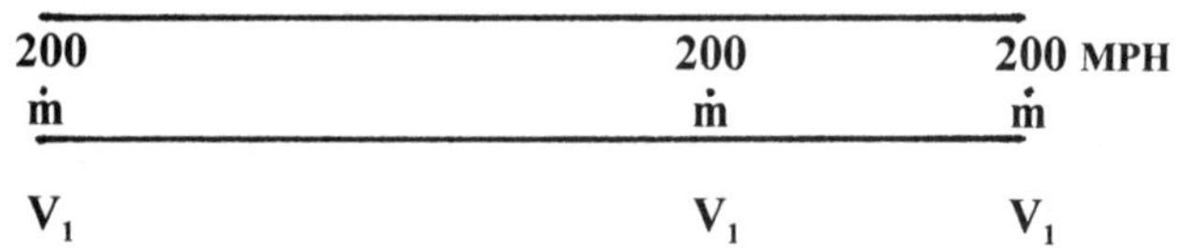

A prop airfoil makes a lower pressure in front, a slight pressure behind, thus when it's working, making a Δ**V**, making thrust, half the ΔV, $\Delta V/2$, happens as ***inflow***, from *<u>a slightly larger diameter</u>*, then $\Delta V/2$ <u>again</u>, as a faster ***outflow***, the *stream tube shrinking* a bit more per Bernouli, and at the final full ΔV that makes thrust.

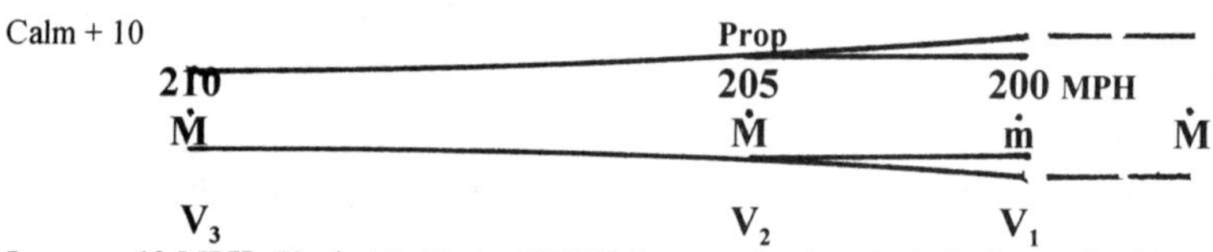

Leaves a 10 MPH ΔV wind behind. 5 MPH <u>Average</u> is a loss to turbulence, back to calm.

The $\dot{m}$ is increased only a small percentage to $\dot{M}$ in cruise. **A big $\dot{m}$, or $\dot{M}$, is hard to throw fast, but <u>that's good</u>, because half the ΔV, $\Delta V/2$ becomes a loss settling back to calm**, and only a **surprisingly small ΔV is needed to make thrust at cruise**, with a **big $\dot{M}$**. It's ***<u>axial</u>* <u>loss ratio is $\Delta V/2V_1$</u>** in math form.

We **calculate $\dot{M}$ <u>at the prop disk</u>** where **only half of ΔV, $\Delta V/2$ has happened**, ***pulled in from a larger diameter***. The ***$\dot{M}$ does <u>not</u> increase downstream***, the <u>second $\Delta V/2$ diameter shrinking</u>!

Why? Pressure drops, the tube shrinks as it speeds up --- per Bernouli --- just like pressure drops over an airfoil!

$\dot{M}$, slugs/sec, can be calculated as simply the area of the prop disk, in square feet *<u>times</u>* the proper axial ($V + \Delta V/2$) velocity at that point, in ft/sec, to get cubic feet/sec --- then multiplying by rho, ρ, the mass density we taught you early in the book, .002377 slugs / cubic ft <u>at sea level</u>. Mass is a pain when you first learn it, but it ***gets the correct answer***. (The nominal air density is .076 #/cubic foot, simply $\rho \times g$, .002377 x 32.174 = .076477 using exact standard numbers.)

You can check yourself -- a full calculation of thrust is on page 31 II, *next*.

A Picture with Actual Numbers

Simple Case -- No Slowdown

In the graduate course we'll show you stream tubes with slowdown --- Recognize the prop flys through <u>*calm air*</u> and <u>*leaves only* ΔV *behind*</u>*!!!*

70" prop / 12"/ft = 5.833 ft.

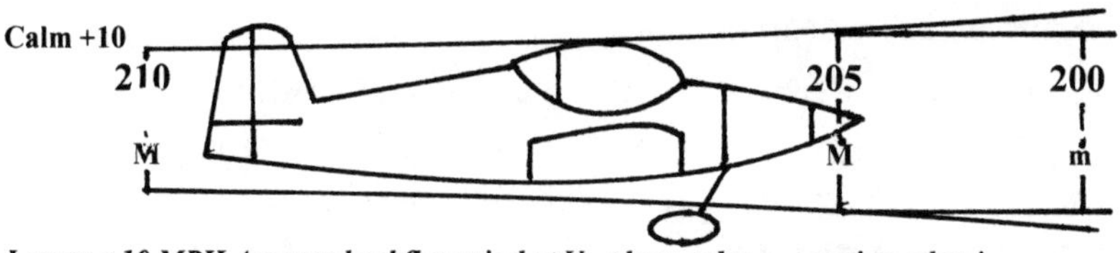

Leaves a <u>10 MPH Average</u> backflow wind at V_3 above calm --- creating a loss!

	V_3		V_2	V_1
MPH	200 + ΔV (Far Back)		200 + ΔV/2	200
MPH	**210**		**205**	**200**
Ft/sec	308	(MPH x 22/15 = Ft/sec)	**300.66**	293.33

Thrust Calculation with No Slowdown for Simplicity

(At the Prop Disk)

Ṁ =	**V_2 ft/sec**	x	Disk Area ft^2	= ft^3/sec,	x	(Mass Density ρ Slugs/ft^3)	=	(Slugs /sec)
Ṁ =	**300.66**	x	$(70/12)^2 \pi/4$	= 8035.4	x	.002377	=	**19.1***
ΔV =	10 MPH	x	22/15	= <u>**14.66 ft/sec**</u>			x	<u>14.66</u>
Thrust =							=	**280.1#**

*** Note that 19.1 slugs/sec x g, 32.174 ft/sec^2 = <u>614.5 #/sec!!!</u>**

ρ, <u>mass</u> density at sea level, .002377 x g, 32.174 = <u>.07647 #/ft^3</u>, air density at Sea Level.

Sure, most pilots are afraid of tackling Physics, even a simple little basic Physics formula, T = Ṁ ΔV. (It <u>*is*</u> confused by the <u>Ṁ in slugs/sec, *divisions by two* on the ṁ to Ṁ increase</u>, and the energy and efficiency loss). Forget that <u>*if* it gets above you</u>. Just <u>grasp the T = Ṁ ΔV</u>, <u>get the horse sense of it</u>, that big speed gets big Ṁ, small ΔV, small loss, good efficiency!!!

High Speed = Higher Pitch, Big Ṁ, low ΔV, Low Loss, High Efficiency!!!

Soon, we do the Basic Chapter 2 II on props, <u>actually easier</u>, and there is a wealth of valuable easier insight. And in this basic chapter, we show <u>*great conclusions that go way beyond the technical difficulty of the chapter*</u>, so you have <u>*a great chance to learn more than you earn*</u>. I hope you can get it, because it *can give you great insight, <u>finally, now in print</u>!*

***If you can grasp what's here, <u>let it soak in</u>, you're apt to find you can get all of the graduate course!* <u>This *Intro* is hardest</u>!**

Summing Up NEWTON'S KEYS TO PROP LOGIC

Solidifying your Basic Grasp of Newton

Here's how to understand the key basic, **$T = \dot{M}\,\Delta V$**, ***completely***. **Thrust in pounds = Mass Flow Rate, $\dot{M}$, (M dot) times ΔV,** Delta V, in ft/sec., not MPH. $\Delta V/2V_1$, increases the *natural mass flow rate* $\dot{m}$, through the prop disk, a small % to **$\dot{M}$**. (Air pulled in from ***a bigger D***).

Imagine we can fly fast with the prop removed!!!!

1. *Without the prop even there*, there is a big, surprisingly ***heavy natural mass flow rate*** we'll call $\dot{m}$, (m dot) through the ***prop disk area***, ***before the speedup***, just ***from the plane's speed.*** **That's important.** It gives the prop **a *big* mass flow rate $\dot{m}$, to throw!**

2. The easiest, minimum math way to see how a prop *Thrusts* is to recognize that it **pulls in and throws back *an already heavy* "mass flow rate" of air! - Fast planes have a big advantage!**

3. Now, that's very important, because you can see from that simple formula that if you have a **big $\dot{m}$**, and final **$\dot{M}$**, you only need **a *small* ΔV for a given thrust!!!!** That's very important - ***ΔV creates an extra wind loss - but only ΔV/2 counts as a loss!***

Why? It's the average speedup, ΔV/2, that sets the distance traveled and energy loss!!!

4. Now, it turns out that the **increase from $\dot{m}$ to $\dot{M}$ is small,** ($\Delta V/2V_1$), because the **ΔV** is small. The big thrust comes from **a small ΔV *times* a big $\dot{m}$ and final $\dot{M}$.** It is ***not*** the small increase from $\dot{m}$ to $\dot{M}$ that makes the big Thrust, **but rather the multiplication of a big $\dot{m}$ increased just a little to $\dot{M}$, *times* a rather small ΔV. $T = \dot{M}\Delta V$. Get that and you're home free!!**

5. Now, **mass flow rate $\dot{M}$** is **like** Pounds flow rate per sec. **$\dot{W}$, except $\dot{W}$ is divided by g,** the *acceleration* of gravity, 32.174 ft. per sec. per sec., (ft./sec^2), thus **a smaller number of bigger units of mass, slugs, and here slugs/sec.** since it's a flow rate. That little drill makes the Thrust answer come out in **pounds,** and numerically correct, not 32 times too big. **Mass** is a pain -- but it **gets the correct answer!** (Why? See M & $\dot{M}$ in the Ch 1 II definition list)

Go ahead, be brave --- Read this key page again --- Get it!

GET ALL THAT *early*, AND YOU'VE ALREADY WON -

A Little Review Narative with Overall Insight

Props, full of Surprises -- *different logic, insights* than you expected!

The **few, key, great, classic books and papers on propeller theory**, are **all heavily into getting the complex mathematical analysis correct**, a few basic characteristic graphs and formulas for key coefficients, but **what has been missing is - *an incisive clarification of what all the math means, an understandable, technically correct, comprehensively complete, logic structure in words, not impenetrable math!* That is our core objective!**

A propeller is *stuck with* overcoming *the drag of the plane* in all its different flight modes. *Flying at the speed of the plane,* the *RPM of the engine, a fixed pitch prop is at excess angles of attack at slow speeds*. There can be ***a major efficiency loss, over and above the prop efficiency loss, particularally on slow planes requiring extra thrust and power*** as the ***prop interacts with the plane***, often much more unfavorably than is realized. That has never been dealt with adequately by Aerodynamicists, simply because they've *never had a practical flight test method, a valid way of testing for it, separating it, in actual flight*.

(Our Zero Thrust Glide Testing has given us ***real data*** to work with.)

It took a multi-year effort to review all the past experts, figure out what they did do, and could ***not*** do before computers, *what was never made clear*. We did all the complex math, computer programming, analysis, to learn how to ***make it comprehensible.*** A pro's rotating wing blade element analysis, proper geometry, didn't work. Real three dimensional flow, only solved by NACA's chief Physicist, Theodorsen, a true genius gave correct answers. Props are as complex a subject as any of us will ever deal with. They're full of hidden surprises, unexpected logic and convoluted interplays. Fortunately the logic ***can*** be made clear. There are really ***wonderful insights and clarifications*** that make it **truly possible to actually grasp the logic *in a few hours.***

OUR mutual problem is there is *so much new insight available*, ***a teacher is apt to drown you with just too much, too quick!!!***

Because of that, the absolutely necessary way to teach it is to ***first** teach **the basic logic to pilots** - get **each new guy aboard on the core logic*** *that is* ***hiding in the basic Physics**. Then* ***expand his insight, but not drown him,*** *teach what's needed* ***to understand even better, with much greater insight**, finally offer* **great conclusions that go way beyond the basic insight, *min effort, max. smart! But FIRST, we must successfully teach you the key to the basic Physics, the correct CORE logic*.**

Once you grasp this basic logic, let it soak in, it becomes much easier to go on, to read, understand the advanced explanation in Chapter 3 II -- ***for those who wish to get it all***, -- ***as usual - reading the book to the depth you wish and choose***.

We teach you the basic logic with the Rankine - Froude concept, the "**Actuator Disk**", 1889. That's ***Newton's*** $T = \dot{M}\Delta V$, ***the core of the basic Physics. In hindsight, it's simple enough! Just grasp how a big*** $\dot{M}$ ***reduces the*** ΔV ***required and*** $\Delta V/2V_1$ ***loss!***

The game is to simply minimize the ΔV, the extra wind you make!!!

Amazingly, **Betz conceived the ideal prop concept in 1919, and it's still correct**. We clarify his concepts early, and teach his airflow geometry in the advanced chapter, 3 II, to not go too fast. Glauert got the air inflow and rotation pretty correct by 1934, Theodorsen, the genius solution by 1948, and we teach it all as you progress. ***Our task is to get you through the first steps into understandable logic, logical horse sense***, ASAP.

There are just a lot of brand new ideas that we need to get you aware of ***so you know enough to go on***, to start appreciating, understanding enough to start seeing the logic and patterns of it all, so **charge, have faith and trust, and you'll be rewarded**.

Only* a computer can nail a specific design**, we'll teach you a ton about propeller logic, ***but most basically grasp: FAST, low drag, low RPM planes with high pitch angle props, high P/D, are efficient - but low pitch props aren't. Slow, high drag, high RPM planes, **anything that gets you low pitch is bad, *EASY!***

But if you ***gear*** a slow, high drag, high RPM plane like an ultralight for ***low RPM, high pitch, much bigger diameter, you can get back to max. efficiency theoretically.*** (**But, *it's usually impractical, too big a diameter,* too low an RPM, *the math self defeating to get high enough pitch - because as the diameter goes up a lot, the angle of the tip goes down again!!!* Geared fighters, low RPM, have reasonably small diameters --- *only Because They Go Very Fast!***) *Slow, you must and will have* **low power density, low H.P. *per* square foot of prop disk area**, like ultralights have, part way to optimum. ---- You'll learn that **SPEED, RPM, C_L** are all far more important than you'd ever guess, **all 800# Gorillas! It's great understandable logic, *but chock full of real surprises!***

q, Dynamic Pressure, much higher at the tip, --- is a 4th 800# Gorilla, D a 5th!

One of the big surprises is that since it's really ***the plane Speed and drag, and the engine RPM*** that **controls and limits propeller efficiency**, ***prop efficiency is really not prop efficiency at all, but rather*** "**a *traveling prop system efficiency***". All the prop can do is to be accurately computer designed, ***actually meet the max. limit of efficiency imposed by the plane-engine system!!!*** *The prop blade has a very important, but definitely limited role.* **Pitch**, **Advance**, **Aspect Ratio**, sets max efficiency ***because*** the ***prop is not in control of its basic efficiency***, the plane-engine is.

A Luscombe, ~ 85%, a Fast RV, ~89, 90%, ~ 91.5% a Classic Max!!!

There are *four kinds* of basic propeller losses, three relating to the propeller **stream tube**, the ***axial ΔV speedup*** that basically creates the thrust, the ***rotation***, ***worse*** for otherwise efficient high pitch props, (a questionable reason for counter - rotating propellers), the ***radial flow feeding the tip vortex***, and finally *the fourth,* ***profile drag skin friction***, that is always there.

Blade Drag is Penalized *Because* Small Chords give Low Reynolds Numbers, Rn.

We're very purposely going to center this initial advanced chapter mostly on the *axial ΔV speedup*, *simply because that's where the sneaky $\dot{M}\Delta V$ logic and multiple surprises are*. The other three losses are there, add up to more loss than the ***ΔV speedup*** loss, but that's ***not*** where the tricky logic is. You can see they logically exist, so we can leave them until later, to keep this basic chapter with the greatest gain in insight, the easiest picture to grasp, free of excess detail and excess complexity.

One of the greatest new insights is when you realize, grasp that --- the *radial root and tip loss*, *the rotation*, and *half the axial ΔV speed up* are *the induced loss*! *Like a wing*, that loss drops to a low level at high speed, leaving the irreducible *Induced and Profile* about 8 to 9%, a 91+% max. efficiency*, EASY --- *huge insight that helps make it all much easier!*

Like a sailplane, Aspect Ratio sets the final upper efficiency limit!!!

I'm very purposely going to skip a lot, leaving it for the advanced chapter, ***because it will be far easier to grasp after you grasp the key core logic!!!*** ***Grasp the Math:*** **Props make thrust by pulling in and throwing back, speeding up**, *the natural $\dot{m}$* ***mass flow rate of air through the prop disk.*** $\dot{M} = \dot{m}(1 + \Delta V/2V_1)$

A big *free* natural $\dot{m}$ is fed through the prop disk by big speed, with no prop there!!!!

Do not allow yourself to ***get faked out***, because we use basic Physics, the **T** = **$\dot{M}\Delta$V** core logic. **It is *not* brain surgery, *not* deep math**, it is *simple arithmetic*, **$\dot{M}$ times ΔV, that simple, a simple two element logic *where ΔV gets smaller as $\dot{M}$ gets bigger!!! Get that. It's very important***, because **ΔV/2V$_1$ is the loss that costs!! Big speed, big V, which gives you a big *natural $\dot{m}$* and final $\dot{M}$, gives you *lower ΔV, high efficiency*!!! You'll thus learn *a pretty tricky, but simple enough logic that you'd learn and grasp no other easy way*, the real key, the real hidden secret to fairly easily decoding all the 100 year hidden secrets of propeller logic!!!** **That's the most confusing basic.**

We purposely keep saying it slightly differently so novices finally get it.

Propellers can otherwise be a huge frustration and task to learn, *or to teach*, because *done the wrong, long, math way, they are just too much*, something that will drown even pros. We **emphasize the axial loss**, blandly say there are three other losses, *that are actually just as important*, but *it's the axial logic where the secrets are hidden*. Trust me and you'll get it, easiest and *very completely if you go all the way through both advanced chapters*. I expect many will find this Introduction difficult, **it's the hardest part**, but you can end up ***proud if you got it***. Compared to the years props can cost professionals, ***you may end up deciding this is the best time bargain you've ever found.***

 *** Ultimately, we'll show you how to beat the Classic Ideal.**

A Concise Look at Prop Design Logic

You can end up with a *Feel for the subject*, much faster than the Math will teach!!!

A **fixed pitch** propeller can be described pretty simply, so we'll do just that and then expand it, so you can grasp very much very quickly. First, a prop must be **pitched (& twisted) correctly** for **plane speed** vs. **engine RPM**, (+ air inflow & rotation), so each blade airfoil flys at a ***low optimum angle of attack*** at ***design speed***.

But it will be way over pitched at all slow speeds --- runup, takeoff, climb!!!

Next, it needs a **proper blade diameter-area-shape** to give the **required thrust** at the ***Speed, RPM***, C_L, and (alt.) air ***Density***, ρ. Now *get all that*, because it's pretty ***simple, concise*** and ***specific!***

Correct Twist and Shape can make it IDEAL!!! Blade Pitch is held to .1 Degree!!!

When you're designing a prop, the diameter is extra important, because the ***tip is going so much faster***, is so much more important than the root or mid radius, **you cannot disconnect the Diameter, a powerful factor, from the Area--Shape calc!** Secondly, a **larger diameter, higher aspect ratio, reaches out to work on a larger mass flow rate**, which directly cuts the induced drag loss, by **directly cutting how much extra ΔV it must throw back the natural stream tube passing through the prop disk.** Like a sailplane wing, a prop wants max aspect ratio, max span. Diameter cuts induced loss in exactly the same way that a long sailplane wing works on more mass flow rate, cuts induced drag ***by having less resulting ΔV downwash, which is a loss, less disturbance of the static air, huge insight.***

MINIMIZE ΔV, don't make extra wind to leave behind as turbulence, a loss. **EASY!!!**

Blade shape is significant, the faster **outer blade *wanting*** much **narrower chords**, and thickness to ***create "a rigid vortex sheet", helical pitch of the airflow, the ideal foreseen by Betz in 1919***, **huge insight**, (the variant of **Prandtl's perfect elliptical wing, but *not* constant downwash**). Back to pitch and twist. If the twist is set to ***hold a constant*** α^o & C_L, and chord widths are made correct to actually hold that C_L, you can actually get "Betz loading", ***get constant pitch helical inflow***, (***but* then, *blade angles must be reduced, more inboard***, for the fuselage pushing a bubble of air *ahead, slowing the air feeding the prop*). This is ***huge grasp early!***

A propeller ***must*** ***be computer calculated for precise 3D flow, to do a proper detail design***, but the math for that took our smartest analysts almost seven decades, from 1865 to **1934** for Glauert to get a *good approximate* solution, actually until 1948, 83 years for **Theodoresen** to nail ~exact **3 D flow**, with realistic **heavy loading**, and **high advance ratio** to get ***accurate inflow!*** We ***won't*** try teaching you the "rare genius level" math, ***pros drown***.

SO HERE COMES THE CONCLUDING TWO PAGES -- WITH THE REAL MEAT SUMMARY!!!

It's much easier to see a prop makes thrust by ***pulling in, and throwing back air***, per Newton's Second Law which teaches us to ***calculate the thrust reaction force to throwing back air***. The formula is $\dot{M}$, (M dot) times Δ**V**, **T =** $\dot{\mathbf{M}}$ Δ**V#**. The thrust is in pounds, of course, as shown by the # symbol. The *total* ΔV (delta V) speedup must be in **ft./sec.**, not MPH, to get the correct answer. $\dot{M}$ is ***mass flow rate in slugs/sec.***, [sounds like a scary new term, but it's just like pounds/sec. divided by g, (32.174 ft./sec. per second, the acceleration of gravity), again to get the correct answer, not one 32 times high]. That ***two item basic formula quickly buys you huge insight***, bypasses 83 years of wild math that most could not possibly understand in this lifetime. And surely don't want to ----

A **fast** plane has a **huge natural $\dot{m}$** ***through its prop disk*** - **free** - from its speed - **without the prop even being there!!!!** It's **increased just a little to $\dot{\mathbf{M}}$** by the ***ratio*** of *half the ΔV*, ΔV/2 vs. V_1 the *slowdown* speed, ***half*** since nominally half the speedup comes from the prop making a small vacuum in front, half behind from a small pressure increase behind, **half the ΔV at the prop disk**. That ***ratio*** is made much easier to understand next.

WE PURPOSELY DO THE PROP CALC AT THE PROP DISK, WITH HALF THE ΔV SPEEDUP.

Prop Airflow weighs a lot more than people realize, remember 76,000# for a 100 foot cube at sea level!!! **Specific numbers** offer **a feel for the facts**. A **BIG $\dot{\mathbf{M}}$ allows the ΔV to be surprisingly small**. Again, an RV 6 with a small 70" prop at 200 MPH passes **614.5** # of air per ***second***, the equivalent of throwing back **1 1/4 ft³ of steel** every **second!** With that, only a very small **10 MPH *extra* ΔV wind** makes 280# thrust. 19.1 slugs/sec x 14.66 ft/sec.

Now let's **ease** the item that might cause you the most confusion: Take **a ratio of half of** ΔV, $\Delta V/2$, **vs.** V_1, **in math form** $\Delta V/2V_1$, where V_1 is the airplane speed feeding the prop, (or in advanced cases, the slowed speed<u>s</u> caused by the body pushing air ahead.) *That is absolutely going to look pretty scary to many pilots, but let me show again how easy it is.* In the RV6 example we're using with a 10 MPH ΔV, it's simply **5 MPH vs. 200, 2 1/2 %**! ***Grasp how easy it is, because it's <u>the 2nd central factor</u> in understanding props, and <u>once you've got it, props are licked</u>!***

1. It's used to **increase** $\dot{m}$ **to** $\dot{M}$. $\dot{M} = (1 + \Delta V/2V_1)\,\dot{m}$, $1.025\,\dot{m}$. We do that because *we calculate* $\dot{M}$ *<u>at the prop disk</u>* where only 1/2 ΔV has happened, having pulled in the $\Delta V/2$ from a larger diameter -- and $\dot{M}$ does not, *<u>can not</u>* increase behind the prop disk, the stream tube diameter simply *<u>shrinking some more</u>*, (all per Bernoulli) - *<u>speeding up more</u>* - *<u>as the **pressure adds** $\Delta V/2$!</u>*

2. We use the same ratio, $\Delta V/2V_1$, 21/2 %, to measure the **axial energy loss <u>ratio</u>**. You might think of it as $(1 - \Delta V/2V_1)$, 97 1/2%. (There's a really obscure reason for using half the full ΔV here that you can <u>promptly forget</u>. In Physics, the basic energy equation of motion, $MV^2/2$, one V of the two is the <u>average V</u>, V/2, because the <u>distance</u> over which energy is added depends on the <u>average</u> speed up!) Forget that pro reason.

So now, to really see the core logic of props, just grasp:

1 ***Fast planes have a <u>big natural $\dot{m}$</u> & $\dot{M}$, thus need less ΔV,*** ***<u>twice helped</u>*** **with a small ΔV, a big V_1,**$\Delta V/2V_1$ ***<u>shrunk twice!!</u>***

2. The $\Delta V/2V_1$ **factor,** *<u>a lot easier to use than it looks</u>*, is used to **increase $\dot{m}$ to $\dot{M}$**, only a small percentage for a fast plane.

3. And that **same factor,** $\Delta V/2V_1$, is used to measure the **energy and efficiency axial loss,** the **2 1/2 % down to 97 1/2 % here!!!**

4. A **BIG $\dot{m}$ from a fast plane means a BIG $\dot{M}$**, (but only up a small percentage), and **that allows a small** ΔV, and that's good, because it's the **$\Delta V/2$ that becomes a loss! (But think, $\Delta V/2V_1$, the loss factor <u>tries to be</u> ~ <u>4 times bigger,</u> (2 / 1/2) *<u>at half speed</u>* <u>for the same thrust</u> - (*<u>except</u>* a <u>bigger, worse ΔV increases $\dot{M}$ more</u>!) <u>Think it thru!!!</u>**

A LUSCOMBE IS LIMITED TO 85% EFFICIENCY MAX, AN RV CAN GET TO 89% vs. 91 IDEAL.

Yes, this lead in *<u>is</u>* terribly <u>repetitive</u>! <u>Green Novices need time, practice</u>!

GET THIS PAGE AND HAVE PROP LOGIC LICKED ---

The Wise Ole Bird — *IF YOU GET THIS PAGE, YOU'RE OK TO PROCEED.*

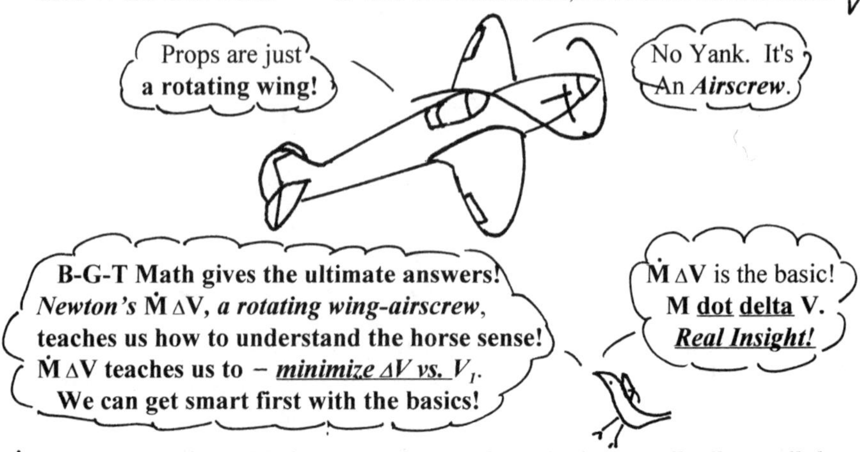

You need ***not*** deal with the *rare genius math*, *we had to!* *We'll tell you all the hidden conclusions*. **This dense summary condenses the core Newton logic**.

WE ALSO DESIGN A PROP KNOWING IT IS A ROTATING WING, AND AN AIRSCREW!!!

1. **Props work on $\dot{M}\,\Delta V$.** ***Thrust = Mass flow rate*** through the prop disk times the ΔV ***speedup***, in ft./sec. They thrust by **pulling in air, throwing it back**, *also increasing* ***the natural mass flow rate $\dot{m}$ provided by airplane speed a little to $\dot{M}$!*** At **twice the speed,** ~ twice the ***final $\dot{M}$***, you get the ***same thrust with ~ half the ΔV!! That's efficient, because $\Delta V/2$ is a loss!***

Not exactly half ΔV, *only because* a smaller ΔV increases $\dot{m}$ to $\dot{M}$ less ---

2. $\dot{M}$ is *like pounds/sec.* **flow rate**, *except* divided by the acceleration of gravity, 32.174 ft/sec^2. That gives ***a smaller number of bigger units of mass***, with a weird name, **slugs, slugs / sec** here, *but* that gives the ***answer in pounds, and numerically correct, not 32 times high, plenty important!***

WE REVIEW THIS CAREFULLY BECAUSE IT'S THE HARDEST SUBJECT FOR NOVICES!!!

3. The key **axial energy loss ratio** in the ***stream tube*** is $\Delta V/2V_1$, ***half*** the ΔV ***ratioed to the slowed air speed, V_1***. V_1 is plane speed V_0, *slowed down* by the plane's nose pushing a bubble of air ahead, *highly variable vs. radii*, (which changes the twist and chord). Later we'll learn about other **3D losses**.

The same small ratio, $(1 + \Delta V/2V_1)$ is used to increase $\dot{m}$ to $\dot{M}$.

4. **High pitch indicates high efficiency and vice versa. High V lowers the needed ΔV, a twice lowered $\Delta V/2V_1$** -- and **high V_1 needs high pitch!!** ***Slow***, gear for low RPM, high pitch, larger diameter, and be just as efficient!

BUT The RPM Tends to Get Too Low, The Diameter Too Big, Heavy Too.

5. *Surprisingly*, ***Plane*** **V and drag, *Engine* RPM set the max. efficiency** possible without gearing. *The prop design job is to actually meet that target*.

BETZ LOGIC, GOLDSTEIN-THEODORSEN MATH SOLVES EVERYTHING ~ **PERFECTLY!!!**

That's more *new thinking* than I'd like to throw at you, but grasp it and *you've won*. *Let it sink in*. BIG insight is based on this *key core logic*!!!

BOOK II --- CHAPTER 2

THE PROPELLER

THE BASIC, ADVANCED PROP CHAPTER

Props are chock full of surprises! You've already seen many in the Primers. Everyone *kind of* understands how propellers work, but *you've seen the real logic is hidden*. Many see 1. *a rotating wing* or, 2. the English concept, an *airscrew*. Let's start there. Think of a *helical pitch*, a *twisted blade*, so *each radius* tries to *screw ahead the same distance* each revolution. **Pitch** is simply *the* ***inches or feet a prop would screw ahead*** each revolution, **with no slip**, like through a weak **solid**, say perhaps, a stiff Jello.

Helical Pitch --- For a 51" P, 71" D Luscombe prop

Tangent β° = Pitch / Circumference at each radii.

(Precise angles easily found on a *trig* calculator.)

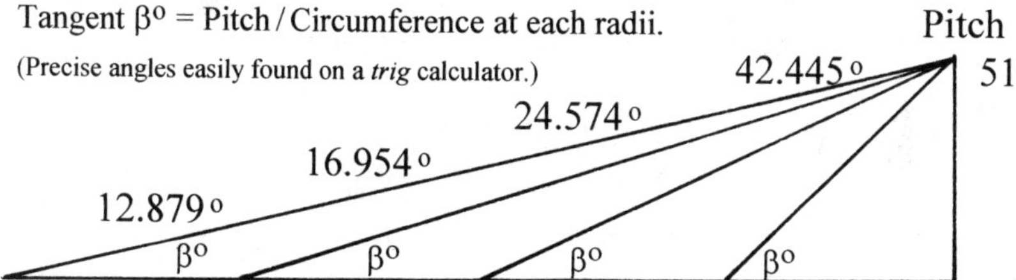

71" x π = 223.05" Drawn to scale, divided into equal parts.

Angles, Circumferences at Several Radii

EXAMPLES:

Dia. Pitch		TIP	3/4	1/2	1/4	Δ deg	P/D	RPM	Max. Speed
71 x 51	**Lusc.**	12.879	**16.954**	24.574	42.445	**29.566**	**.7183**	2545-	116.4 MPH
70 x 79	**RV6**	19.760	**25.594**	35.696	55.165	**35.405**	**1.1286**	2700+	202 MPH
162 x 450.56*	**Reno**	41.518	**49.729**	60.543	74.231	**32.712**	**2.781**	3000/1125	480 MPH

* 450.56" is actually the Dreadnought's Advance, 37.546 ft at 480 MPH, 1125 RPM!!! Pitch is more!

Props *advance* ***less*** *than pitch*, to ***pull Inflow***, there ***is some slip!*** You've seen ***advance ratio"J"***, ***forward*** vs. ***rotational velocity***. ***Important!*** *Realize "J"* sets *and limits* the prop's *max efficiency*! **Great Graphs** will show you that ***J*, Advance Ratio**, *and* **P/D**, pitch/diameter, both tied to ***blade angles***, ***indicate efficiency!!!*** ***TRUTH -- hi speed, lo drag, lo RPM = hi pitch = efficiency!!!***

The reason props appear to ***slip*** is simply that in pulling in and throwing back air, they must be ***overpitched to account for α^o***, and ***inflow,!!!*** So let's *conclude* **Isaac Newton's better way to grasp prop's, their thrust, logic and efficiency**. There are even more intriguing logic surprises. ***After*** *you grasp the horse sense of the technical logic,* ***if*** you wish to pursue propellers all the way, you'll *then* find the graduate level explanation in Chapter 3 II ***much easier*** to understand, with the basics grasped.

We'll want to get you smarter on *advanced logic* soon, but *first*, let's nail the final overview of Newton's basic insight.

A PILOT'S GRASP OF NEWTON'S LAWS OF MOTION

Props ***pull in and throw back air,*** and **"Newton's Third Law"** states what you should now grasp, there is ***an equal and opposite reaction***. Thrust results from pulling air in, throwing it back, just like *you* would be *pushed throwing a big heavy mass*.

Spread your new insight on atmospheric weight! Ask friends how much a 100 foot cube of air weighs at sea level, (if you could actually put it on a scale for honest weight, ***not*** a trick)? **Surprise, shock**, the answer is **76,000 pounds!!!** Air needs to have serious weight or those near million pound 747's would all fall down, and so would *your* plane, and your prop would ***not*** thrust!!! **Wings create *downwash*, throw a serious flow rate *down*, *supporting*, *equaling*** the wing's ***pressure difference lift***. Props ***speed up a stream tube***, *each act as* ***the reaction force!***

There's never an infinite span, so DOWNWASH NEVER GOES TO ZERO!!! √

There *is* a *little* trick there. There are a million cubic feet in a 100 foot cube, 100 x 100 x 100 = 10^6 = 1,000,000!!! Sea level air weighs ~.076 #/ft.^{3}, and that 3/4 of .1 # times a million is a plenty significant ~76,000#. The point is quite valid, ***not*** a trick. You've already seen that a 70" RV 6 prop at 200 MPH is **getting fed 8035 ft.^{3}/sec., 614.5 # / sec.**, *because* the 200 MPH plane is feeding a **big *mass flow rate*** of air into the prop disk, and the prop, you've seen, only has to **add *a small 10 MPH ΔV "speed up"*** to make **a sizable 280#** thrust. ***That's the secret!!!*** √

Newton's Second Law - gives us simple little formulas for calculating the reaction force, thrust. F=Ma, Force equals Mass times **a**cceleration, is for throwing *solids*. F or **T = $\dot{M}$ ΔV**, force, **Thrust**, equals ***mass flow rate*** $\dot{M}$, times ΔV, is for ***flowing fluids***. Mass, in Slugs, is just Pounds divided by g, the acceleration of gravity, 32.174 ft./sec.[2], so mass is ***a smaller number of bigger units***. **$\dot{M}$, slugs /sec.** ***flow rate*** times ΔV in **ft./ sec = T #.** That little drill **makes the answer come out in pounds, and correctly, *not* 32 times high!** If you're not a math person, *don't* let that fake you out, *some will want the full insight.*

That's getting more technical than most pilots would like, but see the **principle is really simple**. **Thrust** is just **$\dot{M}$ ΔV**, **Mass flow rate, times ΔV**, *simple arithmetic*, and **it's the simple key to making props far easier to understand! GO FAST**, get a **BIG $\dot{m}$ and $\dot{M}$**. a **SMALL ΔV**, a **SMALL ΔV/2V$_1$**, ***lo loss***, EASY!

Thrust and Efficiency Think **what's happening at "run up"**. Your propeller is ***not*** acting as a propeller at all, but rather as **a stationary fan**, causing a big blast, a hurricane, *big thrust even with the prop forcing* ***all*** *the $\dot{M}$*, none from plane speed. **But all the energy is wasted, *you go nowhere!*** Grasp that **efficiency** is always ***useful energy or power out, divided by required energy or power in.*** *All the energy that you put into that stream tube is wasted*, useless turbulence, ***finally*** **settling back down to zero, static, a 100% loss** -- so you *must be* and *are* at **zero efficiency!**

Here comes the Super Magic Graph ---

Watch how we've just started you on a path that will give you a really sophisticated, yet simple, grasp of prop logic, that people can otherwise have a terrible time learning. As you start off *on* ***takeoff roll****, your prop is still more a wasteful fan than a prop, churning out a* ***big** ΔV*, ***big thrust*** *you do want for takeoff,* ***but*** *wasting almost all of the energy making a big thrust.* ***Still going slowly***, the thrust times distance, ft. lbs. ***useful energy***, (or T x V, ft. lbs./ sec., ***useful power***), ***is far less than is being put in!*** Making a big thrust, a big **ΔV** becomes a big loss, because that big **ΔV** just falls to static, ***sacrificed energy just to make thrust.***

Lifting off, starting to climb, **extra thrust** is **required** to ***both hoist and fly*** **the plane**. Realize your thrust maximizes, **with less ΔV**! **As you go twice as fast, the natural ṁ doubles, and you get a given thrust with ~ half the ΔV! You're more efficient,** ***because*** the ***plane's speed is feeding a rapidly growing ṁ & Ṁ***, less ΔV required ! Fast planes can easily be more efficient! ***But***, do you see that the **very act of making thrust, forcing a ΔV, making extra wasted ΔV wind, is using energy, lower potential efficiency**. MAKING EXTRA WIND, EXTRA THRUST, especially at low speeds, *low ṁ*, ADDING ΔV, USING ENERGY, COSTS, **lowers** the POTENTIAL EFFICIENCY!!!

But, NOW FASTER ---

Your *prop also is now* ***less overpitched***, ***less stalling***, ***better*** C_L, C_D, ***closer*** to flying at an efficient low angle of attack like a wing. **More Ṁ, less ΔV**, efficiency increases! **But**, fixed pitch, ***still too slow for your fixed pitch***, you're still at *too high an angle of attack*, ***overloading torque***, ***thus less available RPM, and HP!***

That basic logic keeps right on working, right up to and through ***cruise, where the prop was designed to fly just like a wing, at a nice max efficiency, low angle of attack***, and ***lower*** **selected optimum** C_L, now making **a low, efficient** ΔV, a smaller, far more economical thrust, ***designed there, max efficiency at Vmax!***

Realize, efficiency is *good only* near *design speed*!!!

Super Magic Graph Andy Bauer's incisive Super Magic Graph is a key learning tool. It shows how a prop starts at **zero efficiency at zero speed**, starts improving through takeoff and climb, Ṁ, mass flow rate going up, ΔV coming down, better and **better efficiency through the cruise design point**, even **a bit better at *Vmax thrust***, but *drops toward zero again* as you *start overrunning the pitch*, *diving*, **finally back at zero efficiency at zero thrust!** The logic is simple, ***ΔV decreases, helped by Ṁ***, mass flow rate growing naturally with speed. By now, hopefully, you've grasped it's really half of **ΔV**, **ΔV/2** vs. V_1, the ***slowed*** airspeed, **ΔV/2V₁** that shows % **axial** energy loss, ***not hard***. (The Super Magic Graph **accounts for *all* losses**, 3D and Drag.)

Super Magic Graph with Variable RPM, Limited by Engine

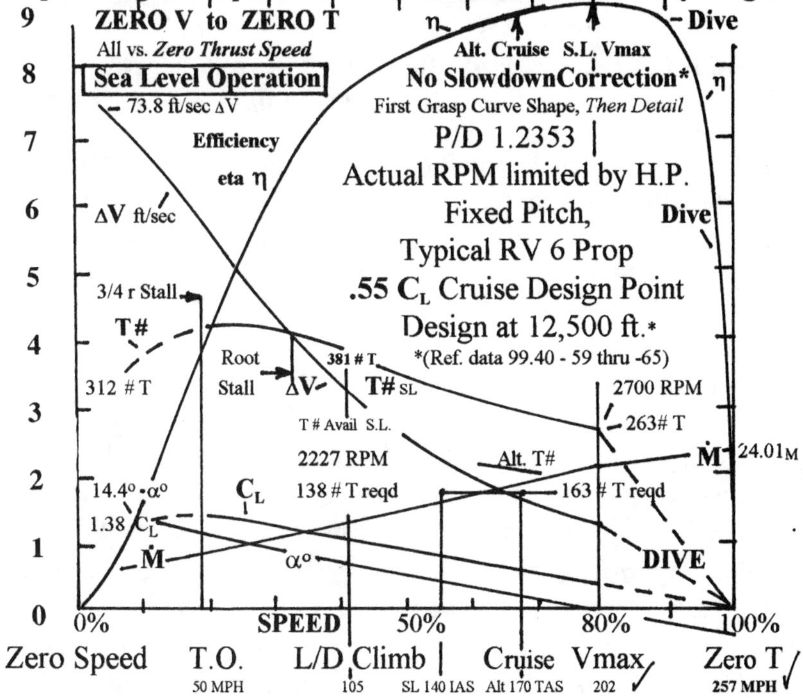

<u>**Sea Level wide open throttle**</u> --- ***<u>actual RPM vs. available HP</u>*** **is used --- at all speeds** from **Zero to Zero Thrust,** (in a slight dive, at 2700 max RPM) --- this shows **Efficiency**, **Thrust**, **Ṁ**, **ΔV**, 3/4 radius α°, C_L, (all vs. *<u>% of Zero Thrust Speed</u>*), **wonderful insight for learning the logic!** Stalled α falls, Thrust peaks, then falls, <u>Ṁ grows</u>, **ΔV** falls. Climbing at a <u>max L/D 105MPH</u>, high prop C_L is needed to both fly and hoist the plane, Thrust, ***<u>above T_{reqd}</u>***, ●*<u>138 #</u>*, *then drops*, still above the *design cruise thrust* of **163#**, 140 IAS-SL, 170 TAS-Alt. ● Vmax, <u>thrust shown</u>, <u>**263#**</u>, ■ <u>requires the rated 2700 RPM</u> --- which is subsequently held as thrust falls to zero. Interestingly, **altitude cruise design**, works ***perfect*** at **sea level**!

The *extra* Altitude Thrust Curve, shows Available Thrust vs. 163 # Design Alt. Cruise

A ***(variable pitch), constant RPM prop***, an ideal, infinitely variable transmission for the engine, *<u>peak HP at any speed</u>*, ***<u>is at even lower efficiency at lower pitch</u>***, that ***<u>allows</u>*** higher RPM, HP, and Thrust. ***<u>Pitch, like J, Advance Ratio, is the eta, η indicator</u>!!!***

 <u>Inner Radii work much better with Slowdown Correction</u>

Andy Bauer's Super Magic Graph offers Wonderful insight! Before we press on, **STOP** long enough to **actually grasp** all the **wonderful basic insight** offered here by Andy's great work, ***the kind of incisive insight you simply cannot normally find!***

LOOK -- DON'T BE AFRAID of the GRAPH --- **IT'S A PICTURE**, **EASIEST WAY TO SEE EVERYTHING**!!!

First, don't be confused. The ***horizontal scale*** is just ***the percent of speed*** vs. the fixed pitch prop speed, 257, where thrust drops to zero diving, pitch too small to make a ΔV, thrust, at 2700 RPM.

An RV Prop would be stalled at T.O., but is far better after Slowdown Correction.

Angle of Attack, α°, shown, is **for the 3/4, 72.3% radius**. Slow, on a relatively high pitched prop, that appears to be stalled. Realize the computer calculates at least 11 different radii, and integrates everything to get the overall answers. Recognize three very important points: ***1. Efficient high pitch are much better after slowdown correction 2. Steep inboard segments stall more than*** **shallow outboard segments**, ***slow****. **3. *As V increases the situation naturally gets better***. Simply, every radius has excess pitch at zero speed, has trouble pulling air in fast enough to not stall. A 51-71 Luscombe prop, P/D **.718**, manages to hardly stall, a 79-70 RV 6 prop, **P/D 1.128**, is more efficient fast, but has to be dragged through the air, at hi α°, slow at T.O. (The **1.235** RV P/D case shown has *excess P/D only because* slowdown was ***not*** used here, ***yet***.)

***Slow, Geometrically, Max stall moves outward, decreases, as Speed Increases!!!**

The Coefficient of Lift, C_L, which **generally grows about .1 for every 1° of α, breaks and falls above about 13 degrees, slow, depending on surface roughness and Reynolds number**. Notice that the C_L plot weakens at stalled speeds. This causes a *major problem calculating propellers slow, because* the exact C_L and C_D are *not* accurately known above stall, so slow speed calculations are naturally more in doubt. We'll look more at C_L.

Wide Open Throttle gives Extra Thrust for Climb. *Note extra at Alt. vs. S.L!!!*

Delta V, ΔV Whereas, it's a wasted hurricane at runup, the very interesting thing to realize is that it's nominally **only 9.2 MPH** at **170 MPH**, **12,500'** ***cruise***, at **163# thrust** for an RV, **only 9.4 MPH** for **263# at 200 MPH**, **sea level**. (At altitude cruise the thin air reduces the $\dot{M}$, so ΔV stays close to the sea level Vmax, high thrust case.)

There are actually **four kinds of losses**, three stream tube, *axial inflow*, unwanted* *stream tube rotation*, (the reason for counter rotating props), the natural *radial flow* feeding the tip vortices, and, of course, the fourth, *blade drag*--which is always present!

***But, you'll learn the air rotation is used to set up ideal helical airflow!!!**

For the graduate course guys, there are two interesting and important basic issues that are worth grasping here. **1.** The prop, like a wing, creates a lower pressure in front, a higher pressure behind. **Inflow $\Delta V/2$ is nominally equal to outflow $\Delta V/2$, each nominally *half* the total ΔV.** **2.** Then, an interesting insight on Physics is that the energy and efficiency loss is based on **only half the ΔV**, ***because*** the energy involved is based on the *extra distance* over which the speed increases, which is based on the *average* velocity increase, thus *half* the total ΔV. *Thus*, pros *only* consider the *inflow* $\Delta V/2$ to calculate the energy and efficiency loss, $\Delta V/2V_1$!!! You, as a pilot, can just skip such graduate fine print extra insight,if you wish.

Notce, only $\Delta V/2V_1$ inflow affects $\dot{m}$ to $\dot{M}$ increase, actually $\dot{m}$ $(1 + \Delta V/2V_1) = \dot{M}$.

If you wish, learn these key fundamentals. ***Axial*** **loss** is **$\Delta V/2$** vs. the ***slowed*** axial velocity **V_1** ***before it's pulled in, at the prop***. **V_1** is the plane speed V_∞, ***but*** *slowed by the plane's nose pushing air ahead*, a bubble, less slow, as radius increases. ***Rotation*** loss, is half **$\Delta\pi dn$**, **$\Delta\pi dn/2$** vs. the circumferential velocity **πdn**, **$\Delta\pi dn/2\pi dn$**. The ***radial*** flow loss affects each calculation, a complex task in Theodorsen's math using the knowledge built up since 1919 when Betz defined min. induced energy loss logic. Theodorsen, solved heavy loading, hi advance, 3D flow by 1948.

We calculate each radius. Small d is *any* diameter, capital D, the full Diameter!

There's a lot to learn here, but *once you've grasped it, **the horse sense is quite simple***. $T = \dot{M}\Delta V$. Making ***ΔV* costs *energy*, *fuel*, *efficiency***. Minimize ΔV, you win, especially if the **V_1** ***slowed speed*** the prop sees is ***high***, you have a small loss percentage, $\Delta V/2V_1$ -- thus high efficiency! *The game is to keep ΔV small but also get V_1 thus $\dot{M}$, up*. ***Grasp that, and you have props basically licked, you've successfully grasped the central logic***.

But, realize, a constant speed prop does *not* prevent this efficiency loss at low pitch!!!

Grasp *why* **low drag, thus low thrust required**, **low ΔV**, ***fast plane speed, thus high $\dot{m}$, $\dot{M}$***, **even lower ΔV**, ***low engine RPM, thus high pitch***, (or gearing), **is what *yields* high efficiency!!!!**

ADVANCE RATIO J=V/nD **The Classic Efficiency Graph**

The **classic Advance Ratio curve** shows **high J, high pitch is efficient**, but **low J, low pitch is inefficient**. ***Efficiency swoops up as advance, or pitch, increases***. (Dividing each side by π, J/π = V/πnD **makes it *easier to see.*** **V** is ***forward velocity***, and **πnD** is **the circumferential tip *velocity***, both in ft/sec.) That's **a *triangle* showing actual tip *advance angle***, (*not the tip angle*). **V/πnD *is the tangent of* the blade tip *advance* angle**. **(D = ft., n**, rev/sec) Below it's labeled **J/π, AAA°**, the ***tip* Actual Advance Angle, (or Ratio)**. The curves *flat peak* about .64, and **J**, the **Advance Ratio**, **simply π times bigger**, flat peaks about **2.0**. **LOOK, to help, *actual Tip advance angles are shown*, *EASY!***

Advance Ratio, J=V/nD, (J/π=V/πnD, is much easier to see)

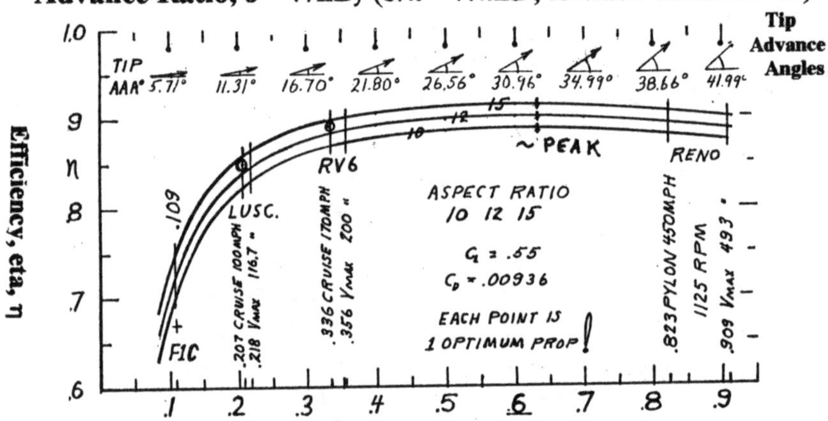

J/π = Actual (Tip) Advance Angle, AAA = AAR = V/πnD

The message is simple, ***Low Advance, Flat Pitch is Inefficient***. Super F1C gas models, 29,500 RPM, 7"D, **3" flat pitch**, are an efficiency disaster, ***below 70%***, but super at getting 1+ H.P. out of a small .15 in³ engine, about **7 H.P./in³!!!** The Luscombe is ***hurt***, ***85% max***, the RV6 in fat city, ***89+%*** without gearing, the Reno Racers more pitch than ideal, would you believe ***a geared 37.5 foot advance at only 1125 RPM***, but in sonic trouble, racing. **Flat pitch causes High Induced Drag and also excess Profile Drag loss, a shallow corkscrew, too far to get to the**

destination. ***High pitch*** **goes with a** ***big V***, ***big $\dot{M}$***, **thus a** ***small $\Delta V/2V_1$*** **stream tube loss ratio**, a proper **low Rn C_D**, **a steep, short, economical corkscrew path to the destination, Profile is ~ 1/3 of the overall loss at .55 C_L**, ***induced ~ 2/3!!! Starting terrible at low pitch angles, ideal efficiency swoops up. Above the RV6's AAA, approaching .4, a J near 1.25, you start approaching optimum efficiency, near 90+%.*** **AAA flat peaks about ~.6+, J ~2+.** ***(Caution - You'll learn using a wrong high or low C_L can and will distort D, thus J, and also P/D. .5 to .55 is best)!***

Do notice that low Aspect Ratio will also limit efficiency!!!!

Now look, clearly we've thrown a bunch of new thinking and logic at you in just a few pages. I certainly expect that you might reread it to really get it. No one will pick it all up in the first reading --- ***but also notice the core is not hard to grasp!!!***

1. Creating ΔV, extra wind, to make thrust costs, ***hurts efficiency.***
2. **Fast V --- increases $\dot{M}$, reduces ΔV, lowers $\Delta V/2V_1$** ***twice!!***
3. **High pitch** ***indicates*** **high propeller efficiency!!!**
4. Surprise, propeller efficiency is really *a "traveling" system efficiency* that ***wants*** **high plane V**, **low plane drag**, **low engine RPM**, all of which **result in** ***high pitch, high blade angles!!!***
5. The computer's job is to **design the prop blade** ***to actually meet the ideal efficiency limit*** set by **Plane's V**, **drag**, and **RPM!**

Remember --- AR also limits efficiency!

It all makes sense: High pitch goes with high V, thus high $\dot{M}$, thus low ΔV, a steep corkscrew path, thus high efficiency!

Thus BOTH INDUCED and PROFILE Drag Energy are LOWERED.

Pitch / Diameter, P/D Ratio -- A Shorthand for Blade Angles. Whereas *Advance Ratio* deals with the *velocity ratios*, forward vs. rotational, the advance angle or ratio, which ties fairly directly to blade angle considering V, ΔV/2 inflow, slowdown, and α^o -- **P/D simply directly nails actual blade angles**. An F1C model prop, 3" P / 7" D = **.4268**, is poor, below 75 % *max* efficiency. A 51" P / 71" D Luscombe prop, P / D = **.7183**, still compromised at 85% efficiency max. ***The RV6 1.128 P/D ratio, (79" pitch, 70" diameter) allows a fine ~89% efficiency.*** **Go for a P/D of 1+** ***~1.3 to 2 P/D maximizes at ~90 to 91%!!!***

Magic Graph -- P/D vs. Efficiency vs. Zero Thrust Speed

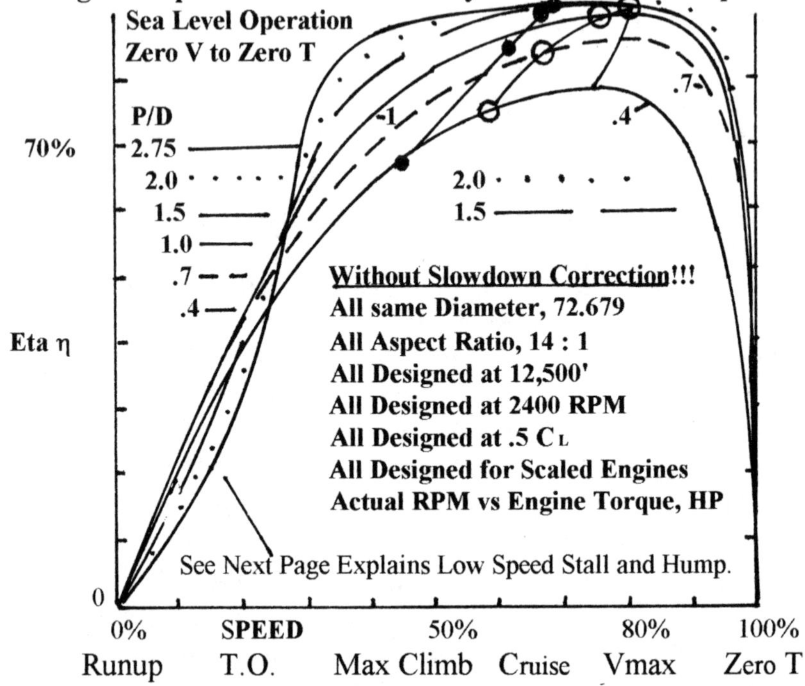

This Graph provides marvelous insight into the <u>Potential Efficiency</u>, <u>and</u> *<u>No Slowdown</u>* <u>Stall Characteristics</u> of normal and abnormally high and lowP/D props. (Like the Super Magic Graph, RPMs are those allowed by engine Torque And H.P, until finally limited by Max allowable RPM --- at each speed, a percent of the Speed where they fall to Zero Thrust in a shallow dive, at 2700 RPM).

Props usually range a little below or above **<u>a P/D of 1</u>**, a **51" P/71" D Luscombe Prop .718**, a **79/70 RV prop 1.128**. Early Spitfires actually used a fixed pitch prop, let's say 2.75 P/D where a *non constant speed* Reno prop would calculate, see p. 51 II, something you'd never do today. A 29,500 RPM, 3"P/7"D F1C model prop would be .428 P/D, for a low range example near .4.

Computer runs show **actual efficiency** and **stall characteristics.** (If, for example, we used a constant RPM, say 2400, **<u>the graph would lie</u> to us at low speeds** because at low speeds where 2400 RPM was *<u>more</u>* than the engine could drive, a false excessively high angle of attack would calculate a false premature stall as speed slowed, *a worthwhile insight to correctly understand what happens here physically*.)

If we were to list the angle of attack at each radial station on the high pitch props, as the plane slows down we would logically find the **high β angle inner radii stalling first**, the stall moving progressively outboard as the plane slowed more, finally causing the **overall efficiency to drop into the S curve you see on the slow end of the efficiency plot**, **worse at high P/D**.

The somewhat impractical, extreme 2.75 P/D case shows a very insightfull characteristic on the computer data. The whole inner part of the prop stalls so badly slow that the **RPM is actually higher at runup than part way into takeoff as the inner blades progressively unstall**. **The attention getting insight is that those very high Beta, β angle blades are lifting, not forward, but significantly counter to the engine Torque, so the engine RPM actually drops as you speed up**, the blades unstalling, fighting the available opposing engine torque ***more!!***

You can see how 1.5 and 2 P/D fit into that logic ---

There, ***slow (but faster)***, the **RPM**, ***is so decreased that the RPM is a much closer match to speed***, lower α°, less ΔV, less thrust slow, but this **allows the efficiency to swoop up abnormally quickly, looking quite favorable** until you figure out that it's happening because **the abnormally low RPM lugging prop** is doing ***a poor job of providing takeoff thrust***. Those early Spitfires would have been dogs on takeoff! Old comments say those **1930's Schneider Cup races were for Seaplanes *because* of ridiculously long takeoff runs, airport runways too short**!

The opposite effect is found on very low P/D props, high RPM maintained at low Speeds *because* the blade angle is *not* pulling against the engine Torque as much as at higher pitch, and RPMs remain much higher as Speed slows down.

The three locus lines at the top of the curves are **the peaks on the right**, the **Vmax 2700 RPM case in the middle**, the **actual design point (at altitude) on the left, (not actually on the curves.) Notice low P/D with greater induced drag and excessively long corkscrew path is penalized operating well below peak!**

Gearing, ***often impractical****, **you can *always* create an ideal 91+% prop efficiency, a high J, if you gear your engine for low RPM, hi torque, hi pitch, **a much larger diameter prop*, more $\dot{M}$, to give the required thrust at the lower RPM!**

With Larger Diameter Prop Tips Fall To Lower Angles, *Counterproductive vs. J!!!

Another way to think of the high pitch correlation with high efficiency is to simply grasp that a prop blade passing can only impart a limited ΔV at a proper lower α^o, C_L, efficiently, and at high pitch, high speed, it's adding less in proportion at a high V_1, another view of a favorably lower $\Delta V/2V_1$, thus a low loss ratio, thus high efficiency, *not* hard!

T he "Twice effect of V", is exactly comparable to the V^2 reduction of Induced loss!!

Yes, this is a headfull the first time, if you've never dealt with science and Physics, but ***if*** you hung in and got it, compliment yourself. Do recognize it is ***not*** possible to strain your brain!!!

We could start right now to teach you what a pilot needs to know ***to operate*** a prop more intelligently, but that is easier to do, and better, *after* you have a good bit of your new insight nailed, so let's wait a bit longer, and you can get much smarter.

Blade Angle Accuracy -- An Example, +/- .1° Required

The key insight is there is ***only .95° difference*** at the 3/4 radius, between a 48P-71D climb prop and a 51-71 cruise prop. That's ***not a tolerance****, but* ***the whole range*** from a *free turning climb type to the opposite kind, lugging the engine!* Thus, only ***.1° is a proper design and manufacturing tolerance to hold manufacturing consistency of thrust and engine torque load!***

I brought my 1951 McCauley 51-71 Luscombe cruise prop to a client's factory to do a granite table "plate check". My 51" Pitch cruise prop had clearly been twisted from a 48" pitch climb prop forging, ~ 48", less at the root, ***a constant 29° in front of the cowl for slowdown***, ~ 51" at the crucial, working outer radii. Checking, I found **McCauley's shop tolerance is only** +/- **.1°**, good reason and need for that accuracy!!! More later --

Blade C_L, An 800# Gorilla -- **A Rotating Wing Calculation**.

We do also calculate propellers as rotating wings -- Area, C_L, q. (Remember q is the ram dynamic pressure, $\rho V^2/2$.) The big complication here is that ***bigger D creates faster tips, (and bigger $\dot{M}$ too). D, and Shape is thus firmly tied to the Area calculation,*** and controlled by your aspect ratio choice! Imagine, if you calculate a prop as a rotating wing, logically ***a definite Area-Diameter-Shape is needed to get a specified thrust at a given RPM, Speed, ρ, and C_L***. Now, ***if*** you **arbitrarily change a .4 C_L to a .5, that's *a big 25% increase*, and you've grossly changed the Area-Diameter-(Shape) required.** -***Realize C_L is an 800# Gorilla!***

Since ***only*** **1° is equivalent to .1 C_L**, prop ***angle tolerance needs to be quite accurate, simply to hold the design C_L and an accurate match to the engine torque and airplane drag combination!!!*** My ***cruise*** prop, **.95°**, more than a climb prop loads my engine enough that it won't quite reach rated RPM, Vmax at sea level, 2545 not 2575. ***.1° manufacturing tolerance is really appropriate***. **Overall angle errors can completely destroy the intended engine-prop-plane design balance!!!!!!**

(*However*, the twist error of a constant speed prop, maybe 5°, ± .25 C_L changing from design to flat pitch *average out, not a big effect on η.)*

Specific numbers: A computer design study, (without slowdown), near the end of Ch. 3, next, in the book, shows a **.3 C_L demands an 83.7" prop** for an RV 6, whereas a **.6 C_L drops the required diameter to only 69.1", AMAZING! A .55 C_L requires a 70.8" fixed pitch prop, about what is usually used.** That was a huge insight for me the first time I saw it!

C_L selection, and holding accurate control of blade angles turns out to be *absolutely fundamental* to *both* hitting a design match for your prop, engine, and plane, and getting consistent results from a production run of propellers. Otherwise, engine load, needed diameter, or speed lose balance! ***How's that for an insight different than you might expect?***

Aspect Ratio -- **Maximizing $\dot{M}$ - Minimizing Induced Drag**.

Just like a sailplane wing, a propeller blade likes to have the ***highest practical Aspect Ratio*** (Span/average chord). ***Big diameter*** has the **least induced drag**, but now you're prepared to recognize **the maximum $\dot{M}$**, thus the minimum **ΔV, the *minimum $\Delta V/2V_1$ and tip induced loss for the max efficiency***. It would seem to be an easy decision to use a high aspect ratio, but **prop vibration can be very complex and *treacherous*,** and can quite literally kill you, so be cautious, no rash moves. We used the outer 90% to calculate Aspect Ratio, **the Luscombe is 14:1**. ***The inner radii are very ineffective***. A spinner reduces core drag.

LATER, WE NOW USE 10, 8% SPACES, 19% TO 99%. A 14:1 A.R. = 12.6 NOW!

Blade Twist - Modifying a Helical Pitch Blade for Ideal Twist

The objective of the classic propeller analysts has been to simply cause ***an imaginary Rigid Vortex Sheet, of constant pitch*** -- by geometrically creating ***constant pitch helical wind inflow angles phi, ϕ, stretched outflow*** at each blade station, ***constant (simple case) prop blade angle of attack, thus C_L, and proper chords*** to match, (to actually make that C_L happen). That takes a fairly complex geometry accounting for plane speed V, the inflow **ΔV/2,** the stream tube and prop rotation. That's all shown in Ch. 3, the graduate course. **But recognize it is *fairly close* to the simple helical blade** shown a few pages back, **just a little different to hold a constant angle of attack alpha, α, simple enough in concept, *not*** hard to grasp, really. (You'll learn in the geometry in the advanced chapter that in accounting for **ΔV/2** and rotation, Theodorsen's, or Glauert's math accounts for the ***half downwash, (inflow) at the prop***, thus the change to effective angle of attack, that lift creates, making it easy to grasp) That needs to be ***modified:*** The pitch needs to be ***unwound more and more as you move inward***, because of the bubble of air the fuselage pushes ahead, causing ***a highly variable slowdown vs. radius***, amounting to **17 1/2 % at the 13" radius** on the Luscombe prop, about **4.64°**, **4.8 % slow at the tip, .597°**

Calculating a Propeller's Area, Diameter and Shape ---------------- **Remember, we're** keeping this the more simple explanation for. pilots, indeed everyone, **to get everyone a good first grasp of the super important logic overview**, purposely avoiding the excess detail that we might mistakenly try to cover before you grasp the basic logic. *This is a subject that becomes much easier to grasp once the light bulb comes on, and you tumble to your first grasp of the overview*. But begin to realize, recognize, ***appreciate the wonderful capability of the modern personal computer*** *that can* ***accurately*** *deliver the complex flow field angles shown in the last section* ***to .1°! It's a modern wonder!***

Now realize this: **The game is to calculate a prop that will deliver the specified thrust, at a given RPM, speed V, altitude density ρ, and C_L Once you select the highest blade aspect ratio that you dare, you have, by that choice, set in concrete the *relationship of needed blade area to diameter*, and, in fact *the blade shape also*, *because* the math will give the planform shape, the cord widths, to give the C_L and *ideal lift distribution which comes from constant pitch airflow!***

Though wildly complex inside the computer, do you see that it. has become something acceptably simple to grasp, to understand? The computer does the wildly complex task of determining *the wind inflow angles, the ϕ's at each radial blade station*, **proceeding just like it was doing a straightforward wing area calculation, *except the desired area-diameter-shape aspect ratio is held***, and all for the twisted wing like prop blade! **Just grasp that you need a certain *"size"* blade to deliver the needed thrust at the given RPM, V, ρ, and C_L -- just like it was easy**, when it's an essentially impossible calculation by hand.

We **understand the logic using $\dot{M}\Delta V$**, because that most easily. allows us to see the hidden logic. Rotating Wing, Blade Element Calculations don't really give correct answers. But we use Theodorsen's 3D precise method. ***Grasp the easy view here***.

A SMARTER GRASP OF PROP OPERATION

Your Engine's Transmission, ***Fixed Pitch vs. Constant Speed***
Have you recognized yet that your propeller is your engine's transmission, and that ***a fixed pitch prop is, in fact, a single gear transmission***, with *a quasi adapting fluid coupling*. But **the variable pitch, constant speed prop**, is, in fact, the near perfect **infinitely variable transmission**. There is a *twist error at low pitch*, ***slow***, ~ 3° to 5°, (presuming a cruise design point).

THERE IS LESS TWIST, LESS ERROR, WITH SLOWDOWN CORRECTION

The ***surprise*** here is that the **constant speed prop *also* has low efficiency, slow**, *because* it's doing an even better job of making **ΔV, extra thrust, extra wind that is going to be wasted**!!!! It's a far better transmission, because *it allows max rated RPM, thus max HP* at any speed, offers *max thrust availability*, but slow, at low pitch, you've learned that ***still gives low eta, η!!!***

AS ABOVE, THERE IS ALSO TWIST ERROR, LESS WITH SLOWDOWN CORRECTION.

Fixed Pitch Recognize from Andy's Magic Graph, that right from runup, through climb, cruise, Vmax, and finally to the "zero thrust RPM and diving speed", **the angle of attack is always decreasing** --- from what can be a terribly overpitched condition at run up, worse on otherwise efficient high pitched props, even worse at the higher angle root radii. Especially on the efficient high pitched props at run up, the air cannot be pulled in fast enough to not stall, and even outboard it's operating at inefficient excess angles of attack at high drag. If you just grasp the horse sense of it, it's no big surprise that you can only turn 2000 RPM, or so, at runup, the engine torque way overloaded, and at a lot less RPM than you'd like in slower max climb. I cruise climb, get more RPM, HP. The prop is more efficient as speed increases, and at more reasonable α° vs. radii.

Our thinking here presumes Wide Open Throttle Operation

You automatically fly at lower efficiency every time you try making extra thrust, because you're making extra ΔV, which gets wasted, and ***fixed pitch***, **slower than design cruise**, you're ***flying the prop at high angles, lugging the engine at poor RPM***.

The Constant Speed Prop. Every engine needs a transmission to match its ***RPM*** and ***torque*** curve to the individual application. The *Constant Speed*, though it ***does not hold low RPM, hi pitch, hi*** η, is the near perfect, infinitely variable transmission, ***from the engine's standpoint***, no confining gear ratios, ***any variable load you need to always allow max rated RPM, thus max H.P.***

The **downside is from the propeller's standpoint**, where **high RPM**, desirable from the engine's standpoint, is ***highly undesirable*** from the propeller's standpoint, ***the related low pitch still inefficient at max thrust*** **for takeoff and slow max climb**!!!! The good news is max power is only used for a few minutes, **the dollar cost of that prop inefficiency negligible**.

Cruise Efficiency The **constant speed prop is still a hugely important contributor to propeller and flight efficiency**, in a way that is ***insufficiently understood in the pilot population!!!*** The low pitch, max H.P. case is ***not*** the really important usage. **The really important need and usage for the constant speed prop is to be able to pull back to low RPM, lower H.P., *higher pitch*, limit the H.P. output, get your engine wide open at the lowest possible altitude, for the highest possible engine efficiency, lean the engine at well less than 75% H.P. --- get the plane and the engine ideally matched, the central objective of this book. Remember, the lower the altitude you can get your engine wide open, the more efficient your engine can be, because altitude hurts engine efficiency, but you must get it wide open and leaned first, and *at the IAS and/or angle of attack for the plane's best speed vs. drag!!!!*** **You don't know actual blade angle, but it's always as steep as the engine can turn. Pitch variable, a .5 C_L gives a bit more diameter!!!**

That's a bit more efficient, but Heavier

Gearing is a Separate Subject. Don't miss the point that **a geared, low RPM, larger, high pitch prop would give *more* thrust at higher efficiency**, **(*but* more weight), on *any* low pitch case!!!!** ***BIG, fast planes need a BIG, geared, low RPM prop to develop BIG thrust, and stay below .9 Mach tips!!!!***

The Propeller, A Different Kind of Machine

Propellers are the unusual case, different than other machines in a fundamental way, and that, you'll see, is a fundamental reason why we have to teach you a seemingly unusual logic of why they work the way they do. In say, an electric motor, there is friction and heating, and those ***peripheral*** losses, that you can't get rid of, account for the efficiency losses, as you pull torque, and rotary power out for useful work. The torque, the power is the product. The efficiency losses are ***peripheral*** to the output.

Of course, more load, more power does generates more peripheral heat loss

In a propeller, the situation is fundamentally different. A **propeller must make ΔV to make thrust, and *that ΔV itself*, the product, *the necessary product to make thrust*, is a *central part of the efficiency loss!!!* That is *fundamentally different* than machines where the losses are peripheral. *You don't see through that looking at it as a rotating wing!***

Next, ***a propeller is quite different, in that, it's the Plane's Speed and Drag, the Engine's RPM that sets and limits the prop's max efficiency*** η, (if it can't be geared to restore the ideal high pitch math relationship). ***The prop is only able to reach the max ideal eta set by the plane-engine. It can't raise the max!***

Only gearing for lower RPM, higher pitch, big D, **or high AR**, can raise the efficiency limit.

When engineers talk to each other, they discuss the math, the very complex leading edge problems on how to get the very challenging analysis correct. **We never got a true explanation of the logic, because that's *not* the challenge to them**. There's another problem. When you've done all the math correctly, made the computer print out correct answers, you're ***not*** done. **The numerical answers processed inside the computer do *not* make the logic clear**. **You have to go far beyond that to define and write out the logic**. That's very difficult because there are a ton of subtle and very complex interrelationships. **It's the *next step harder* to sort it all out, and say it correctly in logic and plain English**. Much of the ***advanced insight purposely skipped here, is in the final chapter and conclusions***.

Overall Propulsive Efficiency -- A Second BIG Loss.

The biggest gap in Aeronautical Engineering is failing to understand and comprehensively deal with the *other half* of propulsive efficiency loss! *But now we're getting to the facts!!!*

Gus Raspet, a somewhat legendary Aero Professor.at Mississippi State, did towed, **propellerless, glide tests** of several private aircraft in the mid 50's. ***He found proper drag numbers!*** *That was **not** favorable, **because** **compared to the power required to fly the planes, it proved disastrously bad propulsive efficiency**!!!* The Bellanca Cruisair, a plane most thought well of, was a disastrous **58%, ducts sealed*, vastly worse than any possible propeller efficiency loss**. Cubs, Cessnas, also did poorly. He provided the needed breakthrough insight, and hard, ***absolutely irrefutable data***. But the industry never picked it up and did anything comprehensive with it. They were *stopped, no way of getting safe, valid glide test data*

* THERE CAN BE ~10% COOLING DRAG, THUS ~ 68% WITHOUT COOLING LOSS!

Andy and I took that on as a challenge in the 80's, first with crude tests, then with credibly accurate analysis of zero thrust RPM's and glide tests, accurately accounting for slowdown. ***Finally, we did very accurate glide testing** accurately sensing the movement of the prop from thrust to drag, as it moved back in the crankshaft bearing **endplay***, with a vibration proof, high natural frequency model airplane wire feeler, lighting a sensor bulb. That solved the most basic, longest unsolved problem in flight, ***how to test for accurate drag of a fixed pitch prop plane?***

There tends to be two losses, that multiply to a **bigger overall loss!!!**

We confirmed Gus's great pioneering work on my Luscombe, with **Zero Thrust Glide Testing**, overall eta, η, from the low 70%'s at low power, to the low 60%'s at high power, an RV6 and a Whitman Tailwind, at the CAFE Group in Santa Rosa, at 80 %+ overall at Vmax, ***much better***, a poor Luscombe 75% η prop, vs 67% overall loss, there can be ***EXTRA SCRUBBING DRAG***. We now need some loose Lycoming engines on worthy planes to carry the initial work forward in a next phase.

Blading Efficiency - ~ 45° β Best - Profile Drag Isolated

For \$2.50 in 1944 Wartime dollars, at age 17, my Dad bought me a copy of Wilbur C. Nelson's "Airplane Propeller Principles"*, a great text, even if it didn't show the essentially final incisive insights that we have here, compliments of Betz, Goldstein, Theodorsen.

A basic Vector Analysis of prop Lift and Drag vectors shows:

$$\text{Efficiency } \eta = \frac{\tan \phi}{\tan(\phi + \gamma)}$$

where: phi, ϕ, is the air inflow angle; gamma, γ, is the drag angle C_D / C_L

Thus, efficiency depends only on inflow angle, and drag angle, γ. A little Calculus, differentiating that vs. phi, ϕ, equating to zero, we find **optimum blade angle is 45° - $\gamma/2$, half the drag angle,** which for **.5 C_L, *.01 low Rn C_D*,** is 50 : 1 L/D, $\lambda/2$, **.02/2 = .01.** Arc tangent of .01 = .573°, thus 44.43° β° optimum, ***nominally* 45°**. Now look at his plot, but imagine a **50 : 1 L/D, just for profile**.

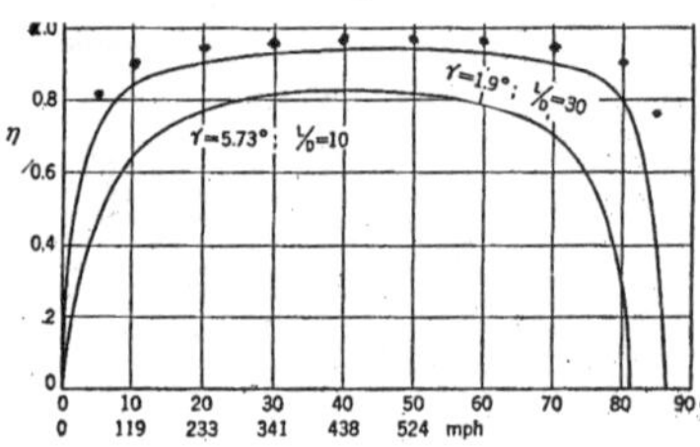

Simple Blade element efficiency variation. - **Imagine Profile *only*, at 50 / 1 L/D**

Where Nelson does a simple derivation, doesn't distinguish between profile drag and Induced loss, thus plots lower L/D's, doesn't specifically include air inflow and rotation, but shows an imaginary, oversize angle of attack, Glauert uses a more complex derivation and final formula **just for profile drag** and gets higher η, similar curves, **valid at a *profile only* 50 : 1 L/D.**

The very ***important key, basic insight*** here is to see that **profile losses are low**, but are ***larger at very lo and hi blade angles!!!*** Andy's accurate program using **low Rn C_D's** works comparably.

 *Proper Credit is Due John Wiley and Sons, great texts!

A **Super Magic Graph** for a **Prop Corrected for Slowdown**?

The Super Magic Graph on page 45 II gives great basic insight on a theoretically perfect Theodorsen prop for an **RV 6** at various speeds, slow, all the way through shallow dives at zero thrust, **2700 RPM, 257 MPH**, the prop running out of adequate pitch that fast. Great insight, but it's **misleading at slow speeds**, because **proper slowdown corrections are large enough inboard** that they **work significantly better**, ***not stalling inboard as early***, with the significant blade angle corrections, greater inboard ---

The first theoretically exact prop, **Triple Ideal,** Minimum Induced, Minimum Profile Drag. Min. Torque, **for the RV 8, designed for 220 MPH, 2700 RPM, 150 HP, 75% power at 8000', corrected for slowdown, it can work much better slow** than the uncorrected comparable RV 6 prop. We planned to give you a slowdown corrected Magic Chart for ***comparing***.

SORRY: We Intended to give you a Slowdown Corrected Super Magic Graph, as our Last Goal Line Task, but the computer program didn't want to run with both Slowdown Corrections and all the "Off Spec. Conditions" of the Graph. With no time to solve that we had to Pass. But below, see the ***RV 8 Slowdown Beta Angle Corrections*** for the Various Radii. This can give you a pretty good feel for the significant changes!

% Radius	Beta Angle °		Reduction °	Final Angle °	
19%	65.888		-17.649	48.239	
27%	57.315		-11.810	45.505	**Remember!**
35%	50.164		- 7.559	42.605	We're trying
43%	44.282		- 4.800	39.482	To **Hold .1°**
51%	39.453		- 3.081	36.372	Blade Angle
59%	35.471		- 2.023	33.448	**Tolerance!**
67%	32.161		- 1.367	30.794	
75%	29.383		- .953	28.430	For +/- **2%**
83%	27.029		- .685	26.344	Consistency
91%	25.015		- .506	24.509	of Product
99%	23.277	**42.611Δ**	- .383	22.894	**25.345Δ**

CONCLUSIONS

We have purposely ***not*** **covered all the material in this, more basic Chapter**, because in market testing, we found it would **drown you with way too much too quick**. If you first *grasp the logic, the new thinking, in this Chapter*, **let it soak in** *until you're comfortable with it*, it is ***then*** *much easier to tackle the* ***Graduate Chapter 3*** with all the **final Professional Insight**. ***Then it doesn't seem too tough***. ***The goal is to make it easier***.

We can give you a great set of conclusions that even go past what you've earned so far, get you quite far with min. effort!

1. A **~91+%** efficient prop can always be designed by gearing it for optimum advance, high pitch, (low RPM, high Torque, big Diameter, but diameters often too big, impractical). Gearing is common on excess RPM, slow, high drag, otherwise low pitch, inefficient ultralights, and large fast planes that need **high H.P., BIG, low RPM, multi-blade props**, for very **BIG thrust, while avoiding sonic tips**.

Very high fighter Speeds, make Prop Diameters Reasonable!!!!

2. **NACA postwar testing**, targeting 1% accuracy, showed **91 to 92 % efficiency**, *even with constant speed prop twist error*, ***until low pitch, or exceeding .9 Mach tips degraded performance***. Significantly, however, ***those tests were with limited diameter, symmetrical drive housings***. **Real airplane tests with large embedded bodies**, unsymetrical, inefficient cowlings, can produce **larger losses** than those predicted by *extra stream tube velocities*, thus, *not* just extra scrubbing drag!

Ham Standard charts show a 91.5% sweet spot!

3. **Fixed Pitch,** without gearing, **maximum propeller efficiency is set *and limited by* Plane *Speed V*, *(and drag)*, and Engine *RPM*.** Propeller efficiency is ***actually a Traveling System Efficiency*** in which the ***blade plays a limited, but finally very important role***.

4. **The Classic Betz Prop and Computer Design of propeller blades Targets the maximum efficiency limit theoretically possible!** ***(But, now we see Elliptical Lift Distribution can do better!)***

5. Plane **V, RPM, and prop C_L**, are ***Three key Unrecognized Giant Gorillas***, of propeller design logic, ***because they exert very powerful, and generally unrecognized, control of thrust, efficiency, and required propeller size --- as follows.***

High q tips, needing counteracting, and Diameter are also powerful Gorillas, 5 Total

6. **High speed, high V, thus high mass flow rate $\dot{M}$**, through the propeller disk area, **directly lowers ΔV, the *extra wind* required to make a required thrust**, and provides ***a larger V_1*** as a ***divisor to produce a lower axial efficiency loss***, **$\Delta V/2V_1$**. Though ***axial loss is but one of the four basic losses, axial, rotational, and radial stream tube losses, and blade drag loss***, ***doubling speed, cuts the ΔV based loss ~ in half, and in half again as the percentage divisor --- dropping the induced loss to ~1/4 if the drag and thrust are not changed, key basic logic*** (*Finally, higher V produces a smaller D, showing the complex interplay!!!*)

(~) Less $\dot{m}$ to $\dot{M}$ increase with a lower delta V, thus *not* exactly in half, and 1/4.

7. **RPM vs. V is *the* primary controller of propeller pitch! *Thus*, it is the primary controller of propeller efficiency**, *with V*. It's thus, ***far more important than is generally recognized!*** **Engines want *high* RPM, for high power at light weight**, but **props want *low* RPM for high pitch, high efficiency**, exactly ***opposite, directly conflicting requirements***. ***Gearing can, might help!?***

Understanding CL, next, is complex, one of the secret insights on props!!!

8. **C_L, Coefficient of Lift**, ***can directly distort required propeller Area-Diameter-(Shape)***. With a change of just ***1 degree α, .1 CL representing the difference between a cruise prop and a climb prop***, ***C_L is capable of great mischief***, major effect, and ***by distorting D, changing assessment of J = V/nD or P/D.*** BUT, **Props act forgiving, --- But Realize a missed design, a low α° and C_L allows, *but demands* more engine RPM, more fuel usage. Excess α° and C_L lugs the engine, less RPM, HP, Speed! Design Study shows a .5 to *.55 C_L is a proper cruise-climb choice,. .5 C_L* Constant Speed**

9. **Blade Accuracy Required, +/- .1°** Because any blade angle errors can directly distort the actual C_L, thus the required prop Area-Diameter, and the engine load, required angle accuracy is much tighter and much more important than is generally realized. ***However in repitching, it's the owner's way to fundamentally change engine load and propeller characteristics, if desired!!!***

10. **<u>Pitch</u> must be <u>accurate</u> to hold C_L** (vs. speed, RPM, inflow, rotation, slowdown). *<u>Then</u>*, **Blade Angle is *<u>the simple indicator</u>* of *<u>potential</u>* propeller efficiency**. (The **ideal is a *45° blade angle <u>near the 2/3 radius</u>*, a J of 2+, a J/π of ~.6+, a P/D of 2+,** *<u>very steep</u>*, ***91% eta, η, but, like the <u>RV6, near the 1/3 r</u>, J/π .34, P/D 1.13, only costs ~2%, then drops as pitch flattens!)*** **On an efficient RV**, slowdown can equal inflow ΔV/2, plus angle of attack. *Thus,* **Vmax in ft/sec can be ~equal to a direct calculation of revs/sec times the needed pitch in feet**!! A slow, draggier Luscombe is ***<u>only 95% of that!</u>***

With the flat top of the Advance Ratio Curve, efficiency gets poorer below a P/D of 1+, a J/π of .33, stays good above

11. **<u>Blade Twist</u>** To obtain a theoretically ***ideal blade loading*, <u>airflow angles</u>, ϕ, are arranged analytically to have <u>simple helical pitch</u>, *a Rigid Vortex Sheet*, to which, typically, a constant angle of attack, constant C_L is added to obtain blade β angles**, and matching <u>chords</u> <u>shape</u> to really get that C_L. ***To account for fuselage caused slowdown, Luscombe 35% radius blade angles <u>need to be</u> <u>unwound</u>* 6.14°**, the tips **.435°**. (Blade angle and chord can be traded, but more drag if wider. ***We teach constant α and C_L as*** <u>the standard simple, and best case</u>.) A constant speed prop's low pitch twist error averages out -- though analysis targeta a ***rigid vortex sheet*, *precise Betz blade loading* vs. radius.** Flight tests can show large interaction losses, covered in Chapter 3 II. Separation can also exist, with stream tubes distorted from large, very unsymmetrical cowlings upsetting flow coupled with energetic blade tip spiral vortices, key insight.

12. **<u>Aspect ratio</u>** As previously described, **the highest safe AR is proper technically to optimize efficiency**, but risk of deadly vibration or structural problems cautions against any ill advised moves, that can, in fact, kill you! **<u>Half the Axial ΔV</u> stream tube loss, and the full tip loss is exactly comparable to wing induced drag losses that sailplanes minimize with long spans, high AR, max M. -- It's fundamental, *<u>but</u>* be cautious!**

Props do have Rotation Induced Loss too, that wings don't ---

13. **<u>Blade Area-Diameter</u>** The proper thinking here, wildly complex inside the computer, is dead simple. ***You simply need <u>the proper Area-Diameter-Shape to deliver the required thrust</u> at the V, RPM, altitude ρ, and C_L intended. Caution on C_L distortion.***

14. **Blade Shape** That beautiful Spitfire elliptical wing has that shape to **deliver *minimum induced drag*, *constant downwash*, *optimum elliptical lift distribution*, *constant C_L's with ideal elliptical chords***. Now, think of turning that Spitfire wing shape into a rotating prop, the ***tip now rotating twice as fast as the mid radius of the wing half*, *4 times the q*** - ***if not also moving forward. The prop-wing will logically become highly tapered***, if it is to have ideal Betz loading, ideal lift-thrust distribution!!! See the highly tapered shape of ideal computer calculated chords on page 147 I, and realize that **broad tips, (high activity factor) are for power absorption, *not* for efficiency**. Broad, loaded tips are thicker, stronger, *more reworkable*, but manufacture excess losses and noise, *almost like reverse taper wings*! Older NACA tests show broader tip props can produce **more static thrust**, ***but they require wasting power to do that***. ***Unwinding a wider tip can hold proper loading, but costs drag, RPM, unless into Mach V's!***

15. **Sonic Limits** Mach 1 on a standard 59°F day, (518.7°R) is 1116.46 ft/sec, 761.243 MPH, and **decreases at altitude proportional to the square root of the absolute temperature**, ***unaffected by pressure***, to 968.09 ft/sec, 660.06 MPH at -69.7° F. at 36,089', the beginning of the supposedly constant temp. stratosphere. Even relatively thin airfoils may have a 30% speed up, which can cause standing shock waves, like you can see on a jet wing surface on an East or West trip when the light is correct, but they are *weak shocks*. **Tests show the drag rise really starts at .9 Mach, 1004.84 ft/sec, 685.119 MPH at sea level, 871.28 ft/sec, at the stratosphere**, and ***the vector*** of forward and circumferential V ***must be kept below that .9 limit***. ***A 6' prop at 2700RPM, 250 MPH, is at a 924.1 ft/sec vector V***.

So how's that for an understandable, penetrating insight, and conclusions on Aero's most complex subject, propellers, without having to deal with the hugely complex math and computer programming, the technical, horse sense logic put to words? There's more but it properly comes in a next step.

TEACHING PROPELLER LOGIC

My experience with explaining propellers has taught me to ***not*** feed people the whole McGilla in ***one gigantic dose***. It can be just too much. Most people **need time to become comfortable with all the surprises and new logic**. I believe you'll find the subject much easier, if you perhaps reread all, or part of this, **let it soak in**, **make it your own**, then do the advanced chapter.

When engineers discuss complex subjects like propellers, they do it in technical papers, with complex math, feel justifiably fulfilled when they conquer, refine one next step facet of the complex riddle, but because of all that, ***almost never deal with the overall explanation***. I believe their necessarily targeted focus is the reason we never got a comprehensive explanation of the overview, because the lesser guys, not at the leading edge never were able to get their mind around the myriad of complex interrelationships. ***We've been through all the expert's work, the complex math, checking it all, dissecting it, to try to nail it***.

For Engineers, the core of Theodorsen's math is in Appendiox T --

There are three reasons you don't see the heavy duty math here. Regular people don't want math, can't understand it, are just put off by it. Second, ***the objective here is to get this never adequately explained subject into words, a few picture graphs, a simple core formula, or so, to nail the heart of the technical logic, make it understandable horse sense logic***, so a regular guy can grasp it, if he's willing to try hard, hang in and actually get it. Third, don't misunderstand, getting it into correct words is just as hard as the math, in some ways harder, because only the human brain can do what even the computer can't, grasp the hidden logic, and put it into words. If any of you engineers out there doubt that, try it. You might be surprised to know that after we got the words nailed well enough, we had multiple instances where the words, corrected the math and the computer! Multiple times the logic showed us the math and the computer were temporarily wrong, an obscure error not caught.

Propeller Computer Analysis - One way of looking at it: **Betz** in **1919** conceived the "**rigid vortex sheet**", optimum logic. You'll learn the highly ***variable axial and rotational air particle motion combine to form a perfect helical screw surface***. The trick is to get the computer to design the **blade angles** β, **chord**, **and angle of attack** α, **to make that miracle happen** for each blade element, to get the **least induced loss**. It's a rotating wing, somewhat akin to ***Prandtl's optimum loaded elliptical wing***. **Glauert**, **1934**, got the ***inflow and rotation essentially correct for somewhat simplified equations (using Prandtl's F Factor)***. It's ***simple helical pitch***, not of the prop blade itself, but rather of ***the air flowing into the prop blade stations***, *at helical angles, phi*, ϕ, logically adding a *constant angle of attack*. With proper blade chords that gets Betz's ideal inflow and loading *(ΔV/2 in, slightly stretching, helical + ΔV/2 out.)* One pro's solution uses sixteen basic equations, algebraically reduced to four, the computer almost instantaneously **iterating a trial and error solution** for the multiple blade stations, finding **the wind inflow angles for** a twisted wing calculation setting α, C_L, to get all the data and integrated answers. ***The angles are then unwound to lower pitch to account for body/nacelle slowdown***. A refined program code yields full data, *separately, an off optimum speed analysis.* **Theodorsen in** 1948 solved **3 D flow**, **axial, circumferential, and radial,** ***for heavy loading***, ***high advance***, **for ideal props!**

But it turns out that only Theodorsen's Math gets really correct Betz Answers!!!

The Axial and Tip Vortex Loss is Really the Induced Loss!

Here is the great insight that brings together the unique prop stream tube logic, and the wing induced drag insight you learned much earlier in the book, because they're really the same thing! Remember that when you ***double the speed V***, you also double the natural mass flow through the prop $\dot{m}$, and as a result, cut the needed **ΔV** in half, ***if drag not increased, ignoring*** $\dot{m}$ ***to*** $\dot{M}$ ***increase*** Now you can see that ***the*** $\Delta V/2V_1$ ***axial tube loss is cut to ~ 1/4 its previous value, because the V divider is doubled as*** ΔV ***is cut in half, cutting the axial loss to ~ 1/4!!!*** That's the secret!

 The Double effect of V is like 1/(V squared) Induced loss.

Now, remember way back, much earlier in the book, that a profile drag curve is a V^2 curve, whereas ***an induced drag curve is a $1/V^2$ curve that drops to 1/4 if you double speed***. The same thing is happening in a stream tube due to prop thrust*, as is happening in the downwash from a wing, ***they're both just the same old induced loss, perfectly equivalent****. In both the prop and wing case, the tip vortex loss is part of the induced loss.

***As ΔV/2V affects the ṁ to Ṁ change, it's not exactly 1/4,** a fair amount different, but similar.

The equally valid stream tube logic and explanation, however, is the way we can most easily get you aboard to grasp all the sequential surprises of prop logic that even smart guys have trouble getting through their head. *Don't feel alone out there --*

Realize, the **rotational loss** is **unique to propellers**!!! ✓

Now that logic happens just exactly as explained, but ***in the real world, more happens***, but the computer program handles it properly. As you go faster the complex algebra also shrinks the prop diameter. Props are a mine field of interrelationships, such that you must be very cautious of extrapolating what we teach you, or you can quickly be awash in wrong conclusions. Don't be afraid however, ***those 15 Summary Conclusions have already given you a wealth of insight on propellers you didn't have a few hours ago, and never would have found in any other book!***

A Summary Insight of the Efficiency Losses. **At higher speeds**, as you can see above, ***the $\Delta V/2V_t$ axial loss effect almost collapses as the key propeller energy and efficiency loss***. A *lower level of both axial and vortex loss at reduced ΔV* is married with *an increasing stream tube rotational loss* associated with higher pitch, and a normal surface friction drag loss. ***The characteristic Advance Ratio efficiency curve results, rapid improvement, leveling off to a very flat peak!!!***

The beautiful final way to look at this, is that *the induced loss collapses with speed*, all losses finally totaling a min. of 8 to 9%, which means about 91% ideal efficiency vs. AR!!!!
THAT IS A FANTASTIC INSIGHT - APPRECIATE IT!!!!
Realize, you only hit your peak at optimum speed, worse slow! √

A real world example of all this logic is that an RV6 won't go twice as fast as a Luscombe at equivalent drag, but it will do altitude economy cruise at 170 MPH TAS, 140 IAS at not too much more drag than a Luscombe at 100 TAS, 85 IAS, and with a more efficient prop, and much better overall propulsion efficiency, **at better MPG, 27.4, vs. 26.6, slow in a Luscombe**. An incisive insight here is that the **160 HP engine** uses only a ***70" D*** prop vs. the 85 HP *71" D*. ***Notice that at much more speed, much more HP is required at almost equivalent drag. A much bigger engine is required to put that traveling energy in at a faster rate!*** **You've seen required prop diameter decreases at higher speed, so here's this much bigger engine driving a 1" smaller prop than on the Luscombe, much faster, more efficiently**!!! Props are an intriguing challenge.

Chapter 3 II Takes You to an Advanced Level

The Prop Primer, through Chapter 4 II, Propellers, are separated as Book II, to not complicate the basic Airplane Book, and **the book's objective of teaching how to match and maximize the Plane - Engine combination**. Chapter 3 II has a wealth of additional insight valuable to anyone who wishes to go for an even broader, deeper grasp of props. **The whole geometry of Betz loading is explained early as a major objective**, so you understand that key fundamental. **The whole McGilla seems too much to most people if given in one giant dose, but if you've grasped this more extensive basic presentation, review it, help it to soak in, really make it yours. The Graduate Course may then even read fairly easy to some**, because it will be filling in many of the details we purposely skipped in this basic chapter, and is apt to answer many questions a sharp guy will have seen, and be wondering about. ***To really become a propeller expert takes a considerable effort, because they really are as complex as any subject you will ever deal with. But see how much insight you've gained here in a very short time. It's our job to get you a great start!***

ADVANCED PROPELLER LOGIC - AN INTRODUCTION

Propellers are **a rotating wing**, ***but imbedded in a stream tube***, ***a very complex case*** **needing world class math to get usable detail design answers** ***to accurately get*** the **engine loading torque, thrust, all the key performance characteristics and shop lot control intended**. Since there is nominally **only one degree pitch difference, .1 C_L, between a cruise and a climb prop**, ***you can see that a necessary design and manufacturing tolerance is .1°, point one degree*** for a proper controlled result. An **inadequate program, inferior math**, flawed, inadequate logic ***will get you poor design answers***, *not* ones that will give you what you intend - ***more likely misleadingly wrong!***

Once you've set a prop's ***Size***, **Diameter, Area, Chords-shape, if you don't have the inflow angles correct**, a proper solution to the **complex 3D, heavily loaded flow**, including an **accurate correction for body caused slowdown**, you can **miss your intended Thrust, and engine Torque Load** by a bunch, back to cut and try. **But an accurate program and a** ***wonderful modern PC*** **will print out an** ***ideal prop in just seconds!!!***

The problem is that doing all the studying that you need to do to truly, accurately, understand all the advanced, very complex math, all the technical logic and insights is a real brain bender, right out at the limits of intelligence of the most brilliant, rarely occurring analysts we've had in the 20th century. It took **a multi year hobby challenge effort**, two very experienced lifetimes of advanced engineering knowledge to get you the **wonderful insight** that you get here **in just a few hours - a huge bargain**.

Teaching pilots ***the math is not even close to a real option***, and you would not want to be drowned in it. ***But we can teach you the incisive, professional level insight and conclusions that come out of it all***, **again in just a few hours, a huge bargain!**

In hindsight, you might *now* be better able to see that what we did *after* the **Primers, in Chapters 1 II** and **2 II** was to **decode** the ***tricky, basic*** **propeller logic --** ***then*** **teach professional level conclusions on much more**, actually far more than you had to earn of your own personal sweat, a *minimum of frightening math.*

I'm sure that many novices were either ***unnecessarily*** scared away, or temporarily confused by throwing you into $\mathbf{T = \dot{M}\Delta V}$, something they'd think was scary basic Physics, a strange Greek letter and scary math. Those who saw through it OK are apt to see in hindsight that the equation is only *a simple enough, two item formula* ***that shows*** **the efficiency game is no harder than getting the speed up, the natural and final mass flow rates, $\dot{m}$ and $\dot{M}$ up,** ***so the ΔV extra wasteful wind can be small, the loss small, the efficiency high***. The logic it decodes is tricky, but once you see through it, the math is just ***simple horse sense***.

You didn't actually have to do the math, just grasp the logic. We did **a basic RV6 example**, so those who got it OK, **could see the details**. It turns out that *in hindsight*, that's ***the easiest way to teach you the tricky little basic logic of props***, that ***the loss gets small if you just go fast***. **BUT -- If you gear for Low RPM, High Pitch**, **Bigger Diameter**, on the slow, excess RPM cases, ***you can theoretically get ideal efficiency***. **BUT -- diameter** ***can*** **get excessive**! Like a wing, **the downwash, tip loss, (and rotation) - is the induced loss**. If you just ***go fast, that loss collapses***, and your prop is more ***efficient with less induced loss***.

Planes teach us Induced Loss is a $1/V^2$ type loss, V lowering Induced TWICE!!!

The second little formula that fell out of it all, $\mathbf{\Delta V/2V_1}$ again would at first look pretty scary to a new "non math guy", but since things are always far easier in hindsight, hopefully you can now see that it's just ***a ratio*** **of half the axial speedup (done to the stream tube, by the prop), to $\mathbf{V_1}$**, the speed of air coming at the prop ***before*** the prop suction affects things. That ***fraction*** establishes the ***decimal or % axial energy loss*** -- ***and*** the prop caused increase to the natural mass flow rate ***$\dot{m}$ to the final $\dot{M}$***.

Now, if you got all that, all the **insights in the prior chapter**, the **15 key Summary Conclusions**, you should have a pretty good grasp of basic prop logic, and be ***prepared to attack this advanced chapter***. To a large degree this chapter will simply be expanding your grasp and knowledge with many insights that, if given to you in one chapter, before the basic conclusions, just comes out to be *too much in one dose*. **Time helps it soak in!!!**

We're going to presume that you got the basic logic and move on. We'll, explain constant dT/dQ, its significance, give you ***a lot more insight into the vector diagrams that show you in picture form how the air flow works***, *how thrust, torque and power are distributed vs. radius*, ***ideal blade loading, vs. shape***, **the shape changes vs. advance ratio, steepness of pitch**, ~45° **best**, insight into the **slowdown caused by the engine cowling, interference losses**, very important *graphs*, ***pictures*** that show you that you can **tell a huge amount about a prop by simply seeing what its P/D ratio is, if it's over 1, or not, and by how much**, give you **greater insight into the stream tube**. The **blading efficiency graph** teaches the ***small drag changes on blades over wide blade angle ranges,*** but *worse at very low and high angles.* **A BIG Surprise, design studies**, show how **C_L can be swapped for diameter, at near equal efficiency!!!!** We cover ***thrust required***, **activity factor, airfoils, blade thickness**, better skipped at first, **but necessary as you expand your grasp.**

The airflow geometry drawings, the vector diagrams are particularly important to the basic insights offered in this advanced chapter so they'll be an early and very important part of this advanced chapter. *I expect you will find they'll be the most mentally challenging part of this advanced chapter*, just as you probably found T = ṀΔV to be challenging at first. ***You can of course skip the challenging geometry*** as you could skip T = ṀΔV if it's too much for you, ***but*** it's very much at the heart of really understanding props, so it's *very worthwhile if you can get it*. ***Go for it***. **As usual we'll help, key's nailed, emphasized.**

Please understand the risk we take in this book, ***and why***. It was tempting to do the cartoon filled, "dumbed down" broadest market book, which I knew would be the biggest money way to go, but we *purposely* didn't do that. We didn't care about money, profit was not the game here. We wanted the book that **gets you the straight, factual, professional insights**. There are plenty of books, some actually very good and helpful broader market books that properly go after the biggest market, the most dollars. The problem is that there are **no other books that take you right to the really correct technical answers, the real truths**, the true heart of the matters that the few best pro's know --- and in some cases past what pro's know, because we've dug deeper, **tied more together**, like ***an integrated grasp of plane, engine and propeller logic*** ---- and *explaining better what the professional papers don't come close to making clear*.

In skipping much, assuming much, not digging, not tieing everything together adequately, not making everything clear, the pros can miss a lot too, not understand many important points themselves that get missed, not really grasp it all well enough. Since we've dug through it all, tried to see how to get the real meat and truth over to the pilot who really wants to get it right, it's been ***our objective to write the book with the real truths***. If you have to sweat a bit over $T = \dot{M}\Delta V$, or the coming geometry, it's worth it and I'm betting you can get it. If you can't get it all, hang in, because you can still learn plenty as you go, and in the conclusions. **Our objective is to let you go deep or shallow, as you wish, but always lead you to great insight, clear conclusions** *that everyone can understand, specific, clear!*

This *is* the Advanced Chapter, *that goes for a pro's insight*.

Many pros will gag at the super emphasis, the purposeful repetition, but they're ***not*** the central targeted audience. We're ***trying to help the deserving guy who wants to learn, but needs max. help***. In fact, ***pros waste huge time on the too esoteric papers, and they too can find the real truth here much faster!***

THE WISE OLD BIRD

Rigid Vortex sheet. Constant r tan phi. Constant dT/dQ vs r. Betz Loading. Simple helical pitch of the air inflow, outflow. What's all that stuff? Can I understand it?

A ***necessary*** game in teaching a professional grasp of propeller logic to learners is to *not* drown the new guy with way too much too fast --- and to find "easy enough" ways to teach a subject that has math and interrelationships that took our most brilliant analysts from 1865 to 1934 and 1948, one of the world's most complexly interrelated 3D analytical problems. **Then**, we feel the best way is to **take him right to the heart of the matter**, so he **immediately sees the real core of the logic**. Then, it's the time to go back, take him through each key step of the subject, with enough *chances and time*, super *emphasis* and *help*, that he really has a chance, the time to get it. ***Great incisive conclusions*** insure that *even guys who have to struggle can get real insight*.

You can *best help me help you*, ---. by figuring out how to best read this book --- *at your level*!!!

In this chapter there is a **fundamental change**. We're going to be dealing with **heavy loading**, where there is a **helical vortex sheet** ***moving significantly faster in and out of the prop***. The ***whole game is*** --- ***to get the air inflow and rotation correct!*** Betz figured out the proper objective to design an ideal prop in 1919, in the time of Prandtl - Munk, but it took their wing theory for Glauert to *get a good approximation of the air inflow correct*, by 1934. Theodorsen achieved his ~ exact analysis by 1948.

BOOK II --- CHAPTER 3 ***The Graduate Course***

ADVANCED PROPELLER LOGIC

PROPELLERS FOR THE GUY WHO WANTS TO SEE IT ALL
- The Explanation That Has Never Been Offered Before -

Like Babe Ruth pointing to the left field seats, **let's challenge each other**. -- I'll go for the ***most important, incisive, core pages ever written on the never adequately explained subject of propeller logic***. -- **You go for it**. I'll help you **actually get it**

The simple, specific technical objective of ideal prop design, is to configure a prop to have **constant dT/dQ**. That is, simply, **a constant ratio of Thrust vs. TorQue (or HP) at every radius**!! That's a -- **constant efficiency at every radius**, ***every radius as good as any other radius****, max efficiency. It's a huge **Surprise**, **a huge insight**, quite *counter intuitive* if you thought tip loss killed the effectiveness of the tip area, or saw very steep inboard blade angles hardly pull forward, steep airfoil angles appearing to produce more resisting torque, but quite weak forward thrust.

*That assumes the ***simple case*** --- ***with no blade drag***, -- **but we add low Rn drag*!!!***

Betz saw in 1919 ***that is possible***, and ***how to do it***, a ***"rigid vortex sheet"***, **a constant r tan ϕ_1 at each radius** --- **simply a *perfect helical pitch inflow of the wind*, (a perfect screw surface), *in, then out*, *stretched by the second $\Delta V/2$ to ϕ_2's*** -- by **proper loading!** Phi, ϕ_1 is just the angle(s) *of the wind* vs. *the prop plane* that the prop blade (***at each radius***) sees coming at it. ***Helical ϕ_1's + $\alpha°$*** sets prop airfoil angle(s) β, vs. the approaching wind. You'll learn it ***combines*** the ***air stream rotation*** with the ***weak ability of the inboard radii to move air axially***. **Easy enough, it sets up the *correct math***

The tricky part: Helical pitch of ϕ is real, makes the math correct. *But the axial ΔV varies vs. r!!*

Betz Loading is *achieved*, by **setting blade angles and chords** to get **Wind** inflow vectors, **W**, **with simple helical ϕ_1 pitch**, **same pitch at all radii**, adding, **a constant angle of attack**, the simple std. C_L case, --- shaping the blade with the **proper chord at each radius** to actually give "**Betz Loading**" -- **which makes helical airflow**!!! Thus, understanding the geometry of the **air inflow**, **stretched outflow is extremely fundamental and important**, so it should be, and will be, an early key and fundamental in our explanation.

Helical airflow is just a perfect ϕ_1 screw surface at all radii - ***stretched by a second $\Delta V/2$, ϕ_2's***

Now you can imagine that the 3D math to get that air inflow and outflow precise to < .1 degree, took genius level math that had challenged our greatest analysts for 83 years, from 1865 to 1948. It's far more complex than our incisive explanation. We can skip the wild math, but show you the **marvelous insight** it offers. **Constant Thrust vs. Torque** (or HP) - ***from proper loading, twist, chords, thus shape, helical airflow, is the core objective!***

In the initial Chapters for pilots, we used the Rankine-Froude "Actuator Disk Concept" to teach you Newton's Laws and the *tricky little central logic of props*, "**go fast, get a *Big* natural $\dot{m}$ and final mass flow rate $\dot{M}$ through the prop disk, thus a small required ΔV, thus a low $\Delta V/2V_1$ *axial* loss, thus high efficiency**. The **Actuator Disk Concept teaches the logic, *but* Assumes Constant Axial Velocity vs r**. **It's *not* that simple**!!

Forget $\dot{M}$ ΔV, Actuator Disks, *for awhile*, while we learn 3 D Rotating Wing Analysis.

Now, we're going to **think of doing a rotating wing analysis**. We'll soon **learn to get helical pitched inflow correct**, so we can **get angle of attack correct, *at the prop***. The ***twisting axial velocity* will vary vs. radius** -- but it forms a ***"perfect screw" shaped "Vortex Sheet" that moves back faster than the rest of the stream tube***. So we'll ***only get back to the "whole stream tube" after learning Theodorsen!!!*** Glauert and Theodorsen both recognized that the key to ideal analysis is to ***analyze for Betz's helical pitched airflow sheets, far back*** --- (even if it does ***not*** exist in perfect form far back). **That gets the math correct, all the energy put in at the prop!**

Theodorsen solved it for exact 3 D Flow, feeding the tip vortex, Heavy Loading, Hi Advance.

But, see Tibery and Wrench Kx loading above .5 λ in 1964 vs. T T's Voltage field tests.

Theodorsen finally got it essentially perfect, (if you don't try analyzing it *off its ideal design point*, or consider the fuselage forcing air forward, causing a slowdown). **Separate analysis does those.**

There are some imaginary parts about it, but the math stays correct, you'll see! .

Now the final initial key to your professional level insight *will be* to **understand what "Betz Loading" really is**, and to also understand that vs. the big surprise that the thrust vs. radius plot ***will*** ***show you the inner radii of the prop can be near useless!!***

A picture plot of Blade Shape and T# vs. r" will show you "Betz Loading", p.147 I.

Confused? Now, in case this explanation, which seems fantastic to me does ***not*** seem great to you ***yet***, don't worry too much. Here, **we're purposely showing you where we're going**. As usual, we'll come back and explain it in more detail, so you'll have more opportunity to catch on. We anticipate that we have to say each item a few times, a tad more and deeper each time to give you time and extra chances to catch on. Hopefully, you'll get it, maybe with some gaps, either new terms, or concepts, and the gaps can go away with more chances to get it. We ***all*** need time and repetition to catch on to new concepts. *We repeat to help*. If you never do catch on to *some* parts of it all, don't worry, we always lead you inexorably to understandable conclusions -- so everybody can learn plenty, by just hanging in.

Theodorsen, who scares almost everyone because his book is complex and difficult to understand, *even for pros*, is, in fact, much easier to use than is generally understood, simply because he uses "**K(x)** ***circulation blade loading factors*****", vs. radius and Advance simply from his charts, to load the prop to get "Betz loading"**. His **mass coefficient factor Kappa**, κ, relates the actual ***average*** axial velocity of the ***final tube***, to the axial velocity of ***his faster, heavily loaded, higher advance, vortex sheets***). -- His math, which is essentially 1929 Goldstein math, properly considers the stream tube and vortex sheet "far back" - includes a small change in diameter, higher pitch for heavy loading. *Both Goldstein and Theodorsen were very, very smart!!*

(Kappa Shows as Chi, χ in Andy Bauer's Analysis, **Theodorsen Greek Letters easily confused**)

Tibery and Wrench at David Taylor Model Basin, in 1964 got higher Kx Blade Loading, above a .5 λ, than Theodorsen's 1948 Voltage Field Answers, most probably correct.

Glauert ***can*** get "**Betz Loading**" ***only if*** his math is set up to get **constant r tan ϕ, *which is just,* helical inflow, constant helical pitch of ϕ_1's --- *then stretched helical outflow, stretched ϕ_2's!* He uses *Prandtl's "F" (fraction) Factor* to account for the blade's interaction with the streamtube, *prior passes* at any point on the blade radius** (**forming the rigid Vortex Sheet**) and **tapering the calculated thrust vs. radius for tip loss!!!** P**roper chords and C_L, thus angle of attack, $\alpha°$** do it. ***Sketches show the geometry***.

Geometry looks at the prop blade like a wing half. Forget the stream tube for now!!!

Is All The Complex Math Worth it? Yes, Really YES!!!. High pitch props can calculate to be 90+% efficient. NACA's best postwar tests with small drive housings confirmed 91+%. I was after the truth of the subject, and I'll ***show you the truths!*** I'll show you a stream tube can quickly become a real mess behind a prop, especially on a poorly designed very unsymetrical nose, which makes it look like these guys were dreaming with "a rigid vortex sheet, far back". But they were very smart guys and if **you put all the energy in at the prop, you get the right answer, *if* you've done the math precisely correctly, even if the tube becomes a mess after it leaves!!!!**

The Tip and Root Radial Flow actually breaks into TWO Separate Vortices!!!

There can be big *extra* propulsion efficiency losses, especially on the old, slow, poor plane designs. Gus Raspet tried to teach us that in the 50s, the same truths we're finding with zero thrust glide tests. **But the idealized math works, and teaches us the logic of what's happening**. We're just getting you started on a pro's insight here, just quickly getting you to key central truths of the big picture. Soon we'll go back, ***start teaching you the airflow geometry*** that can give "***constant simple helical pitch of the wind inflow ϕ_1's, then stretched helical ϕ_2's outflow***", thus ***proper blade loading vs. radius*** ---- **the best prop for any application** --- with the ideal twist, chords, thus blade shape, a **constant Thrust vs. Torque at all radii**, (not considering blade skin friction drag). **And, we'll teach you a lot of practical insight and great conclusions as we did in the initial chapters**

Betz's *Rigid* Vortex Sheet, Just Pure Helical Airflow Pitch.

You'll soon see the Rigid Vortex Sheet concept is tricky, because the inboard ***axial*** velocities look too weak for the "Rigid Vortex Sheet" to hold together! ***We'll explain soon how it twists into a perfect helix.*** But **I'll bet a picture of *what this example of simple helical pitch looks like* will help you better visualize it. It's *simply 2, dual, merged, helical screw surfaces*, from tip to centerline**, ***stretched, +ΔV/2***, ϕ_2's. It's actually ***two merged*** Archimedes' Screws, **one from each blade**. He died in 212 B.C., and they're ***still*** used in a pipe to pump irrigation water in Egypt. You can actually make one in glorious 3D Technicolor, by simply bolting up a stack of ice cream sticks. It makes quite a nifty piece of modern art, varnished, the natural wood color variation making it look quite nice, a creative 3D art form better than you see in many museums. We have an interesting little art collection and, laughingly, this became part of it. *Few will realize what it is.* **We're simply setting up the math to do this. That's valid, *but* more happens, it also breaks into tip and root vortices, but the objective of the math is simply to set up the air inflow *at the prop* this way, ϕ_1, but next, the stretched ϕ_2's, you'll see. Interestingly, notice how the *dual trail* from *two blades* makes it *look like the pitch is half what it really is -- LOOK, realize an RV prop pitch is greater than its diameter!!!***
You'll soon see *it really isn't rigid*, the inner rotation faster than at the Tip, at a smaller radius and circumference!!! √

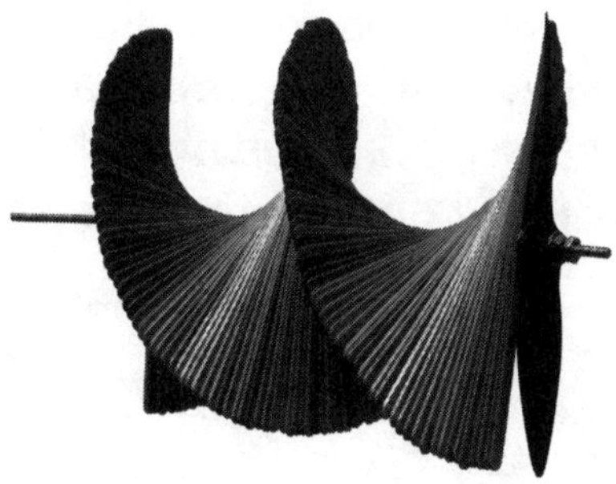

It's precise SHAPE and *Twist* --
That sets up pure Helical Flow!!!

fig. 1 An RV prop model and its theoretical *Dual* Rigid Vortex Sheets.

Betz's Rigid Vortex Sheets --- **An imaginary way to set up the mathematics.** Think of an instantly curing helical plastic sheets with downwash from *each* blade's trailing edge, or **a thin wire of prop diameter size, rotating forward at plane speed --- but *with extra pitch, slipping, feeding the 2 blades a ΔV/2 extra velocity, throwing two* Archimedes' Spirals rearward, *adding the second ΔV/2*!**

Grasp this concept, because it helps you visualize the *two bladed* Archimedes' Screws in the stream tube. That **works mathematically, even for Real World Heavy Loading, where the spirals are moving back behind each blade faster than the rest of the stream tube, a wild proposition you can see!!!**

Later we'll teach you how Theodorsen's *Kappa* gets *an avarage ΔV* from Heavy Loading!!!

I'm ***not*** going to emphasize the "Vortex Sheet term" too much, because it can confuse some, because you'll learn the axial velocity vectors are *not* equal at all radii, so the famous (2) ***rigid?*** vortex sheets don't *seem* to stay together. I'll explain later how the rotation twists it into a helix. Instead, I'm going to emphasize that we want to ***set up simple helical inflow of the airstream, at the prop, in, then out, stretching the helical pitch, the real objective* --- *easy enough?***

(We *will* explain *how* the airflow twists into a perfect helical screw!!!)

Helical Pitch: Helical pitch is so fundamental, we simply must repeat the easy, basic **helical pitch *blade angle(s)*** layout, $\beta°$, that modelers learn to ***advance all radial stations*** of the prop ***equal pitch***, inches or feet, through a soft solid, vs their ***circumference***. Once grasped, ***a calculator with Trig accurately nails angles***. **It's simple: All radii,** (circumferences) **simply go up the proper *ramp angle* to *all* reach the same pitch, or advance, in a solid.**

There is "no slip" moving a Blade Helix through a solid.

Helical Pitch **(Shown for a 71" Diameter, 51" Pitch Luscombe Prop)**

Tangent $\beta°$ = Pitch / Circumference at each radius

Don't fail to get a feel for the real examples, real numbers shown on the Ch. 2 II examples. on page 41 II

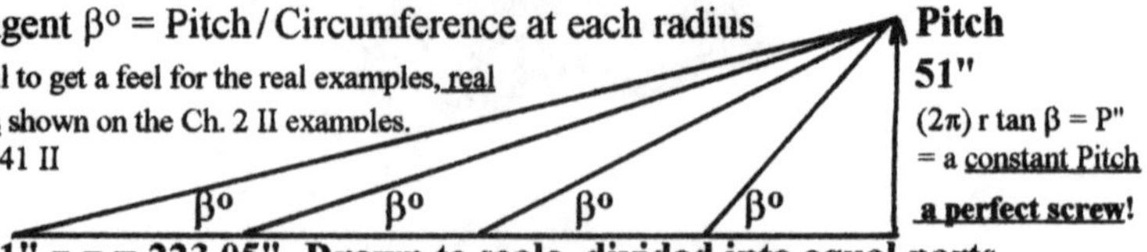

71" x π = 223.05" Drawn to scale, divided into equal parts.

fig. 2 **Angles, Circumferences at Several Radii.**

Blade angle, Beta β, Angle of attack, Alpha α, Wind angle, Phi ϕ

Aerodynamicists, logically set up **lift and drag force vectors**, (vectors have an angular ***direction***, and a ***magnitude***, like pounds or speed), **perpendicular and parallel to the airstream**, which naturally is usually considered to be horizontal. But **prop airflow angle** is called **phi, ϕ**. (Look at the following propeller airflow geometry fig. 3.) An airfoil is typically at some **angle of attack alpha, α,** to that airstream, so a **prop blade angle beta, β, is simply the addition of alpha and phi**. $\boxed{\beta = \alpha + \phi}$ That, of course, is **really basic to propellers**, get it, remember it. **Look at the sketch**.

Of course the ϕ and β angles vary at each radius.

You must already recognize that *the basic propeller game* is to get a helical phi, ϕ, a simple helical flow of the airstream, *not* the blade, *not* the blade β°, beta angles. With a small constant angle of attack, **Alpha,** α°, like 1.6°* **for a .55 C_L, or ~ 1.1° for a .5 C_L** on a prop blade, the blade is close to helical, but that different, on a simple case Betz prop, **1.6 degrees steeper pitch at all radii on a Theodorsen RV 6 prop blade, or 1.1°!**

*** Airfoils develope ~.1 C_L per 1° angle of attack, but typically start at ~ - 4 degrees α!**

Here Comes The Tricky Part, Understanding The Airflow

In the Basic Chapters the game was to get you to understand, and get past $T = \dot{M}\Delta V$, so you could see through the tricky way basic prop logic worked. Subsequently it got easier, professional insights and conclusions, without more brain strain. Now in this Chapter the game is similar. **Prop Airflow geometry** starts easy with the basic chapter's simple helical pitch layout β°, then a very important **vector diagram** of air movement, **the airflow direction and speed** to grasp what's really happening. *There are some tricky parts*, **that's why we teach it**, so you can really understand As before, *don't get scared away, we'll help*. The challenge will soon pass. You're far better off if you get it, but *if you don't get it all, there is still* ***plenty of great, easier insight*** The whole game in this book is to not be afraid. If you keep trying, we'll be there with help. Everything is easier in hindsight.

If you've never dealt with vector diagrams before don't be afraid. The **Vectors just show the *directions*, angles, of all the separate elements, the *speeds* here represented by the vector lengths, to scale.*** **So it's a picture** --- **much easier to grasp** than math. We'll explain the tricky insights in words, and how it all works. You'll see it is nifty, tricky, fundamental, great insight.

*** Vectors, usually to scale. a and a' are exaggerated here so you can see them.**

Blade *Inflow* Geometry:

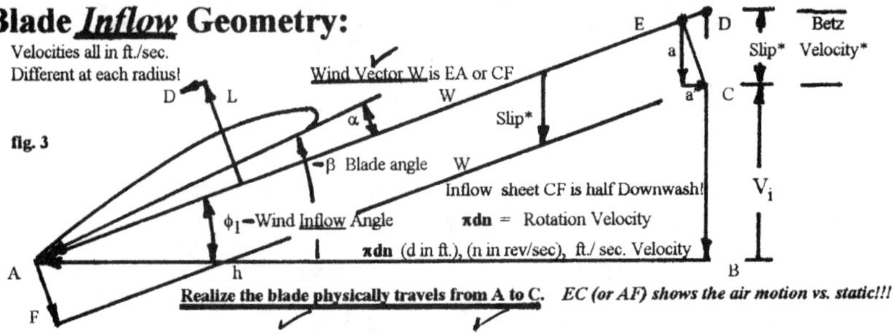

Glauert correctly established the axial and rotation inflow of the **stream tube**, a hugely important step in the developement of Classic propeller logic, using his fundamental ***a and a' factors.*** Think of Glauert as **calculating a rotating wing with a and a' to get pitch and angle of attack correct!!!** **He uses Prandtl's F *fraction* Factor to take care of *both*, 1. The interaction with the streamtube and the prior passes of the blade, and 2. Tip loss, tapering the calculated Thrust along the blade to zero at the tip.**

a is a ***ratio*** of **half** ΔV, to plane speed, V_0, or V_1 the slowed inflow speed. **a' is a *ratio* of half the rotational ΔV to the props rotational V.** Thus, **a** = **ΔV/2V_1** --- and **a'** = **Δπdn/2πdn** **That looks complex but they're just simple little ratios of the axial *inflow only*, a, to the V of the air coming at the prop, before the prop affects that speed, and a' the *inflow only rotational speed ratio* - *to the prop rotation speed*. Note *everything is different at every radius*, and *with heavy loading* a and a' are *both faster* than in the MΔV concept with uniform flow velocities!!!**

Yes, this is complex at first, but noodle it out. It's the key to the Real World of Props.

Grasp how the **inflow ratio a times V_1**, and the **rotation ratio a'** times **πdn**, combine with **the plane's velocity** V, (or the **slowed** V_1), to set up the **relative wind inflow velocity vector W** -- it's **angle ϕ_1, vs. the πdn rotating prop plane. Setting up** ϕ_1 is **the real key to ideal computerized propeller design**. The computer ultimately does a Theodorsen solution of ***everything***, Phi, ϕ's at the core of it all **Realize that we're still setting pitch angles here**, and that next we're going to set up **helical pitch for ϕ's**, rather than the blade, but *now* we're **doing it with velocities**, ***not*** *in." or ft.' pitch.* (Recognize a and a' are ***velocity ratios*** to the V_1 and circumferential velocities, but simply multiply out to small velocities.* Grasp: **a** is like ***our*** *Newton's* **$\Delta V/2V_1$** ***inflow ratio***, (***but faster***, ***heavily loaded***) ***The hidden secret is, required thrust demands and sets the required inflow, and thus the needed a and a'.*** ***You'll see that after we add the second $\Delta V/2V_1$ the geometry shows us the downwash angle ϕ_2*** ***Big insight!!***

Grasp: RV 6 ΔV_{avg} only 5%,** ***inflow 21/2%. But, Heavy Loaded a, a' bigger, still small.

Downwash, let's **understand it.** Mentally draw in the line from ***A to C, the actual prop path***. Notice that establishes a wind line with **no inflow or rotation, no thrust**, *because* it collapses our a and a' triangles to zero, goes right to the top of the **V_1** velocity vector. ***Realize that a symmetrical airfoil set with no incidence in that wind line busily revolves, but at zero thrust!***

Soon we'll show you **zero thrust is ϕ_0, aligned with AC.**

Now look at our wind lines with thrust, which **have *inflow*** and ***rotation*** **increasing the ϕ_1 angle of the two parallel wind lines**. Those set **half the final downwash angle increase to ϕ_2!** The interesting thing to see here is that there is a **real downwash**, and **an angle formed**, ***because*** **there is inflow and rotation**, and the resulting ***wind line W sets half the downwash angle***. ***(Surprise,*** **to the airfoil** now set at an angle of incidence to W, it **looks like it's *flowing up a tad* at the trailing edge!)** Notice that to a cambered airfoil, which can deliver lift or thrust at a small negative angle of attack, that effect could be hard to see.

Now there's one detail that our graduate students, indeed everyone, needs to know to really understand this crucial sketch. Aerodynamicists set everything up perpendicular and parallel to the wind line, so this sketch is done with EC, **the resultant** of the a and a' vectors ***perpendicular to the W wind line***, and you can study the geometry and see that *positions the W vector and ϕ_1*

But actually, EC is a tad off perpendicular, *with drag*, and 3D math, a detail we'll explain later --

We use the term "Tippy Top", Corny, but it helps people "Get it, Remember it", *Important!!*

Slip: Notice how a and a' jack up the wind line W, (it ***is*** EA or CF), to where the "***tippy top***" at **D is above E. DC then is the SLIP of the Wind Line W, now *overpitched* to pull wind in at ϕ_1!!!**

We're going to call that Slip the Betz Velocity. It's held constant!

You can better appreciate this geometry sketch when you realize it took 69 years, needed Prandtl-Munk to finish the wing vortex theory, and more. Glauert understood it, applied it to prop logic, finally got inflow and rotation ~correct, an 1865 to 1934 toughie.

Now we're going to move on to a very important sketch that shows how an outer radius and inner radius prop station interrelate, with insights of Extreme Importance, *Really Basic!!!*

I first saw this great sketch in John Roncz's Voyager prop SAE paper.

Constant Helical Pitch of the Wind *Inflow*, ϕ_1 **(constant height to D)**

fig. 4

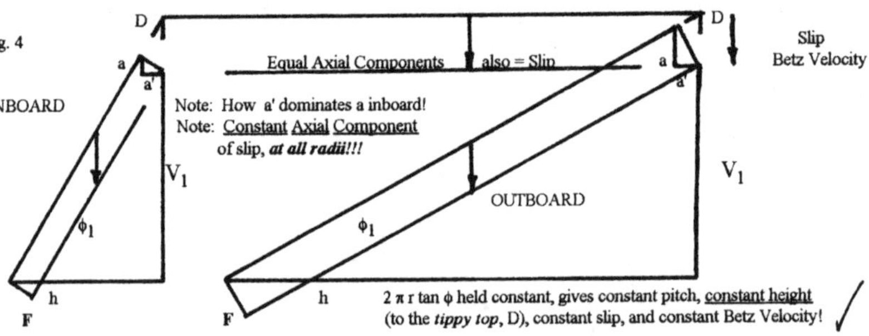

The Geometry *that makes* Betz Loading happen - Ideal Props

This is the ***"Rosette Stone Sketch"*** that ***makes the central objective clear***. I've simplified it, leaving all extraneous labels off for clarity, so you can see **the simple key point, the SLIP, Betz Velocity, is constant at every radius**. That results from **constant simple Helical Pitch of ϕ_1**, that is ***set up by a and a'***. **Notice that a is very weak inboard, but a' is very powerful inboard, reversing roles in jacking up Phi, ϕ_1, due to the geometry, to hold helical pitch of the airflow, with enough excess pitch to pull in a and a' the needed amount to get T#!**

Math shorthand holds r tan φ constant. That times 2π is constant *actual* pitch of ϕ_1.

This *constant slip*, *constant helical pitch of phi*, ϕ_1, *when made to happen* gives *Betz loading*, an *ideal prop*, *min induced loss!*

It's that simple guys!!! If you load a prop correctly, per Betz, you produce simple helical flow of the air inflow ϕ_1, (*and* you'll soon see, ***the outflow*** ϕ_2, because all the energy is already into the prop. The slight pressure behind produces the **second** $\Delta V/2$, the final downwash ϕ_2) **All the wild math done by everyone for 140 years is finally aimed at that simple result**, so ***study these basic sketches, grasp them!***

Of course it's hard for you, everyone, at first. — Me too!

THAT SETS THE MATH METHODS UP CORRECTLY.

If you load a prop correctly per Betz you get simple helical pitch of the airflow --- and if you actually have that simple helical pitch result, you have loaded the prop correctly.

Constant pitch of ϕ means constant height of the pitch triangles to D — the tippy top.

Realize the huge importance of that insight, because if you know that's the simple objective, you don't have to know the Doctoral level math, you only have to realize that's what they do to get an ideal prop, with minimum induced loss!!! EASY!

For any given Thrust, Aspect Ratio, Diameter case:

Helical airflow is **the result** of a correct Vector Diagram, the **correct angles**, with ***correct chords loading*** to a ***correct*** C_L, (meaning a correct angle of attack). The ***simple standard case*** Betz or Theodorsen ideal prop would **use a constant angle of attack, $\alpha°$, thus a constant C_L**, and the **math then provides the proper chords** to **give the correct Betz loading**, and thus the ***simple helical pitch***, all matching the thrust required, (at the speed, RPM and altitude density applicable to the case). Think it through and you'll see it's ***logical, something you can grasp***. ***Recognize when you've set the chords you've set the blade shape*** (You can use a little **less** C_L matched with a little **more chord**, ***but more drag***, or vice versa), but **let's learn the simple optimum case *first***.

If you have trouble with the concept of chords controlling the load, *help is coming!*

The concept of SLIP can be helpful. *Recognize* that the prop must screw ahead fast enough to **actually go airplane speed**, ***but also*** be **overpitched to pull in air at the needed $\Delta V/2$, and throw it backwards at a second $\Delta V/2$, now *at the full ΔV to make the needed thrust*, and Betz's *rigid* vortex sheet. Once you get it, it's simple enough. Hang in, Learn the Diagrams**.

The Downwash Vector Diagram, ϕ_2 Throwing Air Rearward

IMPORTANT Realize the Prop Blade is Pitched vs. Point D, ϕ_1, LineAF, w/2, + α°.

If the **game here is simply to use a precisely overpitched prop to pull in and throw back air, in a precise helical manner**, it can be very **helpful to turn our vector diagram around** so we can **more easily visualize the air flowing in, at nominally half the downwash angle**, (aerodynamacists say to the quarter chord), ***then speeding up a second ~ equal amount***, going ***out, past the trailing edge***, **finally achieving the full ΔV, the final downwash vector W_2** ***and angle*** **ϕ_2 downstream!** (A wake velocity rake test showed it all occurs by ~1/2 diameter downstream.)

ϕ_0 represents zero thrust

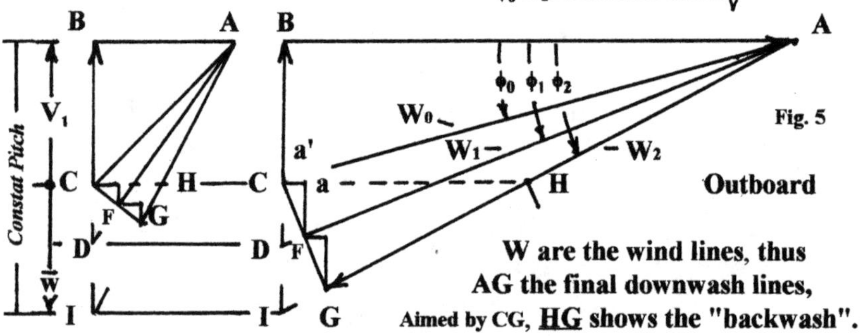

(Theodorsen's w is important, *a more accurate than* 2x Betz velocity, heavy loaded)

Recognize that the plane velocity, (finally modified for slowdown, V_1), is represented by BC. You can see that the airflow below the horizontal line at C shows the **downwash angles**, ϕ_1, then ϕ_2, **and speed, HG**, --- to an expanded scale so you can see the details. **LOOK at how the "a factor ratio" inboard is far smaller than that outboard!!! Notice that** ***inboard*** **the a' becomes much more important than the a in jacking up the phi angles, ϕ_1, ϕ_2, to set up constant helical pitch, as shown before inboard!**

IMPORTANT Realize the Prop Blade is Pitched vs. Point D, ϕ_1, LineAF, w/2, + α°

The **airflow is** ***deflected*** **to establish helical pitch**, by the **combination of a and a'**, as before in our other sketches, but is that helical flow maintained as a ***Rigid*** Vortex Sheet? Look. The **vector direction is correct**, but *the magnitude is just too weak inboard* to make a ***Rigid*** Vortex Sheet real - ***but that's OK***.

Is the Vortex Sheet ***Rigid***, Real? Not Without Brain Strain!

The fine print for the brave souls who want to look close, get it all.

I dislike, deemphasize the Rigid Vortex Sheet concept, because it's just too hard for many people to see through. Simply setting up **helical Pitch in, stretched out**, is the **easiest, least confusing way to understand**. Let me explain **what's happening physically to the air particles. Here goes ----**

We, of course, normally think of the prop and plane roaring through air, or the air roaring into them, like in a wind tunnel. In fact, **the prop is twisting through *static air***, and **leaves *2 relatively slow screws moving back!*** Look at **vector CG in fig 5, ~ *perpendicular to the wind line, the resultant of the a and a' inflow caused by the prop, with nominally twice that flowing out deflecting* Vector HG**, **the *downwash, backwash flow***. CG is exaggerated and thus the *deflected* HG and its angle are exaggerated. CG is rotating more inboard than outboard, but lagging axially inboard, so the outflow wind lines W_2 are all moving axially and also *wrapping up into the two perfect screw surfaces*, finally at a stretched constant pitch as the full ΔV is implemented, constantly being formed, with just the right geometric combination of the a and a' components to stay a perfect screw.

Theodorsen's Ao shows inflow is not quite equal to outflow. Bigger if Low Pitch

It really isn't a *rigid* sheet, because point G outboard is moving axially back faster *physically* than inboard, but point G inboard is *rotating faster*, and at *a smaller radius and circumference*, than outboard to form the perfect screw surfaces!!! In other words, the axially slower inner radii are continually spinning ahead fast enough to keep up with the earlier outer axial flow so it keeps on perfectly forming so it looks just like the double Archimedes screw photo. Not Rigid It looks Rigid. It can be hard to see without good spatial visualization, so I'd rather go with the sucker stick Archimedes Spiral photo, the concept of a perfect screw surface, constant pitch, formed by the geometry drawings.

It's argued the air *actually does rotate, not move tangentially*, guided by adjacent air!!!

!!! HG is the ***airflow left behind the prop***, **flowing backwards and rotating --- at *only* a small percentage of prop velocity. HG's are *not* physically moving even close to propeller speed,** because (double) **a and a', and HG** are small percentages of the forward and rotational prop velocities --- **even heavily loaded, and the stream tube average is much less!** **p. 94 II has real data ---**

The vortex sheet looks real, looks rigid, but is twisting, stretching to look rigid!!!

Recognize the Pitch to point I stays constant!!! *Forget all the 3D complexity if you wish*, if this gets too deep for you. The Easier way is: Just get the concept of a *perfect screw surface in, stretched, out*!!!

Bringing Everything Together (Forget The Vortex Sheet)

Everything is always more difficult to understand the first time you see it. **Try to get the Geometry**. **It is reasonable enough after you have time to let all the new insights soak in**. The model builder's *helical* β *prop blade pitch angles* sketch **simply flies a perfect helix screw *through a soft solid*, no slip**, all radii going circumferentially up their perfect ramp angles, to move forward exactly the same pitch distance, **EASY, once you see it**

Only go as deep as you wish, Theodorsen solves it all ---

The **air inflow geometry here simply is *aligned*** at the prop, **to do the same thing**, aimed, to **move Betz's helical phi, φ angles of the wind** through **air, *with slip*, *throwing* a second step of steeper pitch as the second half of the ΔV happens** behind the prop blade to arrive at the final pitch, **throwing air rearward, in two steps, HG**. That satisfies Betz's **constant r tan φ, phi**, just his math way to say the **pitch triangles** to the ***tippy top, point I*** (corny, but people understand) are ***the same height*, *the same pitch at all radii*. It's a very simple concept, once you get it, and that's what sets up the wild math correctly!!!!**

Slipping, throwing helical airflow is only a bit more complex.

We need to do it in **two ΔV/2 steps, *because* Glauert calculates everything at the prop**, at ΔV/2 where the α° **is set**, the **blades being designed for the inflow angles there**, all the **energy going in there**. Theodorsen actually considers his math far back, but that's due to more complex math reasons. ***Note we design for blade, inflow only !***

Grasp that HG's are the backflows, twisting into a helical surface!

Of course the airflow is a step more complex, an airfoil moving air perpendicular to its wind line, drag free, weak a factors inboard, *but rotating more to hold constant pitch*. The math handles all that complexity and more. *Catch on the easiest way at first, even if you're an engineer*. The game is simply ***setting up, getting constant helical pitch of the airflow, a perfect helix, a perfect screw surface***, and you have an ideal prop. ***It's helical going in as inflow, stretches to a steeper helical pitch behind***.

Making Some Minor, But Basic Details Completely Clear

Be sure that you realize that the **final downwash angle** is the **difference between ϕ_2 and ϕ_0**, the downwash ***at*** **the prop is half that, at $\phi_1 - \phi_0$** feeding the prop as inflow. There's no downwash, of course, at zero thrust, ϕ_0. Don't misunderstand. When we say that downwash nominally doubles, we sure don't mean that ϕ_2 is twice ϕ_1!!!! Just the difference from ϕ_0 doubles.

We show **Actual Downwash Angles** on **Data Page 94 II**

When we showed you the first airflow vector diagram, fig.3, we told you that an *airfoil moved air perpendicular to its wind line W*, so you could grasp the geometry, easily enough. (And Aero's logically set the vectors up perpendicular and parallel to the airstream). ***(But then told you in fine print that it wasn't exactly perpendicular, to be accurate).*** Realize that *drag cocks that vector a small angle*, about 1° and the ~ exact interactive 3D math can alter the exact alignment. Also, in our greatly exaggerated geometry sketch, you can see that CG *cannot* be precisely perpendicular to *both* vector AD and AG, that an insignificant difference in small angles, a *nit*.

Theodorsen uses his $\overline{w}$ **factor, ~ 2 Slip -- for exact heavy loading** -- *"Far Back"*

Theodorsen's A₀, precise math shows that the *first increment* of *both velocity and rotary change*, fig. 5, are *not* exactly equal to the outflow, ***a tad more or less.*** **Theodorsen's *heavier loaded* $\overline{w}$ *is nominally two times the "inflow only" Betz Velocity. The Vortex sheet* behind the prop blade *moves back faster than the rest of the stream tube, at higher pitch than average!!!*** Trust that Theodore T. did it all precisely correctly, 3D, because he did. Let the computer handle the really nit picking details --- so we don't overly complicate the explanation, but let you know you can trust Theodore Theodorsen's great work. It's proper to bring up nit picking details like ***not*** **perpendicular**, ***once***, or it confuses some guys smart enough to see worrisome details. Trust T.T, he's very, very good! His **K(x) loading charts are listed to *three significant figures***, and his work is precise math. ***If you get thrust, velocity and angle answers equivalent to 1, 1 1/2%, three significant figures, you've broken the bank at Monte Carlo!!!!!***

Realize T. T. did a Votage Field for Kx above a .5 λ, Tibery and Wrench, 1964, got Higher kx

The Concept of Loading by Chord Length

Does the idea of loading a prop by chord length confuse you? You know what **Betz loading** does, it **makes helical airflow** **into then out**, *stretched, just behind the prop blade*. That **makes a constant dT/dQ, max efficiency, constant efficiency,** **If Drag Free!!!** ***Betz loading is set up by chord length,*** ***(as well as*** C_L***, naturally)!*** ***Does that confuse you?*** ***Don't let it, -- but there is a Surprise!***

Betz Loading is established by Twist and Shape, meaning α° and Chords!!!

Think of a wing or a prop **replaced by a wire of zero chord**. Obviously that's not going to make any lift or thrust. But if we start adding an airfoil with chord length, we do start generating lift or thrust, a little bit with a narrow chord, lots more with a wide chord, just like adding wing area. Then, the more C_L, the more lift and downwash. **Do you see that we can generate more lift with** ***either - more chord, -- or more angle of attack!***

Surprise: Angle of attack, α°, thus C_L**, increases the a and a' pulled in, and thus downwash and lift or thrust created, but more chord, Span, same AR, increases the lift, thrust,** ***BUT not downwash***! The wing formula: DW angle = $C_L / \pi AR$ Radians, shows the C_L *as well as wing* aspect ratio controls downwash, ***zero DW at infinite AR***, but that ***cannot*** *be reached*! *A key insight!!!*

At Infinite AR, D.W. goes to Zero. Real Wings can't have Infinite Span, or Zero DW !

That's tricky enough we'd better stop and think a minute. A 747 size elliptical wing, at the same V, AR and angle of attack will have the ***same angle of downwash*** **as a ~Spitfire wing --- but it's affecting a lot more air, gives a lot more lift. Tricky, but it can make sense once you grasp it. Props are wild. That's not my fault, --- but I can sort it out for you.**

We'll learn a Betz prop ***will NOT have constant DW, Geometry gets involved!!!***

Prandtl's Perfect Wing, Elliptically Shaped and Loaded.

A perfect elliptical wing has ***constant*** C_L, split half ***elliptical loading***, loaded by chord length, **lift at any span proportional to chord length α° and** C_L**, falling to zero at the tip, zero chord length!!** An ideal wing has **a constant downwash angle, controlled by** C_L**, and AR, min. induced drag.** ***Variable q props are different.***

Betz Loading --- Peek a prelook at the next few pages ---

Think of a perfect elliptical Spitfire wing. But convert it into a rotating prop, with **twice the rotational velocity at the tip,** vs. the mid span, **four times the q**, (if it wasn't also moving forward.) It's going to become **highly tapered**, even pointy, trying to maintain a perfect lift distribution, with that **big V^2 effect on q**. You'll see next, ***a standard Betz loaded prop***, with a **constant angle of attack and C_L** does have **a highly tapered planform**!

q Ratio - Luscombe 25 : 1, RV ~9 : 1, Reno only 2.275 :1 !!!!

The clearest and easiest way to give you a grasp of **what Betz loading is physically**, is to show you next, a plot of the **thrust vs. radius loading** and the **blade shape** that gives us that Betz thrust loading, and **simple helical flow in and out of the prop**, the **constant dT/dQ**, or HP, the constant thrust to torque ratio, **no inferior radii, max efficiency,** an ideal prop. Peek that look.

Remember constant dT/dQ *only* happens at Zero Profile Drag, (a few % off with it.)

We'll first look at **standard simple case Betz loaded props**, that is, a **constant C_L, a constant angle of attack added** to our **simple helical ϕ airflow**. There are **interesting insights, incisive ways to grasp what Betz is physically, functionally**!

The examples purposely don't use slowdown, to get pure answers.

Betz does ***not***, *cannot*, have a **constant downwash**, $\phi_2 - \phi_0$! It does ***not***, cannot, have a **thrust vs. radius that is constant, or proportional to chord length.** You can see those are not about to happen with a **highly variable rotational Velocity, Prop V = $\{$(Plane Velocity)^2 + $(2\pi r\, n)^2\}^{1/2}$**, (the Sq Root), a **huge variation of q,** dynamic pressure, **vs. r**, based on **the (Prop Vector Velocity)^2**!!

We have the **blade thrust**, and **shapes vs. radius** you see next. ***NOTICE!!!*** That ***downwash, dT/dQ, blade width, efficiency, maximize ~where ϕ_1 is about 45° on lower pitch props***!! *See p.146 I*
The ***max chord moves inboard on low pitch***, ***low advance*** props, ***more outboard on high pitch***, ***high advance*** props, **(but pulled back from the tip.)** The ***thrust maximizes at 70 to 80% radius***. The ***a factor is max at the tip, but the chord is narrowed to get the proper Betz T # vs. r. to size that a!***
The max chord moving inboard on low pitch props, makes them more highly tapered. LOOK!!! Page 146 I helps make this more clear!!!

<u>Shapes</u> and <u>Resulting Loading</u> of Constant C_L Betz Props

(These examples are *not* modified for Slowdown)

Luscombe 1400# GW	**RV 6 1600# GW**
73.8" Diamater vs. 71" normally	70 " Diameter, (or 72" is good)
151.5# Thrust (140# T at new η_2 OK)	**164# Thrust,** (156# T at η_2 OK)
.55 C_L 1.612 degrees alpha	**.55 C_L** 1.612 degrees alpha
2280 RPM 48.15 H.P. Alt Cruise	**2400 RPM** 82.9 H.P. Alt. Cruise
100 MPH TAS 85 IAS	**170 MPH TAS** 140 IAS
10,500 ft. Density Altitude	**12,500 ft. Density Altitude**
Ref Analysis 99.36 - 21-23 & 99.75 - 67,68	Ref Analysis 99.36 - 24-26

Realize Betz SHAPE, thus LOADING vs. radius are shown here.

Thrust vs. Radius #/ft. **Blade Planform** Chords"

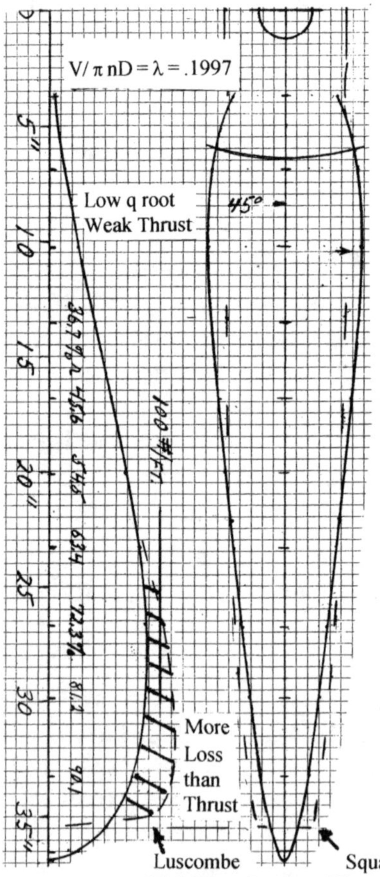

Optimum, High Pitch Propellers, a V / π n D = λ = ~.63 Have Symmetrical, ~Canoe Blade Shape, Max Chord 50% r., not excessively pointed tips, a pleasant, ***Nature's design.***

Thrust vs. Radius #/ft. **Blade Planform** Chords"

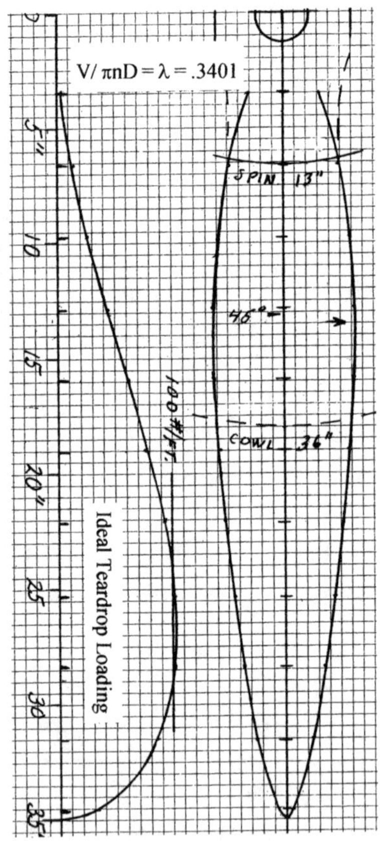

Square Tip Loading, Thrust won't happen, Losses will!

High Tip q, but Zero T# result *wants* Outer Chords Narrow for efficiency!!!

Ultimately Realize the Good and Bad of *half* Teardrop Radial Loading --

It's <u>Precise Betz SHAPE</u> and twist that gives <u>Ideal Loading vs. r.</u>, min Induced Drag Loss

We later design for Less Luscombe Thrust, thus a little Less Blade Width, Chord.

 * Remember, for Century 21, we can force Elliptical Loading, do a few percent better!

A Look at Near Equal Prop Thrust, Fast and Slow

The **required 163# thrust RV prop looks rather elegant**, I believe, **like something that ideal math, nature designed**, and noticeably different than the props we use with much wider tips. (Wider tips add profile drag, increase torque, get less RPM, less Thrust!!!)

Later, p.114II, we discuss final, lower Luscombe Design Thrust, not 151.5#, ~137# ✓

It's most interesting that due to poorer η, interference efficiency η_i, and scrubbing drag, that a Luscombe at a slower 100 MPH altitude economy cruise, with a ***gliding drag of 123 #, at a 1400# GW***, requires about 137# thrust, while a ***1600 # RV6 at 170 MPH*** economy altitude cruise, **only needs 163**, (148/.9039η_i). (You'll learn there's **more scrubbing drag, more interference efficiency loss η_i**). *Slow, narrow tapered chords*, the Luscombe Theodorsen prop prefers a little extra diameter, ~ 73.5" vs the normal 71" *to avoid extra blade width*, or high C_L with the highly tapered ideal blade. A wide tip normal McCauley prop, (see outboard dashed outline), with slowdown, not correct loading, highly variable C_L's, loses maybe ~ 10% extra efficiency.

Notice: **Thrust does *not* equal Gliding Drag -- significently more!!!!** ✓

We ***pay attention to blade width to minimize metal prop forging weight, and high AR is most efficient***. The wide chord inboard is a max width of ~ 6.8" wide for Theodorsen blades. My McCauley prop actually quits helical pitch at a 13.3", 37.5% radius, goes to *a constant 29º within the cowl width.* We'd narrow the ~useless, excess width blade inboard of **3/8** r, hardly any thrust loss, you'll see on the thrust plot. *And thrust is ~useless blowing **at** the cowl face.*

On p.117II the broader tip, mis twisted actual Luscombe prop has irregular C_L's a poor Design!

Hardly Any Thrust Inboard! Look at the dT/dr, ***thrust #/ft.*** plot for the **fairly high pitch ideal RV 6 prop**, the accurate result of a comprehensive Theodorsen analysis. It's **another major surprise** on propellers, **very low thrust inboard**, **even lower at high pitch.** ***Wow!!!*** Many of us might have guessed that a prop's inboard thrust was weak, *too slow, **low q, blades not pulling forward**, **less circular area***, but I'd bet this weak is a big surprise to almost everyone. The fun of all this for me is ***nailing the real specific facts, all the surprises, after 140 years now!!*** ✓

Key Technical Data, Theodorsen RV 6 and Luscombe Props

(But with *no* slowdown, thus Exact Ideal. Done as a study for comparison.)

RV 6 Prop -- 162.543 # Thrust .55 C_L 2400 RPM Cruise Alt. 12,500'
Diameter 70" Pitch 86.617" P/D 1.237 Area 2.0448 A.R 13.479
dT/dQ, drag free = .93341, see below HP = 82.891 Q = 180.717 ft#
J/π Design = .3401368 C_T = .054102 C_P = .065034 AF = 76.186
Eff. eta, η = .88895 (Loss .11105) Zero Drag η = .92601 Thus Drag = .037057
Velocity, V = 170MPH x 22/15 = 249.33 ft/sec ΔV = 9.239 MPH
Theodorsen $\overline{w}$ = .1566 x Kappa, κ, .34704 x V, 249.33 = ΔV = 13.550 ft./sec
A₀ = .078005 $\dot{M}$ = 12.2179 slugs/sec. (Calc. T = $\dot{M}$ ΔV = 165.5578 #)
(with drag)

% r	r"	T#/ft	C"	β°	dT/dQ	η%	DW°	a	a'	ΔHel
✗ 10 %	3.50"	3.772	3.053	76.383	8710	.8641	2.1212	.00538	.06901	**.81673**
✓ 18.9	6.615	13.169	4.931	64.384	.8947	.8876	3.3984	.01633	.05887	.32559
27.8	9.730	27.072	5.970	54.492	.9008	.8937	4.0126	.02841	.04769	.07737
36.7	12.845	43.917	**6.342**	**46.637**	**.9022**	**.8950**	**4.1641**	.03897	.03791	**.00535** ✓
45.6	15.960	61.762	6.243	40.464	.9016	.8944	4.0654	.04731	.03019	.03272
54.5	19.075	78.474	5.829	35.591	.9000	.8928	3.8548	.05364	.02432	.10686
63.4	22.190	92.056	5.217	31.699	.8978	.8907	3.6067	.05840	.01991	.19837
72.3	25.305	100.690	4.485	28.545	.8954	.8883	3.3570	.06200	.01658	.29249
81.2	28.420	**101.116**	3.627	25.952	.8927	.8856	**3.1213**	06475	.01403	**.38247**
90.1	31.535	87.367	2.574	23.792	.8898	.8827	2.9054	.06689	.01205	.46561
99.0	34.650	32.456	.799	21.968	.8869	.8798	2.7104	.06857	.01050	**.54117**

100 % 35" for **70" diameter** prop Ref. analysis 99.36- 24, 26 1-25-01 Prog. THEODT28

(at .85 η)

Luscombe 8E - 151.47 # T, (140# *final*) .55 C_L 2280 RPM Cruise Alt. 10,500'
Diameter 73.8" Pitch 57.01" P/D .7725 Area 2.1680 A.R. 14.13
dT/dQ, drag free = 1.4601, see below HP = 47.579 Q = 109.206 ft#
J/π Design = .199767 C_T = .042434 C_P = .031367 AF = 63.434
Eff. eta, η = .8490 (Loss .151) Zero drag η = .89693 Thus Drag = .04793
Velocity, V = 100 MPH x 22/15 = 146.66 ft/sec ΔV = 11.658 MPH
Theodorsen, $\overline{w}$ = .21476 x Kappa, .54282 x V, 146.66 = ΔV = 17.098 ft/sec
A₀ = .1100146 $\dot{M}$ = 8.86344 slugs/sec. (Calc. T = $\dot{M}$ ΔV = 151.546 #)
(wiith drag)

% r	r"	T#/ft	C"	β°	dT/dQ	η%	DW°	a	a'	ΔHel
✗ 10 %	3.690	3.954	4.725	67.433	1.3941	.8564	4.1960	.01846	.08652	**.41211**
✓ **18.9**	6.974	13.178	**6.481**	**51.295**	**1.4105**	**.8664**	**5.5007**	.04606	.06176	**.00533** ✓
27.8	10.258	26.076	**6.827**	**40.313**	**1.4103**	**.8663**	**5.4175**	.06701	.0429	**.00545**
36.7	13.542	40.697	6.510	32.865	1.4056	.8635	4.9130	.08040	.03096	.14140
45.6	16.826	55.574	5.947	27.644	1.3993	.8595	4.3629	.08882	.02340	.29860
54.5	20.111	69.176	5.279	23.840	1.3920	.8551	3.8715	.09427	.01852	.44232
63.4	23.395	80.299	4.580	20.969	1.3844	.8504	3.4156	.09793	.01524	.56546
72.3	26.679	87.454	3.864	18.734	1.3765	..8455	3.1082	.10048	.01295	.66912
81.2	29.963	**87.538**	3.082	16.951	1.3684	.8406	**2.8176**	.10232	.01130	**.75629**
90.1	33.247	75.443	2.165	15.497	1.3604	.8356	2.5727	.10368	.01008	.82999
99.0	36.531	27.970	.667	14.291	1.3523	.8307	2.3645	.10471	.00915	**.89280**

100 % 36.900" for **73.80" diameter** prop Ref. analysis 99.36 -21, 23 1-25-01 Prog. T-DT28

Later we design for19% to 99%, 11, 8 % Segments, and Less Luscombe Thrust.

Real Data -- Let's Look Close, Get Some Great Insight!!!

There are strong parallels and differences between a **Prandtl wing vs. a Betz prop, elliptical loading vs. the T# vs r. shown, (weak root, strong tip loading), elliptical chords vs. tapered tips,** ***each perfectly shaping the loading; cutting to zero at the tip,*** **constant downwash vs. Betz geometry controlled variable DW.**

Let's Look

Betz Loading The **T#/ft column** shows the data for the *Thrust vs. Radius Plots* shown --- **the best way to grasp what Betz Loading is and looks like, vs. radius, a bit different for high vs. low pitch. It maximizes between the 70 to 80% radius** on the high to low pitch props, far out, **heavily loaded tips, the 90+% radius still high, zero at the tip!** An excessive tip load would be lost rolling into the tip vortex, a very **good reason to** ***not excessively load with wide tips, with extra drag, and torque, with loss of thrust!!!*** **It is a revelation that a precise Betz loaded prop can maintain constant dT/dQ out there**, (*if* drag free)!!! **The third line of data shows the** ***ideal drag free*** **dT/dQ, the sixth column, the dT/dQ at each radius, with drag -- fantastic insight, the decimal loss vs. ideal clear!**

See next, that blade twist and shape are basic to ideal blade loading design.

Betz Constant C_L Chords We purposely teach the simple case **constant C_L, constant α^o, so you can see the more tapered profile** ***that*** **ideal design math produces**. If you want a wide blade tip you can unwind the blade twist and C_L, the math will still hold Betz loading, still ***quasi ideal***, **but** it will have more skin friction, profile drag, ***less RPM and Thrust***, no longer ideal!

The columns show **chords**, **dT/dQ**, **efficiency**, and **downwash maximize ~ 45° ϕ_1!!** Note the ***blade shape, high vs. low pitch!***

Math keeps the max chord from moving too outboard on a Reno prop with a near 45° tip!

Blade β Angles, ***Not Helical*** Column 5 shows the **β angles**, at each radius, **1.612° more than the helical ϕ_1 wind angles** for the .55 C_L RV and Luscombe props. **Compared to a tad higher helical base at 45° ϕ_1**, the ***root and tips twist to a slightly higher angle at each station*** tangent about 45° ϕ_1, *shown in Column 11*

The Blade Angles, β, and Twist, is, of course, the Air Inflow Helix, ϕ, plus 1.6 degrees α.

DW°, The Final Downwash Angles. Clearly the downwash angles are ***not constant***, as ***the Prandtl wing is***. If you study fig.5 you can see that **$\phi_2 - \phi_o$ cannot be constant**, simply because the **geometry controls it, (max at ~ 45° ϕ_1)**, Look! At **zero radius $\phi_2 - \phi_o$ falls to zero, as it does at an infinite radius**. Just mentally moving the radius in and out, you can see that the angle between $\phi_2 - \phi_o$ is forced by the geometry to change. Trigonometry controls it. **The thrust required sizes a and a', and the trig sets, and limits, $\phi_2 - \phi_o$!**

See next that blade drag is about 1/3 of the total loss at .55 C_L, a bit more at lower C_L's!!!!

Efficiency The **overall efficiency, vs. the zero drag efficiency on line 5 -- gives easy insight into the percent loss due to drag shown** -- along with the **local eta, η, at every radius in column** 7 --- the max eta moving outboard one plus stations on the high pitch RV prop! Figure the math and see the **.037057** drag loss is **33.36%** of the ***.11105 RV total prop eta loss*** --- the higher **.047935 is 31.74%** of the **.151007 Luscombe loss, a slow, low pitch, high induced drag prop!!!!!!**

Notice Planes fly with more Profile Drag than Induced., Props Opposite, ~2/3 Induced!!!

a and a' Factors Just like in the vector diagram **the a factor maxes at the tip**, but the tip **thrust vs. radius is contoured to that shown by the *highly tapered cords*** working at the ***ever increasing dynamic pressure as radius and velocity increases***, **the inner radius thrust cut by low a, low q, excessively steep** β

Our Theodorsen calcs do not use or generate Glauert's a factors, so we synthetically generate them, ***but with*** Theodorsen's ***heavy loading*** they're a bit stronger than Betz approximate ***factors!!!***

Realize -- With no slowdown these examples have higher pitch than reality!

An Overview of the Fairly Complete Data Supplied Thrust, Diameter, Area, RPM, TAS Velocity, C_L, Altitude Density, AR, Pitch, P/D, Airfoil, etc., etc., are all supplied so the props are well defined, --- as well as quite incisive resulting data, even more on the coming actual Luscombe prop, you'll see. An Aspect Ratio, (based on the outer 90% of blade radius), of about 14:1 is used, since the Luscombe prop is safe for vibration there. For design studies, an NACA 4412, 12% thick airfoil, much like a Clark Y is used, that used on so many early props. Structurally thicker sections at the root, of little consequence, are ignored for studies. A variable thickness RAF 6 is used on the coming *actual* Luscombe prop analysis. C_T, C_P, and AF, taught later are there for reference.

 A Major Look at Thrust, H.P., etc, is in Slowdown Section.

HEAVY LOADING -- understanding w bar, $\overline{w}$, and Kappa, κ

Theodorsen's w (or $\overline{w}$, a ratio of V_1), (Ref. fig. 5). This, w or **$\overline{w}$ is the basis of Theodorsen's calculations** -- **it is the axial velocity of the vortex sheet, to the tip of the vector diagram, point I,** *(It's ~ twice the "inflow only"* **Slip, or Betz Velocity** *of Glauert's similar, geometry), both heavily* **loaded**, *Theod. more accurate.*) **It indents the stream tube, as explained previously.** Notice that the $\overline{w}$ is **.1566 for the RV6,** ~ 15 2/3 % of V_1, ***bigger***, .21476, ~ 21 1/2 % **for the slower Luscombe.** This makes good sense because the **faster planes need relatively less ΔV** than the lower mass flow rate, $\dot{M}$, slow planes. *Get the horse sense of that, easy enough.*

At .55 C_L both blades have similar loading, except q and T are bigger faster---

All props for fast planes have less $\overline{w}$ and ΔV, because they're fast, ***but Theodorsen has about 5% more pitch, vs. average loading***, you'll see, ***because they're heavily loaded!*** Theodorsen's ***math accurately has the prop indenting the stream tube a lot***, you'll see, --- where Glauert does pretty well using Prandtl's F Factor, ***(but his pitch is about 1.5 % less, low, for a given Thrust, you'll see)!***

Now here comes an Amazing Insight into Heavy Loading!!!!

The **Small Kappa factor, κ, shocks you into how *very* significant heavy loading really is, because it's a LOW** NUMBER!!! **Its function is to knock the Vortex Sheet velocity $\overline{w}$ down to the *average* of the Stream Tube, so $\overline{w}$ (V or V_1) κ = ΔV,** of an $\dot{M}$ΔV calculation. ***A small Kappa, κ, the vortex sheet is a lot faster!***

Note - the $\overline{w}$'s are 15 to 20% ballpark, a lot bigger than the overall ΔV's need to be!!!

LOOK, the **Kappa**, K, is **only .54282 for the slow Luscombe, an even lower .34704 for the fast RV**, which says the **Stream Tube Average is a lot slower than the Vortex Sheet, the heavily loaded Vortex Sheet a lot faster than the average of the Stream Tube!!!!** But Surprise, in both cases the ***heavily loaded pitch is about 5% extra*** *at .55 C_L*. The math's a bit tricky: The pitch is based on ϕ_1 ***thus half $\overline{w}$.*** **The trick is to find the Δ pitch difference *between Theodorsen and the average stream tube*!!**

RV. $\overline{w}$/2 = .1566 / 2, = .0783 extra pitch x .34704 = *.02717 avg. vs .0783* = .0511, ~5%V_1Δ

Lus $\overline{w}$/2 = .2147 / 2, = .1074 extra pitch x .54282 = *.05829 avg. vs. .1074* = .0491, ~5%V_1Δ

So heavily loaded Pitch is ~ 5 %V_1 more *at .55 C_L* - than average loading!

Glauert --- Using Prandtl's F Factor, vs. Theodorsen

Glauert was the <u>first man in the world</u> to get his *<u>axial</u>* <u>and</u> *<u>rotation inflow</u>* <u>math basically correct</u>, using the ***<u>geometry taught here</u>***, and **<u>using his *axial* a, and *rotational* a' factors</u>**. He calculates the prop, *<u>more as a stream tube</u>*, *<u>and</u>* as a **Rotating Wing**, using **<u>his equations</u>** for **<u>Thrust and Torque vs. Radius</u>**, in math terms **<u>dT/dr and dQ/dr</u>**. Of course, they account for **D'**, the **number of blades**, B, **chords**, c, α^o's, thus C_L**'s**, all at the appropriate **q's, based on V^2, and** ρ, Density, as well as the proper **trig, all oriented in the axial and rotational directions**.

His math was approximate, a necessity before computers, lopping off the minutia ends of long complex basic equations, and approximate trigonometry for small angles where that is often quite sound. His answers, not perfect, are actually quite good, just *<u>not</u>* the perfect ***<u>constant</u>*** **theoretical drag free dT/dQ**, the perfect twist and blade shape of Theodore T.

Glauert's math does *<u>any</u>* prop, does *<u>not</u>* give an *<u>ideal prop</u>* unless constant r tan ϕ is set up!

The ***<u>second key</u>*** is to use ***Prandtl's F (fractional) approximation***, actually a complex *exponential* equation -that gives you a ***(<u>Decimal</u>) Fudge Factor***, **somewhat elliptical, below,** to ***<u>knock down the calculated Thrust at each radius</u>***, **shaping it vs r., zero at the tip.** That makes it ***much more sophisticated***, ***close to accounting for real three dimensional heavy loading***, *like Theodorsen.*

Prandtl's F Factor

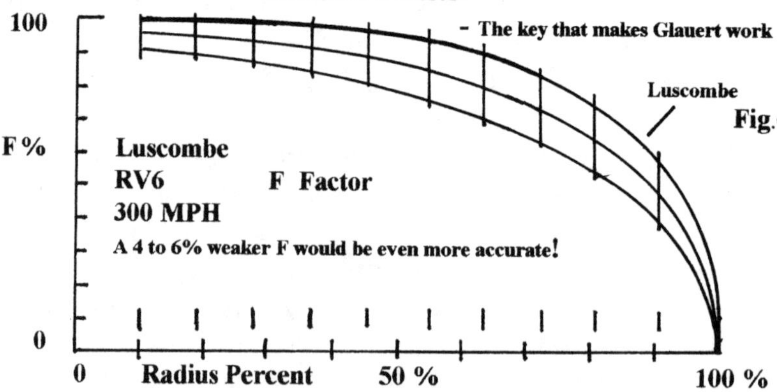

Fig.6

Looks ~ Elliptical? *<u>NOT</u>* with steep inner Blade Angle Trig, q vs. Radii.

F Factor --- vs. **Percent Radius** --- Luscombe, RV, 300 MPH Case

r 10%	18.9	27.8	36.7	45.6	54.5	63.4	72.3	81.2	91.1	99% r
.99267	.98860	.98227	.97243	.95711	.93323	.89591	.83709	.74248	.58074	.19886
.96106	.94864	.93224	.91054	.88177	.84346	.79202	.72190	.62332	.47405	.15783
.90953	.89007	.86631	.83722	.80145	.75716	.70167	.63081	.53704	.40267	.13200

Note, faster, Steep Pitch, gets less Thrust, less ability to move the Streamtube!!!

Glauert, (all the early pro's) started with **(a streamtube)** Multi Blade **Light Loading.** He gets **Heavy Loading** ***using*** **Prandtl's F Factor** *fairly elliptical* **(accounting for ~ 3D flow), shapes the *Thrust fall off at the tip***, calculating it somewhat like a wing, **with real inflow and rotation, downwash, backwash** ***(that is the real vortex sheet, if set up ideal, constant pitch of φ).*** Don't tie it to a stream tube yet, *Theod. will!!!*

Weak a factors inboard, and the F factor blends Glauert to ~ Theodorsen's Kx factors

Theodorsen accurately handles everything and finds a huge effect, the **Vortex Sheet ΔV is 2, 3, 8 times faster than the stream tube average ΔV!!!** Kappa, κ, you've seen, cuts that down to an average imaginary stream tube speed for a $T = M\Delta V$ *calc check*. **The prop really is screwing through the air like an English airscrew, *blowing the whole stream tube LESS!!!*** The **F factor *starting at ~ 90%*** in the 300 MPH case above, shows ***ever less effect on the tube at very high pitch and advance!!!***

Realize the F Factor shapes ~ ideal T # vs radius, *and* the fall to zero at the tip!!!

Prandtl, who with **Monk,** literally invented wing theory, and the ideal elliptically loaded elliptical wing, created the **F factor as his best shot at a good overall correction factor of the *embedded* rotating wing in 1919**, an Appendix of Betz's paper. **Theodorsen** accounts for full **3D flow** feeding the tip vortices too, **heavy loading, all radii affecting each other,** ***a wonder of sophisticated analysis, an 1865 to 1948, 83 year challenge!!!***

If F were made even *weaker*, at high advance, the results could be even *closer*!!!

Comparing Glauert computer runs to ideal Theodorsen Luscombe and RV prop runs we find they produces 6 and 11# **extra thrust, 4 and 6%, ~1% too much efficiency. Untwisting only 1/4 to 1/2 degrees,** (.2842 and .43185°), each about ***1.5% pitch error***, we match the target thrusts, ***a significant error*** if you're trying to analyze ***to .1 degree***, but ***pretty good, really, for 1934***.

99 II *Simply* ~ **Elliptical F Factor, x increasing q = half teardrop loading.** !✓

Theodorsen -- The Ultimate Ideal Prop Analytical Solution

Theodorsen created essentially ***the ultimate, exact, 3D airflow, ideal prop analytical method, by 1948.*** [With modern *light speed fast* computers, able, with iteration --- (that is, *homing in* trial and error solution capability) --- to do more complex mathematics, attacking actual flow patterns, including vortex formation ---they can simulate more. ***But more*** **analysis won't really be better in any meaningful way**, as far as the prop itself is concerned, because ***an essentially exact method is already available here***, in a form that is pretty reasonable for a pro engineer to use.] (We'll address slowdown and off peak soon.) More fancy analysis is apt to miss some of the wonderful capability already created here by 1948. **Theodorsen**, and his analytical forebear, **Sidney Goldstein**'s work showed they not only had **genius**, but the caliber of genius that only rarely comes along. **Let's make it understandable!!!**

A fancier analysis **that didn't get heavy loading correct would go backwards!!!**

Theodorsen's method is really based on (extending) Goldstein's $K_{(x)}$, circulation, or lift, **blade loading factors for heavy loading** and **high advance ratios** -- by simply considering the imaginary vortex sheet ***far back***, thereby correcting for the tube diameter and pressure changes applicable. Significantly, he solved the very complex differential equations ***for complex cases***, before computers, by a ***clever, accurate "voltage field" simulation!!!***

Theodorsen's Charts read to three Significant Figures, accurate to ~.01, 1%. *

The core of the mathematics is based on a **full 3D potential flow solution** of the correct differential equations. What that means is that we have a mathematically accurate simulation of full 3D flow, **axial, rotational, radial flow feeding the tip vortex** with **all radii affecting every other radius**, a wonder of mathematical accuracy, just about **as good as it gets!!!** That is very, very significant, far ahead of earlier approximate methods. He gets the radial flow feeding the tip vortex correct. That plus rotation, the downwash are **the total induced drag loss** -- which is the **largest**, the **most highly variable**, the **most important prop loss**, the basis of our **$T = \dot{M}\,\Delta V$ axial loss logic**, the central key to understanding what is happening physically, logically!

***Tibery and Wrench found Higher Kx above λ of .5, replacing Theodorsem High λ Kx.**

Blade **profile drag is nominally 1/3 of the total loss**, **at .55 C_L** the **circumferential loss**, unique to propellers, (ie. wings don't have that), at high pitch ~ **20 %.** **of induced,** **less at low pitch.** It's crucial to get the **large, axial, (and maybe ~equal) radial tip Vortex** *induced loss correct*. **We know that if we go fast, that induced loss collapses, just like it does for a wing. The max efficiency at optimum J/π is about 91%, depending on how high we get *aspect ratio*, and C_D!**

A coming Design Study Surprise is the somewhat flexible swap of D for C_L.

Here is a capsule insight into how a computerized Theodorsen analysis works in a practical sense. Theodorsen's vortex sheet velocity **w** to the second tippy top of our geometry, (or $\bar{w}$, w bar, a decimal ***ratio*** of ***that*** air velocity to the second tippy top -- ***to V_0***), represents the ***extra velocity to get to*** **any point I in fig. 5**, p 86 II. **It is the *imaginary* constant pitch *final* Vortex sheet *axial velocity*,** defined by the constant axial velocity of that **second tippy top, $\bar{w}$, caused by ϕ_2 - at *any* radii**. Silly, ***tippy top*** gets understood!

V_0 is airplane speed in Theodorsen!!! ---- We deal with Slowdown, V_1, later! ***

The analyst ***adjusts*** **that $\bar{w}$** (to the second tippy top), **to get the thrust required**, at the D, RPM, Speed and altitude Density, **rho, ρ,** which, of course, sets up the pure helical flow in and out of the prop, at the prop, ϕ_1, ϕ_2. It's built like ***Glauert's similar geometry, but built from ~ twice Glauert's inflow only a and a'***. (I say similar, because Theodorsen is more accurate.) ***The Heavy Loading Theodorsen (and Glauert) math*** produces ***a taller velocity schematic*** *than the average stream tube*, and that means the comparative *a factors* are roughly 5% bigger, p. 97 II, than average ***a*** factors, $\Delta V/2V_1$ **for an $\dot{M}$ ΔV** calc. **Look at the $\bar{w}$'s on page 94 II. They're ~15+% to ~21+%, the Kappa's which converts to the average stream tube velocity, are ~1/3 to 1/2! Remember our schematics of a and a' are exaggerated a lot, so you can see them, a few %, *inflow only*, forming a helix ~ to w / 2. They're SMALL!!!**

That ~ 5% extra Heavy Loaded Pitch, or V, is valid at a .55 C_L only, zero at zero C_L. ✓

Behind the scenes the computer is actually referring to a **Chart of K(x) "circulation", or blade loading factors *interpolating*** vs. the proper **advance ratio** and **radius**, properly twisting and shaping the blade chords for the C_L selected at every radius, creating tapered blade tips if a constant C_L is selected, and creating the precise helical screw surface ϕ's for the *imaginary*

Rigid Vortex Sheet -- **in actual fact, simple helical flow,** ***at the prop*** **in, then stretched out!!!** Thinking in terms of Glauert's **a** and **a'** factor geometry, the ***weak inboard a factors*** have enough blade width to meet the small inboard axial thrust demand of Betz loading --- simultaneously narrowing outboard chords to ***not*** make excess thrust, again meeting ***Betz loading vs radius***.

Theodorsen's K(x) Factor Chart controls the Thrust vs. Radius Shaping

Theodorsen uses a similar geometry schematic, but everything is based on his w or $\overline{w}$ to the tippy top of the velocity drawing, see fig. 5. He doesn't use Glauert's approximate **a** and **a'**, but , ~ **double their** ***effect*** *accurately.* ***Vector*** $\overline{w}$**,** fig. 5, **can be thought of as being in there,** $\overline{w}/2$ **at a ~5 % taller pitch** vs. avarage. The individual air particles are actually moved ~ perpendicular to the phi angles by the airfoils, just like Glauert, *because the schematics have the same objective*, the **pure screw surface flow in and out at the prop**, ***vector HG, backflow***, the result. **Theodorsen simply uses his** $\overline{w}$**, w bar, to do his math, but remember he's actually working** ***far back*** **at** ***97 - 99% D*****, because that math concept is valid for heavy loading and high advance ratios**!!

No axial velocity ever actually moves at the tippy top velocity, the plane and prop moving forward at V, --- the imaginary vortex sheet moving back at w, or $\overline{w}$, formed by the combination of ***twice*** a and a', in our geometry at least in imaginary form, ***ignoring the tip vortex formation***. **The w or** $\overline{w}$ **is just the best, simple math basis way to do the very complex math**.

When you can follow this banter, you've got it!

To sum up, the beautiful, and key insight that you can come to appreciate and understand is: ***Theodorsen's Kappa, κ, knocks w, or*** $\overline{w}$ ***down to the average effective ΔV velocity of the stream tube for a basic*** $T = \dot{M}\,\Delta V$ calculation of thrust. *Simple, right in tune with what we taught! Due to heavy loading, a significantly faster Vortex sheet, Theodorsen needs a* ***higher pitch***, done precisely, Note that the ***w bars,*** $\overline{w}$***, are 2, 3, 8, times the average ΔV, the Kappa's SMALL***, ~ 1/2, 1/3, 1/8, page 94 II, **the Vortex sheet ΔV, is 2,3, 8 times faster** ***than stream tube average*** **ΔV*!***

You wouldn't want to attempt the genius math in Theordorsen's book, but **the picture we give you here is right on track**. For any **engineer** who reads all this, **gets the great level of understanding these four chapters can provide**, you'll find it can **easily save you a year**, if you try going back through Betz, Prandtl, Goldstein, Glauert, and Theodorsen. If you can **start, knowing, understanding, it is hugely easier** to conquer props.

Engineers can attack Theodorsen's math, App. T, much easier grasping all this!!!

The computer concurrently calculates all the many factors of interest for a given thrust case vs. RPM speed and altitude, also with the drags, at first left out as in the basic theory --- Thrust, Torque, Efficiency, $\overline{w}$, Kappa, M. dT/dQ, proven constant drag free, dT/dQ with drag, off ~3% at .1 r., (RV), to 4 to 5% outboard, (Luscombe), where **tip drag increases more**, as shown on 94 II, --- a, a', dT/dr, thrust vs. radius, chords, α^o, C_L, ϕ, β, etc., etc., **a long list** of insightful factors ***overall, where appropriate***, or shown at each of **11** (to 20)**, or more radii, depending on your specific program code**. Modern computers are just wonderful, yielding a **magnificent Xray vision** into what was essentially impossible to see in such detail and accuracy earlier, **a propeller's physical characteristics and physical operation laid open for our understanding!!!** The joke is, our now old computers bought November 1993, terribly slow, only 33 MHz, ***only 33,000,000 operations per second***, gobbles up and displays all the ***answers, almost instantly***, after the design task numbers are typed in.

Two comparable parallel programs ---- modify, retwist the ideal design for *slowdown* --- *or* show results off the ideal design point, slower, for takeoff and climb, where the numbers change hugely, **fantastic insight!** *See Super Magic Graph, p.45-II*

So now let's move on, away from the basic geometry, all its implications - and learn everything else we can about props with minimum brain strain, after this full dose of basics. We only spent so much time on airflow and insight to help!

Is a Triple Ideal Theodorsen Prop actually Better?

On our **Best Test, one Prop had near Twice the loss of the Best, ~81% η, not 90+%.** .

We saw tapered props in the old days. A great TV show on **Lindberg** last night, showed his in **1927**, amazingly, ***Ham Standard way ahead***, probably not telling everything they knew, proprietary. The great companies usually have very able engineers.

In WW II, Huge Engines, <u>the focus was on Power Absorption</u>, and Mach, not η.

Why don't we see narrow tips from the manufacturers these days? There are comments in old NACA prop reports that broad tip fixed pitch props are ***<u>better</u>*** for ***<u>static thrust</u>***. **Logically, at runup, slow takeoff speeds, the tips would be "least worst", at more reasonable angles of attack** than steeper, overpitched, middle and inner radii. Are compromise props actually better?

A small diameter, narrow tip Long eze prop, badly stalled inboard - took off poorly!!!

I would not expect the smaller propeller manufacturers to understand, or have a clue of how to design an ideal Theodorsen prop, may not have even heard of him. Our biggest Constant Speed Propeller firm got into Theodorsen in the 70's. Their latest prop, which works best, clearly better than old designs, is almost exactly correct, almost Triple Ideal, still hanging on to a little extra tip area, better than those who got into the math in the 70's, but never understood the Triple Ideal Prop - lots of noisy props.

NOISE is the TELLTALE of UNNECESSARY TIP LOSSES!!! - LISTEN!!!

The 71"D 51"P, Klip Tip, non ideal shape, non ideal twist Luscombe prop can lose 10% more efficiency, 75% not 85% and with interference loss, only 67%, ***<u>at economy Cruise!</u>*** **Bad, we need ideal!**

With Slowdown, the square tip Luscombe Prop has extra losses, loses tip thrust!

Will broader tip fixed pitch props help in climb and takeoff? The computer can look for us! All you have to do is ***<u>not</u>*** use the standard, simple, constant C_L case, we taught for Triple IDEAL. ***You can simply unwind the tip, lower the C_L, swap it for more chord, and still have the same thrust vs. radius.*** But, for an RV prop, dragging more tip surface area, you lose 40 RPM in climb, 2.2 HP, 5# less thrust. ***The hoped for option of an ideal cruise prop that can climb better too, just does not seem to be real. You'll learn Takeoff is the key problem***. <u>We need more good professional testing now, to check the computer.</u>

The Strong Vortex - A Smoke Tunnel Propeller Photograph

This great photograph, with huge insight, was taken in the 50's by F. M. N. Brown at Notre Dame, shown in Eugene Larrabee's excellent Scientific American Article on propellers, July, 1980. It's only a ***very low pitch*** model airplane propeller driven by a spherical gear box, ***not* an ideal prop**, but it still gives great insight. Yes there is ***vortex formation at the root of the blade***, not just at the tip. There are **supposed to be two per blade**. It's ***not*** just turbulence from the gear box, but also a weaker inner vortex. The ***strong tip vortex takes over the real stream tube!!!***

The ***ideal* a factor axial velocities are max at the tip, *but* the ideal thrust profile, "Betz Loading Profile"** is **set up by the narrowing tip chords**. The central issues here is the tip ***vortices engulf the theoretical vortex sheets, but that's OK***, the **energy and math are still correct**. ***Weak axial flow at the low q, steeper angle root is stronger here with a very low pitch model prop***
This broad tip **prop goes for extra thrust at the tip, but you can see it's *swallowed*, killed by the strong tip vortex. Excess tip thrust is a loser! An *ideal*, Betz loaded prop blade is best**.

Fig. 7 **Flow Insight from a Model Propeller**

SLOWDOWN --- The Surprise that Changes Everything!

Now that we've ***taught you how to understand*** and design a theoretically ***ideal, Betz Loaded, Goldstein Theodorsen Prop***, ***with essentially exact math***, we're going to show you that the **subject of slowdown blows everything right out of the tub!!!**

Realize that as a nice, sleek, symmetrical Aerodynamic body moves through the air, the air must speed up as it goes around the fattest cross section, *incidentally dropping the pressure*, per Bernoulli's Law, just like the airflow over a wing airfoil, --- then slowing, restoring pressure as it moves to the tail. We'll skip the *ever thicker*, more stagnant boundary layer, and any separation for now. Most people don't realize that ***a body nose, maybe an ugly engine cowl, signals, pushes a bubble of air ahead***, and if you mount a propeller on its front, the ***prop plane sees a highly varied average slowdown speed vs. radius, perhaps ~17.5% slower at the .367 radius, just inboard of a Luscombe cowl's width***, **~5% at the prop tip, as shown in the accompanying sketch.**

See Luscombe data, p. 117 II

Since that can force, perhaps, a 4.77° angle of attack change at that 3/8 radius, a .61° change at the tip, a .47 and .06 C_L change, respectively, there is **a huge change in calculated thrust**. An ideal Theodorsen prop had to be designed for 106 # thrust at low C_L to hit an early, 151.5# (old) thrust target when dropped into a Luscombe's slowdown, and then, of course, **all the *ideal angles were wrong*, the targeted *C_L now highly variable*, the ideal *chords and Betz thrust vs. radius distribution in disarray*, much more thrust inboard, where the angle of attack, α°'s *increased* the most**. ***Untwisting, with more chord can almost get the ideal thrust vs. r., at a lower q, a small deficit inboard!***

Adding chord can elminate the small deficit, a tad less AR thus efficiency.

Typically the body nose is highly unsymetrical, and of course, the blades ***can't*** articulate, change their pitch, like a helicopter can, to adapt to that precisely. A pro, using cross sections, *volumes*, models an ***average symmetrical body of revolution***,

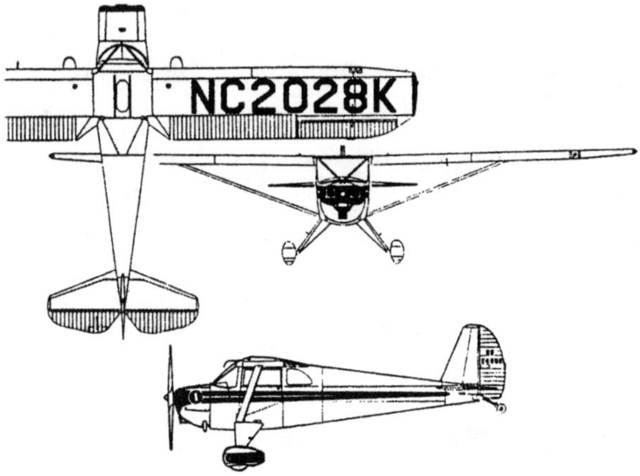

Fig. 1 Three-view drawing of the Classic Luscombe 8E.

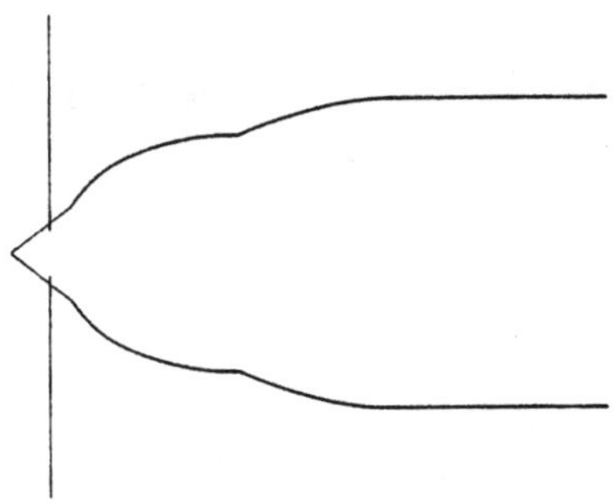

Fig. 2 Propeller plane and the Luscombe 8E fuselage profile.

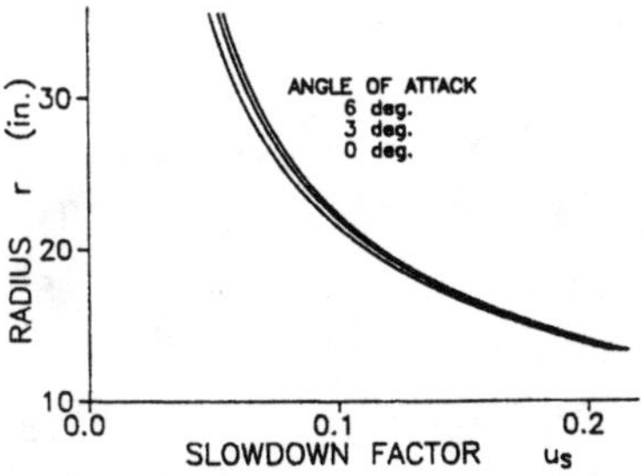

Fig. 3 Luscombe 8E airspeed slowdown factor at the propeller plane.

as we've done here for the Luscombe, he's done **as good an adaptation as is possible**. Using "source, sink analysis", a pro can accurately analyze the ***average*** ***velocity*** ***at each radius***, and we have, ***an orderly basis for a precision reanalysis, retwist,*** but ***not*** one that deals with it better than the ***average at each radius!***

A Source Sink Analysis to Model Nose Flow Velocity

Prandtl and Tiejens, teach the **Source Sink Math Model** method**,** amazingly credited to **Rankine**, the first Prop Analyst in 1865, who also created the Steam Engine Thermodynamic Cycle, the Rankine Absolute Temperature scale needed in Thermo calcs, and Mach Speed vs. Absolute Temperature, **an early genius!!!**

Briefly, if **an imaginary pinpoint air source** expels a certain **flow rate, cubic inches of air per second** it will force its way into a **100 in./sec. flow stream** and create **a theoretical bullet shaped insert in the stream,** soon converted to a mathematically proper diameter. It turns out that there is **a stagnation point** on the **nose of the bullet** where the flow stream will be stopped dead, faster and faster at bigger radii, finally back to stream velocity far **back at the full D as taught in Appendix SSSS**. Interestingly, the focal point, or source location, will be **1/4 D back** from the bullet nose, and the **90° location** will be the square root of 2, 1.414 times the radius, or **.707 times the diameter**.

Interestingly, since the *surface area of a sphere is simply 4 times the area of its circle*, *the mathematically effective source velocity* in **in./sec**. at any point, is just the *source flow rate in cubic inches per second* **divided by the surface area of the sphere, $4\pi r^2$**, square inches, thus **in./sec. velocity**, really simple once you see it, catch on. Logically, the source velocity **slows** the stream velocity **forward of the source**, then only pushing outward at 90 degrees, but weaker at ever bigger radii, speeding up behind the 90 degree position, but that **speedup dropping to zero back at infinity**, the air theoretically never mixing, both at 100 in/sec far back, the source now having inserted itself in the stream flow, at D, ***clever!***

See Appendix SSSS for a full explanation of this great Modeling Method

Now, beautifully, by simply using a combination of appropriate flow rate sources and sinks, located appropriately, we can mathematically model any reasonable embedded body, even pull it back in to model a tail cone, (not usually necessary far back), and ***get a decimal, or percent slowdown at any radius, any stream velocity!***

The Stream Tube: Realize V_1 is **at the prop, no prop there!** Then V_2 is added --- With basic axial changes vs. radii, (and slowdown), the tube **acts like a set of concentric shells**.

fig. 1

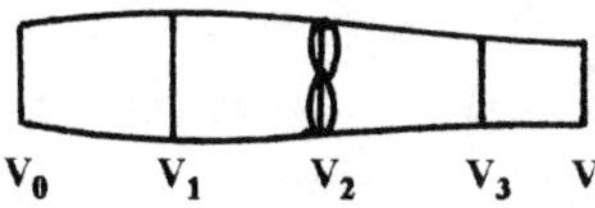

Grasp this: We use it next for Thrust reqd. **But, with Heavy Loading Betz's Vortex Sheet Speed $\bar{w}$ is Faster than the Stream Tube Avg.** - ***but Theod's Kappa gets the avg!***

V_0 = The free stream Velocity, at atmospheric pressure, **P_o**

V_1 = **V_o**, ***slowed*** by the body *pushing air ahead*, **V_1 at the prop, P_o** up to **P_1**

V_2 = the **V at the prop -- now *also* seeing the inflow to the prop** ($\pm \Delta V/2$)

V_3 = the V due to the prop outflow (thus + the second half of ΔV, + $\Delta V/2$)

V_4 = the final speedup to drop pressure from **P_1 at V_3 back to P_o**

Mr. **Bernoulli *shapes*** the stream tube. When you have a body or nacelle in the stream, it sends a pressure wave ahead, the air slows **a variable amount vs. radius to V_1** as it flows into the prop, and *expands* in diameter as the pressure increases to **P_1**. Then you have the ***natural shrinking*** of the speeded up stream tube as it is sucked into the propeller, thus **V_2**, **half the ΔV in front** at a reduced pressure, **P_2**. The pressure jumps back up behind the prop, the tube is thrown back, **half the ΔV behind**. At **V_3** all the propeller caused speedup has occurred and the pressure has returned to P_1, but above atmospheric pressure. It now speeds up more to **V_4**, shrinks more, now back down to **P_o**. If there were no fuselage, no slowdown, the first and last section does not occur, no V_1, V_4, **P_o** existing in front and behind the tube.

In the basic chapter, learning Newton's Laws, $T = \dot{M}\,\Delta V$, we **assumed constant axial velocity at each radius**. In the advanced concepts here the **axial velocity varies at each radius, as you learned in the airflow geometry, *also* due to slowdown varying vs. radius, *concentric shells***, ***finally with heavy loading, Theodorsen, varies velocity around the circumference, as the prop rotates, WOW!!!***

But Theodorsen's Kappa factor relates $\bar{w}$ to the average!!!

With 3D flow, V varying vs. radius, then with Theodorsen's heavy loading, different behind the prop, they had a genius math task. We'll learn the basics, -- then Theodorsen's method handles all that, **gives an ideal prop.** ***Separately***, we'll deal with slowdown, highly variable vs. radius, repitching the blades, for a *symmetrical* body of revolution, since ***the blade can't wiggle*** $\beta°$.

Gus Raspet Found Terrible Overall Propulsive Efficiency???

In the 1950's **Gus Raspet,** was the **multi facetted**, **creative**, **maverick** head of the Aerophysics Laboratory at Mississippi State, highly valued by his associates and many students. **Bruce Carmichael**, later an expert on Laminar Flow, thought he was great, **Dick Johnson rebuilt his milestone RJ5, 40:1 Sailplane, the real start of Advanced U.S. Soaring. Gus** challenged all -- got students to go try tests, anything that got them into doing real things, getting their hands and brains engaged, challenged to put their education to real use, to go solve real Problems. I've seen the same characteristic at Cal. Poly, San Luis Obispo, that gave us the brilliantly creative (former modeler) **Burt Rutan**, and such great young Engineers as **Brian Hobbs,** of Edwards, my right hand on **Voyager**, intelligently helping resolve complex flight data for writing **Voyager,** ***the World Flight***, the official flight report, to provide an **incisive X Ray of what was done, how it was done!**

Personal planes, Sailplanes, hands on, brains on, were a real basis of learning

Gus's creativity in the 50's led to **towed propellerless gliding drag tests of key private planes, notably George Lambros's *Bellanca Cruisair*, found 58% overall Propulsive Efficiency** vs. a sealed plane, that was otherwise well regarded. How could it possibly be that bad, *1/.58 = + 172%, the power required?* One of our core objectives way back in the 80's was to get better, specific insight into that ***grossly bad result***, not picked up in Aero.

We needed some real Data, FACTS on DRAG!!!

A Windmilling, dragging, power absorbing Propeller, makes it impossible to get an accurate glide, thus drag data, ***never solved!*** So I invented **Zero Thrust Glide Testing**, a simple, insulated, 1/16" inch, stiff, (thus low frequency vibration proof), wire feeler that lit a bulb as the propeller started to transition from thrust to drag in the ~ .016" crankshaft bearing axial slop, ***comically easy, once seen***, the first solution to that conundrum, 86 years after the Wrights.

Necessity *is* the mother of Invention! I wanted to nail that!!

We got marvelously, reproducible data, gliding in dead air, five miles out to sea at dawn, a little more risky than I preferred, but great results, the L/D vs. Gross Weight converted to ~***Pure Drag***. Dawn, dead calm, E - W, W - E Speed-Power runs over

Camarillo's *former* F 102, 11,100' runway, compared to Gliding Power got us fine **Propulsive Efficiency Data**. Excellent data, all Reproducible, I felt we had numbers as good as we'd ever get.

The surprise was a somewhat jagged plot of the core objective, Propulsive Efficiency, Drag vs. Thrust Required, actually, more specifically, Gliding Power vs. Power required, a slightly wiggly final line. Feeling all smooth, reproducible data was as good as we'd ever get, I felt pretty confident that we were looking at how it really was, that some variable separation, from a pulsing ΔV on the Classic Luscombe, the world's first all sheet metal, small private plane, would be no surprise. See the actual plotted data next. (& Airplane book, p.109.)

It was harder to get good data at the famous, Santa Rosa CAFE Test Organization because we found tight new Lycoming Engines that also swallowed their Axial end play, because of the differential expansion of the Aluminum Crankcases vs. the Steel Cranks, and surprisingly long front bearings on the 4 Cylinder Lycoming Engines. We did feel that we got good Drag Data on an RV 6 and Whitman Tailwind, smooth, it looked just about right. I plan to run matching Speed Power tests as soon as this book is done, but with good overview speed data, we already have good insight.

So how do we explain the terrible overall Propulsive Efficiency, η_P, the multiplication of Interference Efficiency, η_I, and Propeller Efficiency η? The first important insight was to see that the really **bad results** ***preferentially occurred* on the slower light planes*****, with propellers that could be seen to be** ***farthest from ideal Shape***, needing the far more tapered blades required on the slow props, with the **biggest tip vs. root q amplification, (& more ΔV)!**

*** 11 MPH ΔV at 100 MPH, is 11% An RV 9 MPH ΔV at 180 MPH is only 5%!!!**

The next important insight was that a Good test showed a foreign RV 8 Prop could be as bad as 8 to 10% off the BGT optimum Efficiency, quite a lot really, our first solid insight that poor props could be that bad, and that was for an Efficient RV 8 class Prop. A lower η class could be worse!

The Luscombe Overall Propulsive Efficiency η_P, **was 67%** with the cooling ducts open, of course, since it must fly to the ZTGT test. With **cooling drag usually understood to be about 10%,** the sealed ducts **Bellanca and the Luscombe are ~ equivalent**. Undoubtedly, the tightly cowled, and tightly sealed great modern homebuilts, with pressure recovering, slowing, expanding inlets, are doing much better. **Maybe we have a nominal range of ~6 to 10% loss for cooling**, perhaps a wider range for the terrible, to excellent, most diverse examples. (I know of no new hard data there.) We **probably won't have any *actual thrust* cases like the P 51!**

We realized that **with the thrust producing ΔV we would surely get extra scrubbing drag once we turned on the fan.** We had done a lot of looking, calculating, arguing back and forth on how the whole picture might finally fit together. **With the slowdown velocity profile at the nose of an embedded body, the prop is running in slower air, can produce more thrust for a given H.P., - but a *counterbalancing pressure* is created on the nose, which *demands an equivalent Thrust Increase* to get the *original Net Thrust, at the original Power, a WASH, a surprising trick of Physics, that can look like free Drag!*** There was a lot of arguing about that one, but ***it finally is a wash***, ***no cost!***

We have real Airplane Drag vs. Speed, we know there will be some extra Scrubbing Drag, we think we see evidence of Separation. We have what should be a very accurate calculated H.P., η, and Thrust. To get more assurance, we'd need a Wind Tunnel and its accurate Scale. I'd love to do that sometime. We must, for now, trust the Elegant BGT Math. for Efficiency, Thrust, and Power, look at Probable Scrubbing, Separation, **look at how the whole jig saw puzzle** *most correctly*, **fits together!**

If a low pitch Luscombe Prop, a 51"P/71"D, a P/D of only .718, a lot less than the desired 1++, can have **a max** η **of .85**, but is far from the desired almost triangular shape, twist off too, *perhaps* it is actually **down an extra 10% to only .75** η, per our best test! √

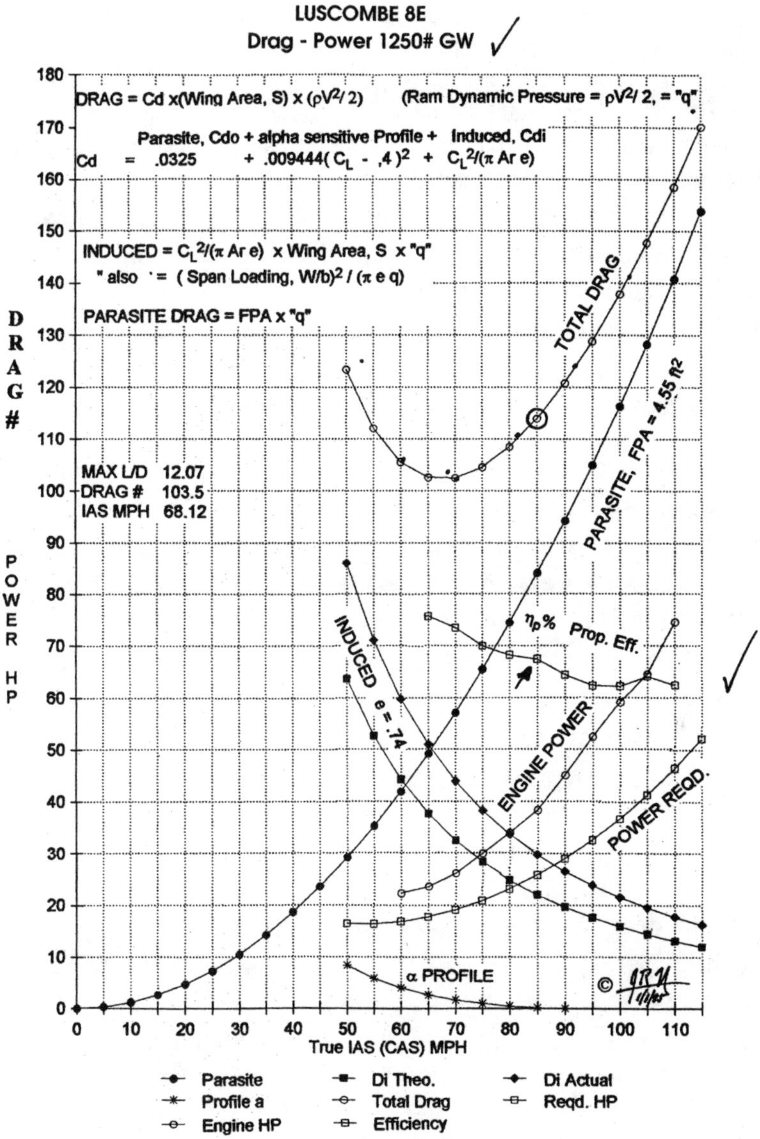

If that is the case, we only need a **.9** interference, η_I, for a **.67** η_P! Notice we have an η_P that **degrades with more power**, shows a **jagged line, quite probably variable separation**, happily **<u>an up .67 where we do altitude cruise</u>, 85 IAS, 100 MPH TAS. It is <u>great to *finally* have some real facts, from ZTGT</u>, *<u>to noodle out</u>*!**

Hubris is never acceptable in the technical business, *pride precedeth the fall, a good rule in life*, but within that admonition, I'd like to take a little personal pride in that all encompassing graph. As far as I've been able to find, that is the first time that complete Aerodynamic Data, ***including accurately measured propulsion efficiency*** has been achieved on a flying propeller driven plane. It took some serious thinking on the subject, the necessary invention, hard work, a little risk, and serious technical care. I did everything I could think of to get spot on, incisive data, and I believe it genuinely is, a little personal challenge nailed. I felt for years that ***Gus Raspet's creativity*** that produced **such insightful data on just how bad propulsive efficiency could be**, **deserved a proper response**, no one else attacking it.

Normally traveling at **1400# G.W**. the drag curve moves up to **~123#**. At the **49 H.P**. I fly at in Altitude Cruise: At **75%** η, that would produce **137.81** pounds of thrust, **14.81# above 123**, **to explain.**

(49 H.P. x 550 ft #/sec./H.P. / 146.666 ft sec) x .75 = 137.8125 # T.

123# Drag / 137.81# Thrust = .8925, η_i, OK to justify .9 η_i

A **T=$\dot{M}$ ΔV** calculation shows that $\dot{M}$ & **137.81# T** creates an **11 MPH ΔV**, and that ΔV *can* cause **~9.5#** of extra scrubbing drag*, thus **5.3#** of variable separation drag, **14.8 + 123 = 137.81** - ***all* worse as more power is applied***, **as we see on the η_P plot,** until we get to full power at sea level, where the degradation, levels out, possibly the faster air somehow conforming better.

***100 MPH TAS + 11 MPH ΔV = $(1.11)^2$ x 41.5# scrubbed profile = 9.5Δ#.**

Of the 123# drag at 1400# G.W., 40 is induced, leaving 83# of Profile Drag, perhaps up to half of it, the messy half, body, tails, landing gears, cowling, wing, windshield, body intersection exposed to the slip stream, not sophisticated analysis yet, but perhaps able to produce an extra 9.5# of scrubbing, as above:

11 MPH ΔV at 100 MPH, is 11% An RV 9 MPH ΔV at 180 MPH is 5%!!!

Obviously, that is not a sophisticated Aerodynamic Calculation, but what we're trying to do at this point is rough out how the jig saw puzzle might actually fit together. The clean, fast planes can have very small propulsive efficiency degradation, little separation, and an RV has ~5% ΔV vs. speed - a Luscombe ~ 11%, a big V^2 effect.

Keys, Slow, more misshaped, high q magnification props ***even worse*, more ΔV loss too.**

NOT EXACT, the CALCS ON THIS PAGE, are QUITE CLOSE TO THOSE ON p. 117 II

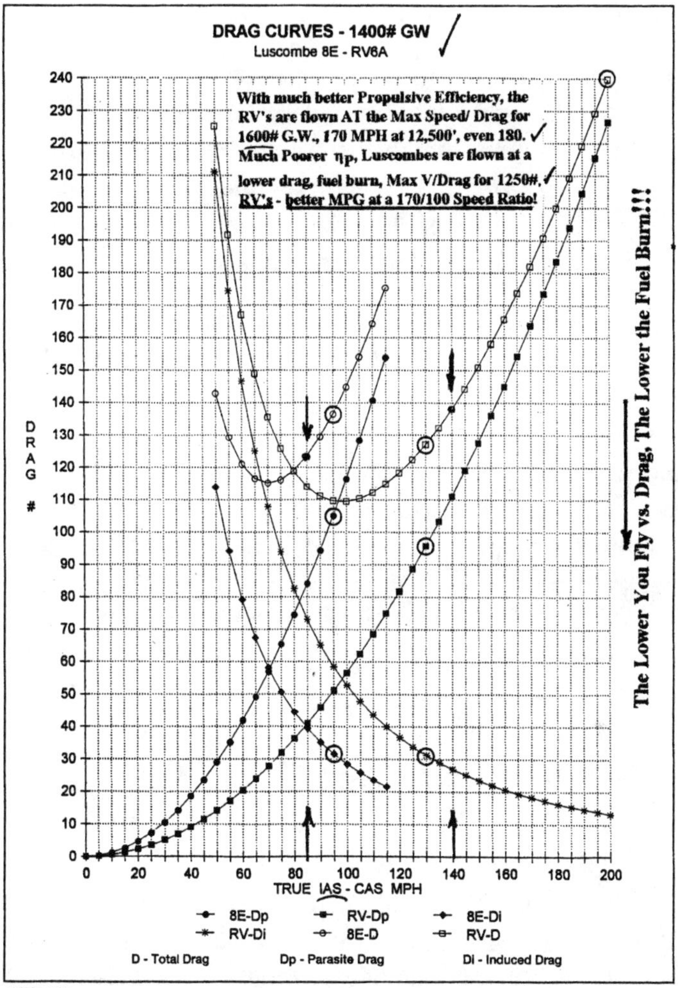

46 APRIL 1995

In the **Overview on Propulsive (in)efficiency**, it appears to be that **~6, 8, 10% Cooling Drag is to be expected, depending on the caliber of installation, almost none to ~10% due to bad prop design, from ideal, and 3, to ~10% Interference loss, η_i!**

We need to run accurate Speed Power data before we can nail RV propulsion efficiency.

The interesting composite drag curve here **confirms Luscombe drag at 1400#**, shows the low A.R. Hershey Bar wing RV has no more Induced loss than a Luscombe, **the RV FAST!** With much better propulsion efficiency, better MPG at a ~180/100 TAS ratio!

An Appraisal of the Luscombe Propeller

Adapting Theodorsen math, for a ~ Perfect Luscombe Prop with slowdown, we then calculated my McCauley 71" D, 51" P prop using exact Luscombe dimensions, actual angles, chords, twist, shape. ***No longer ideal***, the math would lie about efficiency, thrust, power. Our best insight saw η dropping to 75%, not 85%. We made the best rational adjustments, ***checking every way possible!***

GLAUERT CAN GIVE A PRETTY GOOD ALTERNATE CHECK. WE CHECKED EVERY WAY POSSIBLE!

A careful "granite plate" inspection, like the best shop pro's do, taught me that it is easy to get very precise angles, much closer than .1° on an RAF 6 airfoil by simply clamping a 12" steel scale on its flat bottom. It became clear that the prop was in essence a 48" climb prop, retwisted to be ~ 51" pitch at its outer stations, fading a bit right at the tip, and having a constant 29 ° β blade inboard of the 3/8 radius, *a crude accommodation of slowdown.*

NOTICE HOW THE C_L's RISE AS YOU MOVE TIP, TO INBOARD.,shape not right.

The final results on p. 117 II are fascinating! In 1947, not designed for slowdown, the α and C_L are out of control high **above a .55 C_L** target inboard, mid radius to 1/4 r, undesirable, **closer to stall, slow at takeoff** - until the **constant 29 ° takes over killing (the already weak) thrust in front of the cowl**, OK.

Outboard, the **square *Klip Tip* prop forces the math to give a wrong answer, a squarish thrust vs. radius at the tip, which simply can't happen physically**, **the false thrust rolling off the tip into a bigger vortex loss, *just what Betz wanted to avoid***, with his *teardrop* thrust loading vs. radius. It's a bad tip, more drag out there at max radius, bogging down the engine R.P.M., thus available H.P. spending energy to make thrust, losing the Thrust into an excess tip Vortex, making noise, bum engineering.

REALIZE BELOW, MORE SPEED DEMANDS POWER CUBED. I USE A CONSERVATIVE 2.5 POWER.

Our biggest problem here is that **the Elegant Math *goes for perfect***, so **we have no *comparable way* to calculate how bad the bum ones are**. We saw that on a very good, and fair RV 8 test all props flown on the same plane, the worst prop lost more than 8% η, by going 8.2 MPH slower than the best prop, close to Theodorsen perfect, 90+% for a fast, high efficiency RV 8 prop!

Actual Luscombe Propeller - McCauley 71"D 51"P

V_o = 100 MPH, 146.66 ft./sec. **RPM = 2280** **Thrust = **136.216#** **HP = 49** **Eff.** η = **~.75

Alt. = 10,500' **P/D** = .7183 **A.R.**= 13.99

Ref. ABB 99.33 - 39.3 and 40.3, 7/13/200 as modified7/9/05

% R	r"	T #/ft	C"	t/c	α°	C_L	β°	dt/dq	η%	Pitch	a	a'	V_1	V_3	V_4	T_R	Slo dn
N.A. 10.0	3.550!	**-1.236**	4.115	.4609	-4.62	-.0607	29°	-6.566	-2.161	**12.364**	.0222	.0476	78.57	82.07	148.57	.5441	**.53575**
18.9	6.709	**7.691**	4.618	.3186	1.52	.5411	29	2.315	.9237	**23.368**	.0394	.0300	95.26	102.76	151.64	.6638	**.64947**
27.8	9.869	**22.594**	4.976	.2181	3.38	.7234	29	2.033	.9397	**34.372**	.0509	.0218	110.37	121.60	155.29	**.7682**	**.75256**
36.7	13.028	**49.270**	5.188	.1589	5.30	.9127	29	**1.737**	.8804	**45.376**	.0577	.0165	121.00	134.96	158.38	**.8391**	**.82499***
45.6	16.188	**65.196**	5.257	.1318	4.04	.7883	25.394	**1.632**	.8736	**48.284**	.0642	.0126	127.78	144.18	161.16	**.8835**	**.87123**
54.5	19.348	**71.951**	5.183	.1116	2.39	.6260	21.786	**1.572**	.8693	**48.589**	.0695	.0098	132.03	150.39	163.38	**.9109**	**.90022**
63.4	22.507	**78.936**	4.966	.0962	1.46	.5351	19.224	**1.518**	.8566	**49.314**	.0730	.0078	134.76	154.43	164.92	**.9281**	**.91884**
72.3	25.666	**84.920**	4.607	.0850	.91	.4809	17.270	**1.466**	.8386	**50.138**	.0751	.0063	136.60	157.11	165.94	**.9396**	**.93136**
81.2	28.826	**87.418**	4.107	.0773	.52	.4428	15.654	**1.414**	.8165	**50.755**	.0762	.0052	137.90	158.93	166.59	**.9476**	**.94025**
90.1	31.985	**83.868**	3.463	.0724	.19	.4105	14.232	**1.359**	.7905	**50.974**	.0767	.0043	138.88	160.17	166.97	**.9535**	**.94689**
99.0	35.145	*(**72.403***	*2.695)*	.0693	-.13	.3794	12.947	**1.300**	*(.7603)*	**50.765**	.0767	.0037	139.64	161.06	167.19	**.9581**	**.95207***
100	35.500	*(Did 40% Tip)*															

*** Text refers to~ 17.5% and ~ 5% Slowdown. ***

Thrust vs. Radius Distribution .
dT/dr #/ft

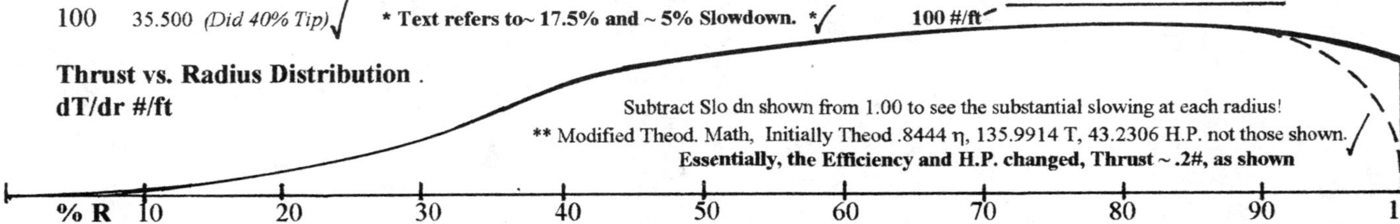

Subtract Slo dn shown from 1.00 to see the substantial slowing at each radius!

** Modified Theod. Math, Initially Theod .8444 η, 135.9914 T, 43.2306 H.P. not those shown.

Essentially, the Efficiency and H.P. changed, Thrust ~ .2#, as shown

**At Lower Efficiency, Higher H.P., Tip Loading will Actually be Lower, Higher at 3/4 Radius and Inboard, more Teardrop, actually wanting to be more Elliptical, the Core Reason Props Lose More Than Wings, Lower Efficiency, -10% to 25%, not 2, to 3% of a 50:1, or 33:1 L/D Wing!!!

Not the 5% of a 20:1 L/D wing, where 1/20 = 5%

To be accurate and fair here, we must give the devil his dues, **and show what's favorable about this prop,** ***but see it's still a loser!*** The high C_L, mid radius, and inboard do have the favorable effect of pulling the lift, Thrust inboard, good, doing a favorable thing! Likewise, the low C_L's outboard, also tend to deemphasize the bad, excess blade widths out there, also pulling the loading more inboard from the tip, also good, it could seem, at first.

But there's **a double negative there too, unfortunately**. You just don't win in the overall game, because excess blade width out there, where max q is happening, and **producing much more profile drag proportionaly, and at max radius, max lever arm, the excess, completely unnecessary torque addition, bogging down the engine RPM capability, thus available H.P.!** Not a big thing, some might say? Well, ***there are only two kinds of losses, profile, and induced***, and **we're hurting this prop's capability, both ways**. **Excess drag, at max lever arm, *creating excess torque load***, bogging down the engine, is hurting, not helping, just not the smart move, **because ease of rotating the prop, for max lift, thrust, is what the game is all about! Outboard thrust is wholly against the proper objective here!**

The gross insult to good sense here, though, is not only excess outer blade width, but **a square tip, massively creating excess induced loss,** trying to create high lift where it just can't be well retained, dumping it out the side door, manufacturing an ***excessively big, and wholly unnecessary tip vortex***, **NOISE**, the worst kind of induced, direct cost, with the ***worst posible payoff!***

And a **slow, low pitch prop with the worst possible q amplification, is the worst, most harmful place to do it**, ***most likely to lose the full extra 10% efficiency loss we've just used!*** **Theodorsen, on his page 33 shows that even with the Betz half teardrop loading done perfectly, *the radial tip vortex loss can be bigger than the axial we must produce to make thrust!!!* Radial flow, Excess Tip Vortes Induced Loss. *Profile there too*, is the CORE, WHY props are less Efficient than Wings!!!**

Are we designing for Sea Level or Altitude - Cruise, or Vmax?

Some Insightful, Bottom Line Conclusions on Prop Design

***First*, You just want to know *the core bottom line* Answers!**

For the SPEED, RPM, POWER AVAILABLE for the given Application :

1. **What *SHAPE* should the Prop be?**
2. **What *TWIST* is correct?**
3. **What PITCH is correct?**
4. **What SIZE -- *Diameter and Blade Area* - for available HP?**
5. **Do I want Big Diameter, Narrow Blades, or stubby Blades?**
6. **Will a Super Airfoil make a Super Prop?**

You'll make CHOICES: C_L, C_D's for low Reynolds Numbers, ASPECT RATIO.

If we **just use Betz Logic, Goldstein-Theodorsen Math, *no tricky changes,*** with an **optimum** α^o **and** C_L, we get a **Classic Triple Optimum, Min. Induced, Profile, Torque, Min. Area Precisely Placed, Ultimate Classic Logic Prop,** ***Norris's 6th. Law!***

IT'S ALL ABOUT CORRECTLY LOADING THE BLADE VS. RADIUS!!!

That fits you right into the B-G-T **Family of Classic Shapes**, vs. Advance Ratio, the **Precisely *Correct Twist***, for ***Constant Pitch***, ***Constant Slip***, of the **Air Inflow, that accounts for the** α^o **and** C_L **you choose - for heavy loading - where the** ΔV **at the blades is faster than the stream tube average, and heavy loaded Stretched Outflow --- sizes the Diameter, Area - Shape, that are *inter-dependant, can't be separated* -- to use the overall max. Aspect Ratio you chose, and judged sound for vibration.**

WE HAVE THE COMPUTER, AND RARE GENIUS, ELEGANT MATH TO DO THE WORK FOR US!!!

We've learned the **prop EFFICIENCY limit is set by the Plane SPEED,** and **Engine RPM, which sets Pitch, and Advance Ratio, which sets the Max Efficiency Limit, our core goal**!!!

We've learned that we had to **add a drag subroutine** to the **Classic Drag Free Math Program**, and with blades having the speed of Jets, but the small Chords of Model Planes, **C_D's must be higher than normal, for Low Reynolds Numbers, often below 1 Million**, and **thus we'll not get super low drag from exotic Airfoils, especially** with **bugged or eroded leading edges**.

FINALLY, WE'LL ADAPT THE BASIC DESIGN FOR SLOWDOWN!!!

We'll ***adapt the basic design*** to work in the **Embeded body** nose velocity profile, **shallower angles**, ***especially inboard***, **wider blades**!

REALLY IMPORTANT INSIGHTS & CONCLUSIONS

To get a computerized propeller design really correct it is necessary to recognize and do *three things* that are *not normally understood*, in many cases, even by professionals.

1. To have a **correct Thrust and HP design target** it is necessary to recognize that the required **Thrust** and **HP** can be from ~**3%, or more to 10%+, higher than the plane's gliding, or accurately calculated profile and induced drag**, due to the **interference efficiency,** η_{i}, caused when the extra **ΔV over the embedded body**, creates **extra scrubbing drag**, and possibly **separation drag**. Inefficient, misshapen ***low pitch props*** with more ΔV, typically ***worst***. Separately, cooling drag can be 6% to 10%.

THE Power Loss can be *QUITE LARGE,* especially considering COOLING DRAG!

2. The **truly accurate Theodorsen 3D design mathematical method** accurately accounts for ***heavy loading,*** which can require ***typically 5% higher pitch*** and in addition supplies an **accurate ideal Betz radial loading design**, more accurate, **a little lower thrust and efficiency**, by accurately accounting for 3 Dimensional flow, **axial and radial induced losses, as well as stream tube rotational losses**, - (also *low Rn blade drag* when added).

3. ***Significant repitching is required to account for Slowdown,*** the body pushing air ahead in the prop plane, perhaps 3 to 4% at the prop tip, ~12% at the half radius, 25%+ at 1/3 r depending on the prop and cowling size, and the prop location. *Prop extension?*

Use Theodorsen's Design Method. It is essentially Exact.

The essentially exact Theodorsen math method, which took 83 years for history's smartest analysts to finally conquer, which comprehensively takes care of ***heavy loading*** and ***high advance ratios***, when **.1 degree design or manufacturing tollerence** is proper, makes it a bad idea, simply unnecessary for amateurs or professionals alike to try their own less comprehensive methods. ***A greenie doesn't have a prayer, and all but the most rare, unique and brilliant modern professional would not either!!!***

C_L Logic --- Selecting a Design C_L vs. Diameter

You know that α° **and** C_L ***always increase as you slow down***. The key is **the opposite is also true**, α° **and** C_L ***always decreases as you speed up, even up to and through Vmax level***, then in a shallow dive to and **through zero thrust at Max Rated RPM!!** There is *great significance to that, how everything works out,* not clear until you study everything, see through it.

Here Goes! CL is Powerful --- But **SELECTED** ALT. CRUSE **RPM** is a **KEY $$$ Decision, Also!!!**

You'll want to **design your prop** for your **optimum altitude cruise condition, where your fuel money is spent, *at your best overall*** C_L, **(.5 to .55) right up as close to the peak of Andy Bauer's Magic Graph as it will calculate!!!** THEN, **at the max RPM S.L. Vmax condition, you'll move *to even higher* efficiency, eta, η, even *more efficient at a lower*** C_L, ***ΔV, ever lower induced loss, the key!*** YOU'LL SEE, **both efficiencies can maximize!! Conversely, if you design for Vmax, you're not designing for the condition where you spend the most time, spend the real money!**

Let us now show you a ***Cruise*** **design study *at Altitude*** that I asked Andy to run, that **shows how** C_L**, Diameter-Area, and Efficiency, all relate, *then* how they perform in *Climb at Sea Level and Altitude* - *Vmax at sea level,* and *Takeoff at sea level the problem!!!*** Andy's program is **modeled for actual engine power and torque, *therefore models a real set of tests!***

(Next, 96.8 is max HP available at Altitude, 2400 RPM, *not* all used, Throttled.)

Design Study: 160HPSL RV6 CRUISE, at **12,500'**, **170 TAS**, **140 IAS**, ***163.7# required Thrust*** = (152.24 # drag /.93 *extra* η_i ***conservative***, ~ 82.8% overallη_P)
2400 RPM, **Aspect Ratio 14 : 1** (based on outer.9r) Study Airfoil, NACA 4412
Do a **Study of five moderate** C_L**'s to find the optimum case**. ✓ **Drag %**

C_L	Dia."	Area ft²	η	HPreqd	J/π	Pitch 3/4r	β°3/4r	C_D 3/4r	% of Loss	ΔV ft/sec
.3	**83.705**	2.815	.88315	84.625	.2844	74.524	20.74	.00900	**62.35**	10.07
.4	**77.308**	2.401	.89290	83.549	.3080	79.443	23.61	.00900	48.86	11.54
.5 ✓	**72.679**	2.122	**.89319**	**83.346**	.3276	84.120	**26.21**	.00917	**38.40**	12.79
.55	**70.788**	2.013	.89111	83.435	.3364	86.434	27.44	.00936	34.32	13.34
.6*	**69.103***	1.919	.88840	83.615	.3446	88.713	*28.64	.00962	31.03	13.86

Note: Pitch here appear larger than normal, since study purposely is *without* slowdown. --- ***next***

*** Diameter shrinks, exaggerates β 3/4 r and Pitch 3/4 r increases. Nominal .3 Cl, 3 deg. Beta Increase is ~4 real, shows as 8 Degrees! WHY? There is ΔV and radius change too ---**

It is obvious that C_L **changes the diameter grossly!** ***LOOK how small the efficiency change is!!*** But .3 C_L loses the most, and ***requires an excessive diameter***. **.5** C_L **is best, but only narrowly. Thus you *could* choose .55** C_L **based on weight.** The efficiency change is small, **because blade drag lowers enough with rapid diameter reduction to balance the increased stream tube induced loss!!! Note drag goes to 62% of the total loss at .3** C_L**!**

It would be easy to misunderstand that C_L and Diameter-Area are not critical, can float around with little effect. **WRONG!! Once a diameter is selected, the prop *has to be pitched accurately* to *get the design C_L* or you lose your prop-engine-plane-RPM* match, *a Speed vs. RPM change!!***

***Less Pitch, RPM up, More HP, $ Cost, Plane Faster — More Pitch, Less RPM, HP, V, $ Cost.**

Sea Level Climb Performances -- ***of props designed at altitude cruise.***
At 1600# max GW, 105 MPH max L/D speed, (above min power, max climb).
With ***extra*** η_i **loss, 130.23# gliding drag/.9412** η_i **needs 138.3# thrust *to fly*.** ✓

C_L	RPM	η	Thrust	HPAvail	T#climb*	Climb*	J/π	$C_{L3/4r}$	$C_{D3/4r}$	DW°
.3	2019.7	.8276	381.9	129.21	243.66	1407.2	.2088	.680	.0097	4.65
.4	2114.2	.8062	**384.8**	133.67	**246.60**	**1424.2**	.2159	.853	.0115	5.86
.5	2181.9	.7851	383.7	136.83	245.42	1417.3	.2226	1.008	.0144	6.98
.55	2218.2	.7728	382.3	138.51	244.06	1409.4	.2248	1.085	.0165	7.58
.6	**2251.4**	.7614	380.8	**140.04**	242.53	1400.6	.2268	1.159	.0187	8.15

*** This Shows Max G.W. Climb. Light it's ~ Inversely Proportional to G.W., MORE!!!**

Here we see two conflicting logics! The prop design with an overly large diameter, an excessively low .3 C_L, has a new **3/4r C_L *not*** too big, but with high drag turns less RPM, less HP, but at good efficiency, fair thrust. **The RPM, HP winner is the *small diameter*, *high .6 C_L* prop, that can be dragged through the air well, now at *even higher* C_L, but less thrust, lowest η! Max available thrust for climb, max climb is the .4C_L prop.** The .5C_L prop is close, ***the .55 prop 1% less***, but ***lightest, a tradeoff***. **A narrow tip .5C_L Betz prop has ~ 2.7 inches more D", *vs 70"D often used*.**

A Broad Tip Prop Calculates to a Smaller Diameter, 70" for an RV 6, at lower η!!!

Altitude Climb Performances -- ***of props designed at altitude cruise.***
At 12,500 ft. Cruise Alt., 1600# max GW, 105 MPH IAS, 127.18 TAS
At same IAS, 130.23# gliding drag/.9412 η_i **still needs 138.3# thrust *to fly***

RPM	η	Thrust	HPAvail	T#climb*	Climb*	J/π	$C_{L3/4r}$	$C_{D3/4r}$	DW°
2119.4	.8629	221.6	87.04	83.30	582.42	.2409	.5213	.0099	3.29
2184.0	.8530	223.9	89.00	85.70	599.14	.2531	.6726	.0108	4.25
2235.5	.8397	**224.3**	90.55	**86.06**	**601.68**	.2630	.8144	.0121	5.20
2257.0	.8325	224.0	91.20	85.71	**599.24**	.2674	.8826	.0128	5.67
2276.6	.8254	223.5	**91.78**	85.21	595.72	.2716	.9495	.0137	6.13

Again we find the original .6 C_L prop has the most RPM, the most H.P, but the .5 C_L with **more Diameter and efficiency yields the most Thrust**, the **most Climb**. But the climb rates are so close, you'd never know the difference, and we **could still choose the .55 C_L based on weight**, with **only 2.4 ft / min. less climb rate.** (***Actually***, the program, a tad low on HP vs. RPM at altitude, **.55 climb ~ equals .5 !!!**) **Interestingly**, notice that the **available RPM is higher at Altitude Climb**, because the **127.18 MPH TAS** is **quite a bit faster than the 105 IAS-TAS Equal Drag case at S.L!!**

200 MPH Sea Level Vmax Performance -- *of altitude cruise props.*
247# gliding drag, with high η_i loss may need **260#. to 268.8# T at 200 MPH**.

RPM	η	Thrust	HPAvail	J/π	alpha 3/4r	C_L 3/4r	C_D 3/4r	DW°
2743.1	(.8812)	267.4	**161.84**	.2928	-1.3363	.2587	.0090	1.45
2732.0	.8975	271.5	161.37	.3183	- .4201	.3508	.0088	1.94
2717.5	***.9019***	**271.8**	160.75	.3404	.5039	.4414	**.0088**	2.44
2709.0	.9018	271.2	160.39	.3506	.9602	.4861	.0089	2.70
2700.4	.9008	270.3	160.02	.3602	1.4138	.5305	.0090	2.95

Notice how the efficiencies *here* have gained, (not .3 C_L), *both* Cruise and. Vmax optimizing, -- with Vmax climbing closer to the peak of Andy's Magic curve, -- than the cruise case can get, slower, more draggy C_L's, J/π's, than the Vmax cases. It's wonderful to get insight this incisive.

• • • (72.3 % radius data is used for 3/4 r data -- on these extra charts)

Here you can see that with all the angles of attack and C_L's falling due to higher speed, the original 72.7" .5 C_L prop delivers the max thrust, highest efficiency, **but the original .55 C_L, 70.8" prop is still a good compromise, only .6 pound less thrust, with a tad less efficiency.** The RPM's shown are the limits, set by engine torque, the RPM's shown the max attainable, (but some like **Vmax to overrev.**)* **All these props can overrev in a dive.**

* **An overrev design** gets you a faster cruise, **but costs cruise fuel $!!!!.**

We've consistently emphasized that **props want to be a very precise device, pitch held to .1 degree** as McCauley does. With everything floating around as it does here, obviously there are choices, but be sure you grasp that ***once a selection is made, if you miss the pitch, your pitch no longer matches the diameter and blade, and you've lost the combination, though it may only change RPM and Speed a little!!!* A 1° error is big!!**

1 degree, .1 Cl, at .5 Cl, is a 20% error, a big change to RPM vs. Speed match!!!

Sea Level Takeoff -- of high Pitch RV props *designed at altitude cruise.*
At 1600# max GW, 50 MPH RV 6 minimum Takeoff speed. you want all the Thrust you can get for acceleration, but these Props are STALLED*!

RPM	η	Thrust	HPAvail	R_n .9r	J/π	alpha°	C_L 3/4r	C_D	DW°
1862.2	**.5585**	**509.6**	121.64	963,330	.1078	7.658	1.145	.0180	10.36
2011.9	.4967	479.9	128.84	913,420	.1081	9.639	1.337	.0240	12.11
2147.4	.4407	446.9	135.23	883,530	.1077	11.43	1.491	.0295	13.77
2207.7	.4049	419.1	138.03	871,950	.1075	*12.27*	*1.485*	*.1171*	14.58
2262.5	.3690	388.9	**140.55**	861,230	.1075	*13.08*	*1.159*	*.1379*	15.34

All props are *Stalled* - to the 36.7% radius on the .3 Cl Design Prop. 45.6% at .4 Cl, 63.4% at .5 Cl, 73.2% at .55 Cl, to the 81.2% radius at .6 Cl. This is the Hidden secret on Props for high performance homebuilts.
It's particularly crucial on the otherwise great Long Eze which has ***a too small* ~65" *restricted prop diameter, lacks good slowdown design*** at the rear.

* We'll soon see Slowdown Correction Helps a lot!!!

Realize here, the marvelous Insight available with the modern, speed of light personal computer, with the Elegant, rare genius level math that we have inherited. It's available, the product of 83 years of Historic Technical Work! The *relative* ease of use of **Theodorsen's Math**, the ability to **learn Source Sink Analysis**, makes the use of **BGT logic, and Math**, **for Ideal Prop Design, a task that can be broadly used** to **get Props out of Cut and Try Black Art**. **Andy has the program solved, available!**

Incisive Insight into the Blade at All Radii -- Climb, T.O. Stall

Our Design Study gave great insight into all props, at all key flight conditions, but in a simple format could not show what happens at *all* radii in Altitude Climb, and Stalled at Takeoff

CLIMB at 12,500' - .5 C_L case *ALL* Blade Radii Analysis

105 IAS, 127.12 TAS, 2235.5 RPM, 90.55 H.P., 224 # T, η .8397

% r.	β°	φ	α°	C_L	C_D	Rn_{10^6}	DW°	Eta, η
10.0	75.18	70.88	4.300	.8138	**.0162**	**.2450**	3.159	.8531
18.9	62.78	57.39	5.391	.9220	.0146	.4460	5.783	**.8594**
27.8	**52.70**	**47.12**	**5.574**	**.9401**	.0142	.6345	7.014	.8520
36.7	**44.80**	39.40	5.402	.9231	.0136	.7915	**7.187**	.8466
45.6	38.66	33.54	5.123	.8955	.0130	.9016	6.839	.8431
54.5	33.86	29.03	4.830	.8664	.0125	.9621	6.301	.8405
63.4	30.05	25.49	4.555	.8391	.0122	**.9728**	5.733	.8382
72.3	**26.98**	**22.67**	**4.306**	**.8144**	**.0121**	**.9345**	**5.198**	**.8357**
81.2	24.46	20.38	4.084	.7923	.0121	.8365	4.719	.8327
90.1	22.37	18.48	3.886	.7729	.0124	.6518	4.298	.8286
99.0	20.61	16.90	3.711	.7557	.0165	**.2205**	3.931	.8106

Notice Here: The angle of attacks, C_L's and C_D's are reasonable, the 127 MPH TAS better than S. L. Climb IAS, but the C_D's are up somewhat, **above .01**, (at 12,500'), the **Reynolds numbers below 1,000,000**, down in the 200,000's at Slow Root and Narrow Tip. We've Highlighted the ~3/4 radius data, 72.3 % r, used in the Study. **45° φ, - half the Total drag angle**, (**Profile *and* Induced**), is normally the **best radius at the design point**, max. efficiency, ***but*** off the design point here things get distorted a bit, as you can see. For insight, the .55 C_L prop has higher angles, ~1 degree, more ΔV too.

S. L. TAKEOFF -- .5 C_L case *ALL* Blade Radii Analysis

50 MPH IAS/TAS, 2147.4 RPM, 135.2 S.L.H.P., 224 # T, η .4407 *!!!*

% r.	β°	α°	C_L	C_D	Rn_{10^6}	DW°	Eta, η	S.D° *new°*	α° *new°*
~~10.0~~	75.18	*21.98*	*1.045*	*.4267*	**.1792**	11.31	.2797		
18.9	62.78	*22.32*	*1.040*	***.4329***	.4149	20.40	.2879	- 17.6°	4.72°
27.8	**52.70**	*19.65*	*1.095*	*.3757*	.6825	**22.73**	.3009	- 11.4	8.25
36.7	**44.80**	*17.17*	*1.232*	*.2715*	.9278	21.74	.3472	- 6.9	10.27
45.6	38.66	*15.20*	*1.340*	*.2013*	1.1143	19.69	.3763	- 4.0	11.20
54.5	33.86	*13.66*	*1.421*	*.1540*	1.2302	17.51	.3970	- 2.5	11.16
63.4	30.05	*12.43*	*1.479*	*.1211*	1.2726	15.51	.4131	- 1.6	10.83
72.3	**26.98**	**11.43**	**1.491**	**.0295**	**1.2419**	**13.77**	**.4960**	**- 1.0**	**10.43**
81.2	24.46	10.60	*1.425*	.0274	1.1244	12.29	.4953	- .72	9.88
90.1	22.37	9.89	1.492	.0259	.8835	11.04	.4983	- .49	9.33
99.0	20.61	9.29	1.200	.0268	**.3008**	9.97	.4863	- .19	9.09

Look at all those wild ***stalled*** α°'s, C_L's, C_D's in *Italics* **It's worse slower, and at .55 C_L +~1° stalled** one station **farther out.** Notice how high the C_D's go, post stall, leaving **a big separated wake**. **But LOOK: The Slowdown Correction, eliminates the excess, on the edge of stall, but all radii unstalling just as Takeoff starts, even with this Fast, High Pitch Prop, even at a slow 50 MPH takeoff speed, for such a fast plane. A Slowdown Corrected prop is *fundamentally fixed,* unstalled for Takeoff. Of course it's a mess at Runup, and rolling to T.O.**

Looking at some Key Insights Visible in the Study.

Variable Pitch Constant Speed Props: We're purposely studying fixed pitch Props. They're the ones to learn on, so you understand the tough parts, everything. A variable pitch prop is the easy way out on a stalled root problem, on the otherwise hi efficiency, high pitch props, and clearly the way to go if you have a fast plane that justifies the expense, weight, complexity, especially if your plane has a T.O. problem!

The Cost of Missing your Pitch: We've been saying that Pitch must be precise, within .1 degree, either as **a manufacturing tolerance** or **a proper precision needed for a computer design** calculation. The easy way to understand is that there is only **.95° difference** at the 3/4 radius between a relatively free turning 48" P, 71" D Luscombe climb prop vs. a relatively lugging 51" P Cruise Prop -- which ***limits the altitude cruise RPM to about 2300 RPM***, (more economical than a climb prop that would turn about 150 RPM faster), **slower like you'd use a Constant Speed Prop in economical cruise at a lower RPM!**

To yield a consistent manufacturing product McCauley properly uses a +/- .1 degree blade tolerance. **Remember the blade has to screw ahead at plane speed, but then be accurately overpitched *only* enough to account for and pull in *only* a small ΔV/2, *only* a few percent of V, plus a degree or so angle of attack.** **Δ1º error is:** 16.9 º(3/4r) vs. 15.9 º is **1.06289**, a 6.29% ballpark speed loss, ~ **6 MPH at 100**. or **2444.6 RPM vs. 2300** an increase to get back to the same speed, on **a simple ratio basis**, **purposely skipping the precise computer to keep it a simple calculation. You can easily think and see through that yourself --- no computer!!!**

DEAD SIMPLE, the Ratio Method is only a bit off the exact calc!!!

Cruise Design RPM: **A Key Cost vs. Speed Decision** Excess pitch, (or too big a prop), lugs the engine, less RPM, less H.P. less Speed, it probably can't reach rated RPM and H.P. But too little pitch, extra RPM, H.P. and Speed, **overreving, at Vmax what the go fast guys do on purpose** --- but you see **that costs you economy every hour you fly. The economy game is to lope along, wide open, leaned, the engine perfectly matched to the plane, High, getting free TAS, (but not excessively high where your engine loses too much efficiency). You're not going slow, you're going Vmax at altitude,** , but you're **using the Prop and its design as though it was a costly Constant Speed Prop. Realize, fixed pitch, *you never get or use Rated RPM*, and H.P. unless you're doing a buzz job, on the deck, and who cares, *or can judge your exact speed, V?*** **Going nonstop, you beat the go fast guys!!!**

Selecting Cruise RPM is one of the most important choices in designing a prop. Make it low enough and you have a good economical cruise prop, wide open, a best chance at good mixture distribution, but can't go as fast. *Higher RPM, you have the go fast option too, **but throttled may have uneven mixture**.*

Digressing to the Merits of Electronic Fuel Injection.

We've been to the Moon, but we're flying with 1930's technology. Electronic ignition, Leaning, ***Fuel Injection covered in the Engine Chapter can bring exact Science to economical flight***. A carburetor can be an amazingly subtle, sophisticated device created by masters, in the best cases, but it's also potentially a crude device, and **a canted throttle plate** can be a busy **liquid droplet separator**, ***assuring you won't get an even mixture*** distribution, unless you get both Smart, and get Lucky.

If we can inject ***even***, ***lean*** fuel shots, with **knock proof variable advance electronic ignition**, we can fly with modern Science. Till then try to get ***even fuel distribution***, ***wide open***, ***lean***.

Because we can fly at altitude we can do what autos can't, *reduce our power, without throttling*, avoiding the pumping loss sucking air in from a duct throttled *below atmospheric pressure*.

Weight: A really interesting insight is that if a **constant Aspect Ratio** is held, the Diameter, Chords and Thickness go up in proportion, **a cube of the ratio**! Look at the study .3 C_L vs. .55. 83.705" D. / 70.788" =**1.18247**)^3, =**1.6553**, a **65% weight penalty**! .5 C_L vs. .55, it's 72.679 / 70.788 = **1.0267**)^3 = **1.0823, 8.2% penalty! Clearly weight is a proper consideration, .55 C_L a possible alternative, shown purposely, but do consider stall, of course**.

RPM, HP, Thrust *Slow* -- Big Dia., Low C_L vs. Small D, Hi C_L

For years I've wondered whether it was best to have a Big Diameter, Low C_L, prop ***Slow*** -- or vice versa -- from an RPM - HP - Thrust standpoint?? Well there's the answers, sitting right there in front of us, just begging to be understood. With all the complexity and interdependence, there is no way any of us mere mortals will see through that one cold. **Fantastic Insight: In slow climb, the High C_L Small Diameter wins the RPM and HP race, *but not* the thrust race!!! LOOK!!** But, ***conversely* the *BIG* D prop has max RPM and HP, at Vmax, *but* the middle .5 compromise is best!!** ***Insight you just can't find!!!***

Really Understanding the Aspect Ratio Subject

Aspect Ratio High AR props are more efficient than *low* AR props, ***1 to 2% better***, going up from ***12 to 14:1***, ***low to high J****, λ. A friend, whose bright comments I value, cautioned me about pushing that concept, arguing that it was really the diameter increase that caused the improvement. Smart people will regularly show you it's really lowering Span Loading on wings, by increasing Span, that knocks down wing induced drag.

*LOOK AT THE ADVANCE RATIO GRAPH, p. 48 II, TO SEE THE GOOD EFFECT OF ASPECT RATIO.

Smart guys know even more good fundamental reasons why a big D, should be the driver here, a big D, a bigger Disk Area increases $\dot{M}$, mass flow rate, lowers the demanded ΔV stream tube Speed Increase, a lower $\Delta V/2V_1$, less induced loss, more efficient. Additionally, pros know the Thrust coefficient is a D^4 function, a D^5 for the Power Coefficient, the modeling tools a pro uses to scale props, small test props to big, at NACA, for example. So to a pro ***D the driver*** seems a pretty pat conclusion. **All the smart guys know D is an 800# Gorilla, in Prop logic**.

But look what the Study just taught us: D vs C_L can swap, cancel each other, **relatively little efficiency or other change at *constant AR***. **More Diameter got you more blade area, drag went way up**. Thinking back to the Advance Ratio Graph, calculated with all up math **Aspect Ratio is *itself* a key driver** at constant C_L, constant Advance ratio! ***What's going on here?***

You only need so much prop *Size*, (Dia. Area, Shape) to meet your **Thrust requirement**, at Spec. conditions. As AR goes up, yes D goes up, but **now narrower tips keep Drag loss in check** and **we get the efficiency gain** we didn't get swapping D vs, C_L. Yes, ***D is important, like we learned***, but ***AR is also a true basic***

How all this works out, the fundamentals and subtleties, the true interactions that you can learn from this study, is one of the most important Insight Opportunities you will ever find on Props --- because it is as close as you will ever get to a grasp that is all encompassing!!! ***Think it through, Soak it up!***

A Bottom Line Insight -- Ideal Betz Props vs. Black Art Props

We've been teaching Ideal Theodorsen-Goldstein-Betz Props for 3 Powerful Reasons:

1. **There is a Precise Design Math** developed by Rare **Genius** Historic figures that allows us to get creditable ~ **exact design point answers, only adding Profile Drag, for next level insight**.

2. **Betz Logic** provides the **magnificently orderly, logical insight** basis we find **in Nature**, **Science**, that **makes one of the most complex, convoluted problems in Engineering, explainable, understandable**, all from **simple helical flow**, and a **constant dT/dQ**, **every radius equally valuable, efficient**, that **cuts through the gross complexity like a Magic Knife.**

3. The concept of **Ideal Minimum Induced Drag Design**, with **Practical, Helpful, *Easy Enough* Math**, offers the hope of finally **getting propeller understanding and design out of the Black Art Category** it's been **stuck in for 135 years** at the Millennium --- **Getting Propellers Properly Efficient!!!** --- even win **proper recognition** for the 7 **Intellectual Heros** who over 83 years **won us an Essentially Exact Analytical Solution**!!!!

Rankine, Froude, Betz, Prandtl, Goldstein, Glauert, Theodorsen!!!

It's always proper to recognize the Giants, rare Genius Work!!!

If you are maybe beginning to **realize the initial Mental Jungle** that **Propeller Logics were, before they were rendered orderly** using our 7 heros, you'll know how indebeted we are to their past genius!

So Now, **What is the GIANT BOTTOM LINE to all this?**

If a normal **Clean Wings** can have a **20 : 1, to 25.:1 L/D**, that's only a **4% to 5% loss**, a **95% to 96% efficiency**. Now, if a perfect BGT prop normally falls into an 85% to 90% efficiency band, 10 to 15% loss, ***or worse, if Black Art Props***, 2 to 3 times worse, it's time we **understand WHY**, ***get everyone aware***, ***get them right,*** a full half Century late, look to how we can **get more of an Elliptical Thrust Distribution** for Century 21, maybe get 92% to 93% η. TIP LOADED WINGS ARE BAD, DUMB, **Do Better!**

It makes absolutely **no good sense** to **expose excess tip area to the high q, high profile drag acting at the longest possible radius arm against engine torque** --- **when the productive thrust you hope for is just going to DISPROPORTIONATELY roll off the tip, lost, but its cost paid**. It's **just not possible to maintain a high thrust at the tip, it just rolls off** and at the high q cost, **proportional to the high tip (velocity)** ^{2}. **(Go look at Prandtl's F factor plotted on page 98 - II to grasp how strongly a calculated thrust is knocked down at a tip. It's riveting, thinking of a square tip)**

Once you see the **modest inboard Thrust vs. Radius Loading of Betz plotted,** like p.92 - II, **realizing that tries to stall slow, (OK with S.D correction??) that area might be narrowed to cut weight!!**

WANT TO ENLARGE THRUST -- OUTSIDE THE COWL TO ~75% RADIUS IS THE PLACE TO GO!!

So **the world makes broader chord outer radii props, going for a better combination climb-cruise-takeoff prop???** No sweat, we mistakenly thought at first, we can **just unwind the blade angle to a lower** α^o **and** C_L, add chord, shrewdly **maintain Betz's ideal Thrust vs. radius loading. unwind it ideally. WRONG!** We promptly got the excess drag you could predict, **lost 40 RPM and 5 # Thrust** for Altitude Climb!!

Cutting diameter, we restored RPM, got Thrust, but started getting **more induced drag** naturally, **lost efficiency, required more HP**, could at least play with it, **not lose too badly, but ALWAYS LOST**. It's positively **amazing how smart Betz's insight proves to be, a 1919 wonder. (What we did gain was the ability to help the gross inner blade stall problem** we just demonstrated in the study, **on otherwise efficient high pitch props. Slowdown makes very big angle corrections inboard, especially in front of the cowling, and as shown in the angle corrections in the last column on p. 125 II, there's lots of help**.)

The point is IT ALWAYS COST US!! We learned how to fix a problem intelligently, neatly, go backwards the least!!! The basic point that dawned on us pretty quickly is that **not only is a Betz prop best, a constant** C_L **with the min. area drag wins!**

An Important Grasp of the Overview, The Final Results ---

We've been teaching you the Logic of Propellers using the **Betz** Minimum Induced Drag concept, and the **Theodorsen-Goldstein** Math, specifically because Betz has a great ***understandable, explainable, learnable logic***, and Theodorsen Math gives us a credible, ***essentially exact solution***, a proper base for learning. **That is reason enough for using the method** --- **learnable**!!!

Now the obvious questions is **are these props actually better, should I be trying to use one on my plane?** When I originally wrote this in 2001, I had full faith, but said we needed to test them to prove them. No real technical Pro would answer differently.

With a lifetime of technical experience, and my products all over the world, a pro is always cautious, **makes all products prove themselves**. There is nothing in this business quite as good as actually making products prove themselves, *shake out the unforseen*. Jim Rust, WhirlWind Propellers, El Cajon, San Diego jumped at the opportunity to prove our 10 year's work. His new 200 RV Two Blade, and 151, Three Blade beat all his prior Products.

On a valid RV 8 test the WhirlWind 200RV beat all other props, except Hartzel's excellent latest, the only other basically Betz Theodorsen Prop, with all of Hartzell's best tricks, airfoils, a little extra tip area, **208.9 MPH**, 2500 RPM cruise, vs. the 200 RV **208** BUT, I'd expact **the *.9 MPH Δ is just what I'd expect forever*.** **A metal prop should always beat a composite prop by a tad, *finer edges*, but the Composite prop was a BIG 17# lighter, no vibration vulnarabilities, a BIG Plus**!!! ***BGT beat all here!!!!***

A HUGE test proof to me here, the worst prop was Europe's best, 200.7 MPH. 200.7/208.9 =.960746 $)^{2.5}$ = **.9047**, inverted = **1.10529**, **91/2 % Δ.**
A Speed increase is normally said to take power cubed. I conservatively used 2.5.
That was a three bladed prop, which normally **costs 1/2%** η. So lets say that this high pitch prop that was expensive and looked great, **cost 9%**, say 81% η, not the 90% a BGT RV 8 prop target. *How much can non BGT Props miss by?* ***This data*** **8 to 10, 9%.**

Just How Bad can a Non BGT Prop Really Be -- and WHY?

All of the Prop Math that everyone has worked on for 140 years, now, has been on how to design a prop Right, Correctly, doing a **poor, to good job,** (BGT), **of recognizing the *real* size of the losses**, and as we come to the goal line here, wrapping up our long project, to me that is clearly **the area of weakest insight!!!**

Flip way back to page I 18, and I 19 of the introduction, to me, among the most important pages in the book --- because they expose the core, insightful fact, that **a propeller is really, in effect, two outer wing halves, at best, ~no center wing, even with BGT loaded too far out, a big mass of excess radial flow, greatly excess tip vortex induced loss**, ***mathematically nailed on p. 32, and 33 of Theodorsen***, as **BIGGER than the AXIAL induced loss that we must have to make thrust**. To me that is the **fundamental reason an 85% to 90% prop efficiency is so much worse than a 20 to 25 L/D, 5%, to 4% loss wing!!!!!!!!!**

Most prop design, unaware, has no understanding at all that is the core problem, fully able to klutz it up, make it worse!!!! To me, that makes the 1 legitimate data point in Van's RV 8 test above hugely valuable, a good honest test that shows a good looking professional prop can rationally, conservatively calculate to 9%, or more, worse than a BGT prop, almost twice the loss, ~81% overall, vs. ~90% overall. (An old pro, I'm not about to lose my head over data of 1 point, but **that is hugely more important than none!!!**)

NASA, cutting funding on aircraft, Langley is now deemphasized, (Lewis actually did the last turboprop work), but we need some serious testing to get specific answers on real prop losses, specifically how bad wrong props can be. Bureaucracies do become useless. **NACA should have nailed Raspet's interference loss, got us real data, this level of explanation --- a half Century ago!!!**

If I could see we had no prop explanation as a kid, waited a half Century, finally had to do it with Andy Bauer, where was the Research Establishment? **I've never heard one word that propellers are fundamentally flawed, try to tip load, two outer wing halves, no center, BAD - - a HUGE hole!!!**

With ***five*** **C_L vs. Diameter Design Cases** laid open, then **explored for Sea Level Climb, Altitude Climb, Vmax,** and **Takeoff**, you are never again apt to get an opportunity with such all encompassing specific insight. Don't miss your opportunity to see and grasp the inner workings and interactions of all the real outcomes of propeller logic. **The insight is marvelous, and quite different than you might expect!!!**

C_L vs. Diameter -- Induced Loss vs. Profile Drag Loss *(Two Pairs)*. The **interaction *within* each pair**, then **the interaction *of* each pair** with the others **is extraordinary** --- seeing that large variation in the powerful basics of props produces ***only small differences in efficiency, and each performance characteristic***.

Diameter ***is*** *a powerful 800# Gorilla* in Props, but **see how C_L, another 800# Gorilla makes such large changes in Diameter**! Likewise, **see how induced loss** and **Profile Drag** loss **interplay**, obviously **as diameter goes way up induced** loss **drops**, but **profile expands startlingly to counterbalance induced** loss, **becomes 62% of the loss**, but **only relatively small changes in efficiency, thrust, H.P. amazing insight.**

Vmax, $\dot{M}$ up, ΔV, C_L fall, efficiency improves, logical after you think it through, as explained in the text, **both altitude cruise and Vmax efficiency as good as they can be, *winning 2 ways!***

Takeoff tries to be the problem case. Those props with a **P/D**, Pitch / Diameter Ratio **above 1**, ***the only ones that can be truly efficient are the ones that have a substantial Stall Problem*** at Takeoff. But, see on **p. 125 II how *Slowdown really helps stall!***

Notice the Study fine print offers Extra Insight!

Climb, ETC. Look at **how altitude affects H.P.**, **how Speed effects RPM, thus H.P., thus Climb**, more RPM at Altitude due to a higher TAS. **Marvelous! A treasure trove of Insight!**
Look for yourself. See what else you can see, EXPLAIN???
Marvelous! Am I too enthusiastic? You can't buy this $$.

Learn from your Friends. With different backgrounds. and interests, your friends, and associates, over a lifetime, are apt to know all the things you don't. Listening, learning, growing can be the secret to maybe even getting wise. And ***friends insights, queries, taught me many things I needed to clarify in this book***

An ace Engine man ***quizzically*** said, **"Propellers seem to be pretty forgiving"**, a casual, but **very perceptive truth at several levels**, but exposing hidden lies. I instantly knew I had to **show how and why there was both truth, and lies there too.**

1. We learned early that it was **the Speed that determined the Mass flow rate, the RPM that divided the Speed into a High or Low Pitch, or Advance Ratio, thus the steepness of the corkscrew path of the prop Tip, that sets the cap on efficiency!**

The final steps to Max Efficiency hide in the final BGT details!

2. We've all seen **highly variable shapes, and they all seem to work OK** -- but the Truth is Hidden. The **Hidden Secret is that a Speed Increase requires Power³, cubed,** (or at least a 2.5 power). **A Careful Test is Required! Airplane Math hides Poor Props.**

5% more efficiency only gets you ~2% more Speed! $(1.05\ HP)^{.4}$ = 1.0197-- 10% only 3.9%!

3. We just showed you in our comprehensive study that TWO 800# Gorillas of Prop Logic, **Diameter and C_L simply swapped with each other, facilitated gross changes in Diameter with amazingly small effect on efficiency, all the key performance specifics**. **Profile drag, normally a lesser ~ 1/3 loss, jumped to 62% --- as induced dropped --- as Diameter got quite big, (but too heavy) - *startling***, and ***vice, versa, a small D, a balancing wash!*** It starts looking like nothing matters, but of course, that's a lie.

You can match Torque, get your RPM, but configure a poor Prop!

4. Another friend, **nailed a *Correct* objective, but caution here. The first job of a fixed pitch propeller is to load the engine correctly, get the correct RPM, get available HP**. ***Simply one must get torque vs. RPM matched*** That's absolutely true, basic, ***but*** economy argues we really do ***not*** want to design for an excess cruise RPM, or you'll be burning a lot of fuel, and not going much faster. **You want to be able to do economical**

cruise, with the engine as efficient as possible, more subtly, matching the plane where it's efficient, *"max speed vs. drag, at Cruise RPM*". My Luscombe *with only* 67% overall propulsive efficiency, BAD, can fly 800 Miles, at 100 TAS, LA to McCall Idaho on its 30 gallons, only 33/4 GPH, 26.66 MPG, at 2280 RPM. Its Cruise Prop ***can't reach 2575 rated RPM***, only **2535 Vmax on the deck**. **A big shortcoming? I couldn't care less**, a bit less climb. ***I'm cruising just like I had* a constant speed prop!**

You set Economy, Range, Vmzx, climb, when you set Wide Open Power vs. Cruise RPM

5. We've learned another key insight that looks fairly close to another wash, but isn't. **Because of Item 1. there are only small efficiency differences, a few percent, with some undesirable moves, and that's a trap. It turns out efficiency is not the whole story, you have to look at performance specifics to see if you lost there**. If, for example you hang extra tip area out there at less α° to try to counteract a stalled root, you drop RPM, also you don't want thrust at the tip where we just took it away, get less, lose more profile and induced. Just use Betz Goldstein, Theodorsen as the logic intends to maximize efficiency, results.

6. Here's the Bottom Line, a FACT, a TRUTH. In 1990 at a **CAFE 400 Comparative Aircraft Flight Efficiency contest, Dick Van Grunsven, of RV Design Fame ran a test of all the supposedly best props for an RV 6 --- and there were significant performance differences, design does count**! Many designers had substantially **missed the RPM target, missed on SIZE, Diameter, Area, Shape, or PITCH, RPM simply too high or too low**. But there were **several designs** that got that correct, but then **went on to work better or worse, so you'd buy one, not the others**! ***There were winners, losers - off target.***

7. **Design Does Count, Betz has True Merit: That narrow chord tip gets you More Thrust for Less Drag, *BGT* Optimum. Slowdown correction makes the excessive inner Beta Angles of efficient high pitch props much lower for slow speeds, a big help. *Want to do even better, try an Elliptical distribution!***

The Speed of Sound, Mach Number Effect -- Mach Limit

As a body moves through the air it sends a pressure signal wave ahead **at the speed of the molecular motion speed of the gas**, and thus related to its temperature vs. **Absolute Zero, - 459.7 F°** **where all molecular motion stops**. The Speed of Sound is quite logically the speed at which that pressure wave propagates. If that speed is exceeded, the air can't signal ahead, piles up in a pressure wave, and wave drag is created that acts very much like the bow wave of a ship. **The relationship is proportional to the square root of the Absolute Temperature** in **degrees Rankine, R°, 459.7° + the degrees above Zero° Fahrenheit**. It is **a function of temperature only**, surprising to some, at first, is **completely unaffected by pressure**. On a prop, logically, it's **the combined vector velocity of the forward and rotational velocity**. **(Remember school, the hypotenuse of a rectangle is the square root of the sum of the squares of the sides.)**

Sonic Limits Mach 1 on a standard 59° F day, (518.7° R) is 1116.46 ft/sec, 761.243 MPH, and **decreases at altitude proportional to the square root of the absolute temperature**, ***unaffected by pressure***, to 968.09 ft/sec, 660.06 MPH at -69.7° F. at 36,089', the beginning of the, **(in theory)**, constant temperature stratosphere. Even relatively thin airfoils may have a 30% speed up, which can cause standing shock waves, like you can see on a jet wing surface on an East or West trip when the light is correct, but those are *weak shocks*. **On Prop tests the drag rise really starts at ~ .9 Mach, 1004.84 ft/sec, 685.119 MPH at sea level, 871.28 ft/sec, at Stratosphere** temperature, and ***the vector*** of forward and circumferential V ***must be kept below that .9 limit***. ***A 6' prop at 2700RPM, 250 MPH, is at a 924.1 ft/sec vector V***.

Now, that was the party line at NACA long ago, but John Harmon in his Rocket ran careful tests and went no faster above 2500 RPM, at 2550, with a 7 foot, 84" prop at 256 MPH, at a standard **~59 F°**. That's ~Mach .88, so we give you a graph, with John's ~.88 test point. **I'd thus be a bit conservative on .9**.

Understanding the .9 Mach Graph **It shows allowable V vs. RPM**

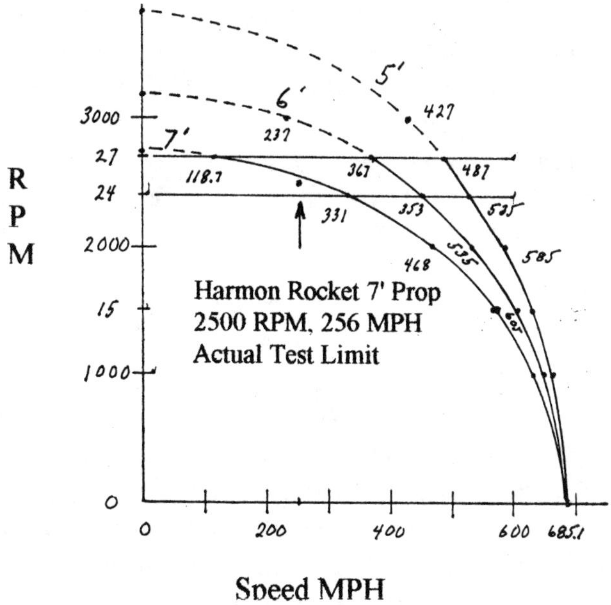

Realize it's better on a hot day, worse on a cold day, as explained vs. 518.7 R°, a standard 59 F° day, in Rankine temperature!

You can see that a **6 foot prop stays out of trouble even at 2700 RPM and 300 MPH**. John Harmon's 7 **foot prop gets into trouble above 2500 RPM at 256 MPH**. An **8 foot prop gets into trouble above 250 MPH, even at 2200 RPM, starts needing to be geared down if it's on a really fast plane.** The Furias and Dreadnought at Reno have a very low 3/8, 3 to 8 gearbox from a Boeing C 97 Tanker, only 1125 RPM vs. a 3000 RPM P&W 4360 engine -- would you believe **advancing 37.5 ft. per revolution at 480 MPH --- 4 1/4 Tons of air per Second!!!**

At 480 MPH, the **13.5 foot prop** turning that 1125 RPM, **18.75 rev./sec**. has a tip speed of **1062** ft./sec. / **1116.46** ft./sec. is **Mach .951279!** But on a 100° F day, $(559.7\ /\ \mathbf{518.7})^{.5}$, the Speed of Sound goes up a factor of 1.03877, so dividing the tip Mach goes down to Mach .915775, probably the reason that the Dreadnought goes about 480 MPH top. Can you see that OK?

Activity Factor vs. Aspect Ratio Activity factor is a Calculus based characteristic based on the following formula that ***pro's use to understand the "Power Absorbing" characteristics of a prop blade's shape*** --- where wide chords outboard are big power absorbers. Calculus Integration ***"Integrates, sums up" the effect of the blade chords, shape***. To understand, notice $(r/R)^3$ gives very heavy weighting to outboard radii chords, ie. $(.8)^3$, a much bigger radius ratio, gives a lot more weighting to outboard chords, because it's a bigger number, than $(.3)^3$, an inner radius ratio, a small number, gives to inboard chords. Do you get it? an r/R of .8 is out at the 80% radius, and that number cubed is much bigger than .3 radius cubed. (weights outer chords more)

$$\text{Activity Factor} = 100{,}000/16 \int_{.2\,R}^{1.0\,R} b/D \cdot (r/R)^3 \cdot \text{derivative}(r/R)$$

We don't want to teach you calculus so just see the "Integral sign" just sums up all the chords from the 20% radius to the 100% radius and the $(r/R)^3$ gives more weight to outer radii. The factor 100,000/16 just jiggles the numbers to make the answer come out to be from roughly 60 to 110, **100 nominal** for a wide blade at the outer radii, which absorbs a lot of power. It's all just a way to tell pros if you have a wide tip, a power absorber, or a narrow tip. ***Betz Props*** with ***narrow chords*** come out ***about 70***. ***Fast, Hi H.P. Mach Limited Planes use Hi A. F.***, ***but that hurts efficiency*** (I sure didn't want to get you into Calculus, but you should know about A.F.)

Thrust, Torque, Power Coefficients. NACA used to run small Test Props and Scale Data up or down using Thrust, Torque, and Power coefficients, and they are used in Theodorsen's Math.

$T\# = T_c\, \rho\, n^2 D^4$ $\quad Q\text{ ft.}\# = Q_c\, \rho\, n^2 D^5$. $\quad P\text{ ft.}\#/\text{sec.} = P_c\, \rho\, n^3 D^5$

These are quite useful and with the exponents show how powerful n revs/sec. and Diameter ft. are and how to scale data up or down **for geometrically similar, propellers, and similar test conditions**. For Thrust, one D^2 may be thought to come from *Disk Area,* a second D^2 from the V^2 *q effect* of radius, thus D^4. For ***torque*** we need a lever arm effect for the drag, thus D^5. Power = $2\pi Qn$ so a bigger **D** *and, or* **n** increase **Torque, or RPS** so Power is equal to $P_c\, \rho\, n^3 D^5$ --- vs. the Thrust Formula form.

We have No Bias, Not Selling Betz. We're After the Truth!

This Technical Business Works, because we are able to stand on our Predecessors Shoulders, and in fact, we did just that, and we expect those who follow to improve and refine our work.

With modern computers, we might have charged off on our own, confident that with modern tools, we could do better than any old fashion worker. Two lifetimes of experience taught us to respect that we had 83 years of **genius level work here**, and that we would be **better advised to first learn what those who came before us had learned**. As is so often the case in this business, **we found that brilliant analytical work has been done, long before its time**, that was **just too poorly understood, recognized and used**, and that **the task that needed to be done was to fully understand, evaluate, check, appreciate, explain and use it** --- to get propellers out of the too long Black Art status.

We've tried to do good professional work here, to add as much significant contribution as we could, so that we could offer more to those who follow, who can check and refine, and carry forward what we have tried to contribute. **We've tried to add to the body of knowledge and understanding, and subject to the review of those who follow, we believe we have, *real insight***.

In inventing **Zero Thrust Glide Testing**, **we got the first valid glide tests of drag** on *flying fixed pitch propeller* planes. We were able to do **repeated glide tests in dead air out over the ocean at dawn, and got excellent repeatable drag data**, which, when **compared to the higher powers required**, **confirmed Gus Raspet's incisive 1950 discoveries of very poor overall propulsive efficiency, using propellerless glide tests**. We've carried that forward to find **significant interference efficiency loss** compounding propeller efficiency loss a lot, **potentially explaining the phenomena, a combination of scrubbing drag, and a rear fuselage variable separation drag loss, caused by the higher ΔV imposed by the prop,** all worse on slow planes.

In *explaining the logic of propellers*, and of the **Betz, Goldstein, Theodorsen analytical work**, and **the conclusions to be drawn**, we believe we have contributed significantly to potentially **broader appreciation and understanding** of their benchmark, historic analytical work, by the flying public, as well as **engineers, who can gain a far quicker understanding of this complex subject**, in **easier preparation for any work they may wish to pursue**. ***If you already understand, it's easier***.

In view of the substantial depth and breadth of the explanation here, based on Theordorsen-Goldstein-Betz math, and the enlightening insights presented, we hope that we have met our original objective -- which was **the first propeller logic explanation in words, not math, that is both understandable to a broader audience, and technically comprehensive, based on an exhaustive review of what this historic body of work -- really implies and means**.

That's what we tried to do. If you are a Pilot just looking for a basic insight, this may well have been far more than you'd ever want. But **remember, several times, we admonished everyone to only go as far as their interest carried them**. With 650,000 pilots in this country, we have the full spectrum audience. **For those who want the whole explanation of propellers, it's all here, and for all those in between, as you wish.**

I've purposely written this book in a more conversational tone, if in bold, with a lot of Emphasis, because there's more real technical meat here than most of the audience would normally see in a lifetime. I want to both help the new guy, and not only not be "put-offish" but **have it as conversational as possible**. ***If*** I think an insight is ***marvelous***, ***I say so***, even if some may see that as less professional. **I've been looking for some of these insights for a lifetime**, and it's ***great*** **to finally nail the *Truths***.

We'll cover several additional subjects that need to be covered, but **then** do **a whole Chapter of Summation and Conclusions.**

Additional Basic Classic Subjects ---- **If You Want It All**

There are so many things to teach you about props that we don't want to extend the matter, but there are **a few more classic subjects** that need coverage for a really complete job, and a few **last items that shout for clarification** --- so we're offering them.

Angle of Attack, α° --- **Trickier than you may think**!!!

Modern Airfoil data, α° vs. C_L vs. C_D is taken in a Wind Tunnel with α°, **quite naturally measured from the horizontal**, the **direction of the airflow**. The tricky part, not clear to a Novice, is that the **wing section runs from side wall to side wall**. That makes it act like an **Infinite Span, Infinite Aspect Ratio** - which has **no downwash theoretically and actually**, **otherwise impossible!!!**

In a **limited span wing or propeller there *is* an inflow**, **ΔV/2, Theodorsen's w/2, Glauert's a factor**, **from the front of the propeller,** --- **and the top of the wing**, which is never mentioned on the wing, **half the final downwash angle**. There is an ***effective* angle of attack *decrease*** --- **that Pro's correct, on a wing** The formula gives the angle in Radians, **so we multiply by 180/π, to give it to you in degrees.**

Effective Change to wing α° = 57.29578° C_L / π Aspect Ratio.

Since the **air is flowing in from the top** the α° **is LESS** that you thought, *a decrease to the effective α° measured from horizontal*.

Now to really muck up the subject, if you've seen a smoke tunnel flow into an **airfoil it starts below the airfoil, and heads *up* too it**, simply because it's attracted to the lower pressure field above the airfoil. In the same vein the **whole mass of air *above* the airfoil is also attracted to the low pressure** surface, and the **net effect**, vs. the infinite span wind tunnel data is a ***decrease* in the effective angle of attack**, just as we've explained -- **honest injun!!!**

Now it is very important that we make this clear for the sharp guys looking close, because **our propeller geometry drawings account for the inflow and rotation**, ***no correction needed!!!*** **Glauert's a and a' factors**, **or Theod's w/2, account for the inflow** -- so we **correctly define the *actual and effective*** **α° vs, ϕ_1,** the wind **inflow line W**, **see fig.3, p. 94-II** --- **No Correction Needed!!!**

The Nitty Gritty of Propeller Aspect Ratio. **Also inflow ---**

Props like to be High Aspect Ratio, as high as is sound vs. vibration, as **long and narrow** as the case allows, generally higher than the AR of the wing of the plane. The problem is that it acts as ***two separate*** **rotating wing blades**, **affecting air 180° apart in the stream tube**, even has **a tip vortex at both the root and the tip, the one at the root weaker, of course.**

However, the hub is structural, hardly a wing, and it makes no sense to include it in the wing calculation. We had to make a command decision on how to present it all. Andy chose to use **the outer 90 % of each blade**++. Thus **when we design a 14 : 1 AR prop it's really 14 :1 vs 90 % of the diameter, and it really is two separate 7 : 1 blades**. It seemed misleading, maybe confusing to novices, to call a long skinny prop 7: 1.

To help your insight -- if we used the α° change formula above for a prop we'd **use a 7 : 1 Aspect Ratio, that of one separate blade**. Thus for a .55 C_L ***if the prop blade were a wing***, **which it is *not*,** the correction would be:

$$\alpha^\circ \textbf{ correction} = 57.29578^\circ \times .55 / \pi \times 7 = \mathbf{1.4329^\circ}.$$

The **final downwash** would be **nominally twice that**, **2.8659°**

***You can see Actual, Variable *Bigger* PROP Downwash vs. radius on p. 106-II**

BUT Surprise: Theodorsen's $\bar{w}$, is Big, cut down by his kappa factor, to maybe ~1/2, 1/3 or 1/8 to find the average ΔV, half that for inflow, ΔV/2. Props have more inflow, *because they don't move the whole stream tube*, and also vary vs. radius.* BUT now realize, the correct inflow is already built into our geometry -- *No correction is used*, this just to teach you all the nitty gritty at work here! Only wings are corrected.

I'd sure prefer to not have to teach you so many things, but all this is why props never did get comprehensively explained before and **this *is* the advanced chapter intended for those who want to go all the way**. A lot of things are clarified and pinned down here that you would have a hard time finding any other place. If you're ***making a game out of this you're right on***.

++An RV having an 18% Spinner, we later used only 81% of the diameter for Aspect Ratio calculations, our 8% stations at 19% to 99%. Thus a 14 : 1 Aspect Ration is equivalent to 12.6 A.R..

A Final Look at Advance Ratio, Efficiency, and P/D Ratio

Advance Ratio, J = V/nD, mystifies people because they can't grasp what it means physically. As you've learned, if you simply **divide both sides** of the equation **by** π you get **$J/\pi = V/\pi nD$** which is **still valid mathematically**, both sides now **just smaller**, **still equal**, and it's easy to see it's **just the corkscrew AAA**, **actual advance angle of the prop tip**, or **AAR**, **actual advance Ratio of the prop tip**, because it's easy to see **V** is the forward velocity of the plane and **πnD**, is just the circumferential velocity of the prop **tip** --- and when you **put those two velocities together at right angles** you get the **AAA, or AAR, or J/π** in the simple little sketch below. Now those are **velocities**, but obviously they have a **close relationship to Pitch, Pitch Angles, P/D Ratio** in inches, feet or even angles, because the **prop blade has to be at a little steeper angle to account for inflow and angle of attack α^o**. **A prop appears to slip, accounting for inflow** + α^o.

V/nD = J **Advance Ratio**

n = revs/sec. D = Diameter

$V/n\pi D = J/\pi, \lambda$

AAR

AAA°

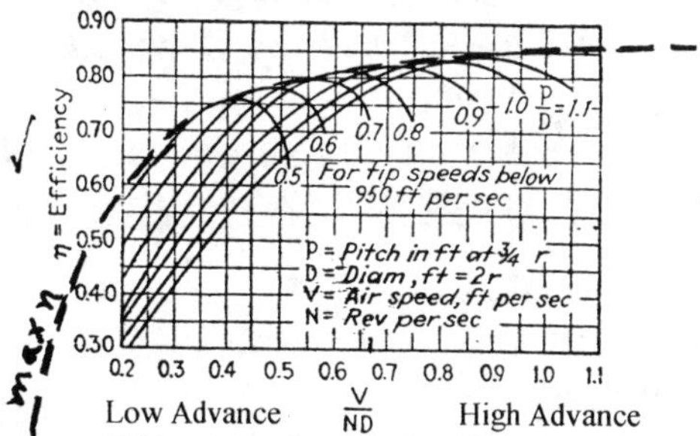

Efficiency of a Series of Two-blade Propellers.

P/D Ratio I hope you caught that the **Advance Ratio Graph** p 48 is made up of the **max. efficiency points** of a **jillion Ideal props**. Now, this old graph in an early McGraw Hill, Marks, Mechanical Engineers' Handbook, does a great job of tieing together the **individual efficiency curves vs V/nD, *Advance*** of a varied **P/D** range of props and **how they form the Advance Ratio Curve!!!**

Very old, the efficiency is low, not accurate, vs the real J Graph, p 48 II. All left ends must go through 0,0 on the horiz. V/nD, and vert. efficiency axis, but nails how they all fit together. Grasp how all props start at 0,0, but the high P/D props, a huge speed range --- the inboard stalls slow*!!

Blade Element Analysis - **vs. Actuator Disk** - **vs. Theodorsen**

A few pages back we went through the drill of showing you that **wings need an angle of attack correction**, but that **props do** ***not***, *because* the **inflow is already accounted for in prop geometry and math**. ***Then*** for very good reason we showed you that if you try the wing α^o correction on a prop it doesn't work right, **and some insights on why not**, the **real prop inflow and final downwash are bigger!!!** That could seem confusing, but we're ***helping you grasp inflow, outflow is faster, higher angles.***

The Newton Actuator Disk --- gave us very important major **insight into the tricky basic logic of props**, ie. Go Fast, Get a high mass flow rate $\dot{M}$, thus a **low ΔV** for any required thrust --- only a **small $\Delta V/2V_1$**, a **small axial energy loss**, thus **High Efficiency -- only a High Pitch, High Advance Ratio prop can be efficient** -- Low pitch has high induced losses, too long a path to the destination, thus high profile drag energy losses also!!! We **tend to Lose Efficiency if we Increase Required Thrust, Go Slower, Reduce Prop Diameter, or Raise RPM, (a longer path). Great -- but it didn't deal with the prop geometry needed, was quite unrealistic in assuming constant axial airflow, useless for actual design**.

Blade Element Analysis --- Is the kind of analysis engineers love and some very smart guys spent decades trying it, but **never got really correct answers!!!** It considers a radial sliver of the prop chord and calculates its thrust --- and HP absorbing torque --- at the proper RPM, radius, thus velocity, thus q, dynamic pressure, all at the correct angles, correct trigonometry, desired C_L, then sums it all up with calculus. Thousands of Aero's were taught that way. **We'll use it next to get some interesting insight**. The flaw is that it ***assumes*** an α^o, C_L, **doesn't calculate, solve the accurate complex actual air inflow -- or ideal shape and twist**.

Theodorsen, **of course**, is our technical hero who got the complex, **heavily loaded**, 3D flow correct, accurate, finds a higher velocity helix moving back through its stream tube, the **only way you get real accurate inflow calculations real usable calcs**. **More Soon!**

Do Wildly Steep Blade Angles Really Work ??? Key Insight!

Think of a prop sliding *up* a steep, ***greased*** **60° Wind line, W** incline, pulled by a *horizontal rope.* **With low *Profile friction* drag,** *only* - a 50:1 prop airfoil, **"moves farther forward than the *rotation* rope moves"!! Though it isn't pulling forward well, actually pulling against engine torque, it's moving forward more than the engine rope that's turning it is moving, so it's *using energy efficiently*. Realize, *drag free*, we get constant dT/dQ!**

Per this Sketch: Bottom Line - A Prop *IS an* ***Airscrerw: That works OK!***

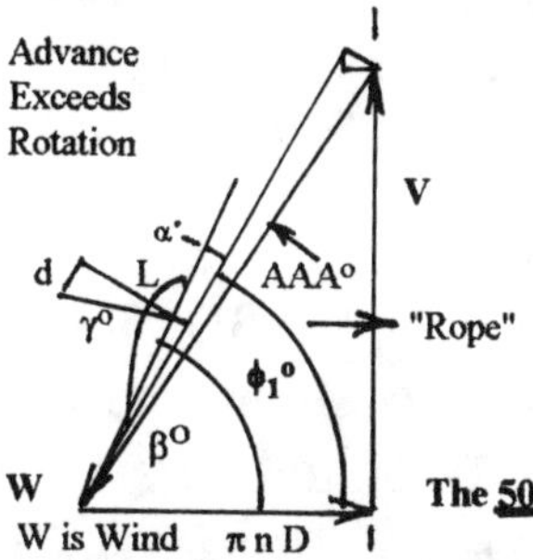

W is Wind   πnD

Vector diagram of a blade element

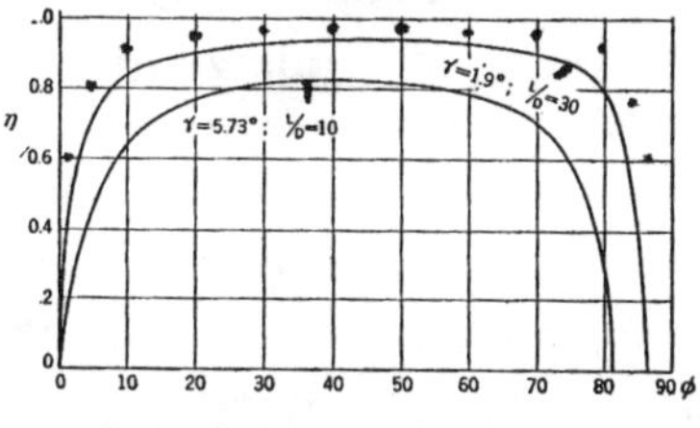

FIG. 1·9. Simple blade element efficiency variation.

The 50 : 1 L/D Graph ••• is for Profile Drag, SKIN FRICTION **ONLY**

Graph is of Actual inflow ϕ_1°, not Blade, β, not AAA)

Blade element efficiency vs. inflow angle ϕ_1°

(fig. 3, p.94-II, shows how stream tube inflow sets the angles φ and α° to get β) Also see ~ p. .60 II

Blade Element Analysis, using Algebra and Trig., with all the forces and angles **(even if it can't calculate air inflow correctly)**, calculation, can easily enough look at the **efficiency** of **a screw at any angle**, and yield the plots above, and for various blade airfoil L/D ratios.

The key Graph insight is that a phi$_1$, ϕ_1 **inflow angle of (45° - γ/2) is the most efficient** -- the blade angle at α° steeper. (γ is the drag angle, like L/D, Lift vs..Drag in angle form). Per the graph it shows the **efficiency of any and all inflow angles at all reasonable blade airfoil L/D's**.

Now, the great insight gained here is that **efficiency drops moderately as you vary from a ϕ_1 inflow angle of (45° - γ/2), but is hurt most at very shallow or very steep angles. Now that is some quite instructive, practical overview insight** ---

For Insight, If a Prop were a Wing! For a low N_R, **profile drag C_D of ~.01** -- and a C_{Di} for Induced drag* for a .5, or .55 C_L and a blade Aspect Ratio of 7 we'd get a C_{Di}, **of about .015, .025 total, and vs. a .5 C_L that's an L/D of maybe 20+ : 1, a ~5% loss**!!!!

*(Remember, a Wing $C_{Di} = C_L^2 / \pi ARe$ -- where e is Oswald efficiency, ~.8)

Now that becomes a hugely incisive insight on the subject of propellers!!! We know that a pretty high pitch, efficient RV prop can be 89 to 90% efficient, 10 to **11% loss** -- using an airfoil that on a wing would have a ~ 20+:1 L/D, **5% or less loss**. **WOW!!! Use an airfoil on a propeller and you get *at least* Twice the Loss, at a comparable C_L and Aspect Ratio**!!! Even with BGT half teardrop Loading we get excess tip loss, can do even worse with poor design. **Props, *better than Ducted Fans*, are the best, most efficient Propulsive unit, but *at least* 2 to 3x loss**!!!

A Low Pitch Luscombe Prop, at 85% η has 3x loss, 5x if at 75%, if bad Design!

Furthermore, only Theodorsen Math copes correctly with the **heavily loaded faster inflow**, faster movement of the pure helical inflow, stretched outflow, gets the complex 3D flow correct, **so you can get what your calculations show**, and it **gives Betz's Ideal Minimum Induced Drag Twist and Shape**.

Blade element Analysis, like a Newton Actuator Disk, can teach us a bunch about basic logic, but designing a prop with anything but the real math is just not correct, just kidding yourself, because you only have it right when you accurately calculate airfoil inflow. That's a pretty simple Key Insight!!!

But there's more!!!. We learned from Gus Raspet that there can be a big interference loss, maybe another 10% loss on slow planes with a bigger ΔV, less on fast sleek planes with less relative ΔV vs. V. Zero Thrust Glide Testing, a Classic like the Luscombe, we found 67% overall η_P, with a ~75% prop η, a ~ .9 η_I interference efficiency. We don't have final overall numbers yet on RV's because we must do Power tests, but they can be quite good maybe 85% overall with an 89% Prop. That's key Insight --- never available before!!!

The Real Induced Drag Loss. We've purposely been treating Induced Drag like it consists of **Downwash** and **Tip**, **and Root Vortex loss**, like it is on a wing, to not confuse learners, **Rotation Loss separate**. But **Think**, obviously **Stream Tube Rotation loss is just another type of Induced loss**, part of the reason **props have ~ twice the loss of a wing, or much more**!!!!

Get a Professional Grasp of Theodorsen's w factor. It's BIG

Go back and look at Theodorsen's **$\overline{w}$, w bar** on p. 94-II, the **Total Extra Velocity Speedup Ratio** of Betz's Helical Vortex Sheet. **See that it's .1566**, a ***BIG ratio of plane speed V_l*.** **At 170 cruise, on an RV, it's 26.622 MPH, much bigger than the average 9.239 MPH ΔV we've been talking about!!! Why?** Hopefully, you already know. Theodorsen's **Kappa Factor, κ which knocks the $\overline{w}$ down to the average ΔV is** .34704 for the RV on p. 94-II, so the **effective ΔV is 9.239 MPH**. But you see the prop has to be **pitched for nominally half the full w, 13.311 MPH** above 170, .0783, 7.83% faster, .0511, 5.11% faster than the .02717, 2.7% that would fit if the whole stream tube moved at the average ΔV!!! **That confirms the ~5% extra pitch rule we told you about earlier that seems to work at ~.55C_L!!!!!!**

Theodorsen's A Zero Inflow Factor, A_0 Actually, Theodorsen's **A_0 Inflow Factor**, (.078005), accounts for whether the inflow is a **little more, or a little less than half the $\overline{w}$**. This guy is Amazing in his thoroughness and mathematical insight, a true genius. Of course when we're setting pitch we have to **account for angle of attack, alpha, best if its simply a constant α° added to the helical phi's of the air inflow -- because that's what sets up a constant C_L** for a minimum Ideal Profile Drag, as well as Ideal Minimum Betz Induced Drag. **Now we're ending with this little drill here so you both understand the extra pitch that Heavy loading demands and conclude that Theodorsen is the only sensible way to go about designing a Propeller**, because **he deals with all the complexity and gets it right**, all else an amateur drill, not worth wasting your time on.

Theodorsen's Epsilon Factors ε Theodorsen actually divides his 3 axis losses into 3 axis Epsilon Factors and makes it possible to take the losses apart, shows that **the radial flow loss is very big, bigger than the axial loss, the big loss, never understood! In Appendix T we concisely deal with Theodorsen's math methods and equations, so you can see his methods.**

Airfoils Profile Friction Drag For Decades props simply used a basic, old **Clark Y** airfoil or an **RAF 6**, a special prop airfoil that actually had its **bottom cut away**, which gave it an **easy to measure**, **absolutely flat bottom**, not all bad if we must **measure to .1 degree Beta,** β^o. Tests showed the **RAF** was better for **Takeoff**, the **Clark Y best for Cruise, a lower C_D**,.

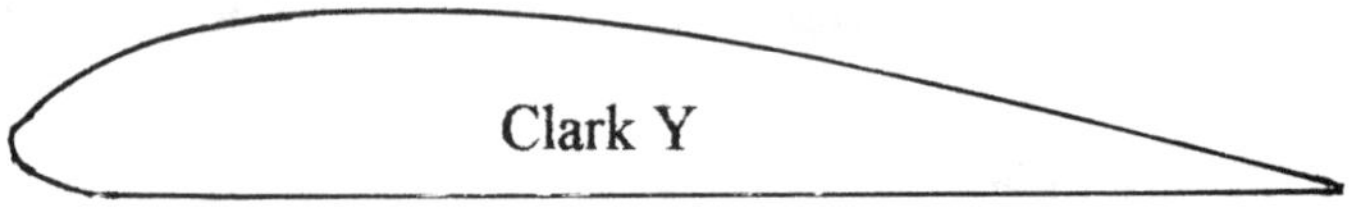

Pretty Unsophisticated!

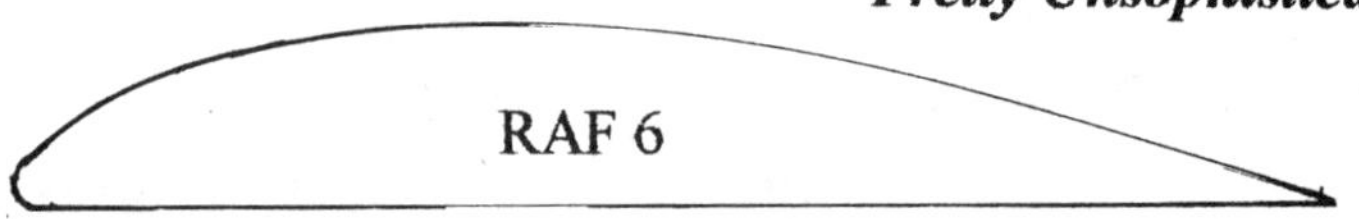

Yes, we want **low *PROFILE DRAG***, but with **eroded**, **bugged**, and buggered leading edges and **low Reynolds Numbers**, betting on **low C_D's is not realistic**. Conservative is appropriate. The game in laminar flow is to **Hold Laminar at Reynolds Numbers of several million, hold a ballpark C_D ~.00 5**, not let it jump up to the **significantly higher Drag of turbulent flow, ~.008.** The problem is that as you drop to R_N's of a **million and below, even laminar C_D's raise to .01 *and above***. At a .5 C_L, 1 million R_N an **.01 laminar C_D ballpark is a proper conservative nominal**.

On the other hand an ideal airfoil with really good leading edges can't hurt. John Roncz designed special optimum airfoil sections for each station on the special **Hartzell around the world Voyager propellers**, and they worked great, **compliments to John Roncz**. **Compliments to Hartzell too**, because those props **they contributed saved the Voyager Program** when a **controllable pitch wood prop failed**, a blade pulling out of the hub!!! The really nifty 1986 trick was John designing the special props on his computer, loading them on **a floppy disk**, Hartzell creating an NC Machine tape directly, the Numerically Controlled machine milling the blades out of a forging, untouched by human hands in minimum machine time - **ready for final hand finishing**.

Low Reynolds Number Drag.

Below is a standard Graph for Profile Drag at a Wide Range of Reynolds Numbers, both for a lower range below 1,000,000, down to ~200,000, then above, to 2, or 3,000,000, common ranges. Low Reynolds Number Drag Data is pretty limited with rather inconsistent answers. Basically, see that there are large changes below a 1,000,000 Rn, from below .01 to well above at 200,000 Rn. For our Low Reynolds Number calculations we conservatively assumed turbulent flow, used .01, rising above that down to 200,000, and also rising with greater angle of attack. Actually if one uses a conservative overall C_D of ~.00937 overall results can be comparable, so we used our private chart only where we were looking for significant effects, because Reynolds Numbers easily get down to 200,000 at slow inner radius chords, and also fast, but narrow outer tip chords.

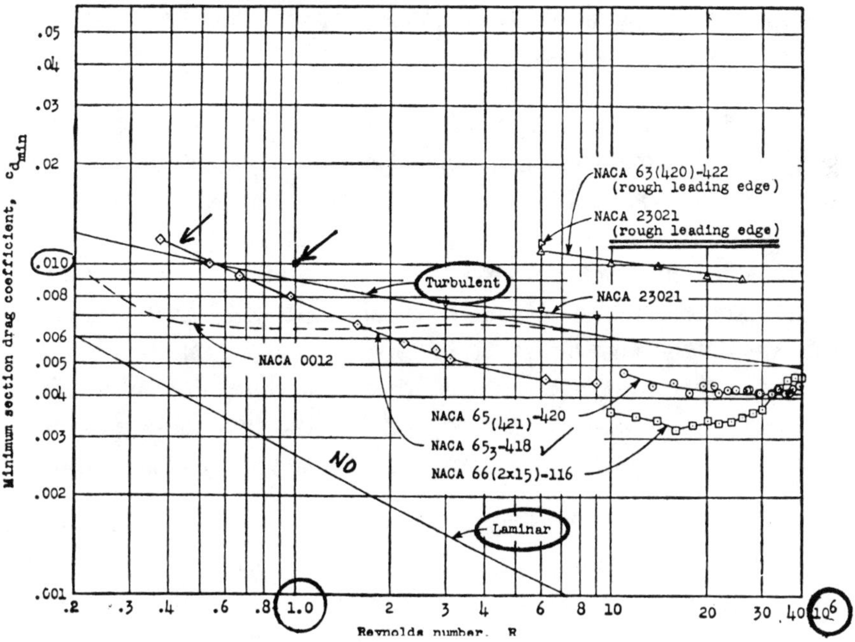

FIG. 66. Variation of minimum drag coefficient with Reynolds number for several airfoils, together with laminar and turbulent skin-friction coefficients for a flat plate.

A Goal Line Personal Comment, and Evaluation of all This

Understand that I'm a very Senior Aerospace Technical and Top Management Consultant, long ago financially retired, *exactly* how Airplanes work, ***a lifelong serious challenge hobby***. I've been watching, waiting for the serious Propeller explanation, for 60 years, since I was looking for it as a kid, winning the National Model Championships two of the three years I was there to compete in 1946, 48, 107 of my Moon Program Spacecraft Small Rocket Maneuvering Controls in the Milestones of Flight Gallery, the Central Hall of the Smithsonian, National Air and Space Museum, July 4, 1976, the day it opened --- about as qualified as one can get to finally, Technically, practically explain Propellers.

For decades, I've thought it was ridiculous, unacceptable that we never got the Professional Level, but Incisive, Understandable Explanation of Propellers, especially since Theodore Theodorsen, maybe the top technical professional at NACA, nailed the genius level Mathematics almost 60 years ago in his 1948 book.

Understand, that I'm neither the technically flaky top management guy, or the narrow vision technical guy, who lacks the no fault allowed Top Practical Grasp. I spent a lifetime in key, core flying products that had to be on the Technical Leading Edge, but absolutely could not be allowed to fail, or we had a National, Front page disaster: First, big, tri-axially Stressed Landing Gears, the Absolutely must never fail for *ANY* reason, Primary Jet Flight Controls, that your, and a few hundred other lives depend on, the Spacecraft controls that had to, did live through the Apollo 13 explosion, brought it home safely, maybe man's greatest adventure!

The core point that I'd like to make is that Theodorsen's work, was ***not*** the flaky, too far out theotrician's analysis, but the ***Tough Minded*** all up ***Final Professional Work***, that accurately accounted for **all the 3D losses** after 83 years, the smartest rare geniuses the industry had produced, giving the final most conservative, most correct technical work. **I work at the, *it must be correct level!!!***

The 7 Historic, Genius Pros - The final really correct Answers

Propellers are, and should always remain, our most efficient method of rotary propulsion, especially important as inexpensive fuel disappears, but almost laughable, the dumb things are **trying to work inside out, trying to make max Thrust, and Drag at the Tip, where Thrust *must* fall to Zero**, Dumb as a Stump. That was **the first thing we taught you**, by far the ***easiest, most fundamentally important basic grasp in the book***.

The **Ultimate Technical Swamp**, the reason Andy and I chose it as **a Life's *Give Back*, Goal Line Intellectual Challenge**, they were such an unexplained morass, that no one saw through the **Aero 101 insight**, that they were trying to operate inside out, and that was **the main thing we had to recognize and fix -- Theodorsen's magnificent technical method**, as a **SIMPLE BLADE LOADING CHART**, sitting there, ***waiting an Extra Half Century*** for us to get Computers, Wake up, and put a great example of an ultimate Technical Answer to Work. I hope old Theodore is up in Heaven having a big laugh --- **we finally got it!**

It's hugely Proper, that we clearly honor Rankine, Froude, Betz, Prandtl, Goldstein, Glauert, Theodorsen, already Technically Historic figures - or their Rare Genius work in our behalf, clearly deserving Historic Recognition and Status - providing us with a ready to go, and use, sophisticated solution to probably the most complex fundamental problems in Aerodynamics --- a practical, understandable, accurate analysis of Propellers - *amazingly, a simple Blade Loading Chart*

Doctor Andy Bauer, my lifelong friend and Collaborator in all this **clearly deserves to be Historic person #9**. It was he, with a lifetime of top Technical Experience, who wasn't snowed, put off by all the Partial Differential Equations in Goldsrein, ***smart enough to see the final answers were in easy chart form***, seeing that **with his final 19 page program we could nail everything. Andy earned, deserves a proper place in Propeller History**

Gus Raspet, that marvelous Creative Mississippi State Maverick, **clearly deserves to be # 8** for showing us all in the 50's **how really terrible overall propulsion efficiencies could be**, a good looking, Classic Bellanca Cruisair, **a terrible overall 58%**, ditto Cubs and old Cessnas - the marvelous insight that got my attention, got me interested and started on the whole project, essentially got me qualified to be Technical Director of Voyager Mission Control, by its 1986 date, all sharpened up ready to go with a Steel grip on all Flight Fundamentals.

I absolutely want to name **Paul Lipps of Arroyo Grande CA**, as **pro # 11 in Nailing Props**, because ***thinking outside the box***, a highly intelligent RCA Radar Technical Pro from Vendenberg AFB, his mind not cluttered with Party Line Gross Aero Technical Detail, but very bright, creative, he was **brilliant enough to recognize that perhaps a prop should go right back to Prandtl's ideal Elliptically Loaded Wing, as a rotating Wing**, **a genuinely top level technical perception** that Betz, or Prandtl themselves might have seen a Century ago - saving us a Century of gross, noisy tip overload losses, if we mortals were perfect. His Elliptically Loaded props, that look wild and crazy, seem to perfectly correctly do the next better step compared to the Betz, Goldstein, Theodorsen Half Teardrop loading, vastly better, quieter than the ridiculous tip overloading of too many of todays products after 140 years. ***Good Work Paul!***

I'll leave room for myself as #10. Hell, at age 79, I earned it the old fashioned way, in addition, a lifetime of creative products, that absolutely had to work, soundly, reliably, lots more than you heard here, creative, simple, technically astute Nuclear Plant Seismic Protection, good stuff all over the world, ***great satisfaction***. To me, it was totally unacceptable that something as fundamental to flight as propellers, went unexplained for 140 years, answers I was looking for as a kid of 18, 19, 20! I wasn't ready to quit at the answers that Andy and I had years ago. **I wanted technically sophisticated answers - simple, clear!**

The Wise Old Bird

This Crazy Guy decided,
as a challenge, he was going to
do **the real propeller explanation**
that **never did get written** before,
- **making the ultimate morass**
Understandable ---

The secret insight is: **All of nature, no matter how complex, is always orderly, logical** --- and that is the way our brains work, in order and logic, and thus if someone can figure out the logic, we can understand anything in nature! **That's what Science is**, the first guys **figuring out the order, the logic**, so it can be **universally understood and put to work for the good of man**.

For those of you who know of Stephen Hawking, who have seen him on TV, or read him, you may realize that he points out that if we can finally understand, we ***amateurs*** can know the mind of God.

Computers are **a specialized tool** that we can put to work for us because they can **process numbers** at essentially the **speed of light,** even do it in accordance with complex mathematical programs we can set up, **without stopping to think**, giving us almost instant evaluations, numerical answers, often even put out in non mathematical form like this text, or movie special effects. **Our minds can't work *fast* in math**, **so we put computers to work!**

So you see, **it's imperative that we decode the order, the logic of propellers** and ***teach it to you that way***, **because that's the way you can understand it**, have it make sense to you, all interrelate, **make it possible to remember it, *because* it does make sense!**

Our minds work in layers of logic, remembered if something does make sense, relate in logical ways to other things that make sense to us. **So the *Summary* you're about to read is the first, comprehensive, orderly, logical explanation of propellers.**

BOOK 2 --- CHAPTER 4

PROPELLER SUMMARY

THE CHAPTER with the **GREAT HINDSIGHT SUMMARY**
JUST ENOUGH DETAIL to help an ACCURATE GRASP

With Fan Duct Losses, and potential 21st Century gains in Propeller Efficiency, the Propeller is, and most likely will remain our most efficient method of creating Thrust from a Rotating Engine. It is Amazing that it has taken until now, 2006, to get a Comprehensive, Understandable, Explanation of the Fundamental Engineering Logic --- recognizing that Rankine, Froude, Betz, Prandtl, Goldstein, Glauert, and Theodore Theodorsen, all Genius, Technical Historic figures, labored 83 Years, from 1865, to 1948, but in that time had created Betz's Classic Propeller Theory-Logic, and Goldstein, Theodorsen, essentially exact Math Solution of Betz Logic! It is Amazing that such a Fundamental Aerodynamic Device has been allowed to remain an unexplained, technical Swamp this extra HALF CENTURY. WHERE WERE the PROS? I saw this gaping hole as a young model builder in 1946, 1948!

The most fundamental insight that I want you to realize is that all this time, all too unattended, the Propeller has remained, essentially *not* understood, a rotating wing, *not* with the easy 4, to 5% loss of a basic, clean, nominal Aspect Ratio wing, but rather the much larger nominal 10, to 15% loss, *or a lot worse*, of a device with excess radial flow tip vortex induced loss, losing paid for thrust, and all too often, excess profile wide tip losses, at max lever arm, bogging down the available engine torque, losing H.P., losing 4 ways!

It is particularly Amazing, that this excess loss situation was never adequately grasped, since Theodorsen was such a respected figure in the Aeronautical Engineering Community. He showed on his page 33 that the radial tip Induced, Vortex loss could be *Larger* than the Axial Induced Loss Necessary to Produce the Basic Thrust!* We normally fly a wing, at more like a 25% ballpark induced loss*. *Induced flys ~TWICE Profile* on Props.* What a BIG FAT HINT!! Clearly, everyone in the Research Community, running off to Jets, abandoning this Most Fundamental Aerodynamic Device, no one was looking!

***REALIZE TIP VORTEX IS JUST A ~25% *part of* INDUCED LOSS ON A WING!!**

The basic Betz-Goldstein-Theodorsen Propeller, in it's pure form, the Family of Ideal BGT, Triple Ideal, Minimum Induced, Min. Profile, Min Torque Propellers, MINIMUM AREA, PRECISELY PLACED, Norris's 6th Law, should have been the basic design Standard in the Industry for the greater part of the Last Half Century. The Research Community has been, unfortunately, Missing in Action!!!

The first appropriate step is to get some approximation of everyone involved to understand, that Propellers can be "easily enough understood", just get the radial loading correct per BGT, by a move, no more difficult than getting the **SHAPE, and *Twist*, correct**, ***of course*** Pitch, and all other Size, Area, Diameter, and all other optimum, ideal characteristics. **That's easy to Grasp**! BGT math has been available for over a half Century, since Theodore Theodorsen's 1948 book. **We have the Computer Programs running, checked, can teach those manufacturers who don't have the knowledge and resources -- learn how to use them.**

Understand Prop Engineers, Good Guys, Deserved Help Here*

We have to be smart enough to know, as customers, that we won't get them, unless we are smart enough to go after them, because old manufacturers have a half Century of old design props, all approved, and will only slowly get modern, unless we get smart. *Research gave prop engineers no help since 48! In these days of ever more costly energy, we need our props as good, efficient, Quiet as they should be! It's time!!

The Areo 101 of Propellers - Fantastic Summary *x-ray* Insight

We've learned that a 7:1 Aspect Ratio Wing can nominally, have a 20:1 L/D, thus a 1/20 total loss, 5%, 95% efficiency. But even an optimum BGT private plane prop can only be 85% to ~90% ballpark, nominally, and poorly designed, may lose another 10%, **2, 3, 4, 5 times as much loss** - ***Clearly*, a big wasteful loss here!**

Theodorsen shows us the basic problem on page 32, and 33 of his book. Even for an efficient high Pitch, .5 Lambda, λ, prop, the radial loss, the **Tip Vortex Induced Loss**, **is even a bit Bigger than the Axial Induced loss, fundamentally needed to make the Thrust.** Rather than a nominal 25% extra Induced, common on a wing, **the extra, auxiliary loss** you'd hope was only a fractional loss, **is Bigger than the Needed Loss to make the Thrust! With poor design, trying to get thrust at the tip,** where the Smart Game is to Get both Profile and Induced Away From the Tip, **it's possible to kill another10%, create a 4x loss on a 90% case, 5x at low pitch, at 75%, not the basic 3x loss!**

It's our job to understand, know how to Professionally Counteract Excess Prop Tip Loss!

You just don't tip load *any* wing, exactly what the fast, very high q prop tip is trying to do to itself -- where the Aero 101 of a wing shows us the lift, thrust must fall to zero, pretty much elliptically, into a ~ 25% extra Tip Vortex Induced Loss, with a .8 Oswald efficiency factor representing a nominal tip Vortex loss, over and above the needed theoretical loss to make lift. Of course, the prop has rotational *induced* loss too, *but the Tip is worst!*

Now knowing that Betz-Goldstein-Theodorsen always pulls the radial blade loading back into their "**Half Teardrop Loading**" we see ***that high loading is still unfavorably close to the tip***, the reason radial tip loss is as big as the axial. **YOU DO NOT TIP LOAD ANY WING, EVER - BAD, WRONG, COSTLY! The core ball game in prop design is being smart at the tip!!!**

With essentially exact Theodorsen Math, and Source-Sink nose profile correction, we have *the* accurate, pro way to go, to design. With Heavy Loading, BGT Pitch is nominally ~5% higher, a significant error, if you use less accurate methods.

1. The new *Century 21*, Lipps, Elliptically Loaded Outer Prop

Betz and Prandtl, *or a host of others*, **anyone calculating**, *might* have seen that BGT fundamentally pulls the blade loading back away from the tip, into **a half teardrop radial loading**, to prevent unnecessarily excess tip radial flow tip Vortex, induced loss, and unnecessary tip Profile Drag, at Max lever arm, bogging down the available Engine Torque, increasing cost, reducing available H.P., **but I've never seen it before this text**.

Paul Lipps a very smart Radar pro, and hobby plane improver, saw, **without knowing of BGT classic Theory and Math**, that ***we could, should move the loading back even more inboard to Prandtl's Minimum Induced Loss Elliptical Radial Loading***. It makes a pretty crazy looking, concave outer blade shape prop, but it ***simply does what BGT does, just a little more inboard***, to ***an elliptical tip loading***, from the BGT half teardrop. As such it ***does have the potential to do even better than the BGT loading!***

Winning Races at Reno, more testing, refined math, and slowdown design, Truth soon!

The two pages, way back in the Introduction, though quite brief, do a good job of showing Lipps basics in easy picture form. so I suggest that you go back and read it again, ***see the core logic of it***

See pages I - 28, and 29 - II

Now, if you ***reread*** the Introduction on Lipps Props, or if not, you should be able to easily enough, by now, see that the very **fundamental flaw of the basic prop**, and unfortunately still the Lipps prop, is that **all rotary props effectively have No Center Wing**, effectively have **two separate outer halves**, **almost only two separate outer wings**, **connected to the engine by two ineffective, low q inner halves. That's fundamentally Poor!**

ONLY **TWO OUTER WING HALVES, IS MOST UNFORTUNATE!!!**

If you think about it, the spectacular 60 : 1 glide ratio modern sailplane, **1.666% Whole Plane loss, 98.33 percent efficiency**, can do that because it has a **huge Center Wing, far removed from tip loss**, all very high Aspect Ratio, long narrow tips, themselves far removed from the tip, ***cutting the entire induced loss to near an absolute practical minimum!*** That cuts the induced from the basic lift, and 60: 1 **slashes down tip loss too!**

So, with the Lipps Elliptically loaded *outer half* Prop, what we're potentially doing is *practically* opening up the *possibility* of getting Propellers to their Ultimate Possible Efficiency!!!

CAUTION: FIXED, HIGH PITCH, IT MAY TAKEOFF MORE POORLY!!!

If a fast, 200 H.P. RV 8 has just over 90% η BGT potential, let's just say 90% max for the best of the non all out speed merchants, time, more testing, refined math, may get us **a bit above that with a Lipps** prop, ***never as good as a 95% wing!*** Trying is ***a legitimate piece of normal Engineering Progress*, a proper objective for this 2nd Century of Flight!!!** We'll certainly make more, prove conclusively if they really can be better, try to get math equivalent to BGT, if possible. **GO!**

---

2. Thrust Required The **first thing** to **understand** about props is that **Thrust does *not* equal Gliding Drag!** There is an eta$_i$, η_i **interference efficiency loss** once you turn on the prop throwing back a downwash, backwash ΔV wind to make the thrust. That causes **extra** $(1 + \Delta V/2V_1)^2$ **scrubbing drag***, and ***possibly***, some separation. **In Slowdown, a pressure force on the imbedded body does *not* cause a new loss, a wash actually, the prop now running in** slower air, at less H.P. But there is a need there, to generate a bigger Gross Thrust, for the same old Net Thrust to balance the pressure, a wash, at the original H.P.!

***On a Clean, Fast, low ΔV Plane, the extra loss can be a Small Percentage, RVs Good!**

As taught in the Advanced Chapter, 3, if you know the ***accurate gliding, or calculated drag***, and the ***interference efficiency***, η_i, you can get the **necessary design thrust** by **dividing glide drag by** η_i. If you don't know those, but know the actual H.P. and speed, V, since Horse Power is T # x V ft/sec / 550 ft #/sec per H.P. you can easily find the **Available Thrust** for 100% Propeller Efficiency, and then decrease that by the proper (high) prop efficiency, to get the proper (high) thrust available - Somewhat above glide drag, (as related through the interference efficiency, η_i, eta$_i$). **Ideal** (high) prop efficiency can be found from Advance Ratio.

That is taught in the Advanced Chapter 3, because it is one.of the final sophisticated caps on propeller knowledge and one that has not been understood, only clarified herein where we have **actual gliding drag** and **interference efficiency test data, η_i.**

3. **Newton's Laws, The Actuator Disk** Everyone can see that a prop is ***kind of*** **a rotating wing**, but very much of the basic logic can best be grasped by understanding that it can also be seen as **pulling in and throwing back a surprisingly heavy mass flow rate of air**, $\dot{M}$, (M dot), **mostly from plane Speed** only **a surprisingly low** ΔV, (delta V) speeded up wind required, **to make any Required Thrust, maybe only 9 to 11 MPH** ***average***.

$\dot{m}$, $\dot{M}$, is called M Dot, just like it looks.

Newton's 3d Law, is the familiar "equal and opposite force reaction" to throwing a mass, or flowing fluid mass flow rate. Newton's 2d law, T # = $\dot{M}\,\Delta V$ calculates the actual Thrust, Force. The concept is called the Rankine - Froude Actuator Disk, because it **assumes a constant axial flow rate**, which is a great oversimplification, the 3D flow being quite complex, but it **teaches and decodes the following Tricky Basic Logic of Props**:

Speed, V is the Fundamental Controller of Prop Efficiency, because it produces a ***high natural Mass Flow Rate***, $\dot{m}$ through the prop Disk, (only increased a little to $\dot{M}$ by the factor **$\Delta V/2V_1$**), thus **only a small required ΔV wind speedup** *for* **Required T#.** That is efficient because ***the axial efficiency loss*** is ***also*** **$\Delta V/2V_1$**

Only High Pitch Props can be Efficient because they go with a **High Speed,** big $\dot{M}$, a **small ΔV - a big V,** *a twice small* **$\Delta V/2V_1$ and that** ***small axial loss***, **foretells** ***a small total induced loss.***

Only sufficiently **Low RPM Props can be efficient**, because **RPM vs. Speed**, V, controls whether you have High Pitch, a **low $\Delta V/2V_1$** thus **low Induced Loss, High Efficiency**. But also:

High Pitch equates to ***a steep corkscrew path*** to the destination which equates to **less profile drag energy loss - a shorter path**.

The poor Prop is *not* even in control of its own Efficiency!!!

Thus **High Speed** and **Low RPM** equate to **Hi Pitch,** ***both*** **Low Induced Drag Loss**, and **Low Profile Drag Loss**. Thus **HIGH SPEED** and **LOW RPM** become the **fundamentals** of **HIGH Efficiency**, the **simple key** to the **Tricky Basic Logic of Props!**

BUT: Slow, Hi RPM Props can be geared down for Big D, Lo RPM, Hi Pitch, Hi η!

---

4. An Orderly, Logical Way --- to Look At --- ANY PROP

We give you an Orderly, **Logical way to look** at **any prop** --- for a given **Required Thrust**, **3 Spec Items**, **3 Decision items**, **3 Design steps** -- **9 Orderly Logical steps**, **easy to remember.**

Design Thrust, Greater Than Gliding Drag -- can be thought of as # 10

3 Spec. Items: **SPEED**, **RPM** are the two **basic controlling** items in propeller **efficiency**, **Altitude DENSITY, rho,** ρ, is the basic that defines the media the prop works with, ***fundamentals***.

3 Choice Items: Coefficient of Lift, C_L which basically **controls** Diameter, **Aspect Ratio, which affects efficiency**, and controls the **proportion of Area vs. Diameter, (shape preordained, *if* an ideal prop)** and **Coefficient of Drag, C_D** is set in the computer, as graphs of multiple curves, dependent on Reynolds numbers, Rn, and α°. **From our Design Study we select best choices for each!!!**

3 Design Steps: **(Realize Shape, Twist, is Preordained if it's to be an Ideal Prop)**

- **Sizing: Area, Diameter, Shape are interdependent**, can't be separated*, but with **Aspect Ratio** selected, that **sets the proportion**, for any given shape selected -- and the prop simply **scales up or down vs. the C_L selected and Thrust Required**.
- **Pitch:** Speed vs. RPM control Pitch + Inflow, and α° required

With 1 degree error, .1 C_L representing a 20% error on a .5 C_L prop, **+/- .1° tolerance is required to hold a** +/- **2 %** manufacturing tolerance, or design Analysis Target tolerance.

- **Twist vs. Shape: Controls** *whether or not* a **constant C_L** and an **Ideal, Minimum Induced Drag Betz Loaded Prop results**.

There are other minor, only secondary variables covered at the end of Ch. 3

With this **relatively simple outline**, for a very complex subject, it is possible to come to understand the Physical Parameters of propellers in an Orderly, Logical grasp. Most Significantly, as a result, one comes to realize the valuable **insight** that it is **simply Twist vs. Shape that determines if a Prop is Ideal --- or *not!***

4. *Highly Variable Dynamic Pressure q The highly variable **Vector Speed of rotation and forward V** causes **a very big V^2**, ***q difference***, ***root to tip***, **makes ideal props very different than wings** in ***how they are loaded and understood!***

5. **Betz, 1919, defined the Minimum Induced Loss, Ideal Prop**.

There is the Classic *Betz Logic* that we hang the whole prop explanation on, a wonderful, fundamental Ideal Prop Logic.

- **It is First Done with Zero Profile Drag, drag added *later!***
- **It yields an Ideal Minimum Induced Loss Prop by**
- **Creating the exactly ideal Thrust vs. Radius Loading by**
- **Holding a perfect helical air inflow, stretched outflow by**
- **Setting the Inflow Geometry for Constant Pitch, thus Slip!**
- **That yields a Constant dT/dQ, Thrust/Torque, or H.P. Ratio**

Constant Airflow Pitch, Slip, acts like Constant Inflow, when Axial Inflow is Not Constant!

This concise, incisive outline makes this subject much more **clear**, **much easier to grasp**, than you will find any other place, but it is the **marvelous technical work of our forbears** that allows this **Orderly, Logical insight and grasp**, and the **equally, or more marvelous following math solution**, possible.

Crucial to our explanation for Pilots, or professionals, is the fact that the **ideal simple, pure helical screw surface Air Inflow, and Stretched, still perfect helical Outflow can be visualized as a Simple Picture**, (p.79-II), not the wildly complex rare genius level math that it took to solve the total problem.

Blade SHAPE, & *Twist*, counteracting q (: V^2), Controls Radial Blade Loading

Furthermore, the **Ideal Blade Loading, Root to Tip**, and **Ideal Blade Shape**, *Twist*, the **characteristic variation in loading and shape-vs. Pitch**, or Advance Ratio, **can be seen as pictures** p.147 I. **Thus,** either a technically inexperienced Pilot, or Pro Engineer can **quickly, easily grasp the heart of the subject, Pure Gold!**

Furthermore, the **constant dT/dQ**, simply **a constant Ratio of Thrust to Torque -- at every radius -- every radius equally efficient,** (if first considered drag free), **not only makes the concept simple, it proves to be THE MAGIC BULLET** that ***cuts through the impossibly wild complexity*****:** heavily loaded, 3D, 3 Dimensional Airflow, --- compounded by all the trigonometry of the twisted blades and forces, the drop off in the shape of the thrust loading from max, to zero at the tip, + tip vortices, **Pure Gold!**

6. Instant, Easy Computer Answers, Theodorsen's *Solved* Math

Goldstein, in **1929**, solved the Partial Differential Equations for the Betz *Rigid Vortex Sheet* concept, but he believed only for the Multi Blade, Lightly Loaded, Low Advance Ratio Case, where the helical downwash outflow from the prop moved back uniformly in the stream tube, at the Same Speed as the Tube.

Early workers called that Multiple Blade Light Loading, not a final solution

Theodorsen, 1948, published his book, brilliantly recognized that **Goldstein's solutions were valid for full heavily loaded, 3D, high Advance Ratio flow, if one simply considered the 3D flow, and a reduced effective diameter existed far back!!** Interestingly, the **reduction is slight, ~ 99% Diameter**, except for **very** low pitch, **low Advance Ratio Props**, maybe **~ 97 %**.

Theodorsen used Goldstein's simple **Charts of Kx circulation Blade Loading Factors**, **vs. Advance Ratio and Radius**, which the computer can easily and accurately **interpolate to 3 Significant Figures, for answers of ~ 1%, error!!!**

The Kx factors are what control the Ideal, Min. Induced Drag Blade Loading, Root to Tip!!!

Theodorsen's work was no less brilliant, because he **accurately saw through the whole problem**, produced **other factors related to 3 D flow,** most notably the **Kappa Factor**, "κ" that related the rearward speed of the "Rigid Vortex Sheet" helical backwash, **w**, **(or w, a ratio of Plane Speed)**, that showed that with normal heavy loading the imaginary helical backflow moves back nominally **2, 3, 8 times faster than the whole stream tube average**, ***a very important, incisive, X-Ray insight***.

Andy Bauer used Chi, χ, pronounced Ki, in his Program, odd Type in T T's Book

Furthermore, Theodorsen provides a **full set of Engineering Equations** that make it **easy enough** for an Engineering Pro to program and use his method, **especially recognizing that the Kx Charts** are **easy to set up** for the computer to **use**. Additionally, he **solved advanced, complex, multi blades, counterrotation, before computers using a voltage field simulation**, as an analog.

Later, 1964 Tibery, and Wrench work got bigger Kx loading above .5λ.

The **Hugely Important Point** is ***with the math solved*** we ***don't have to learn it***, but can **use the computer for X-Ray Insight**!

7. X-Ray Insight into an Impossibly Complex 3D Problem

Realize Constant dT/dQ is Betz's Simple Outcome to this Technical Swamp

Let us make clear just how prohibitively complex the 3 D flow problem is, compounded by the twisted blade and force math, a swamp so complex no one could possibly see through it - evaluate.

Air is Pulled in and thrown back to create Thrust. Since the Blades are **at an angle the air is also rotated**, progressively **more at the steeper angle root radii**. With **higher pressure behind the blade, lower pressure in front, radial flow and tip vortices are generated** with **opposite flow directions on the front vs. the blade rear**. Interestingly, there are **root vortices and radial flow there also**. With **Heavy Loading**, the **Helical** Backwash is *moving back through the Streamtube* ***2, 3, 8 times faster than the average*** rearward motion of the streamtube. The Disk Diameter shrinks from Bernoulli pressure reduction too.

And Remember, Theodorsen Considers a Reduced Effective Diameter - Far Back

At the same time, the **ever more twisted blades inboard** not only **don't pull straight forward**, they seem to **lift more sideways, seem to pull more and more against the available Engine Torque, lugging down potential RPM, H. P. and Thrust. Actually it works perfectly as a pure Helical Screw!**

Thinking of the ~ elliptical lift distribution on a wing vs. span we **recognize that it is *not* possible to have calculated high thrust loadings near a tip, the thrust just rolls off the tip into a tip vortex, Thrust vs. Radius Loading another complexity!**

Now Recognize that the **Theodorsen - Goldstein 3D Potential Flow Solution**, (meaning every air particle affects and is affected by every other particle in a complex 3 D flow field), **handles all this, in addition to all the trigonometry and force computation,** but **done initially drag free, drag then added in, easily compared.**

Betz's CONSTANT dT / dQ is the SIMPLE OUTCOME — of this Quagmire, See Appendix dT/dQ

With this wildly complex **3D flow** and force geometry you could wonder if the basically **2D flow geometry** we teach you (p. 94 to 98-II) is valid. The 2D picture is easier to grasp, the 3d D radial flow is just perpendicular to the page, ***Theodorsen's math handles it all, 3D***.

8. Understanding the Pure Helical Air Inflow, *Stretched* Outflow

To Start: First, we teach simple **Pure Helical *Propeller* Pitch**, (p. 51-II) like a competition model builder might learn as a young teenager, because it can teach a Pilot exactly what propeller pitch is, quickly, pretty easily, all radii going up their proper circumferential ramp angle to arrive at the same pitch, inches or feet, ***with* zero slip, moving like a perfect screw through a solid**.

That's Prop Pitch, p. 51-II, *NOT* the AIRFLOW, Phi, ϕ

One must be careful though, because if we purposely **varied C_L**, it would **no longer be a perfect screw**. Also, if a prop designer properly accounts for ***slowdown***, **much slower inboard** the actual prop **moves further from a pure helix**, ***a less twisted blade***. Thus, **pitch is usually defined as that at the *benchmark* 3/4 radius**.

Figure 3, p. 82-II shows Glauert's ***a factor***, ***axial inflow***, a small **ratio** of the air velocity at the prop **V_1, at each radius, (V_1 as if no prop were there**), and his **a' factor** showing the ***inflow rotation***, a small **ratio** of the prop rotational velocity ***at each radius***. Studying that sketch teaches **how inflow works**, how **it jacks up** the **air inflow angle, phi$_1$**, ϕ_1, to account for **inflow to the airfoils**. **Adding angle of attack, α^o, sets Blade Angle, β^o, Beta**.

Helical Airflow Photo on p. 71 II

Figure 4, p. 96-II is the Rosetta Stone Sketch, that teaches that **if you simply set up Constant Helical Pitch** of the **Air Inflow**, **constant Slip, at each r.**, you create pure, simple Helical Pitch Airflow, **create Betz's ideal, minimum Induced Loss Prop**.

Figure 5, p. 98-II, *turns the flow diagrams around backwards*, adds the second $\Delta V/2$, thus shows the **downwash, backwash, still helical, now stretched**, "**Betz's Rigid Vortex Sheet**", (just our perfect Helical Screw, Stretched Outflow) **Helical Airflow Photo is on p. 71 II**

It's important to realize the **Prop Design is Pitched for the $\Delta V/2$ inflow only**. ***Everything Happens At The Prop!!! All the Energy is Put in There, the stretched helical 2d $\Delta V/2$ backwash* is already set up, maybe happens 1/2 diameter behind, from a slightly higher pressure**. Theodorsen's marvelous math even **defines if the inflow or outflow is a bit bigger, his A_0 factor!!!**

9. Tip Vortex Swallowing the *Imaginary* Pure Helical Flow
The **Smoke Tunnel photograph** on p. 105-II makes it abundantly clear that the ***tip and root Vortex swallow the theoretical pure helical outflow*** from the propeller, make Theodorsen's **Analysis *Far Back*** seem ludicrous, but if the **math is done scrupulously**, as it is, it is **perfectly valid, all the energy put in at the prop!!**

10. Computational Accuracy The **Goldstein-Theodorsen Kx Blade Loading Factors,** .xxx, created before computers, are basically accurate to **2+ Significant Figures**. One in 100 is 1%, but one in 999 is .1%, far better! *Thus, the basic method is held to be good to 1% **or better***, the calculation at that twice as good as the +/- 2% that manufacturing a prop to +/- .1 degree tolerance gives you, the good work that McCauley's shop offers.

Ribner-Foster do correct, improve some of Theodorsen's secondary numbers, and Hi λ..

Ribner and Foster of the University of Toronto, working in the 90's **confirmed Theodorsen's math using modern computers**, a valuable confirmation so no one misunderstands that this final math of propellers that took the smartest professionals 83 years to create is in any way inferior. A review of their work shows **at worst a 1/2% difference**, in the Kx loading factors, locally. Solving math as complex as propeller math to 1/2%, < .1°, is as good as "breaking the bank at Monte Carlo", winning the lottery.

Tibery and Wrench, 1964 increased Theodorsen's Kx Blade Loading factors above .5λ. Do!

11. Advance Ratio J = V/nD Advance Ratio is the Classic Graph that teaches how Propeller Efficiency works. **(p. 48-II)** If both sides of that formula are divided by π, **J/π = V/πnD** **becomes far more meaningful**, because **V is just the plane's Speed** in ft./sec., and **πnD** is just the **circumferential velocity of the tip of the prop** in ft./sec. **(n being revs/sec, D, Diameter in ft. π being π).** Those two speeds become the **AAA, Actual Advance Angle of the Prop Tip**, **shown on the Graph for easy insight**, the **Advance Ratio/π, J/π.**

The Graph clearly shows **Low Pitch**, **low Advance**, **low angle of advance** props are **poor in efficiency**, but swoop up to high efficiency, leveling off at ~91%, also depending on Aspect Ratio -- Allowing for Slip, **ΔV/2,** and **α°**, ***of course sets Pitch and Blade Angle***.

12. <u>P/D, Pitch/Diameter Ratio</u> Once you grasp what J, or J/π is, it becomes obvious that you can **tell the max efficiency possible** of a propeller **by just looking at *blade angles***, or ***more easily and specifically at the very simple P/D Ratio*** which piggy backs right along, tied to J or J/π, **simply slipping to account for prop inflow, ΔV/2, and angle of attack** α°.

It becomes even easier when you realize that a **P/D somewhat above 1 are the highly efficient props**, and those somewhat **below 1 are the low efficiency props**. It's easy to calibrate your judgment. A **51"P, 71"D** Luscombe prop, **P/D = .718 is limited to 85 % max efficiency**, whereas a **79" P, 70" D** RV 6 prop, P/D = 1.128 is ~89% efficient, getting close to the 91% max you might find on a P/D of 2, **depending on Aspect Ratio**, as seen drawn on the ***Advance Ratio curve done with all up Math, drag***.

A P/D of 1 is ~88%, a 200HP RV8, 1.27, 90.2% η - Aspect Ratio able to raise or Lower η.

13. <u>Andy Bauer's Magic Graph</u> (p. 50-II) If you **plot Efficiency vs. Speed** -- for various Advance Ratio or P/D Ratio Props, they spread out across a wide spectrum of Speed. *<u>However</u>*, if you ***Plot all P/D ratio props vs. 100 % of their Zero Thrust Speed***, (where the props simply **run out of Pitch**, thus Thrust, in a shallow dive at max allowable RPM), **<u>all props plot on the same horizontal scale for comparison</u>** and we have Andy's Bauer's Magic Graph, which offers much more insight than you might expect.

Below a P/D of 1 all props have roughly the **same characteristic shaped hump, obviously <u>low P/D inefficient</u>, high P/D efficient** But notice that **<u>the operating point at low P/D is even lower</u>, <u>far from the peek</u>** at low pitch, **high P/D near the Peek winning twice**, *<u>both cruise and Vmax plotted</u>* **vs. the max η line!**

On the Magic Graph, you see that **all props start at Zero Efficiency at Zero Speed**, *<u>runup</u>*, blasting a **Hurricane in Fan Mode, going nowhere, using fuel**. All props swoop up to their max efficiency as Speed increases, finally running out of Pitch and Thrust, **Zero Efficiency again at Zero Thrust**, **(in a shallow dive) using Fuel, but making no Thrust, *<u>not</u>* helping**.

14. Super Stall **Above a P/D of 1**, the good efficient props, **you get into *more and more stall, slow*, which is going to *hurt takeoff!!!*** Jumping to the **very high 2.75 P/D Ratio**, perhaps like the 400 MPH **1930's Schneider Cup Seaplanes**, or maybe the first **Spitfires** that actually had fixed pitch wood props, we find **Deep Stalls SLOW** at **very high blade angles, *Poor Takeoff*!!!**

This is where you absolutely need a Constant Speed, Variable Pitch Prop

Extreme High Pitch, a very interesting situation, shown by the big stalled S Curve. Props or wings still make substantial lift or thrust stalled, but the **thrust drops significantly** and the **drag goes way up**, especially in a deep stall dragging a **big separated wake**. But what happens is a **Big Surprise!** Before Stall, the very steep blade is lifting sideways, **pulling very opposite to engine torque** and **lugs the engine RPM way down**. Starting Takeoff, **Deep Stall**, but **pulling weakly against engine torque the RPM runs up significantly**, but then lugs down as the prop finally unstalls, limits RPM power and thrust and the prop jumps up to a surprisingly high efficiency, the low prop RPM, low thrust more matching speed, a strange case. As you can see vs. the big stalled S curve, extreme high pitch props exhibit more and more of this characteristic at high P/D ratios!!!

Slowdown Correction, Lower Inner Blade Angles, Helps All Props, Slow

15. Super Magic Graph (p. 45-II) Andy Bauer's Super Magic Graph is also marvelously insightful in that it **shows what happens to all the major variables vs. percent of Zero Thrust Speed.**

Angle of Attack, α^o at the benchmark 3/4 radius **falls as speed increases**, even steeper after max allowable RPM is reached. **Coefficient of lift C_L 3/4 r follows in parallel to α^o, but Stalls Slow.** **$\dot{M}$, Mass flow Rate** naturally **increases with speed**, a bit less drop at slow speeds, higher $\Delta V/2$ slow, tweaking $\dot{M}$ up a bit more there. **ΔV starts very high, slow**, the Prop in Fan mode trying to create a Hurricane, falls, **curves down as Speed Increases, dives, in a dive, the prop now RPM limited, less and less able to make a ΔV.** **Thrust**, blade stalled slow, shows **falloff starts as the root stalls, worse as the 3/4 r stalls, slower.** The **79/70, 1.128 P/D RV** prop **efficiency *humps, like the Magic Graph.***

Far off the design case, expect the non BGT math will be less accurate --

16. <u>THE FIVE 800# GIANT GORILLAS of PROP LOGIC</u>

You come to realize there are FIVE -- 800 Pound Gorillas -- that are **the strongest factors in propeller logic**. Of course the real basic is to **<u>understand how induced</u>** and **<u>profile drag</u>** ***<u>interact</u>***. The **<u>Chapter 3 Design Study</u>** holds some ***<u>Real Surprises</u> for us!***

1. **<u>Dynamic Pressure,</u> q, proportional to V^2**. The fact that the prop tip is going so much faster than the root, *<u>especially on slow, high RPM planes,</u>* and that **q** is **:** $\mathbf{V^2}$, is the fundamental that makes props, their theoretical constant C_L shape, **p.147-I and** ***<u>ideal lift-thrust distribution vs. radius, basically different than wings</u>***

The Plot on p.147 Book I, is essentially the Bottom Line on 138 years of Prop Insight!

2. **<u>Diameter</u>**, is obviously hugely important, because it controls the **size of the stream tube** and the **mass flow rate,** $\dot{\mathbf{M}}$, (also considering speed), the **lever arm of the blade drag**, especially when you learn it's $\mathbf{D^4}$ in the Thrust coefficient, $\mathbf{D^5}$ in the Power coefficient, but surprisingly $\mathbf{C_L}$, the next gorilla, can trump it.

3. **<u>Lift Coefficient</u>, C_L** Surprisingly, a **.6 C_L** shrinks a needed RV 6 diameter to **69.1"**, a **.3 C_L** balloons the needed diameter to **83.7"** proving to be **an even more powerful basic than D*!** A real surprise is that the **profile drag** on a proper .55 **C_L** prop changes from a nominal **1/3** of the total loss to **62% at .3C_L!!!**

***Narrow Tip Chord Betz Props Like a little Bigger Diameter ---**

4. **<u>Speed</u>** is hugely important because it **directly raises the $\dot{M}$, mass flow rate,** ***<u>thus</u>*** **reduces the ΔV** required to make any given thrust required, ***thus*** **reduces induced loss, raises the Pitch** and **Advance Ratio**, thus reduces the corkscrew path length, **less drag** energy **loss, all raising efficiency**, and **<u>the cap on eta, η</u>**. **Speed** also **acts to reduce the optimum diameter required!!**

5. **<u>RPM</u> is vastly more important than realized**, because with speed, it directly **controls Pitch**, **Advance Ratio**, thus directly **controls** the **Cap on Efficiency**, (with Speed), and of course the **induced drag** and **profile drag corkscrew path**. <u>Engines want high RPM for max HP vs. weight,</u> **<u>a direct conflict</u>** <u>with</u> **<u>props that want</u>** <u>low RPM,</u> **<u>high pitch for efficiency</u>**!!! **<u>FIVE Gorillas!!</u>**

17. The Chapter 3 Design Study -- All Key Flight Conditions

If I told you the **Broad and Deep insight available, p. 121-II** etc., is **Marvelous**, you see, I would ***Not*** be **Bragging!** Separate from my day job as a serious Pro, my lifelong Hobby has been Flight, and understanding it better and better. Finding, as a kid, flying competition models, that **the real specific factual insights and help on ideal Diameter, Pitch, Shape, Twist was ~ Zero,** I have to tell you that today is one of those Lifetime Benchmark Smile Days, nailing a personal challenge ***we set our cap to nail!***

Andy Bauer has been a marvelous collaborator, friend, partner.

• The Altitude Cruise Design Case shows specifically how **Lift Coefficient, C_L** and **Diameter** simply **Swap**, as **Induced Drag** and **Profile Drag Swap in Lockstep**, amazingly **small efficiency** and **performance** changes --- ***if*** the ***Pitch matches the case*** to **hit design RPM, intended Power and Speed**, Diameter and C_L counterbalancing over **pretty broad ranges**. Amazingly **drag rises to 62% of the loss on the Big .3C_L Prop** What we have here is specific insight on **exactly how and why props can be pretty forgiving, flexible *if*, --- *if* engine load is maintained,** maybe **only a bit of Pitch change,** ***to tweak it correct***. See p. 121-II

The practical insight to grasp here is that **a prop has to be pitched to screw ahead at plane speed -- then to be precisely overpitched just a little bit to account for *only* a small ΔV/2 and an angle of attack of *only* a degree, or so!!! A 1 degree** miss, **.1 C_L** is a **BIG 20% error** on a .5 C_L prop so we'd be **shaving a wood prop, <1 degree, or tweak twisting a metal prop** to **hit the balance**!!! **I learned that as a kid. It only took a half Century to nail the hard facts!!! Voila!!!**

For years I've heard how ***Diameter*** is the real basic on props, and as you can see that's ***WRONG!*** **Diameter, Lift Coefficient, C_L, Induced and Profile Drag swap in Lockstep,** (but remember, we're comparing ideal props). **Ideal, that takes some Smarts!!! But you see the whole message of our prop Chapters is that's EASY: Betz defined Ideal, Theodorsen solved the Math!!!!**

• **Sea Level Climb** The .5 C_L had been best at Altitude Cruise, the lighter .55 C_L prop quite close, **but here the bigger, heavier .4 C_L Prop climbs best even at significantly lower RPM**. **p.122-II** Notice the **Drag Coefficients have risen well above .01 C_D, low Reynolds Numbers --- worse at higher C_L's, smaller Diameters**. **Weight**, a cube function, the **.4 C_L** is **1.30 times heavier** than the **light .55 C_L prop**, (77.308" D/70.788" D) = **1.0921)^3 = 1.3025**. √

• **Altitude Climb**, 12,500' **Still p. 122-II**. Most interestingly here the **best climb rate returns to the .5 C_L Prop**. Additionally, notice that the **RPM is significantly higher than it was at S.L. due to the higher 127 MPH TAS** at the same drag 105 MPH IAS at Sea Level. The **extra RPM**, of course **helps** the available **H.P., helps keep a respectable altitude climb rate**, though the available H.P. is much lower. Interestingly, **notice that the C_D's fell**, but **only at the higher RPM smaller diameter props**, indicating more **favorable Reynolds numbers** at higher RPM and TAS despite a considerable reduction in air density thus viscosity. There are so many things going on in Propeller logic it **can be quite an education**, going through a study like this, the **math and computer giving us X-Ray insight**, a unique gift. √

• **200 MPH Sea Level Vmax** **p.123-II** As well described in the Text, α, **C_L and ΔV drop, efficiency gets even better at Vmax**. Look close, ***see how everything varies***. ***Can you figure out why?*** √

Nest, Realize Slowdown correction helps a lot, lower blade angles, especially inboard!

• **Sea Level Takeoff** **p. 123-II** The Study lies, an **otherwise desirable High Pitch, High Efficiency Prop *seems to* go Bad**, slow! **The higher the β° and C_L, the more tendency to stall,** ***slow***, the higher the speed it occurs at, extending outboard. It seems a misfortune that **above a P/D of 1**, just where fixed pitch props are starting to get good, quite efficient, they start running into stall at slow speeds, **worse slower than 50 MPH**. But fortunately ***Sowdown correction helps a lot, especially inboard!*** √

An **RV 6 seemingly stalls inboard,** if you tried climbing at **a min. power speed of 80 MPH**, but S.D. correction makes it OK. √

18. Slowdown The Ch. 3 Study was purposely done without Slowdown to give ***pure answers***, But once we teach you how to understand a Betz, Ideal Minimum Induced Drag Prop, we must **show you how it must be modified to account for Slowdown, so it can stay ~Ideal, the same Radial Blade Loading and Drag**.

Realize the STATIC Air is pushed Forward, and Outward, by the Embedded Body!

A Source Sink analysis of the prop plane, at a Luscombe's nose, shows that locally, at the 37.6% radius, the air is slowed nominally 17.5%, but only 5% at the tip. Any body pushes a pressure wave ahead, the phenomena that breaks down at the Speed of Sound. The air has to speed up as it goes around a typical body, but in a Prop Plane, at the nose, you'll find a **highly variable slower velocity profile vs. radius -- p.106-II, App. SSSS.**

Obviously it's **a major correction**, first **untwisting the blade** a precise angular amount, more inboard, but now, **at a lower q**, Dynamic Pressure, **we need a slightly wider blade also**. The **precise T# vs. radius loading can be maintained**, and you ***do not*** *suffer a drag penalty*, because the **q goes down as much as the chord has to go up**. It can be ~ **4.77** degrees at .367 r, **.61°** at the tip, **HUGE corrections if we're trying to hold .1 degrees**.

19. Vibration A **WW II Rolls Royce Merlin**, which sounds so incredibly Smooth is **actually a jungle of many strengths and frequencies of vibration** --- trying to drive a low damping metal prop, which itself has multiple modes and natural frequencies of ***response***. And engine Forcing Frequences change as RPM changes, ***a mess***. **If the Engine frequencies find a propeller mode to excite, it can be a life or death matter!!! Don't try modifying a prop. You might kill yourself -- big complexity!**

See Appendix V, for Basic Insight on Propeller Response Frequencies

20. Off Design Point Performance Analysis To actually do prop analysis you need **Three Separate Computer Programs**.

Also — an Addition to Theodorsen Adds Profile Drag

- **A Basic (Ideal) Design Program** Theodorsen's Legacy
- **A Source Sink based program to correct for Slowdown**
- **A Program to account for any combination of Speed, RPM and ρ changes**. **All of these take serious Professional work --**

Off Design Point, The Farther we get from BGT Math, the greater the Error!!!

21. <u>SUPER PROPS</u> -- There are <u>no Super Props</u> and there <u>never will be</u>! <u>The Max Objective</u> is to get it Right, <u>IDEAL</u>!!!

A Partially Elliptically Loaded Lipps Prop, has the potential to add a few % η.

If you're just reading this as an introduction to prop logic it's OK to wonder if you can gin up a little computer program, or buy one and design a Super Prop for your Homebuilt. If you read and learn what's herein you'll pretty quickly learn propeller design is a good approximation for **one of the most challenging Engineering problems in history**, and that **it took the 7 smartest guys who came along in the industry 83 years to get Theodorsen's really correct math design solution, (Drag Free).**

What that permits a pro to do is to easily enough, after a lot of long, smart, hard work - to **design an Ideal Prop, Drag Free**. It takes a pro's capability to do that, go on correctly, and only if he goes after ***<u>low Reynolds Number drag</u>***, because Propellers have ***<u>Jet Speed</u>* - *<u>but Model Airplane Chords</u>***. That causes the blade to have ***<u>higher than normal drag</u>***, and to ***<u>Stall earlier</u>*** at ***<u>lower C_L's</u>***. Then -- <u>what is the drag</u> for the excessive STALL angles found??

<u>To complete an Analysis</u> one has to **<u>create the next two programs above</u>**, and learn all that is taught herein so you don't take years to figure it all out and get it right. ***That's the bad news***.

<u>The Good News</u> is that we've been through everything with Gun and Camera and **Andy Bauer has the Program that's been worked out over a long time and cross checked every way possible, including some good flight test data that agrees**.

We hope to offer Andy Bauer's Computer Program to Propeller Companies and Propeller designers. After years of development, cross checking, and refinement it would be a huge waste to not have it productively used. It will also **<u>get the fantastic work of Betz, Goldstein and Theodorsen more broadly appreciated</u>, *<u>used</u>*!!!** It's not a Windows Program, but a more Pliable, GW Basic Program where <u>you simply plug in the numbers for the variables you wish right in the Program Lines</u>. It's possible that we may have, and offer. a Windows Program.

22. **Ideal Theory and Good Sense Blade Shape** We know that whether or not you end up with an **Ideal Prop** depends on whether or not you design in the proper **Blade Shape vs. Twist**, to create a **Pure Helical Pitch downwash, backwash outflow**. **It is most Ideal, has the minimum surface area, min. drag if a constant, optimum** C_L is used, ***a pretty simple final concept!!***

Look at the **Thrust vs. radius loading**, and the **blade shape** that produces that loading on page 147-I. **Betz**'s result is a concept that is **Ideal in Theory** and makes Very **Good Old Fashioned Engineering Horse Sense**. The **weak inner loading is practical** with the **weak, lower q, lower potential to make thrust inboard** -- and since it is the **steeper angles inboard that go to** ***excess angles slow***, it is **highly desirable** that it ***not start highly loaded***.

Next, a prop with **high q tips** is going to ***try to load itself highly outboard*** and it's ***wasteful to try to load it heavier*** than the ideal that ***is already loaded heavily at the 90 % radius***, but ***must drop to zero at the tip!!!*** **Narrow chord tips load props correctly!!! To try to manufacture even more thrust out there, at high drag, at a max lever arm --- when the thrust must drop to zero out there --- paid for but lost, is like manufacturing loss.**

Betz has it right, Ideal by the math, and good sense too. It would make no good sense at all to overload it out there ---- Square Tips just don't make any good sense --------

23. **Modifying Blade Shape** ***for better*** **Climb, Takeoff, Cruise?** Years ago NACA tests showed broad tip Props gave **better Static Thrust**, and ever since then we find Broader Tip Props. So we tried unwinding the blade outboard with broader chords to maintain ideal thrust vs. radius, every combination we could think of. Guess what? **We lost every time, just like horse sense says we should, more drag, more tip loss. Ideal is Ideal**.

We could help a badly inboard stalled prop at takeoff with a broad tip, but that **costs in cruise and climb, less RPM, H.P., less thrust!** The Computer says **Performance loses**, every time!

Any Arbitrary Blade Width Increase would be best, ~ 1/2 to 3/4 Radius

Clarifying an Important Unexpected Result

24. <u>Does Wrong Design lose Efficiency or Performance</u>???? Interestingly, as we just saw in the last case, if we hang on **extra, non-ideal blade area**, analysis shows **extra drag causes a drop to lower RPM, less H.P**. is available, and we **get less thrust**, not more. ***But the lower thrust ~ matches the lower H.P., so* we can see a performance decrease as the telling insight, *rather than a BIG efficiency decrease***!!!* That's a **tricky unexpected result**, but **Props Work that Way!!!** Props are **Chock full of unexpected outcomes like that. Remember how that can work!!** ***It's tricky***. Theodorsen Math is ~exact, but leave the ideal loading and the the final, *simple, non 3D math*, **can be fooled, a bit** This subject was a real bear to tame!!!

Interestingly, a key insight is that efficiency really is limited by Speed and RPM. When you monkey up the blade you can lose performance, not efficiency. **Where** we see **we do lose Efficiency is with a Square Tip like the Luscombe Prop**. It has extra Drag, and the computer tries to calculate a "square thrust load" at the tip which can't happen. It appears to ***lose ~ 3 pounds of thrust getting 151 pounds not 154, a 2% loss,*** paid for but lost. With other flaws, early we took the Luscombe Prop to be 82.5% efficient, not the Ideal 85%. It's worse, ~ 75%!!!

25. Mach Limits - A function of Absolute Temperature, R° The Game is to **keep prop Tips below .9 Mach** number, **maybe .88**. We handled Sonic limited Props in Chapter 3 with all the ***key numbers*** and **a graph on p.137-II** that shows **allowable RPM vs Speed -- for typical Diameters, at Sea Level**, so we'll just suggest that you refer to that graph and adjoining numbers.

Wildly, Prandtl, the Father of Wing Theory, Teaches Rankines Source Sink, (Slowdown)!

It's interesting, in passing, that Rankine, the famous 19th Century Scottish Scientist-Engineer won fame for the basic Steam Engine Thermodynamic Cycle, and is credited as the ***first*** water Propeller Analyst in 1865. The Absolute Temperature Scale (needed for Thermo, and the Speed of Sound) is named for him, **R°, Rankine degrees**, Fahrenheit degrees plus **459.7°** above Absolute Zero. He managed to be **first and last**, the Speed of Sound being the last barrier!!! A Standard NASA 59 F° day is 459.7 + 59 = 518.7 R°

Have Pity, Understanding for the Poor Struggling Author

I usually Don't Write in this Style

As a Professional, I had in mind a first class, computer generated, clean, artistically attractive page, professionally written book, incisive, right to the heart of all subjects, all the answers you couldn't normally find. I intended it only for planes, certainly not the gross complexities of the never explained propeller, certainly not all super emphasized, a professional little book to be proud of.

I almost didn't publish it. The pilot, normally an adventurer, not a techie at all, usually avoiding anything that looked like Science, Math, even a graph, which after all is the picture of how things work. With often zero practice on anything technical since High School, I quickly saw I had to give the reader every help I could, and **before I knew it I had backed into super emphasis, repeating, expanding, summing up, everything different than the concise, say it once, depend on the reader for a lifetime of prior knowledge writing style of a pro!**

Knowing full well that propellers automatically had 12 times the gross content and complexity they'd like to see in the friendly, dumbed down, artistic cartooned book, I knew everyone, who would also want it to be written at his specific level, would soon be ***shooting the messenger***, wanting to tell me what I did wrong.

The interesting thing I found, however is that a pilot artist friend could read it and actually get it, which, of course, was the objective: **Put it out there for the guy who wanted to learn how things really work, in flight, his passion, with all the help possible. Damm the Torpedoes --- Put it out there!!!!**

What happened: a guy with no background, who had never heard many of the terms, fairly soon, **if he cared enough, to hang in, take it on as a Challenge, a Game, could grasp what we pros never did!** There's 10 years insight here, plus a lifetime, **in a day**, or so, **the keys highlighted**, not for speedreading! People who care enough can get it, the real insights and inner workings of propellers, all the surprises. **Hopefully, trying, easy enough!**

I'm not the guy who made props complex, they're a bear. I tried for EASY ENOUGH!!

Is The Book Perfect? No Mistakes?

Well, as a Pro I sure tried! Unfortunately, I was out there all by myself, much more than one might normally be. My Lifelong Friend and collaborator was getting old, really no longer wanting the job of pouring through a book for Pilots, technical novices, looking for obscure technical traps, as well as things that the "new guy" reader could fail to understand, miss, or read wrong.

Of course the Book is also for Lifetime Pros too, a Very Demanding Audience!

Milly, the love of my life was a huge help, as always, could see the missing, or wrong punctuation, spell check mix ups, that do occur. No matter how hard I tried, I would soon be mostly reading for content, obscure technical missteps, logic accuracy, that pros might see, not on the same thought train as I.

Invariably, pros will find some items they object to, it would be no other way. And then there will be progress, better future insight. We always stand on the shoulders of our predeceases, and many will invariably stand on mine, on the new insight here.

The hardest task of all is to make the most complex Problems Simple, Easy Enough!

If I have indeed brought the Black Art of Propellers down to Understandable Logic, even the Horse Sense of What is Happening Physically, I'll settle for that as a good contribution to our ability to understand flight, my lifelong Hobby, Avocation. It's a worthy addition to my Spacecraft and Jet Primary Controls. Both are Products that can never be allowed to fail, that must work reliably, no excuses, whether, or not the requirement is in the Design Spec. Lives, and on Manned Spaceflight, our National Reputation were at risk. A long string of products, fun!

As a Kid Modeler, it was great fun to see my Spacecraft Controls go into the Smithsonian!

The Big Funny here, I was really bugged as a kid modeler, a high performance model designer, little F 16's, with more thrust than Weight, that had to fly themselves, without crashing! I eliminated the Spiral Dive by age 19, but there was not one intelligent understandable word in the books on how to get props right, in 1946, 48 when I won the Senior National Model Championships twice. The Big Funny, Theodorsen's 1948 Genius waited a half Century for Andy Bauer and I to sort it out for you!

APPENDIX

Betz	B
Constant Ratio, Thrust to Torque	dT/dQ
History	H
Newton	N
Slowdown	SSSS
Theodorsen	T
Thruat Ratio	TR

Appendix B

BETZ - Classic Ideal Propeller Logic.

An Incisive Review of Ideal Blade Flow Geometry

Betz, Gottingen Germany, 1919, saw the central requirement, and fundamental definition of the **ideally configured propeller, comparable to, but fundamentally different than Prandtl's Ideal Wing**. Prandtl's classic wing definition is simply the elliptical lift distribution, elliptical planform, which yields an ideal 1. constant downwash, 2. minimum induced drag, which results in minimum downwash, min. tip vortex loss, the two sources of **wing induced loss**, the energy cost of holding up the aircraft.

Prandtl, The Father of Wing Theory, defined Circulation, and the Vortex System!

Concisely, Betz foresaw the **ideal propeller** as one forming his now famous **Rigid Vortex Sheet***, and mathematically defined that as **holding r tan ϕ constant**, which we'll soon see sets **a simple rule** for **setting up the *geometry* of all the *air flow velocity vectors***, which means that it firmly sets up the ***direction and speeds of all the airflow at all radii of the prop***, a really very neat trick, controlling a very complex flowfield with, as you'll see, **a very simple math expression**, with a very simple **understandable objective**, **constant airflow Pitch**, **constant Slip**, great consequences, subtle and hidden. **Making all that clear will be the central objective here in this Appendix!**

***The Simple, Pure Helical Archimedes Screw Flow shown on the Book Cover**

But first we must clarify the famous defining term, the Rigid Vortex Sheet*, because as you'll see that is **a name**, easily misunderstood, by the public, really not a vortex as understood by the public, the whirlpool like tornado, like a sink drain, which can cause a lot of initial confusion*. It is not really rigid, since it stretches as the second $\Delta V/2$ happens behind the prop, and another Surprise, the math of propellers says that it does not rotate, but its air does, rotating as the helix moves straight rearward. Let's get a clearer picture, easy enough to grasp, **a simple screw surface!**

The Rigid Vortex Sheet, Better Defined

THE SIMPLE PURE HELICAL ARCHIMEDES SCREW ON THE COVER, SHOWS THE correct PICTURE

Think of an **instantly curing plastic sheet** coming off of each prop blade trailing edge, **precisely aligned with the W wind sheet,** forming **two helically twisted ribbons**, **180° apart**. They're helical because back in our advanced chapter sketch, and in the sketch below, the wind lines are all set up to describe simple helical pitch, just as shown in Chapter 9 --- except here we refer to the incoming W wind lines at helical ϕ angles, (here set up by velocity vectors, whereas in our Chapter 9 example, we set blade angles by inches or feet of pitch and diameter).

Remember, the Vortex Sheets stretch behind the Prop, a pressure, the second $\Delta V/2$

Look again at the cover photo of these helical ribbons coming off at the W wind angles, the ϕ's at each radii. We purposely repeated this here, because it is at once a simple concept, but visually challenging especially at first, and we want this as easy as possible for you, comfortable as you proceed. Visualize, but ignore that the prop tips will have strong tip vortices swirling around them, recognize they're there, we can't get rid of them, but they need not complicate the axial flow we're nailing here.

*** Vectors, usually to scale. a and a' are exaggerated here so you can see them.**

Blade *Inflow* Geometry:

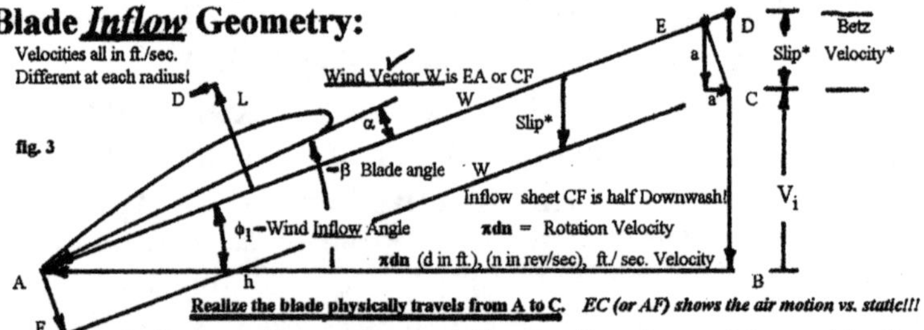

Grasp how the **inflow ratio a**, the **rotation ratio a'**, combine with **the plane's velocity** V, (later the **slowed** V_1[1]), to set up the **relative wind inflow W**, it's **angle ϕ, with the πdn rotation**, ϕ's, *next* for **constant Pitch, thus Slip**, the core key of ideal computerized propeller design, the **computer creating a solution of everything**. Recognize that we're still setting angles here, and that we're going to first set up **helical pitch for ϕ**, rather than the blade, but *now* we're **doing it with velocities**, not inches or feet. (recognize a and a' are ***velocity ratios*** to the axial and circumferential velocities, but can also be understood as small velocities, but remember **a** is simply $\Delta V/2V_1$, the inflow ratio.) ***The hidden secret is, required downwash sets a and a', and thus the vortex sheet that comes off the prop, thus thrust -- real insight!***

[1]Later the Slowdown Design loses this perfection, but after we've designed the Basic Prop.

Constant Helical Pitch of the Wind ***Inflow***, ϕ_1 (constant height to D)

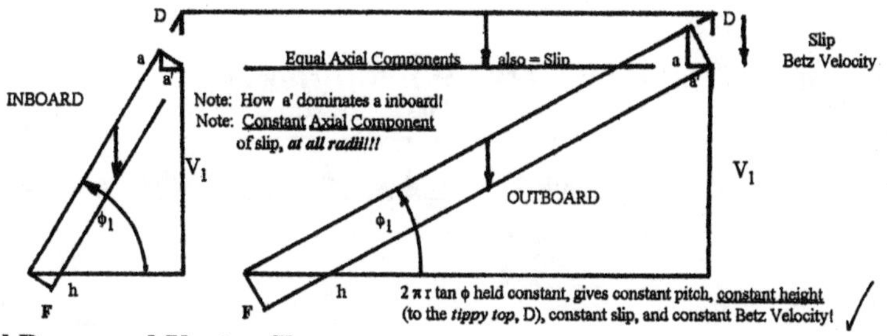

Ideal Props and Vortex Sheets

This is the *"Rosetta Stone Sketch"* that ***makes the central objective clear.*** I've simplified it, leaving all extraneous labels off for clarity, so you can see **the simple key point, the axial component** of the Downwash Sheets, the Vortex Sheets coming off the prop blades are **equal velocity at all radii**, a theoretically **constant axial velocity stream tube**, no unbalance to create extra radial flow, optimum. The downwash sheet, the Vortex sheet comes off the trailing edge of the prop at the required velocity to give the needed a and a' and ΔV and thrust, all tied together. It might help you to think of a set of "concentric glass barber poles", all with **equal (*height*), (Slip)** red stripes, all moving up at equal velocities. For all the guys who like math, and grasp it, (Theodorsen has expanded height), but all methods hold **r tan φ constant**, which simply makes each velocity triangle the same height, **constant slip**, makes the φ at each prop station **simple helical pitch, easy once you see it!!!**

There is more on Betz in the book, this a concise core summary---

A Complete Outline of *Classic* Betz *Min Induced* Prop Logic

Betz Core Objective: Getting Ideal Minimum Induced Loss!

(Basically Achieved by Pure Helical Inflow, *stretched* helical Outflow)

- **It's purposely First Done on a Drag Free basis, the simple case first ---**
- **Drag can then be easily added, Min Induced, *then* Profile Drag** *(separated)*.
- **It's an Archimedes Screw into, out of each Prop blade** (see Cover Photo).
- **Achieved by designing for Constant Slip, *holding constant Pitch Inflow*.**
- **Slip is the difference between inflow speed and Airplane Speed,** *constant.*
- **(Mathematically that is done by holding r tan φ constant. (φ inflow angle).**
- **The Geometry teaches how *highly variable* inflow and rotation combine.**
- **Amazingly, a Constant dT/dQ results, a constant ratio of Thrust vs. Torque.**
- **That means a constant efficiency at every radius, *steep root angles too!***
- ***But remember, it's first done Drag Free, then Drag Added,*** separated.
- **Bottom Line: that results in "Ideal Betz Loading", Thrust vs. Radius.**
- **Betz Loading is like a half tear Drop,** weak root, stronger, hi q outboard.
- **A Family of Ideal Blade Shapes *and* Matching Twist *results, if constant α^o, C_L***
- *Most Significantly*, **Tapered Tips** *vs. Advance Ratio*, **Pulls Back Excess Tip Loading.**
- *Not Allowing* **Greatly Excess Tip Induced and Profile Drag at hi q Tips.**

That final Hidden Imperative is the SECRET, Never Adequately Seen!.

(If the Fine Print is Unreadable -- it is Larger on the Appendix Last Page)

Constant Helical Pitch of the Wind *Inflow*, ϕ_1 (constant height to D)

A Velocity Diagram that Shows Directions of Lift, L, Thrust, T, Torque, Q at ANY RADIUS, Inner, Outer

fig. 6

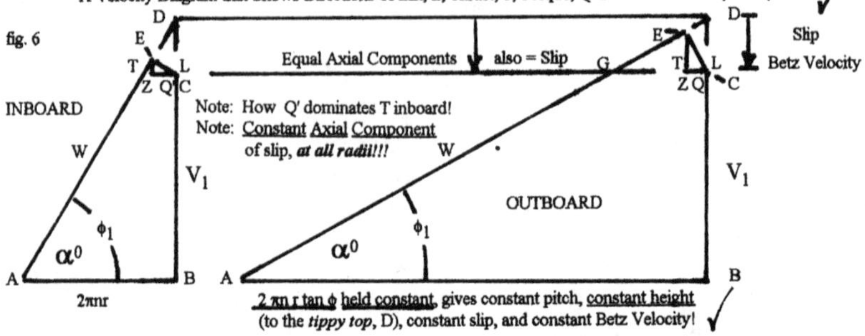

The Geometry *that makes* Betz Loading happen - Ideal Props

Also The Basic Geometry to Prove a Constant Thrust / Torque Ratio, dT/dQ

4 Triangles - ABD -- CED -- EZC -- GCD -- All Similar

Notice: The Four Triangles on each side, (Right and Left,) 4, are Similar, ~ Triangles, *to each other,* that is, exactly the same angles, and all sides proportional in length, just bigger to smaller size -- but that the **triangles family on the left**, for *much steeper Blade Angles*, (2nd 4), have a much smaller Thrust component T, a much bigger Torque component Q, in proportion. (This is the same Rosette Stone Sketch as on p. 84 II, with more labeling, for explanation – because we want to work with **each item**)

Now the trick will be to **prove that the steep family on the *left*, with weaker thrust, Stronger Torque *Force*, but smaller radius,** has the **same Thrust to Torque ratio, dT/dQ as that on the right!!! ???** In other words the **smaller radius** must exactly **counteract the bigger Torque force**.

Slip x cos α° = EC, Blade Lift, L, perpendicular to Wind Line, W, = Slip x cos α°

$$\frac{T\#}{Q \text{ in } \#} = \frac{\cancel{\text{Slip cos }\alpha}\cos\alpha}{\cancel{\text{Slip cos }\alpha}\sin\alpha\ \mathbf{r}} = \frac{\cancel{\text{Slip cos }\alpha}\cos\alpha}{\cancel{\text{Slip cos }\alpha}\sin\alpha\ \mathbf{r}}$$

Canceling where we can

But, $\mathbf{r} = (\text{Slip} + V_1) / \tan\alpha$ or also, $\mathbf{r} = (\text{Slip} + V_1)\cos\alpha / \sin\alpha$

So, when $(\text{Slip} + V_1)\cos\alpha / \sin\alpha$ is substituted for **r.**

$$\frac{T\#}{Q \text{ in } \#} = \frac{\cancel{\text{Slip cos }\alpha}\cancel{\cos\alpha}}{\cancel{\text{Slip cos }\alpha\sin\alpha}(\text{Slip} + V_1)\frac{\cancel{\cos\alpha}}{\cancel{\sin\alpha}}} = \frac{\cancel{\text{Slip cos }\alpha}\cancel{\cos\alpha}}{\cancel{\text{Slip cos }\alpha\sin\alpha}(\text{Slip} + V_1)\frac{\cancel{\cos\alpha}}{\cancel{\sin\alpha}}}$$

Canceling where we can

Everything Cancels out, *except* **T# / Q in#, = dT/dQ = 1/(Slip + V_1)**

And that shows dT / dQ is a constant, the Same at every Radius.

Appendix dT/dQ

Betz Geometry, Constant dT/dQ ✓

Rankine, Froude, ***Betz***, Prandtl, ***Goldstein***, Glauert, ***Theodorsen***

The **Most Important Basic of Betz Logic** is that he sets up a **Constant Slip**, **Constant Pitch**, **Perfect Helical Screw Air Inflow**, and **Stretched, Perfect Helical Screw Backwash, Outflow,** first considered **Frictionless, Zero Drag.** ✓

It turns out that the Geometry of a **Frictionless, Perfect Screw** does have a **Constant Thrust vs. Torque Ratio at Every Radius**, specifically, because it is **working as a perfect screw**, Advancing, Friction Free, up a Perfect Pitch Screw at every r.

In the derivation page opposite, we want to keep the Math brief, but it's probably too concise for anyone not practiced in Geometry, and Trigonometry, so we'll explain that a bit more.

In geometry, we learn that all triangles have 180 degrees of angles, and with one right angle like our Lift Line EC perpendicular to the Wind Line, W, the other two angles must total 180 degrees. Now look close at how the four, 4 separate triangles on each side nest together. Each may be rotated 90, or 180 degrees from another one, but you should be able to see **all four are Similar Triangles, same angles the same place, sides** bigger, or smaller, but **all in proportion to their like triangles!**

Now in Trig, one learns the **three basic angle vs. side relationship** for **sine**, **cosine**, **tangent**, as follows:
sin=opposite side/hypotenuse **cos**=adjacent side/hypotenuse
tan = opposite / adjacent

All we do in the derivation on the opposite page is juggle those Geometry, and Trig relationships. **Silda Smith** my **11th. grade high School Math** teacher would be pleased to see I got it!

And that shows dT / dQ is a constant, the Same at every Radius.

As a boy, competing, and consistently winning nationally with my model designs, I had a very good grasp of basic Flight, Aerodynamics, really good designs, that worked for the right, well thought out reasons, but with Zero Help from professional's insight on the intricacies of Propellers, frustrated, the reason this book happened a half Century later. A key, core that I had enough insight to see, was that the Inner Blades, at a Steep Angle, **seemed to be Hardly Pulling Forward**, but **seemed to be Lifting, Pulling against available Engine Torque**, more and more as the Blade Angle Steepened! **Smart, but Dead Wrong**, exactly why I was looking to see what the Pros had to say, after maybe the better part of a million Props in WWII, **I found a Barren Desert, Zero Help!**

My lack of Incisive Insight, was to fail to look past the Prop as a Rotating Wing, and see that, acting as a perfect, frictionless screw, those steep inner radii were working in perfect unison with all other radii, all doing exactly the same thing, perfectly, a perfect SCREW!!! I know today that friction hurts **a bit** more at low, and steep angles, but that's not a major loss, the size of the loss easily seen when we run the prop, first without drag, then with drag, all shown in the computer printout if we run it in two steps, the normal one step run showing the efficiency, with, and without drag.

The screw concept is the perfect, necessary insight on why those steep inner blade angles don't hurt, and that teaches us to get past thinking of a prop, only as a rotating wing!

When I first set off to **find out how Betz's Logic had a constant dT / dQ**, I was wondering if I faced fearsome Math. Was I shocked to find that I could prove it in an hour, simply **grasping that it really was a perfect screw**, and that I could derive it, **prove it with High School Geometry and Trig.**

I had this marvelous old Math Teacher, who liked boys, wouldn't let any of us be "Laggards"!!! Thanks Silda ------

And that shows dT / dQ is a constant, the Same at every Radius.

There is one other important detail here that has both technical and human significance. To be creative in the technical fields, one must, of course, be quite intelligent, have your subject well in hand mentally, so that you can see opportunities others do not. But then, in addition, it is quite desirable to have a brain configured in a way that you can think freely, creatively, break the bounds of overly disciplined, constrained thought patterns.

But, in contrast to that, in engineering there is great need to be a very careful, disciplined thinker, because in so many cases the core need is to create a product that will never fail, under any circumstances, whether in the formal requirements, or not. This takes a very detailed, disciplined orientation. The very best Engineers somehow learn to be both, freely think creatively, and then, almost change personalty, able to dig deep, nail the details.

Usually individuals are ***not*** capable of switch hitting, doing both, and if the boss is smart he may realize that he must have a creative team, and a team of disciplined thinker, checkers. If he's fortunate to have a switch hitter he can appreciate it, use him.

There is a case in point here. We used **a velocity diagram** here for a derivation proof, cutting incisively to a compact finale to make it possible to maybe grasp it simply, in concise form. But there is an illegal operation there, because we talk of Thrust **Force Pounds,** and **Torque Pounds, soon Ft. Pounds, ft #,** creating the sin of **mixing apples, oranges, and bananas!**

In a more disciplined way, we can develop Thrust and Torque Velocity into force, by using air density, ρ. rho, Area, and V^2. Now, what Betz, Goldstein, Theodorsen do is create precisely the correct SHAPE, Area, with proper twist, rho being equal at all radii, canceling out, In a complex way the Torque radius accurately factors down the extra strong Torque Force, inboard, with the trig involved, then the V^2 effect cancels because we have the same dT/dQ ratio at all radii. The derivation is finally legitimate, but formally, with many more steps than shown.

This Assures the Derivation Page Fine Print is Readable

Fine Print on the derivation page was necessary to permit a compact derivation organization that was comprehensible -- and place everything in the correct relationship to each other, ✓

However on the " book page masters", that were clear enough, it was seen that if the printing, or paper degraded, in publishing, you would hate me, So here is a solution, that can help.

Sorry for the inconvenience caused by the 6" x 9" page format

All Print here is raised to 10 Point Type

A Velocity Diagram that Shows Directions of Lift, L, Thrust, T, Torque, Q at ANY RADIUS, Inner, Outer

Equal Axial Components also = Slip (The Key to Betz Logic!) ✓

Note: How Q' dominates T inboard!
Note: Constant Axial Component of slip, ***at all radii!!!***

$2\pi n\, r \tan\phi$ held constant, gives constant pitch, constant height (to the *tippy top*, D), constant slip, and constant Betz Velocity! ✓

T# = ~~**Slip cos α**~~ ~~cos α~~.

-------- ------------------- Canceling where we can ✓ ------------------------

Q in # = ~~**Slip cos α**~~ ~~sin α~~ **(Slip +V_1)** ~~cos α~~ / ~~sin α~~ =

Appendix H_p

HISTORY --- of Propeller Analysis

A Brief History of Propeller Analysis, Who, What, When

Axial Momentum Theory: is simply based on "***pulling in and throwing back air***", adding a ΔV speedup, increasing the *natural stream tube mass flow rate* $\dot{m}$ to $\dot{M}$, adding a +ΔV *velocity* passing through the propeller disk, causing *"a rate of change of axial momentum"* to yield thrust. Don't be confused by that. The thrust is simply the ***reaction force from the mass of the air thrown rearward*** --- just like *you* would be pushed throwing a heavy mass! It was initiated by **Rankine in 1865** [1] for marine propellers, and was further developed by **R. E. Froude**, publishing his "**Actuator Disk**" concept in 1889 [2]. **Both show the maximum efficiency limit possible, since all losses are *not* included -- *only the primary axial loss*.**

Though historically called *Momentum Theory*, it is more accurately, and easily understood specifically as a Newton's Second Law concept where **Thrust = $\dot{M}$ ΔV**, *the fluid flow version* of Newton's more familiar F = Ma formula for throwing *solids*. $\dot{M}$ is the final *mass flow rate* ***in slugs/sec***, [just like pounds/sec. divided by g, (32.174 ft./sec. per sec)*, simply a smaller number of the **bigger units of mass**]. The basic formula is very simple, just multiply $\dot{M}$ by ΔV, the total speedup in **ft./sec. = Thrust#.**

The problem with momentum theory is that it tells you nothing about how to design the propeller, *only covers the thrust created and the primary loss, not how the prop needs to be configured!* ***Most important, it teaches that air must be pulled in and thrown back to make thrust, and ΔV is easily seen as a loss --- wasted, settling back to zero, calm.*** **It thus shows that it costs to create thrust --- that efficiency can never be**

100%, even with zero blade drag. *Also*, **hugely important**, a **fast plane with a big $\dot{M}$ needs less ΔV, or a Smaller Prop!!!**

Looking closer, it also teaches the tricky hidden point that although *the full ΔV is used to calculate thrust*, **only half ΔV**, the average **ΔV, $\Delta V/2$** vs. the speed V_1, feeding the prop, before any speedup, **$\Delta V/2V_1$** counts as an energy and **efficiency loss ratio**. The reason for this is hidden in the basic energy equation, $E = MV^2/2$, which involves one *average velocity* in calculating the distance over which the energy is added! *Also*, ***only*** the $\Delta V/2V_1$ ratio increases the natural mass flow rate $\dot{m}$ through the prop disk to $\dot{M}$, *because* the calculation is made *at the propeller disk* where the speed has only increased to half the ultimate. The **air is pulled from a larger Diameter, thus $\dot{M}$, not $\dot{m}$!!!** Subsequently, the V goes up, the second $\Delta V/2$, p drops, the **D. shrinks, Bernoulli, no more increase in $\dot{M}$**! The key to grasp is that ***making thrust requires a ΔV speedup --- and making a ΔV causes a loss!***

Huge insight from a simple formula, tricky details exposed too!

The basic concept is very simple, just $\dot{M}$ times ΔV -- ***the trickier details give great insight for those who grasp the full explanation.***

Blade Element Theory: divides the propeller blade into "several radial airfoil segments" with the necessarily *different twist and rotational speed at each station* for individual analysis and averaging -- for the *constant forward but highly variable rotational velocities, **as well as the extra axial and rotational inflows induced***. **W. Froude** [3] made a first try at Blade Element Theory in his 1878 paper, but **S. Drzewiecki** [4] did the major initial development covered in his book, Theorie generale de l'helice, Paris, 1920. All the early work did ***not satisfactorily understand inflow***, ***and rotation*** which caused wrong airfoil lift and drag answers. Also, ignoring Aspect Ratio didn't work, because stream tube **ΔV losses are, in fact, the induced loss**, which must ***not*** also be charged again in the airfoil drag. Several analysts attempted a proper concept and N.E. Joukowski got close by 1918. Betz, the Germans, started using Momentum Theory to determine inflow, seemingly logical, but only half of the full ΔV happens by the prop disk, and it's actually ***not constant*** **vs. radius**. ***Getting inflow and rotation correct was the big hurdle, and that needed Prandtl's work.***

Betz [5] in **1919** looked at *the vortex system of the slipstream*, assuming a frictionless, ***lightly loaded*** airscrew consisting of many blades, (equal advance *at all blade radii, constant pitch*). Betz first used momentum calculations for inflow, no rotation. Importantly, finally, he determined **ideal load distribution** along the blade for the *minimum loss of energy*, analogous to the elliptical loading of wing theory was --- "**The Rigid Vortex Sheet concept**". That is sometimes described as a **rigid helical ribbon** of flow emanating from each propeller trailing edge, which *conceptually* stay intact far behind the propeller. The hidden secret is that holding **simple helical pitch for ϕ**, phi, the wind inflow angle, with proper math and geometry yields a *uniform vortex pitch* as the excess pressure behind the propeller yields the second $\Delta V/2$, staying as ***slightly stretched rigid vortex sheets, just Archimedes Screws!!!*** An ***equal ratio of thrust vs. torque and HP at every radius***, results before drag is added. (Realistic heavy loading, as ultimately used and proven by Theodorsen, shows the heavier loaded vortex sheets *move faster than the unloaded areas* of the stream tube). Betz is *historically recognized*, for ***the optimum rigid vortex sheet concept*** as the fundamental **Classic Prop Theory, a milestone!**

Goldstein, 1929 [6], in his paper "On the Vortex Theory of Screw Propellers", using the rigid vortex sheet concept, **solved the 3D Potential Flow Math for a lightly loaded, single rotation propeller, low advance ratio, devising his Kx circulation factor**, (**shown to .5λ**). That was used by Theodorsen, **for heavy loading**, extended for higher Pitch, Advance*, 1948. He credits Goldstein's analysis as **the greatest single step in propeller theory** and uses it, bypassing Glauert's 1934 approximate method.

*But by 1964 Tibery and Wrench got **higher**, **High Advance Kx**, **above .5λ.**

Prandtl, in an appendix to Betz, gave an approximate correction to the thrust distribution for few blades, the first try at real loading. But, it was the ***Prandtl-Munk wing theory***, **1918 to 31** [7], that finally allowed ***rational calculation of the induced inflow, and rotation***, the remaining speedup farther downstream, nominally half at the prop disk. Prandtl conceived the ***Vortex Theory of wing lift*** that tied the Kutta-Joukowsky circulation around a curving baseball or a rotating cylinder to the ultimate horseshoe vortex system of wing circulation and trailing vortices. Prandtl taught the **Ideal Elliptically Shaped, and loaded Wing**, ***circulation***, ***trailing vortices***, ***constant downwash, minimum induced loss***. **His wing theory allowed *Blade Element Theory*, and prop analysis to get on to a rational basis**. Before you know the *induced inflow, and its rotation*, you cannot do a correct analysis of the airfoils as a "rotating wing". That is the major milestone that solved that major inflow hurdle, absolutely fundamental to getting it right!

Glauert [8] in **1934**, writing **the propeller section, Division L, of Durand's multivolume classic tome, Aerodynamics**, is generally credited with writing the best exposition of "modern" propeller theory and logic. He makes clear that nobody could get correct inflow answers until Prandtl [7] did his Airfoil theory work, which made it possible to get correct "**induced flow**", the ***inflow***, ***rotation, and downstream stream tube speedup***. Thus Glauert finally got prop logic essentially correct, except for heavy loading. He still assumed light loading, multiblades, ***not indented*** for heavy loading, and advised that all analysis done in that (pre computer) era is of necessity done on light loading. ***Also, Inflow is not equal at all radii, we learn!!!*** Also, reading Glauert shows that, in deference to his computerless era, he makes several math simplifications so that his math could be used and solved at the time. The full unsimplified equations would require iteration, trial and error solutions, essentially impossible by hand. That offers an obvious opportunity for the modern PC which iterates at a speed beyond the comprehension of engineers in the 30's, which obviously offers the opportunity to account for the lesser terms Glauert left out. **But next, T.T.:**

Thedore Theodorsen, a true genius, NACA's Chief Physicist during W.W.II went all the way in his **1948** book [9], including **heavy loading, high Advance Ratio, a 3 D flow example, a**

radial flow feed to the tip vortex, (p 32, 33), **that exceeds the axial flow, AMAZING,** (the source of excess prop loss), and the pressure and velocities far back in the wake. He makes the point that the actuator disk concept does *not* represent the true limiting efficiency case, because (in addition to circumferential and huge radial tip vortex feeding velocities) there are heavy loading axial pressure and velocity differences far back. **Significantly, the wake far back, not the prop, is the key to correct analysis.**

Before computers, he used ingenious electrical field simulation, not the complex differential equations, for the potential flow problems, extended the K(x) circulation function, for λ above .5 to 1, See Below*. He points out that the Betz Theorem, the Rigid Vortex Sheet is a key to heavy loading analysis, along with Goldstein's mathematical analysis, **as long as the final result of the vortex sheets far back in the wake is targeted**. He offers a mass coefficient, κ, kappa, that ratios average ΔV, far back to ΔV at the Prop. A simple Kx blade loading chart allows accurate analysis by all who make the effort to understand his insight and math. H. S. Ribner and S. P. Foster of the University of Toronto carried the work forward in a interesting 1990 paper[10] for frictionless propellers. It includes computer analysis by Kramer, and it compares the losses with the basic actuator disk momentum theory, simulating perfection, for two bladed props through multiblades, quite insightful. *(Tibery and Wrench work[14], 1964, nominally confirms Goldstein Kx Factors up to Lambda of .5, but not Theodorsen's Kx Factors above a Lambda of .5 - if proven correct, the final needed step!

Theodorsen says much more of course, more than we can or should cover here. He was generous in his credit for Betz and Goldstein, who laid the groundwork so long ago, now. Practical, realistic, he hastens to point out that the wake will be unstable, not in the form analyzed far back. (That's true, but OK. since he basically analyzes it theoretically, correctly!)

Thus, though *purposely avoiding all the complex math* in the book text, (it **is in App. T**) - targeted for a civilian, pilot audience, but **aimed at getting the correct sophisticated**

conclusions across without drowning everyone in math, **this book *is* based on all the long history of propeller math**, and theory. We very purposely show the most basic math, like the $T = \dot{M} \Delta V$ *where it is simple enough, and shows great insight* for Pilots who want to stretch and get it, and even more detail in fine print for those who want the deeper grasp. Clearly here, prop analysis has been brought to essentially final fruition over more than 100 years, after benchmark analysts in 1865, 1889, 1919, 1929, 1934, 1948, 1964, 1990, and here. Theodorsen's splendid work, the milestone, Tibery and Wrench[10], the necessary final Kx. Professor Eugene Larrabee [12, 13], MIT, and Adkins and Lebeck[14] advanced, updated and refined the classic Glauert analysis, which can give good results, and is more adaptable to off design point analysis, important. (Larabee designed the optimum propeller for the Gossamer Albatross which made the 1979, 22.26 mile man powered English Channel crossing possible, after the original prop proved unable.) Quentin Wald found the ***Tibery and Wrench Kx work missed by Aeros***, a fine overview, a simplified look at Theodorsen like math.

Work continues with *much needed on the efficiency loss of props and afterbodies.*

1. Rankine, Trans., Inst. of Naval Architects, Vol. 6, p.13, 1865
2. Froude, R.E., Ibidem, Vol. 30, p. 390, 1889.
3. Froude, W, Trans. Inst. Naval Architects, Vol. 19, p. 47, 1878
4. Drzewiecki, S., Theorie generale de l'helice, Book, Paris, 1920
5. Betz, A., Schraubenpropeller mit geringstem Energieverlust. Gottinger Nachr. 1919.
6. Goldstein, S., On the Vortex Theory of Screw Propellers, Royal Society Proceedings (A) 123, 440, 1929.
7. Prandtl-Tietjens, Hydro und Aeromechanik, Berlin 1931.
8. Glauert, H., Aerodynamics, Durand, Div. L, 1934.
9. Thiedorsen, T., Theory of Propellers, McGraw Hill, 1948.

10 Tibery C L, and Wrench JW, Tables of the Goldstein Factor. David Taylor Model Basin, Report 1534, Applied Mathematics Lab. 1964

11. Ribner, H., Foster, S., Ideal Efficiency of Propellers, Journal of Aircraft., Vol. 27, no. 9, 1990.
12. Larrabee, E., "The Screw Propeller", Scientific American, Vol 243, No 1 1980, pp. 134 - 148. (Also the following:)
13. Design of Min. Ind. Loss Props. SAE T. Paper 790585, 1979.
14. Adkins, Lebeck, Design of Optimum Propellers, Journal of Propulsion, and Power, Vol. 10, No 5, October 1994
15. The aerodynamics of propellers, Quentin R, Wald, Science Digest, 2006

Appendix N

Mr. Newton - and Engineering Units.

Engineers always work in **a consistent system** of **physical "units", "foot, pounds, seconds"** in the U.S., until the more recent trend to the less friendly and familiar international units. The reason is simple and the advantages great, among them that you can **get the correct answer**, *as opposed to wrong*! Just as you can't add apples and oranges, to get a correct answer to a calculation each item in a potentially long complex calc ***must*** all be in consistent units. Thus you'll find engineers don't do basic calcs in MPH, but rather in ft./sec., or they multiply by 22/15 to put MPH into ft./sec, since 60 MPH is exactly 88 ft./sec.

A **huge side benefit** is that you can do "**dimensional analysis**", that is, put the dimensions, or *units* in a long string calculation and see "if the units cancel out to" the "***units*** answer" you want, *or* if you made a mistake. A simple example: The weight of a 100 foot cube of air: $(100 \text{ ft})^3 \times .076475 \ \#/\text{ft}^3 = 76{,}475\#$. **Wow**, you just learned **two** important insights! 1. If you want a thrust or lift answer you better be sure your calc does in fact come out in pounds, and you'll see shortly how easy it is to be wrong on that one. 2. Yes air has serious weight, and it has to if "throwing it down" is going to hold up a 920,000# 747!

If the feet, pounds, and seconds do *not* cancel out to give the "correct units for the answer you want" in a long calculation where you'd get lost, you *know* you have to chase through and see where your mistake is. It's a very simple check on a basic that could prevent someone's death. I started this way, because **Mass and Dynamics**, basic to Aero, *is where you **need a check***

It's a very simple concept, that gives you a simple check on calculations that can get wildly obscure and complex, a method and insight that will keep you straight, save your life or someone else's, literally, when you're confused, in need of help. Bridges that *don't* fall down come from Engineers **who check**.

Our modern life started *literally* when Sir Isaac Newton and his contemporaries in the 1600's started ***precisely*** nailing how the physical world works, **the specific laws**, so that precise calculations could be done, to produce a Space vehicle, or just a connecting rod on an old "Iron Horse". His **three laws** that nailed "*exactly* what that falling apple was doing", started the field of "Dynamics", that makes it possible to calculate a wing's lift, a prop's thrust, or propulsive efficiency, a long list of basics.

Newton found the concept of ***Mass***, related to weight, but fundamentally different than weight, **weight #, divided by the "acceleration of gravity, 32.17405 ft./sec^2**. That makes the **units of mass** pretty weird, **# sec^2/ft**. **Rho,** ρ, the **mass density** of air, that is "mass per cubic foot", thus is # **sec^2/ft^4**, even more weird, weird even to non Aero engineers, who sit down and pencil it out to get themselves straight every time they hit it. M dot, $\dot{M}$, the symbol for Mass flow rate, in propeller analysis, "mass per second", mass/**sec**, thus becomes # **sec^2/ft sec** = **#sec/ft.**

Now don't let all that weirdness blow you away, because the whole purpose of this Appendix is to explain it all to you, show you how to handle it without getting all confused and defeated by it. It can really screw you up, or you can defeat it and be the master. I think what we have here is pretty good, if you just dive in, do what I say, learn just --- **that one magic paragraph**.

First, **just** look at, think through, and **understand that one paragraph**, just get it. In all of Aero, grasping "those three terms", plus the little we'll do with Momentum are the three keys to **clairvoyance**.

Second, recognize that those weird units, while tending to complicate the eminently simple subject of dimensional analysis, can be easily tamed by that simple analysis -- because you can **easily see if you have it correct or not**. If you want a thrust answer in pounds, like you do on **p. 136**, you can easily check that you are in fact correct**, as we do there!** Look there, check.

Incidentally, let's be sure we didn't confuse the new guys back in the key paragraph. "Acceleration is a speed up in velocity", vs time, foot per second, (ft/sec), per second or **ft./sec^2**. We don't want anyone lost or snowed by what is a very friendly system of writing units, once you get it. Everything seems hard for everyone when you're learning. Later it's easy, so our objective is to help. **Get that one paragraph, follow it through, word by word, with your finger if you have to**, and you'll see that it's *weird but* not *hard*, and you will have conquered a basic that causes no end of grief to people who never nail it.

Now so that we have a rock solid basis for this book. and our insight into Aero Engineering let's go back and learn what Sir Isaac said **about linear motion**, carry it through to new insights.

- **Newton's First law**: Objects at rest tend to stay at rest, and objects in motion, tend to continue in a straight line (at the same Velocity), unless acted on by a force (that changes either).

- **Newton's Second law**: An unbalanced force acting on a body causes the body to ***accelerate*** in the direction of the force, directly **proportional to the magnitude** of the **force**, and **inversely** proportional to the **mass** of the body. a : F/M. From this we get the famous equation **F = Ma**, that we'll tie to gravity in *a gravity based system* where *force* and *weight* are *in pounds*.

- **Newton's Third law**: Every action has an equal and opposite reaction. (In other words, if we throw back air we'll get thrust!).

Sir Isaac, through Galileo in the "leaning tower" knew that the apple didn't just drop, it accelerated. Later pros finally ended up at **g** = **32.17405 ft./sec²,** which means that the apple at sea level, at 45° latitude, speeds up that much each second, in a vacuum where there is no drag. That's a most important insight, but it also tells us **mass** is ***not*** in **pounds**, as we'll now see.

If it takes 1 pound of force to hold up 1 pound of weight, and if we drop it, it accelerates 32.17-- ft./sec², the following is true.

F # = **Wt** # = **M** ?· **a** ft./sec² --- then the **units of Mass** must be **# sec²/ft** (do you see how it works?), and we know that a = g, the acceleration of gravity, 32.17 ft./sec². If *one* "***Slug***" of Mass, M = W/g, it's a ***BIG*** unit, **more than 32 pounds**, but with **very different units**. That will take some getting used to, but it's really not difficult, and with this **one paragraph** of explanation, you can go back and justify the units of ρ, **mass density**, # **sec²/ft⁴**, and M dot, $\dot{M}$, **mass flow rate, mass / sec**, # sec²/ft sec, in the magic paragraph above, which can be simplified to # **sec/ft,** if you wish. That nails the ***three key terms*** in all of Aerodynamics. Yes, that's some new thinking to do, but if you want to, I'll bet you can cope, ***because*** **that's *all* the meat.**

Mass is simply W/g, numerically, and dimensionally, ***not*** a brain buster. All the mass units that spring from it fit in *one* paragraph and grasping that *one* paragraph will let us go on and do all the basic calculations in Aero. *Learn the units of those three terms*!

M = W/g **M = # sec²/ft** **Rho, mass density,** ρ = M/ft³ = **# sec²/ft⁴**

$\dot{M}$ = M/sec. $\dot{M}$ = # sec²/ft sec. or $\dot{\mathbf{M}}$ **= # sec/ft.**

Weird as those may first seem, nail them and you'll succeed.

Finally: **Recognize** F = Wa/g implies Force # = Mass x accel.

Recognize **"G" load = a/g** - If you accelerate 3 times 32.17 ft/sec² you have a 3 G load!

Recognize **Load # = W· a/g** a simple proportion really!!!!

PROPELLERS ---- For everyone, and in particular those interested in propellers, it is very worthwhile to press on and understand the special and basic relationship between Newton's Second law, F = Ma = Wa/g, and ***momentum***. Propellers are often said to work on Momentum, and there are indeed some momentum concepts that are important and basic to propellers, but the idea that props work on Momentum, is fundamentally ***wrong*** and can be ***seriously misleading***, and we can best clear all that up in the next few paragraphs!

Understanding F = Ma is one of the most fundamental things in Physics, (and it turns out that there is a special nifty version of it that works with flowing fluids, like the prop stream tube that gives us our thrust). First we'll look at it the regular way. ---

If you throw a heavy 15# medicine ball, M, backwards, *hard*, you'll feel the big reaction force. You didn't just throw it back at some speed, or velocity, *you had to accelerate it*, speed it up, let's say to 10 ft/sec, in .1 second, which is an acceleration of 100 ft per second, per second, and that is 100 ft/sec². Got it? Now, we can calculate the actual force with our formula, F = Ma, put it in weight form, F # = W # a ft/sec²/g ft/sec² = #
That's 15#·100 ft/sec² / 32.17 ft/sec² = 46.627# (That's 3.108 G's).

Now here's the nifty trick with a fluid like water or air. The problem is we don't have a solid medicine ball to throw, but a flowing fluid! What do we do about that one with our F = M a? Well a = ΔV/t, so we can convert our formula to F = M·ΔV/t. Now math is flexible and it does not care if the t divides the ΔV, or the M, it's all the same from a math standpoint. But now you see M/t is mass/sec, mass per second, and that's ***mass flow rate***, which we call $\dot{\mathbf{M}}$. Now engineers would do this as follows:

Remember that *acceleration is rate of change of velocity*, ΔV/t.

$$\textbf{Thrust \#} = Ma = M\,\Delta V/t = (M\,V_2 - M\,V_1)/t = (M/t)\Delta V = \dot{M}\Delta V\ \#$$

So now, hopefully, you can see how we tweak the F = Ma, #, formula, which works for solid masses, around to **T = $\dot{M}\Delta$V, #,** which gives us **Thrust** from our flowing fluid *prop stream tube*.

Now momentum is **MV**, Mass times Velocity, which you're now smart enough to figure out has the units of **#sec**, --- Mass, **# sec²/ft**, times Velocity, **ft/sec** = # sec, specifically not # -- but if you *divide by sec*, consider a "*rate of change of Momentum*", we have what we need, *pounds thrust*. The big picture thing to grasp is that **props** from which you want **pounds thrust** work on **F = Ma**, which is equivalent to **T = $\dot{M}\Delta$V, #,** per Sir Isaac, or ΔMV/sec, #, rate of change of Momentum, an equivalent but not very helpful concept, **but specifically *not* MV, momentum.**

Now there is a good simplified example that helps to explain how propeller thrust works that we use on page 135 and page E 18, but it also exposes a subtle, initially confusing little point that we'll expose here.

Imagine a hypothetical plane flying along at say 200 MPH, momentarily, magically, with the prop at zero thrust! There will be a 200 MPH **stream tube passing through the propeller disk diameter at some "mass flow rate", *mass/sec*, $\dot{m}$** rate.

Propellers work by "throwing back air", per Newtons Third Law. When they do that they **increase the mass flow rate** *to* **$\dot{M}$**, and "**provide a ΔV velocity increase**". Amazingly that ΔV **may be only 20MPH, or so, *not* a big number**!!! **But notice** that ΔV ***both*** **increases the mass flow rate a relatively small percentage, and then turns around and multiplies $\dot{M}$ to produce a Thrust in pounds** as we've learned above.

The explanations on pages 135 and App E 18 are fundamental and important, so we didn't want to complicate them with going **from $\dot{m}$ to $\dot{M}$**, and that **ΔV acts twice**, so we're doing that here.

Appendix **SSSS**

SLOWDOWN - Source Sink Simulation

Potential Flow, Slowdown, and the Source Sink Method

When a body is moving through the air, it tends to shove the air **forward** and **outward,** or **slow** an airstream coming at it, and move the stream **outward.** This **creates a slowdown and a radial flow in the plane of the prop.** We can somewhat ignore the radial flow, but we need to **accurately know the axial slowdown** vs. radius, to **correct the angles and chords of the ~perfect helical inflow ideal prop,** for any particular airplane. Since we want inflow angles into the prop airfoils to **.1 degree accuracy,** and these corrections can be several degrees inboard, more than 17 degrees next to the spinner, maybe .3+ degrees at the tip, 3+ times our accuracy objective, even at the tip, this is a hugely important correction at design speed. (Incidentally, notice the correction would be zero at zero speed, greater the faster we go, ultimately important in making **a properly corrected fast prop** **better at slow takeoff speeds**, and **climb**.) √

Now, since a prop can't change blade angles as it rotates, we'll convert our body nose cross section area, vs. axial length, into an **equivalent *average*** *symmetrical* area body of revolution, around the crankshaft axis, like a special shaped bullet - that causes the *correct **average** effect at each radius* of the prop, in the plane of the prop, ahead of the nose. That makes it an easier two dimensional flow problem, not a much more complex three dimensional problem. We're simply after the average slowed factor at each radius, maybe .497, at 19% r, next to the 18% spinner, .645, at 27%, .757, at 35%, inboard, ... .968, at 91%, .973 at 99% out near the tip, **real RV 8 numbers.**

Aerodynamacists use the **Potential flow concept** to be able to deal with 3 dimensional flow, or the mathematically easier 2D flow. The **Potential function,** a capital phi, Φ, with a little calculus, a derivative vs. a location gives **a flow velocity at that location.** just the **capability to produce that velocity,** at that place, like a *strength, more easily handled,* if we have to deal with, or divide up flow into more complex 3D, or 2D flow. Goldstein and Theodorsen work with a genius level of this method to accurately understand the complex 3D flow of our basic propeller solution, and resolve it into a chart of **optimum blade loading vs. radius,** vs. **any *Advance Ratio,*** **low or high.** (J/π, our Lambda, λ, the AAA, Actual Advance Angle of the Prop Tip.)

Rather than talk about *Potential*, or do any Calculus, we're simply going to **First: create** a chosen axial lineup of pinpoint **Source+, and negative Sink Strengths** along the prop shaft axis, to **mathematically simulate our actual *average* body nose**. (Here comes Rankine, a genius on yet another subject.)

Prandtl, Tiejens teaches us Rankine's **Source, Sink concept** to create a way to **mathematically create the synthetic equivalent of an embedded body.** First, think of a **simple imaginary bullet shape** with **one air source** of just the right air flow volume rate, at just the right axial position, (it turns out, ¼ D back from the bullet nose, to mathematically model a bullet shape). A certain cubic inches per second flow rate, CIS, from a point source, has an infinite potential and flow velocity right at the point source, ***but*** its Strength, **velocity quickly drops off, divided by the surface area of an *ever bigger* Sphere,** (literally the CIS divided by the surface area of a Sphere, $4\pi R^2$, simply 4 times the area of a circle, to get **a simple velocity in in/sec, in. per sec, at any radial distance from the source**, **in.^{3}/sec. *divided by* in.^{2} = in./sec).***

THINK, *get it*, that's easy enough, once you catch on!!!

Now, if we have **a uniform flow field moving from left to right**, and our perfect point source flowing out in all directions from the longitudinal axis, it's easy enough to see there will be **a stagnation point on the nose of the imaginary bullet**, where the right to left potential of the source exactly equals the left to right potential of the flow field --- and the flow field is stopped, **a stagnation point**, caused by the **one radial** of the source going exactly straight forward in opposition to the stream potential, the source strength, ***its velocity*** stops the flow V! As above, it's at 1/4 D, the diameter of the bullet. **Of course, the bigger the source, the bigger the bullet.**

***Realize we might use inch, or foot dimensions, but we must be consistent!**

There is some advanced math here, graduate level math based on a partial differential equation solution, but it all comes out easy enough to grasp, **insight that we can understand** OK. We **do get a bullet shape**, with **all the source air staying inside the bullet boundary**, like you see in **figure 1.** a **nice definite mathematically definable boundary, all the flowfield air kept outside that defined boundary**. Now, I'm sure you can see where we're going with this. **With multiple sources and sinks** we **can bump that bullet into any shape we wish to essentially exactly match our actual body of revolution** - and the computer will quickly do the math work for us.

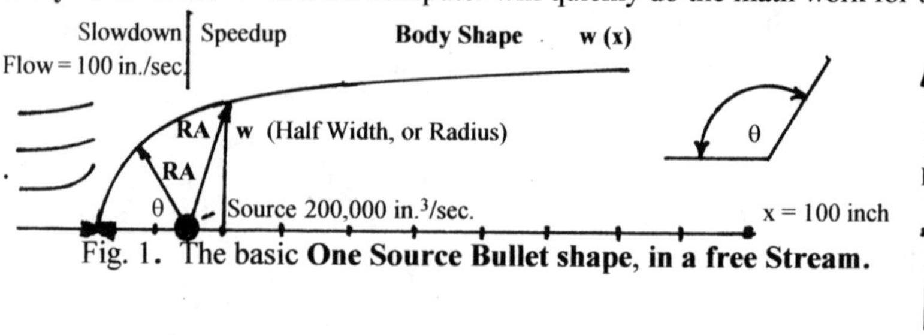

Fig. 1. The basic **One Source Bullet shape, in a free Stream.**

Now, realize that at all x axis locations **to the left of the source**, the **source is slowing down the flowfield** --- but **to the right of the source**, it is **speeding up the flow**, though the flow inside and outside the bullet doesn't mix! Don't worry about maintaining **continuity**, the constant flow rate outside the body, the **outward flow** at the nose ***keeps the flow rate constant***. Now, realize that **in the plane of the prop**, the **axial flow is slowed at *any* radius, a lot inboard, much less as we move out in radius**, past the body.

For the Engineers in the audience who want to understand ***all*** the math, we recommend the book **"Fluid Dynamics" by Victor L. Streeter, McGraw Hill Book Company, 1948**, which does a more complete job of explaining the math than the classic Prandtl - Tiejens, text. Streeter gives the following **two formulas, first for the bullet half width w**, or radius of our **one source** synthetic body of revolution, **at any axial location**, **vs the θ Angle**:

$$w^2 = \frac{m}{2\pi U}(1 - \cos\theta)$$

m = source strength in.^{3}/ sec.
w = body half width, or radius, in"
U = uniform stream velocity in./sec.

The important key to understanding that: the cosine moves from 1 at 0°, to zero at 90°, to minus a minus 1 at 180°, a +1, thus the (1- cos θ) term goes from 0 to 1 at 90°, to 2 at 180° way back at infinite length, the half width, w, forming the bullet shape full $\sqrt{\ }$ **D"** far back, gradually as an asymptote.

Mathematically, m/2πU is a constant, the bracket goes from 0 to 2 as θ goes from 0° to 180°. But we need the square root of those to get any half width **w**, so we have the **square root of a constant, times the sq. root of 0 to 2, Physically**. The source stops the stream at the stagnation point, weakens as the area of a sphere as the radii increase with θ, slowing the flow less, no longer directly opposing it, shoving it outboard, **but weaker and weaker**, as RA grows

This next tricky equation plots the Bullet boundary shape at any axial location, including far back based on **r, the diagonal length, source to point.**

$$r = \frac{1}{2}\sqrt{\frac{m}{\pi U}}\ \text{secant}\ \frac{\theta}{2}$$

r is labeled **RA** on the sketch.

Now, notice that both formulas have **m/U**, the **source flow rate** in **in.^{3}/sec.**, divided by the airstream velocity, **U**, **in./sec.**, which gives us a distance^2, **a distance after the square root is taken**, the units we need, want, plus the nondimensional trig. functions that the complex differential equation math decrees. **Two tricky secant insights there -- hide right before you. A secant of θ, the hypotenuse of its triangle / adjacent side, goes from 1 at 0° to infinity at 90°. Now, with θ/2 controlling the action, the diagonal length, r, reaches way back at θ = 180°, goes to the infinity of 90°, reaching way back, so the bullet shape approaches a perfect cylindrical diameter as an asymptote far back at infinity, just like the last formula. That diameter will be 4 times the nose radius at the stagnation nose, that radius 1/4 D.**

Andy Bauer's Computer Program to do Slowdown, Three Major Steps

1. First: We must measure the cross section area of the cowl - nose at several logical axial locations --- and **create an equal area vs. length symmetrical body of revolution to use as a master to compare, match** vs. the radii, or **w half widths of the math model** we create. **Do realize** that with most noses quite unsymmetrical, **the prop angles of attack will be cycling from above to below our design α^o and C_L, but if not extreme, the linear C_L - α^o** relationship will **keep our math and average result OK,** ***no better possible!***

2. Second: By **trial and error**, we must **find that combination of Sources and Sinks, at x axis locations, that model our actual equivalent body of revolution in 1. above**. The computer will **do the math for us, reading out w, half widths, so we can home in on matching our nose body of revolution half widths, or radii, and contour in 1! More below ---**

Andy's Program prints out all selected widths, w, vs. x's for each source-sink guess.

3. In the **last major step** we will be finding the **slowdown at the various prop radii**. To do this we divide the source strengths **Q*, in.^{3}/sec.**, by the expanding sphere surface area, **in.^{2}, $4\pi R^2$**, where R is each diagonal distance, or radius, from the source to the desired prop radius --- for multiple sources and sinks, to **get a velocity, in./sec**. The **horizontal component** of those **slow the flow field, easy, once you catch on. Notice how dimensional analysis keeps the units correct, helps us to understand what we're doing!!! in^3/sec. / in^2 = in/sec.**

***Andy's Program uses Q for flow rate rather than Streeter's m.**

Now, Andy Bauer's program uses an iteration program that calculates the effect of a long axial string of needed sources and sinks, for **First: matching our synthetic body half widths, or radii, to the symmetrical body of revolution that equates to the actual body nose shape. Second: once we've mathematically matched the real body axial cross sections --- the program uses that assemblage of source-sinks to find the slowdown at each prop radius -- as above, by summing the effect of all source-sinks.**

Realize it's the horizontal component of all those that is the slowdown at each radius.

Andy's program, as it comes to you, has the **actual sources and sinks** that **solve the Triple Ideal RV 8 prop slowdown problem**, so you have an example to learn from and mimic. ***Finding the source-sink array that solves any problem, is trial and error***, easy enough with the ***computer to find the matching*** **w's** for you. Just play with source flow rates, locations to solve it.

Notice the RV 8 model on the next page. The RV 8 has a big source up front, then a sink, pretty typical. That produces a **more blunt nose**, the next sink then **pulls in the bullet shape**, then selected sources and sinks to **bump out and pull in the math model to match your body of revolution model.** Andy's instructions shows the line numbers to plug in your numbers

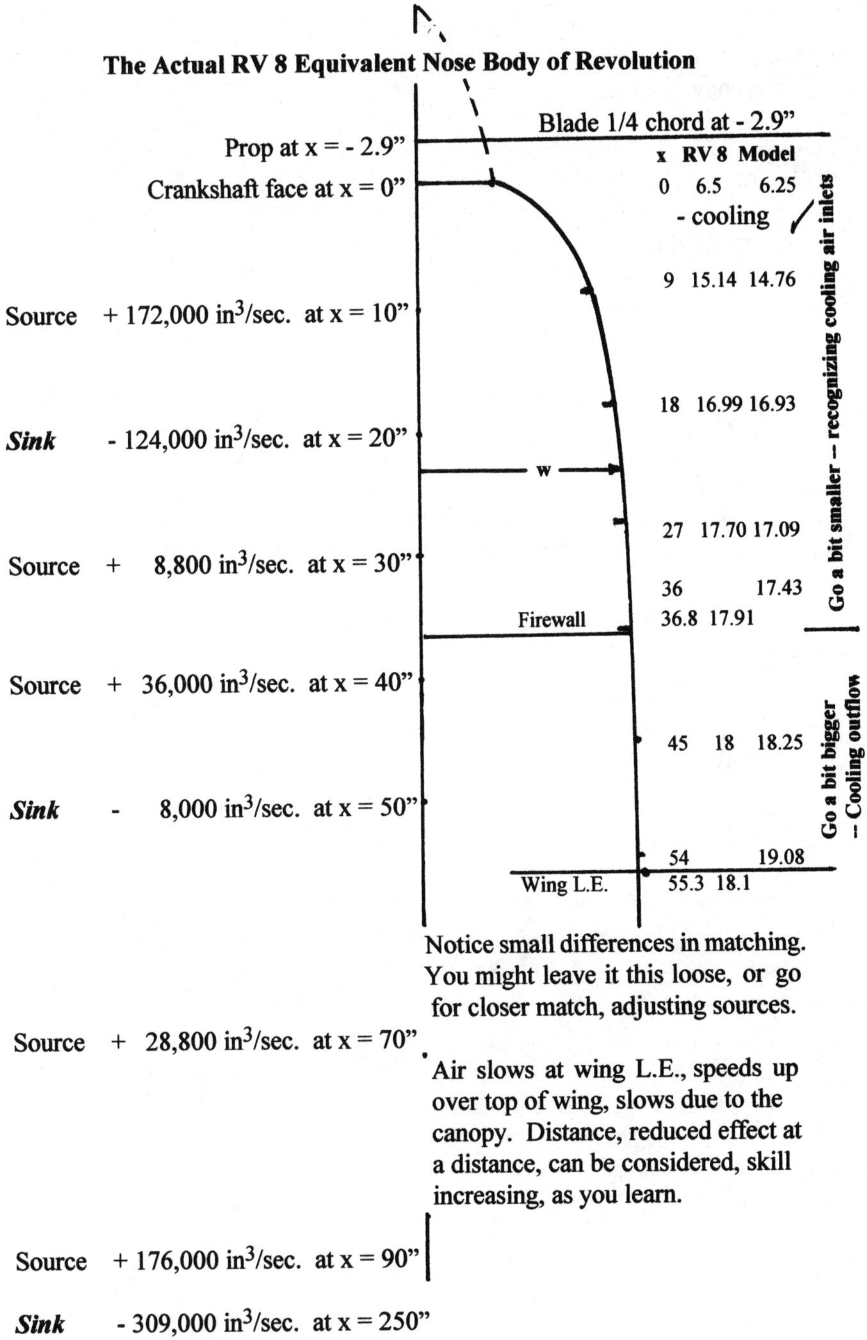

Fig. 2. **RV 8 Nose Equivalent Body of Revolution**

To reasonably account for a tapered rear fuselage: You put a sink at the tail, so that you haven't created, or lost any air, for arithmetic zero. Now, you could duplicate the whole body this way, but if you just model the nose, maybe a little back behind the cowl, including the windshield *if it's close and big*, you'll do just fine. We've done it to also model the circulation about a wing, a slowdown from under a high wing, a speedup tendency from above a low wing, but that's usually not done, a small effect. With the rapid Spherical drop off, distant sources have small effects, soon insignificant. But the big final sink for algebraic zero is big enough, it does change the RV 8 enough to not ignore it. We really don't care about the speedup behind each source, just want the ***composite*** slowdown at the prop.

There is not a more simple way to say where the body boundary is, than Streeter's formulas show, simple enough for how it really works, those final formulas much more simple than their math. More said a few pages forward.

Notice it's simply m/u at the core of both formulas, easy as it could be!!!

Now, I must tell you that Andy's iterative program to handle multiple sources and sinks, spaced out along the prop axis centerline **works a little differently than Streeter**, equivalent math developed more than a decade before we found Streeter's simplified, easier to see single source formulas.

As θ increases past 90 degrees, down to 180 degrees far back, **the cos θ is negative past 90 degrees**, all correct with Andy's math. Naturally we used Streeter's single source formula to check Andy's Math and Program, and they both plot identically, a nice double check that there are no nasty little math or program errors, the scourge of computer math and programming.

I did a lengthy, but simple, rock solid 10 digit hand calculator check of the RV 8 prop slowdown at the 75% radius, and an inner radius, and matched Andy's computer **slowdown calculation** through the 4th decimal place, so we have a solid check that both Andy's Math and Program are correct, also with the identical plot of Andy's body contour plot vs. Streeter's method!!!

For Prop Radii Slowdown use the Diagonal r, each source to each radii!

Slowdown: Sum the Horizontal Components of all the Vr(s) !

My rock solid check method was simply to decrease each sources velocity at the 75% prop radius by **dividing** each source strength **in.^{3}/sec.** by the **$4\pi r^2$ surface** area of its Sphere, for the **correct r**, to get a **slowing in./sec. velocity summation** from all the sources, the fundamental principle here. Their **horizontal velocity component checked the program, essentially exactly.**

Andy uses all inches, not feet, in./sec., but of course, all feet, and ft./sec. can be used.

Andy uses 100 in./sec. flowstream velocity, so the slowdown at each radius can be read like a percentage of any velocity, fast or slow, *smart!*

How and Why Rankine, Streeter and Bauer's Math Works Correctly

In all the explanations in this book it was necessary to **go way past the math** to **understand why and how** the math was working correctly, ***what logic is implied***, why it is correct, to explain the logic of it all, the engineering sense.

Here, we're creating a math simulation of an embedded body in an airstream which slows the approaching flow and shoves it outward. As often happens in a math derivation of a complex situation, a fairly simple algebraic formula falls out, often different, and more simple than one might expect.

When we plot the results of Streeter's math, we find a bullet shape is generated, that the source **slows** the airstream in front of the source, that the flow is **right back to stream speed at 90° above the source**, that the airflow is **speeded up from 90° to 120° behind the source**, that the stream slows ***back down to its initial velocity far back at infinity***, as the diameter slowly approaches its ultimate diameter at 180° at infinity, as an asymptote at max D

1. The Bullet Shape-Size: Streeter's first formula is $\mathbf{w^2 = (m / 2\pi u)\,(1 - \cos\theta)}$ The angle effect parenthesis multiplier goes from 0 to 1 to 2 at 0°, 90°, 180°, as the cos goes from 1 at 0° to 0 at 90°, minus a -1 at 180°, 1 + 1 = 2, at 180°. That *facilitates* the final correct bullet Shape as cos θ goes from 1 to 0 to -1, the parenthesis from 0 to 1 to 2, all smoothly as the cos θ function over 180°.

That's nifty, tricky, Streeter's math making what's happening clearer.

The **size** is controlled by **(m / 2πu)** which divides out to be **a constant**, the source strength **m**, in³/sec. , divided by the stream speed, **u**, in/sec, thus in², but we must take the square root of the whole, (or each parenthesis of the formula), which sizes the half width at each axial station. The π is in there, because we use polar coordinates for the angle in deriving the equation, the 2 is there, because that's what the derivation decrees, (***not*** the spherical area 4π)*. Streeter's second formula works in a similar manner, except it uses the diagonal, or radial to shape the bullet, which plots exactly the same. The **math magic here is that simply m/u is the core *simple ratio* controlling the bullet size, the complex boundary between the internal and external flow, the exact synthetic simulation of the bullet and the flow around it.**

* ***Interchanging u and w²***, **shows u is twice the radial velocity at 90° θ**, tan 1/2, 26.56° flow.

2. Speed, and Slowdown: The source strength, and speed drop off as the source strength **m**, divided by the area of a sphere, of **r** radius, $\mathbf{m/4\pi r^2}$, in./sec., where **r is the radius, or diagonal distance from the source** to the **bullet shape boundary**. But *that* **speed is multiplied by the cos of θ to get the axial, horizontal slowdown effect *at the body* ahead of the source, or speedup behind the source!!!** --- ***But***, finding the **prop slowdown** is a **two step process** where you must **first define all the sources, get the diagonal, r to each prop radius - calculate the sum of all the axial V components!!!**

Using, Understanding Andy Bauer's Source-Sink-Slowdown Program

For those who buy and use Andy Bauer's Ideal **Propeller Design Program**, and his **separate, but included Source-Sink-Slowdown Program** to do the very necessary, very big correction of an Ideal Propeller Design, for any given airplane nose, you need a little additional explanation over and above the basic explanation possible with **Streeter's specific <u>one Source bullet</u>**.

Andy's program works different than Streeter's math but accomplishes the same thing. You could have a tough time understanding just how it works. Andy's program must integrate, or totalize the effect of any multiple combination of sources and sinks **so we can synthesize any required non bullet shaped nose**. To do this Andy plugs in a **20 inch w**, half width, or symmetrical body radius, bigger than the actual, and a **10 inch w**, smaller than the actual, and lets the computer loop, or iterate to find the ***<u>resulting body shape from all the sources and sinks</u>***, and provide the **w's at all the desired check points**. To do a different problem than the RV 8 case, you ***adjust the source or sink strengths, location***s, as necessary, to **home in on matching your body model**. The program works, does all that correctly.

The program will find answers above the 20", and below the 10" limits.

Andy uses **Q**, rather than Streeter's **m** for source strength, thus the formula core **Q/2πu** In program line 1410 and 1420 math similar to Streeter is there, but you'd have trouble identifying it exactly, because it works with the rest of the program math. You won't find (1 - cos θ). Andy uses cos θ, not (1 - cos θ), because he **goes after the slowing effect of all the sources, iterates triangles to solve it**, <u>adds a U, the stream speed</u>, in $(U + VxT)^2$ for Vr in the next, second equation in program line 1420. That confuses this explanation, a little, makes the iterating program harder to follow, but gives you the necessary clue that **Andy programs the math a little differently <u>for the same result</u>**. That clue can save a user from a lot of confusion and wasted time, if he digs to see how Andy's program works, having learned Streeter's more simple one source bullet case to learn the logic. <u>Streeter is easier, as presented</u>, <u>but Andy's program handles multiple sources and sinks</u>. //

Andy provides **a user instruction with the program**. That shows you the line numbers where you must plug in your trial source-sink flow rate sizes, and their location axial dimensions, to match your body of revolution nose model. You'll see Andy uses a series of subroutines to loop, or iterate, find the size and shape of your synthetic math model to match your average nose dimensions, w's, (and diagonal radii the program uses for its **Spherical area calculations). You don't have to dig into the nitty gritty of the program, and I suggest you don't. Just use it, it works just fine. Just get your nose model and source-sinks correct. Spend your time learning that!**

Now, let me make a **hopefully clear final summary** of how this all works, so that I can be as sure as possible that you can understand correctly.

Andy's program, like Streeter's first formula, puts out **radial widths, w's**, at **all axial stations set up**, so you can fiddle with **Source-Sink size** and **locations**, so you can home in on a synthetic math model that matches your body of rotation model of your actual nose. Now, when that is resolved, you have, in essence **plotted your math model body shape**.

Now, you might plot your math model body shape using the equivalent of Streeter's second formula which defines the body as the diagonal distance and angle from each source to the body boundary. But we **don't do that**, don't waste that effort. **The body is already defined.**

WE'RE GOING TO USE THE DIAGONAL FROM SOURCES TO PROP RADII!!!

What we must do, is find the slowdown at each desired prop radius, which is progressively farther out and farther ahead of the body shape as we move out in prop radius. So, **Andy's program does three steps**:

1. It finds the **diagonal** ***distance*****, RA, from each source to each prop radius**.

2. It finds the related **diagonal velocity, in./sec.**, by dividing each source strength, **Q, in.^{3}/sec.** by the area of the Sphere of that diagonal radius, **$Q/4\pi r^2$**, or **RA^2 --- in^2**. But to **get the *axial* slowdown** at the targeted prop radius, Andy's program ***multiplies each diagonal velocity found, by the cos*** θ. and **sums up the axial effect of all sources and negative sinks. Source sizes are set to match the bodies - But we calc *axial* effect AT the PROP!**

3. Finally, ***YOU* Must subtract each summed axial slowing Velocity**, from the **standard 100 in/sec flowfield velocity, u.** That **gives you a decimal, or percent slowed velocity**, like maybe .497 next to the spinner, maybe .968 at the 91% radius, like the RV 8. The computer prints out its answers as Vxt. **You put YOUR slower inflows in line 203 of the Prop Design Program!!!**

Now, aren't you glad **you don't have to do all that math drill,** just let the computer do the drill for you. Learn how to use Andy's program, and you can make **essentially perfect corrections to essentially perfect props,** at **any design point you choose**. You'll find the corrections are big enough that to not do it well is to unnecessarily fall right back to amateur status.

But, see below --- We actually design a whole new prop for the slow condition!!!

Finally, I should repeat the heads up that Andy's program works more with the simple sides of the right triangle, **w**, **Δx**, with **RA as the hypotenuse**, using squares and square roots, rather than trig functions, in case you want to confuse yourself and start digging through the program.

A surprise is that there is a pressure interaction, prop to body, slow -- so:

The text explains that we design a whole new prop for the slowed condition for a *higher gross thrust* - *same net thrust and HP* - in slowed air

Seeing Through the Real Nitty Gritty of Andy's Math Solution Here!!!

The **most fundamental principle in Source-Sink work** here is that of the **Source Velocity V_r dropping** as the **Source Strength divided by the ever increasing surface Area of a Sphere, $Q / 4\pi r^2$, (equals V_r),** (as r increases).

This is the most fundamental and important principle in Source-Sink work!

Now, a **very important insight** is to realize that Streeter's formula that deals with the half width of our model body, *(not the diagonal radius),* divides by **2π**, ***not*** **4π**, the **very simple**, but **key difference** resulting from a more complex partial differential equation solution!!!! The result of combining the algebra in line 1410 and 1420 in Andy Bauer's program for the w, half widths of the model body, does divide by **2π**, ***not*** **4π**, ***correct!***

The basic formula is always valid -- But This Finds the Boundry ***vs. w and x.***

Andy then simply correctly deals with the distance and velocity vector triangles operating within that, (also recognizing the math manipulation in the next paragraph that legally interchanges w and u, etc. to better see through the math). Dealing with multiple sources and sinks, he makes two wild guesses for high and low w's of 20" and 10", and **lets the program, loop, or iterate** to **home in on final correct values for each axial w location chosen,** as affected by all the sources and sinks. Taking the horizontal, axial component, the cos effect of each source-sink he accurately gets the ***slowing*** effect of the sources on the flowfield, the speedup of the sinks, on the body nose --- or all the desired radii of the prop plane..

Looking at Streeter's first formula for **w**, we have $w^2 = (m / 2\pi u) \times (1 - \cos\theta)$, the square root of all that, or the product of the square root of each part. Now, Algebra allows us to solve that formula for u in terms of w, by simply interchanging them, so, $u = (m / 2\pi w^2) \times (1 - \cos\theta)$. Now, at $\theta = 90°$ where $\cos = 0$, the parenthesis falls to 1, and we just have $\mathbf{u = (m / 2\pi w^2)}$ --- which looks just like our basic formula --- *except* we have flowfield horizontal speed u, not the radial velocity of the source, though w is the correct radial distance at 90°. Now, since we're dividing by 2, not 4, we get the **great insight that at 90°**, where there is no $(1 - \cos\theta)$ effect, other than 1, and **u is back to its initial value**, the ***u is twice the radial flow speed*** -- **because we're dividing by 2, not 4!!!!** (As in the text, after 90°, the flowfield is speeded up by the source, maximizing at 120°, then gradually falling back to u at infinity, the source now so far forward that it has no ΔV effect.)

Very Simply, m / u gives Bullet Area, thus D, far back, all at u Speed, Insightful!!!

Separately, if $(1 - \cos\theta)$ goes from 0 to 1 to 2, at 0°, 90°, and 180° and the basic formula solves for w^2, the half width is 0 at the nose, 1 at 90°, and the square root of 2 at 180°, far back at infinity --- or playing with all the math --- **with a nose radius of 1**, the **w is 1.414**, the square root of 2 at **90°**, and **2 far back**, the ***nose radius 1/4 the final diameter, for our basic bullet!!!***

Source-Sink-Slowdown A Terrible Subject, at first, to Teach or Learn!

I expect you'll have a terrible time trying to understand this subject the first time you read it. I assure you it's an equal problem to try to write understandably. It doesn't get easier if written longer or shorter.

Like many involved technical subjects it's easy enough once you start catching on, but can seem difficult the first time you dive in. Let me suggest that you just try to grasp a few key basics, don't try to grasp it all at first, **learn how to grasp the few basic steps to run Andy's program, trusting him and me, and you can work your way into grasping it all!!**

Steps For Running the Propeller Design Program with Slowdown

1. Run the basic Propeller Design Program without slowdown, using desired RPM, Speed, etc. This requires that AAAA be set at 0 (zero) in Line 205.
2. Lines 201 to 204 contain slowdown data for blade stations 1 through 11 for several aircraft. Turn on one of these lines by deleting the ' on the data line that best represents slowdown on your calculation, (or create your own).
3. Adjust AAAA, (ratioing factor), on line 205 to better match slowdown magnitude for your case. AAAA=1.0000, unchanged, may be a good choice.
4. Run the program again and type GOSUB 257 and push the Enter key. A list of eleven negative numbers will appear on the screen. They represent the changed blade angles needed to account for the slowdown effects. Put these negative numbers into Line 1048, replacing the eleven numbers there. Also, remove the apostrophe at the start of Line 1048, to activate it.
5. Run the program.
6. Go to Subroutine 4600 to find the new eleven lift coefficients brought by the blade angle changes.
7. The above is generally not enough to bring the lift coefficients to the desired values.
8. Repeat steps 4 through 7 several times. Generally, about three times will bring all 11 lift coefficients to the design value, normally .5, the value for the no slowdown case. The program will have the new blade angle results in Subroutines 4400 and 4500. You are changing the blade angles to obtain the desired lift coefficients. But, the Thrust and H.P. will fall in the slowed air!
9. Now, rerun the basic program several times adjusting the $\bar{w}$, w bar, up to get back to the desired H.P. and/or Thrust. When achieved, you will have a full 2 pages of complete design data. Significantly, realize you will also have the final correct Blade Chords, and, of course, the final Blade Angles, β^o.
10. Print out, record your final results, your final slowdown corrected Prop!

Appendix T

Theodorsen Calculations (Heavy Loading)

An Overview of Theodorsen Calculation Methods

To make it as easy as possible to use his calculation methods, **Theodore Theodorsen**, the **WW II, NACA Chief Physicist**, intended that we use a basic **Blade Element Analysis**, taught to, or understandable by Aeronautical Engineers, together with his, *simple enough*, **special factors** that **accurately account** for **Heavy Loading**, the quite reasonable, **logical fact that what is happening "right at the propeller blade", is greater than what occurs, *on average*, across the whole stream tube**!

When you think about it, **in the real world, that would be no surprise**, and all Theodorsen is doing**, is simply getting the calculations really correct, especially for such core related issues as *inflow*, *needed pitch*, *advance***. When we see, quickly below, the **quite significant magnitude** of the $\overline{\mathbf{w}}$, vs. average **ΔV differences**, you can quickly see that going off to create your own, less accurate, technically less sophisticated method, is pretty ill advised, since the 7 smartest, special geniuses of the industry worked 83 years, from Rankine's 1865 start, finally Theodorsen's 1948 book ~exactly accurate result. **Theodorsen precisely, finalized Betz, Goldstein, for Heavy Loading!!!**

Betz, of course, **created** the recognized, **Pure Helical Inflow, Stretched, Pure Helical Outflow, Classic Minimum Induced Loss Propeller Theory - Logic in 1919,** with all its insightful ancillary characteristics, **constant dT/dQ**, ***a constant ratio of Thrust vs. Required Torque at every radius***, thus **constant efficiency** at every radius, **constant Slip**, a **Math Basis - all if first considered profile drag free, *purposely, the simple, basic case*, without drag. Drag is readily added in a separate, *segregating* step, better, since each loss can be separated, each effect readily seen**. **Realize Betz Logic** is **created by proper Blade Loading vs. Radius, and Advance, controlled by blade SHAPE, with proper matching *Twist* --- changing vs. Advance!**

Be sure you fully understand the ***Flow Geometry in Chapter 3***, p. 82 II through 86 II, **particularly 84 II, which shows how *air Inflow Pitch is the Same at All Radii*, the core of everything!!** The constant pitch of the air inflow, gives **Constant Slip, vs. Plane Speed, which acts like the inflow is constant at every radii --- when the axial inflow can be seen in the sketch to vary a lot!!! It's the Magic Geometric Combination of Axial inflow, and Air Rotation, greatly bigger inboard, that makes all the Magic Math work out, makes all the nifty, hidden ancillary characteristics happen.**

Mathematically, Hold r tan ϕ constant, to hold Inflow constant.

Be sure you grasp that in the sketch, the "**a**" **inflow factor**, a ratio to the V Plane Speed, **times ~2 is the** ΔV the prop is throwing back. **If you think of that as the Average for the whole stream tube, ~ the same, that's ~"Light Loading". Theodorsen simply substitutes his w, or $\overline{w}$ for the full ΔV. But locally, *at the prop, that's much greater than the average ΔV*. Simply, that creates the way he handles Heavy Loading!!!**

Kappa, κ, the "Mass Coefficient" Ratios the Difference.

Of course, you know by now, that **Goldstein precisely solved the 3D Potential Flow**, Partial Differential equation Math, **for Light, Low Loading of the Stream Tube, in 1929. Then, Theodorsen precisely finished the 83 year *Rare Genius Level Challenge* in 1948, for real world Heavy Loading, quite different, a fantastic, readily P.C. usable, Gift to us all!**

Extraordinarily simple, ***after*** **Goldstein's** rare genius level math, realize **Theodorsen just uses Goldstein's Basic Math Solution, *shown as* Goldstein's Simple Chart of** K(x) **blade *Average* Circulation, or Lifting, Loading, at each radius, for the various Lambda's, λ. But T.T. targets *Heavily Loaded*, constant Air Inflow Pitch, or Advance, at each radius, for low to high Pitch, or Advance, by using his own Lambda, λ! It's IMPORTANT to realize, that to implement his method, T.T. uses a different, special definition of Lambda**, that **includes his w, or w bar, $\overline{w}$, factor**, **to get Inflow, Pitch,**

Advance, really correct. Lambda, is <u>usually</u> defined as the Actual Advance Angle of the Prop Tip, the Plane's Air Inflow Velocity, V_1, divided by the Tip rotational, πnD velocity, simply, $\lambda = V/\pi nD$. (A minute's thought, if you don't already understand, realize that's quite simply, the Actual Tip Advance Angle.) **!!! Notice below that T.T.'s λ *<u>has w, or $\overline{w}$, added, the key, core step</u>* to get the heavily loaded inflow, outflow, really correct**, using his method, <u>realizing that he simply uses</u> **<u>Goldstein's K(x) Blade Loading Chart</u> --- <u>unchanged</u>!**

It's, of course best if you can find a copy of Theodorsen's 1948 book to read and understand, see his text, K(x) Chart, and Equations, but this Appendix T, teaches T.T.'s equations and method Overview, OK if you're a pro who understands. (Also, telling you here exactly how to grasp it, use it, is a huge time saving over figuring it out for yourself in the book.) In addition to considering a λ steepened to account for the full w, or $\overline{w}$ velocity, as just described, he treats his Math as occurring "far back", where there is also a reduced diameter, D_o, simply from the Bernoulli shrinking of the now <u>$\overline{w}$ faster helical sheet</u>. For a high pitch prop, at cruise, that diameter shrinkage may only be 1%, but for a low pitch prop it may be more like 3%, a ***97% D_o***.

What Theodorsen is doing, in **solving his math far back**, is **getting it really correct, (also at the ~ half way point, <u>at the prop</u>, where all the energy is put in, *<u>an "also result"</u>* , <u>all the action happening there</u>.** Using his "Heavy Loading", not our own method, or someone else's less sophisticated concoction, is the <u>difference in getting your design really correct</u>, ***<u>or not</u>!!!***

I know Engineers, an extra dose of Hubris, who feel that with their modern computer, no one from the past could possibly be smarter, do better than they can with their self created program, not a clue, that the rare genius line of Historic figures, were more aware, much smarter in 1919, 29, 48, than they are today!

Frankly, with experience in interviews, and armed with a Psychological screening test, I was avoiding hiring guys like that 40 years ago. Hubris is not a good thing in Engineering or Science. **Checking, knowing what got accomplished before you came along, is always correct, because there is so often, a brilliant man who preceded you, gave you a much smarter starting point.** Checking, then creating, then a lot of double checking, is the proper game. It was often, the rare genius, who accomplished the next significant step in complex propellers, for example, the math here far beyond the capability of the mere, mortal engineer. ***Never skip past rare genius work.***

K(x) Blade Loading Chart - vs. Radius and (Theodorsen's) **Advance, λ.**
Realize x is Radius Ratio, r to Tip Radius, R, simply r/R, (~defines itself).
As above, realize Theodorsen uses his own special definition of λ, for Heavy Loading.

Goldstein and Theodorsen K(x) Blade Loading Chart vs. Radius and Lambda, Theodorsen's Lambda for his Math.

Theodorsen's *Heavy Loading Lambda* $\lambda = V(1+\overline{w}) / \pi n D_o$.
n, rev/sec. Do is reduced Diameter, far Back w bar, $\overline{w} = w/V_1$.

Values Interpolated for Even Fractions

x \ λ	0.1	0.2	0.3	0.4	0.5	0.6	0.7	0.8	0.9	1.0
0.1	0.517	0.291	0.184	0.127	0.092	0.070	0.055	0.043	0.035	0.027
0.2	0.773	0.513	0.338	0.237	0.1735	0.133	0.105	0.085	0.068	0.055
0.3	0.876	0.645	0.456	0.3225	0.2445	0.195	0.158	0.128	0.102	0.079
0.4	0.927	0.732	0.538	0.393	0.299	0.240	0.196	0.160	0.1275	0.098
0.5	0.9475	0.769	0.5825	0.433	0.3325	0.2675	0.220	0.181	0.147	0.115
0.6	0.955	0.782	0.593	0.444	0.3435	0.279	0.232	0.192	0.157	0.124
0.7	0.942	0.750	0.569	0.430	0.337	0.275	0.228	0.190	0.157	0.124
0.8	0.890	0.673	0.504	0.3795	0.300	0.248	0.209	0.175	0.145	0.118
0.9	0.739	0.523	0.385	0.297	0.235	0.193	0.1625	0.136	0.113	0.092
0.95	0.565	0.382	0.284	0.217	0.169	0.1375	0.114	0.0955	0.0805	0.067

Values of Goldstein - Theodorsen K(x) vs. Radius and Lambda, λ.

This chart with even Decimal Values is generally the easiest form to understand. Goldstein's comment was that these values are accurate to 2 to 3 in the 3rd decimal place. That sets the accuracy ~1%, in usual ranges, *as you can see.* Do look.

Realize: In designing a Propeller with Theodorsen's Math, we select a w bar, $\overline{w}$, essentially his greater ΔV at the prop, adjust it to give the required Thrust, or often, the H.P. available, *if* Thrust # is not known!
Quite naturally, we also **select a C_L and proper C_D for Reynolds numbers.**

Note Later Tibery and Wrench Loading Data referenced at the end of this Appendix

Theodorsen's Equations **All units *in* ft. #, second, *Angles in Degrees*.**

Solidity Factor, $\sigma = P\,c / 2\,\pi\,r$ **P = number of blades (2) C = Chord**
Simply, the **decimal of solidity, at each radius,** of blade **chords** vs. ***full*** circumference. Found just above T.T.'s equation 4, in text. **σ Is Sigma.**
It is used in **line 1055** in Dr. Andy Bauer's Propeller Design Program --- ABB Prog. **line 1055**

Beta, Blade Angle $\beta^o = \alpha^o + \phi^o$ **Beta = Alpha + Phi**
Simply, **Blade Angle = Angle of Attack + Air Inflow Angle.**
This is **the standard Angle basis *for* Propellers,** unsaid in the text.

Using Sigma $\cdot$ C_L to find the Chords - With Sigma Formula above.

$$\sigma C_L = \frac{(1+\overline{w}) \cdot (2\,\overline{w}\,K(x))\,(\sin^2\phi / \cos\phi)}{(1+\overline{w}/2)\,[1 + (\overline{w}/2 \cos^2\phi)]}$$

Theod. **Eq. 4, p. 47**
ABB Prog. **line 1040**

x = r/R r/R is simply the radius Ratio, actual r / tip Radius, *used a lot*
Important: realize K(x) in loading chart, is just **Loading at that r/R**

Theodorsen points out that this equation is not exact - has 2 (very close) approximations, Insignificant. Inflow is *not* exactly equal to outflow, Blade lift Vector is *not* exactly perpendicular to Wind Inflow Vector and Phi, ϕ. Both could be made more exact, but it's not worth the considerable complication for very little improvement in accuracy.

Finding Wind Inflow Angle ϕ, from tan ϕ Theod. **Eq. 8, p. 48**
ABB Prog. line 1015

$$\tan\phi = (\lambda/x)\,[(1+\overline{w}/2) / (1+\overline{w})]\,[1 / (1+2\overline{w}Y)]$$

x here, is just r/R, the Radius Ratio, as above, better used that way to avoid confusion.

There is a small contraction in the wake from the propeller Diameter to Do.
It is the convention to show that contraction as a Radius Ratio: Eq 6, p 48.
$\Delta r_o / R = (c_s /\text{kappa}, \kappa)\, Y = \sim 2\overline{w}\,Y$ So, to solve for Y, we need c_s, and κ.

Thrust Coefficient ABB term, DRO means Delta Radius Δr_o..

$c_s = 2\kappa\overline{w}\,[1 + \overline{w}\,(.5 + \varepsilon / \kappa)]$ **So: Now we need both ε and κ** ABB Line 153.
And these are found on **Theodorsen p. 37,** his **Graph,** Fig. 10, **for 2 blades.**

DRO = Y R C_s / Kappa, κ T.T. Eq. 6 p. 48 DRO & DO on ABB Prog. **Line 160.**
The Propeller Wake Final Diameter is: DO = D - 2DRO ~.99, .97 D.

Again, be reminded that **Kappa, κ** was inadvertently called Chi, Xi, in ABB Program, due to unusual type setting symbols for Greek Letters, in Theodorsen's Book.

Including Blade Drag in Theodorsen Calculations

If no drag is included, the **efficiency** is simply the **Thrust Coefficient C_s**, previously defined, **divided by the Power coefficient C_p**.

Efficiency $\eta = C_s / C_p$ where $C_p = 2\kappa\bar{w}(1+\bar{w})[1+(\varepsilon/\kappa)\bar{w}]$

with **Epsilon, ε, and Kappa, κ,** available on **Theodorsen fig. 10 below:**

If Drag is involved we must *add the tangential and axial factors*, t_t and t_a.

Theod. **Eq 18, p. 54** ABB Subroutine 8000, to 8070.

Efficiency $\eta = C_s - t_a / C_p + t_t$ **where:**

Theod, **Eq. 26 p. 55**

$$t_t = (2/\lambda^2) \int_0^1 (\sigma C_D / \sin\phi)\, x^3\, dx$$

Eq. 27 p. 55

$$t_a = 2 \int_0^1 (\sigma C_D / \sin\phi)\, x\, dx$$

Or Theod. **Eq. 14 and 15 p. 49** -- *started* in ABB lines 8020 and 8030.

$$t_t = 2/\lambda^3 \int_0^1 \sigma C_D (\lambda^2 + x^2)^{1/2}\, x^3\, dx \qquad t_a = 2/\lambda \int_0^1 \sigma C_D (\lambda^2 + x^2)^{1/2}\, x\, dx$$

Theodorsen's Graph of Kappa, κ, Epsilon, ε, the axial Part of κ, and ε/κ.

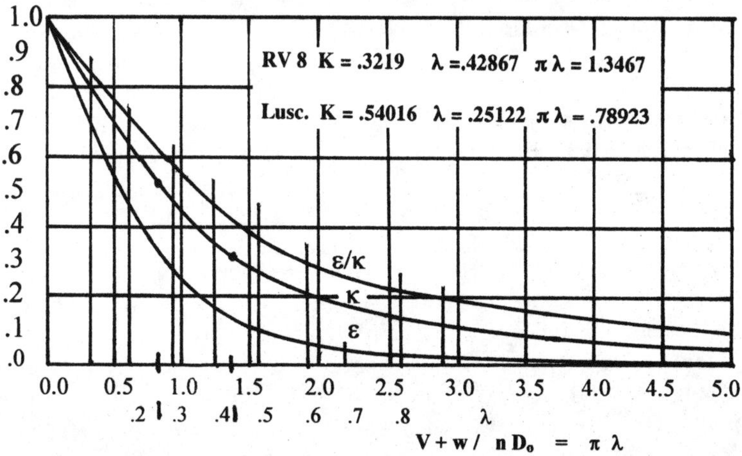

Theodorsen Fig 10 Tentative Values of ε, and ε/κ for 2 Bladed, *Single Rotation* Prop.

The Huge Significance of Theodorsen's Kappa, κ, Factor.

Notice that Kappa, in the Graph here, might be ~.5+ for Luscombes, .3+ for an RV, .12+ for a Reno Racer. The Huge Significance of that is that when inverted, a Luscombe w is ~2x the average stream tube ΔV behind the Prop, ~3x on an RV, ~8x on a Reno Racer, greatly more than the Average ΔV. What is happening **Physically**, is that as a **High Pitch Prop** screws through its Stream Tube, it **has less and less effect on the Stream Tube, uses its mass less well the higher the pitch, the huge insight that only T.T.'s Heavy Loading gets Pitch, Inflow, Advance really correct!!!**

The Milestone Capability, that got us to This Time and Place

Only a sufficiently accomplished, experienced Aerodynamacist, and Programmer should attempt to write his own Program for Propeller Design, a thicket of pitfalls, realizing the core complexity of Propellers. It was only because of Dr. Andy Bauer's lifetime of Knowledge, and Experience, that he was able to correctly sort through that thicket, and **make it possible to find the hugely simplified final insights you see in this book**. The final Program, in Basic Language, is 19 pages long, and was made ever more complete, sophisticated, and correct, for the multiple capabilities finally appropriate, over many years. A very accomplished, experienced, proven engineer, I would not have ever attempted it, without the specifics of Andy's lifetime of knowledge and experience in Aerodynamics, needed to ultimately find his way through Betz, Goldstein, Glauert, Theodorsen, and the total Swamp that propeller math can be when you first wade into it all.

Our mutual collaboration in ultimately getting far past where others had been, in being able to ultimately see through all the myriad of complexities, and **find the core simplicities, for incisive explanation**, took our two lifetimes of experience, but **was only possible, *after* Andy cut through all the best, past Historic work in the Mathematics of Propellers, with an incisive, Theodorsen computer program for broad and specific analysis, World Class Work, that went far beyond where any had been!**

Without any question, Dr. Andrew B. Bauer deserves to be recognized as the 9th Historic figure in the long, now 140 year conquest of Propellers, following those famous Historic technical figures, Rankine, Froude, Betz, Prandtl, Glauert, and the Ultimate Genius, Theodore Theodorsen, who finally got the math precisely correct, after the initial 83 year period, from 1865, to 1948. But, then the job never got finished through a Comprehensive, Incisive, Understandable Explanation, because the core job was just too hard, too time consuming, too daunting. Everyone, even Senior Professionals shied away for an extra half Century, ridiculous, never an explanation.

Not to be missed, if we properly include Gus Raspet, of Mississippi State, with his innate curiosity and creativity, with students, and coworkers at his Old Miss Aerophysics Laboratory, as milestone figure # 8, Andy would be # 9, in that long, and distinguished, Historic List. Gus showed us how really bad overall Propulsive Efficiency could be, only 58% overall on a Bellanca Cruisair, inverted, 178% more power required than that in Gliding Flight, a huge insight, another unanswered challenge by Industry, since his milestone 1950's test work.

The collaboration between Andy Bauer and I began in the last half of the 80's, after the work I did before and during my 1986 work as Technical Director, Mission Control, on the **Voyager world flight**, and started to write **The Logic of Flight**, *The Thinking Man's Way to Fly*. Andy did an Adkins LeBeck analysis of the Luscombe Propeller, and I invented, **Zero Thrust Glide Testing**, and carried out, what was ultimately the **first full Aerodynamic testing of a propeller airplane, including full speed range, Accurate Propulsive Efficiency**, in full flight configuration. It showed degrading, overall efficiency as more power was applied, and a non smooth, accurate curve implying, we believed, some variable separation, which interestingly, seemed to level out, not as bad at the higher speeds. We gave an AIAA presentation and paper at Reno, January 1990, and a final paper, Journal of Aircraft Vol. 30, No. 4, July August 1993. The **final Graph** of **Full Luscombe 8E Aerodynamic Data, EAA, Sport Aviation, Mar. 95**, is used today at NASA Langley as their best reference on Propulsive Efficiency.

A Blade Element Analysis to Create an ideal Propeller Design

To do a Blade Element Analysis of a Propeller, one divides the blade into even radial Elements. We ended up using 10, 8% radial sections from 19 to 99%, 11 boundaries, considering the 18% D. RV Spinner, and the desire to have the last boundary be on the prop blade at 99%, with a real chord, not air, at the tip. The Blade Angles and Chords, are found with the equations above. The whole blade Lift is calculated, Blade Shape curve smoothed, integrated, and resolved into forward Thrust, and Resisting Torque. That can work accurately, if you use **real Theodorsen Heavily Loaded Flow**.

Among all the historic complexities of Propellers, let me show you some **relative simplicities here**. If you can get the genuinely accurate **Chords**, and **Blade Angles,** β, from the above equations, you have Blade **SHAPE**, and **ANGLES**. You have **chosen a target D, Diameter**, or **Aspect Ratio**, and quite simply, **can see** from the chords, **if your objectives there have been met**. Of course you've juggled $\overline{\overline{w}}$ to get the **Thrust** and **H.P.** you seek. Of course, you chose an Altitude and density to start. Of course, you also **chose a C_L**, and **C_D**, the **C_L** probably **.5**, or **.55**, if you're seeking a smaller, lighter propeller, almost as good. In your program you should have set up **a subroutine** to give **proper C_D**'s vs. the Reynolds numbers at each radius, based on a set of curves you selected **vs. angle of attack, and Reynolds number**. With experience, you can guess, and arbitrarily set an overall, higher **C_D** that, though not as locally accurate, gives comparable overall results, satisfactory for a less precise analysis, for example, a quick, more simplified look at overall possibilities, though with modern computer speeds, and capacities, there's no real time difference.

If you accomplished a creditable, sufficiently professional printout of overall results, listed for each radius, **you can have X-ray insight into all local characteristics of the prop at each radius.** Naturally, that depends on whether you are really competent to do the overall job, or not, and how much time you invest. **Our work was done over years**, as Andy created an ever more capable program, that confirmed that a constant dT/dQ actually happened at each radius, what it was after Profile Drag was added, showed **dT/dr, Thrust loading at each radius, before and after slowdown** correction at each radius, **dTT/dr**, the Net of Gross Thrust considering the pressure interaction with the body in Slowdown, **listed Reynolds numbers, Wind velocity, C_L, C_D, Efficiency, η, the LONG LIST of things you really want to know at each radius, finally, a 2 to 3 page data printout.**

We Engineers tend to want to do everything ourselves, start from square one to feel correct, complete, sure we got it all, but where major work was done, that can be less than wise. Only one, far more capable than normal mortal Engineers, is even qualified to repeat Goldstein, Theodorsen. Knowing how much work, how long it took to develop Andy's Basic Language Program, **I'd recommend you simply buy Andy's Program, save your time, more profitably spend your time learning all the capability in that program, and all the insight on real bottom line results that are available, at modern, essentially instant computer speeds, and capacities.** Using the simple (originally IBM) Basic Language, the programming is easier to learn than more sophisticated, later code. The now old GW Basic does have some disadvantages, newer computers don't like to print it out, and you must use some easy enough tricks to print, after the early Windows 3.1, where we started in 1992. We do have a later Windows Program developed by Jim Rust, Whirlwind Propellers, quite slick, but less total capability, flexibility, We hope to have a program available, as time allows, after book publication

A few extra details need to be covered:

A_z A_0 Azero. On his **page 31, Theodoreson** gives an **equation (42)** that shows the **exact inflow, *not* exactly half the total w, or $\overline{\overline{w}}$** delta V at the prop, but quite close, often part of a percent, or so, less, not a biggie. It is of interest in developing the theory, and seeing that precisely analyzed. **The inflow vs. outflow are *not* exactly equal, *not* each half the displacement velocity w, or $\overline{\overline{w}}$. Andy has it in his program,** printed out where he looks at several specific T.T. data characteristics, but we'll not cover more here.

There are actually 3 Epsilons, Axial, Tangential Rotational, and Radial feeding the Tip Vortex, 3D losses = κ. T.T. p. 33, fig. 8 plots 3D losses at .5 λ, shows radial, the tip Vortex loss exceeds the axial at .5 λ, WOW!!!

One Last Time --- Don't think that you can easily do better math *now!*

Betz logic is Elegant, in its ancillary detailed insight, Goldstein's 3D Potential Flow Solution is far beyond the math capability of all but the most exceptional modern Engineer, Theodorsen's insight and work, equally brilliant, a rare, exceptionally tough team to challenge. Ribner and Foster did check the work by modern computer in 1991, generally concurred within approximately 1%, a few less central items a bit more. The chief practical problem there, doing even more on multi-blades, etc. they *plotted* most of the results, inadequate for us to do a digital check.

Of course progress won't stop. We won't close the patent office, though the Administration is severely cutting Langley funding, the surest way to get bypassed. **The clearest path to progress, potentially greater propeller efficiency, a few more percent, is going for elliptical blade loading, cutting tip losses even more than Classic Theory**, loading the blade more inward, **impossible inboard of R/2, without gross chords, far lower q inboard.** Logic, and Math work comparable to BGT, will demand very smart new work.

Appreciate the wildly complex Math that Goldstein, Theodorsen solved.

We're pulling in and throwing back air, not unevenly, but with accurately held constant Pitch of the Air Inflow at each Radius, perfect Screw Surfaces, Classic Archimedes Screws, stretched pitch finally in the outflow, uneven rotation and axial inflow, optimally, brilliantly, but easily geometrically combined to produce that perfect Math Model, far weaker axial inflow inboard, but precisely assisted with much greater rotation inboard, at the steeper blade angles there --- Tip Vortices not seen in the Model, but accurately accounted for in Goldstein's 3D, Potential Flow Math - radial too.

The Radial flow is, of course, outward on the bottom of the outer Blade, inward on the top, but with a two bladed prop acting as two separate, but not independent flows, 180 degrees apart in the stream tube, there are two opposite radial flows on the inner blades, weaker root vortices too at low q.

Theodorsen's brilliant insight that Goldstein's Math Solution, and Blade Loading Chart IS Valid for Heavy Loading, if you just consider the resulting flow Far Back, at a significantly higher Pitch, perhaps 14 or 15% local ΔV vs. V, much less on average across the stream tube, as in the text above, and at a **smaller D_0 diameter**, shrinking essentially as Bernoulli would teach, but with the overall incisive, yet relatively simple math offered by Theodorsen, as used in this Appendix. **I submit that's as Elegant an overall solution as you are ever apt to find and use in all of Engineering.**

Including Blade Drag in Theodorsen Calculations

Our Program <u>does</u> include Profile Drag, and Low Rn when implemented

If no drag is included, the **efficiency** is simply the **Thrust Coefficient C_t**, previously defined, **dived by the Power coefficient C_p**.

Efficiency $\eta = C_s / C_p$ where $C_p = 2\kappa w\,(1 + w)\,[\,1 + (\varepsilon/\kappa)\,w]$

with **ε and κ** available on **Theodorsen fig. 10 following**

If Drag is involved we must ***add the tangential and axial factors*****, t_t and t_a.**

Efficiency $\eta = C_s - t_a / C_p + t_t$ where

$t_t = 2\,(1/\lambda^2) \int_0^1 (\sigma C_D / \sin\phi)\, x^3\, dx$ $\qquad$ $t_a = 2 \int_0^1 (\sigma C_D / \sin\phi)\, x\, dx$

Remember that **analysis** is more **correct** ***<u>if</u>*** the program uses **low Rn C_D's.**

<u>Updated Information</u> Tibery, Wrench - David Taylor Model Basin, 1964

Tibery, C. L. and Wrench, J. W., at the David Taylor Model Basin, Report 1534, Applied Mathematics Laboratory (of the Naval Research Laboratory), did an extensive update of the Goldstein Factor in this 1964 report It is reported for Propellers from 2 to 12 Blades, and **most valuable, is reported in numbers,** not graphically, as Ribner and Foster did in their 1990 work.

Note that the Goldstein Factor is reported for 1/λ, thus requires manipulation

While this work was done much earlier than Ribner and Foster, and before the modern proliferation of computers, we can believe that NRL data of that time deserves to be trusted, and as such, *if correct* can be a very valuable addition, extension of the Goldstein - Theodorsen Tables. Most significantly, where it tends to confirm the original tables up to a (basic) λ of nominally .5, (~ 1%+), at the higher Lambdas it shows that the Kx factors for the higher Lambdas should be progressively bigger, at higher Lambdas, (~ 6% effective, at λ .8), favoring even higher Pitch, and Blade Widths

Thus, at those errors, <u>the implied factor updates, and changes **would not significantly affect normal props**</u>, but **only those of quite high, to very high pitch**. Since we became aware of this work only as we were bringing this book together for publication, we have <u>not</u> changed the book, other than this heads up, or our computer design program. We'll leave the appropriate checking for a next step of work. However if I were faced with a very high pitch prop design, I'd be inclined to use the Tibery and Wrench data, and get busy checking, because David Taylor work is probably correct.

Ref.. The Aerodynamics of Propellers, Quentin R. Wald, Science Direct, Progress in Aerospace Sciences 42 (**2006**) p.85-128 www.sciencedirect.com

A final Overview, and Summation of BGT Calculations

Just Finding the Wald Paper, and thus Tibery, and Wrench expansion of the Goldstein Kx Blade Loading Chart Data, as we were bringing this book together, we did not want to publish without a proper review of Wald's work, and the Tibery and Wrench Tables, did not want to stop our completion of our long project, but most of all did not want to take any chance of publishing wrong, or incomplete information. A few telephone calls to Quentin Wald proved most satisfactory.

1. Even though Tibery and Wrench did their work in the 60's, before the modern availability of powerful Personal Computers, we agreed that working at David Taylor they would undoubtedly have the best of equipment available then, and we should be able to expect that there work had every chance of being correct. It was Wald's view that they had a good math solution of the crucial loading factors.

2. Presuming that the Tibbery and Wrench tables were simply the inverse of Goldstein's Lambda, 1/λ, I had checked their results vs. the original Goldstene Table reported only to a Lambda of .5, and the Theodorsen table to Lambda of 1.0 that we had used as a base, and reported in this Appendix. Simply, T & B seemed to confirm that the Goldstein Math solution was essentially accurate, nominally ~1%, to a Lambda of .5, but that the Theodorsen Kx Loading, obtained only by Voltage Field Tests were not high enough! Since all the props we designed were at less than .5 Lambda, our work, and conclusions should be correct, simply realizing that very high Advance props need higher loading!!! (Thus our blade Shapes, and Loading on p.147 would change to a proper degree, the mathematically ideal Canoe Shaped Prop would occur at a lower speed.)

3. Wald Advised that although he had done new math to clarify Theodorsen, make him easier to understand he did not disagree with T.T's result.

A Bottom Line Overview

1. Betz Classic Prop Theory, 1919 remains Valid – unless we go for Elliptical Loading, to try to achieve a bit more efficiency.

2. Goldstein's Light Loading, Low Advance Ratio Classic basic Solution, and Kx Chart remains valid, subject to a final, exhaustive check vs. T & B

3. Theodorsen's adaptation of Goldstein for Heavy Loading, and his math method, the work that could have won us essentially perfect props a half Century ago remains valid.

4. Presuming a careful review confirms the validity of the Tibery and Wrench Loading Charts, we can convert to that loading, .5 λ, or 0 to 1.0 λ.

Appendix TR

Thrust Ratio - Net Thrust, There's a Loss

Why you need More Thrust Than Drag

Gus Raspet, a somewhat legendary Aero Teacher at Mississippi State was always challenging students to **try the unusual**, **Dig Deep**, **find the Truth**, **understand things better**, the kind of **creative, unique personality that students remember for life**. **He ran a series of gliding tests** towing up popular light planes with the **propeller removed**, sometimes with the cooling ducts taped shut --- **risky enough, few professors would be so bold**.

He did all this in the 50's --- a Half Century ago!!!

To me, he earned the right to be honored as the eighth man in our string of 7 historic figures who accomplished the Benchmark Steps in the Conquest of Propellers!!! He found **correct drags**, but when those drags were compared with the **actual** ***high*** **H.P. required** it **implied absolutely terrible overall propulsion efficiency.** And there was **no way his drag data was wrong**, too low. A plane just won't stay up longer than its drag allows. ***Glide drag errors go high***. A pro, he would have his weight, speed and sink rate right, and he always got the same end result of bad overall propulsive efficiency. A Cub, a Cessna, --- a Bellanca Cruisair had **a terrible 58%**, when he compared **the min. drag taped ducts test with real required Power!!!**

That's hugely different than people have understood!!!

This guy nailed one of the most important fundamentals in Aerodynamics, *a huge missing insight*. The problem, the great injustice was that everyone was off working on jets, the new craze in the 50's, and **nobody seems to have caught on to the fact that this was breakthrough fundamental insight, facts!!**

It was Gus's great work that motivated Andy and I to get into this whole subject, invent **Zero Thrust Glide Testing, see if we would duplicate his results, figure out what causes it all. I want to see Gus get Credit for great Creative Aero Work!!!**

The Book Cover -- Huge Basic Insight!!!

The book cover has really huge insight, right up front, on the incisive Logic of Flight, the Thinking Man's Way to Fly!

Notice the labeled Profile Drag Curve swooping up, a V^2, V Squared Curve, twice the speed, 4 times the Air Friction Drag, 3 : 9, 4 : 16, etc., the Profile Drag we usually hear about.

We're taught that Wing Lift happens because of a reduced pressure above the airfoil, but some hear about Induced Loss, or Drag, the cost of creating lift. What they should tell us is that Wings are also throwing down air, to create a Reaction Force Lift, *Newton's equal and opposite Reaction Force.* Induced is just the Energy Cost, the Loss of Throwing Down Air, Lift.

Look - Induced is a $1/V^2$ Curve, Swoops Down as we go Faster!!!

We simply add those two curves together, and get the "leaning Lazy J" Drag Curve that all planes have, some with more or less Induced, or Profile drag. Profile depends on how sleek, clean it is, also Speed. Induced depends on Aspect Ratio, Span / Average Chord, also Speed. Sailplanes best!

Big Spans involve a bigger Mass Flow Rate, $\dot{M}$, thus slower throwing!

We get Min. Drag, Max. Climb, nominally where the curves cross and are equal! Cruising Faster, Profile Swoops up, Induced Swoops Down, maybe 20, 25% of the Total Drag!

Ah, but now we can get really Smart! At the Tangent Point to the Drag Curve at Sea Level, we have the Max Speed vs. Drag - see, all other speeds, Drags above that tangent line!!!

But now, if we simply climb up, for High Altitude Cruise, in thinner air, we have to go faster to hold up the plane, TAS, True Airspeed, 21% faster at 12,500' vs. Sea Level IAS, Indicated Airspeed, but at Sea Level IAS Drag, Free Speed!

We can also much more safely Lean the Engine at the much lower High Altitude Power, unthrottled, win, win, win!!!! Get all that *"easy enough"* insight and you already grasp the core of The Logic of Flight, The Thinking Man's Way to Fly!

Dense, Incisive - Yes! The Core of Optimum Flight on One Page!!!

THE OBJECTIVE OF THIS BOOK IS TO **SHOW YOU** THAT THERE IS A BEAUTIFUL AND **AMAZINGLY SIMPLE LOGIC OF INTELLIGENT FLIGHT!**

IN FACT, THE CONCLUSIONS ARE SO SIMPLE THAT YOU CAN BEGIN TO LEARN THEM IN THE FIRST FEW MINUTES, IN THE FIRST FEW PAGES!

THE EVEN MORE VALUABLE FACT, HOWEVER, IS THAT WE CAN ***USE*** THE THINKING MAN'S LOGIC OF INTELLIGENT FLIGHT, ***AS A TEACHING TOOL***, TAKE YOU THROUGH THE WHOLE CENTRAL CORE OF AERODYNAMICS, OF ENGINES, OF HOW IT ALL WORKS AND TIES TOGETHER. YOU CAN SEE IT ALL BETTER THAN MOST EXPERTS, WHO NEVER SEEM TO GRASP, TO CLARIFY AND STATE THE WHOLE INTEGRATED LOGIC! IT'S FUN TO GRASP IT ALL!

IT'S JUST A MATTER OF **MAXIMIZING IAS VS. DRAG**, FOR YOUR GROSS WEIGHT, CLIMBING, GETTING **FREE TAS AT HIGH ALTITUDE**, (THAT DOES NOT HURT MPG) -- THEN SIGNIFICANTLY IMPROVING MPG, BY OPTIMIZING ENGINE EFFICIENCY, OPTIMALLY MATCHING YOUR **ENGINE** TO YOUR PLANE, BY SIMPLY GETTING IT **WIDE OPEN**, **PROPERLY LEANED**, AT THAT OPTIMUM IAS, AND OPTIMUM WIDE OPEN ALTITUDE, ***REAL INSIGHT!!!***

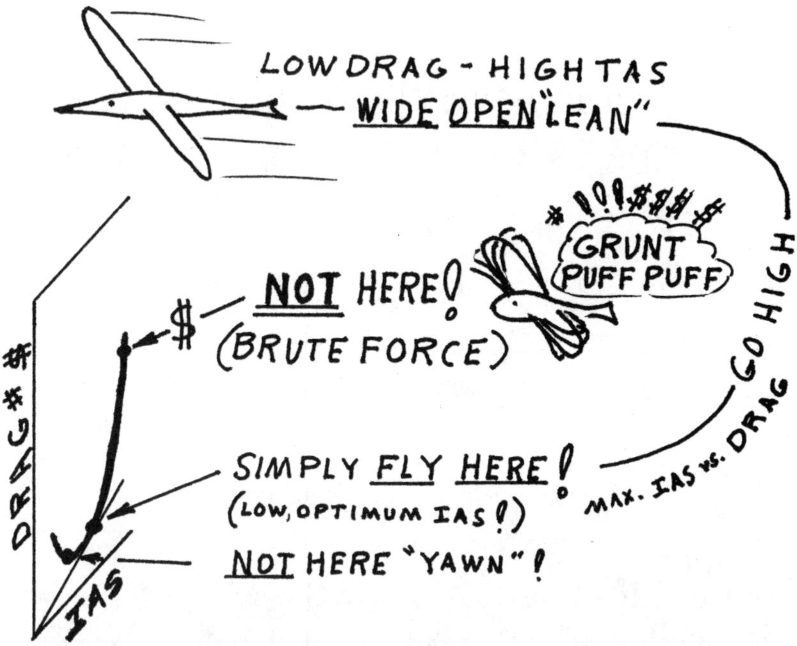

YOU WILL HAVE FOUND THE OPTIMUM L/D, ANGLE OF ATTACK AND "DECK ANGLE", THAT WORKS AT ANY GROSS WEIGHT, ANY ALTITUDE!! YES, THERE IS A "**MAGIC ANGLE OF ATTACK**", THAT WORKS AT EVERY WEIGHT, EVERY ALTITUDE! **IT'S ALL HERE**!

YOU'LL BE AMAZED AT YOUR NEW INSIGHT, YOUR INCISIVE GRASP OF AERODYNAMICS AND ENGINES, HOW LAUGHABLY SIMPLE IT ALL IS IN HINDSIGHT!! YOU'LL SOON REALIZE YOU FOUND THE **BEST LEARNING OPPORTUNITY OF YOUR LIFETIME**!! YOU'LL FIND THE BOOK SET UP FLEXIBLY, FOR YOU TO LEARN JUST AS DEEP AS YOU WISH. IT ALWAYS COMES BACK TO THE SIMPLE CENTRAL THEME, SO **YOU CAN'T POSSIBLY GET LOST**!

THE WISE OLD BIRD

THE LOGIC OF FLIGHT

"THE THINKING MAN'S WAY TO FLY"

INTELLIGENT, EFFICIENT FLIGHT

FEATURING

MAX SPEED / $

BY JACK NORRIS

Jack Norris Publisher

The Logic of Flight
The Thinking Man's Way to Fly

First Edition, Published July 2007
Northridge CA
Jack Norris

All Rights Reserved
No part of this Publication
may be reproduced in any form,
by print, copy machine, microfilm,
scanner, photography, or any other means,
without the written permission of the author.
Price purposely kept low to spread the insight!

ISBN13: 978-0-9620239-1-0

Library of Congress Control Number: 2007904698

Price: $14.95, a Double Book
A $2, $3 Shipping Charge is
encouraged to cover real cost.

propellersexplained.com

11613 Seminole Circle
Northridge, CA 91326
818 360 1105

A personal check, with your
name and address, telephone
number preferred, will serve
as an order. Return shipment
usually 1 day, to one week,
if away, by book rate mail.

Also Available, while supplies last:
Voyager, The World Flight, Jack Norris, Technical Director Mission Control. ISBN 0-9620239-0-6 $15 + $2, $25 personally autographed by Dick Rutan.

PREFACE

Pilots LOVE TO FLY, love planes, the freedom, adventure, the elation of conquering the sky. The novice, eager, learns "to pilot" the plane, so he too can enjoy. The pro, still challenged by flight's many aspects, becomes professionally adept, "at operating" the plane, the machine, learns almost all of flights many aspects. But there is almost always at least one knowledge gap, that causes pilots, to be more "driver", perhaps "master driver" than real master! The pilot seldom learns or understands, the logic of the Aerodynamics that the plane itself senses, flies on, **its logic!** That is unfortunate, and unnecessary!

Without a really good, EASY EXPLANATION of the hidden technical basis of THE LOGIC OF EFFICIENT FLIGHT, pilots are denied a clear insight into how the *airplane itself* really works, what *it* senses, what ITS LOGIC is!

Pilots are "doers", not "students". There is a barrier that is part "technical", part the lack of a good teacher, part the personality difference between an engineer and a pilot, who'll "skip the theory"! HOWEVER, the thought that a truly competent pro, who can, for instance, safely fly into the New York air traffic complex, with terrible weather, with a "Jet load" of people, for an entire lifetime, yet fail to master his subject, for the lack of a really good explanation, is just ridiculous! Our objective is to provide THE ENGAGING SOLUTION, for the novice and the master as well, using the basic *classic plane* as a teaching tool!!

The *intended audience* for this book is ANYONE INTERESTED in INTELLIGENT FLIGHT, pilots, both novice and professional, people from all walks of life that find flight fascinating. Even the technical professional in the Aeronautical Sciences will find *incisive logic* and *insight* in this book that he had not thought to recognize and express!

The unabashed PURPOSE here, is to cut right through the complexity of Aerodynamics, the science of flight, put together the best, most UNDERSTANDABLE, TEACHING guide to THE BEAUTIFUL LOGIC of INTELLIGENT FLIGHT, with the magnificently SIMPLE CONCLUSIONS that lurk, hidden, like a Michelangelo sculpture, in an obscuring granite block.

It will purposely and specifically <u>not</u> be written in the arcane, terse, writing style that technical pros prefer in the obscure technical journal. In every way, using every mechanism, we will REACH OUT, MAKE CLEAR, BUILD the INSIGHTS to help the interested learner come aboard on the magnificent grasp that modern science makes possible. We'll use *italics*, capitol letters, **bold**, to emphasize, make CLEAR! The expert spent years learning, but fails to explain, forgets the learner doesn't "get it" the first time, certainly not the warm, incisive grasp, that *can be built,* by a good helpful teacher leading the way!

The learner needs HELP. Engineering formulas and Greek letters, scare off even intelligent, interested adults, PREVENT UNDERSTANDING! We will teach the beautiful logic and insight that the science exposes. We'll make the CONCLUSIONS CLEAR EARLY, show the INCISIVE PATH, the skeletal framework of the logic, BUILD, REFINE THE READERS GRASP. In the natural repetition, inherent in building and refining the grasp, <u>everyone</u> can "GET IT".

The main game is understanding the <u>big</u> difference, "<u>where</u> you fly on the drag curve", -- "How to <u>also</u> get the Engine efficient", SIMPLE!!! We'll make the INNER SECRETS of the Science of Flight **clear and understandable**, usable for **more intelligent, better informed flight** by <u>any</u> and <u>all</u>! As you come to grasp it all, you will be amazed by your incisive insight. You'll have **a good laugh** over the conclusions, the CLARITY, SIMPLICITY!!

FORWARD - FROM THE DARK AGES
- How We Got To This Point -

We are blessed to live in the greatest country in the greatest time in the history of man! We live in luxury, thermostatically controlled, buy a King's banquet, any product we wish, at the corner "horn of plenty". We have instant news, worldwide communication, entertainment, at the speed of light, navigation by satellite. With portable energy, we zip around in our own motorized chariots, FLY ourselves, FLY 900,000 pound jets. The *Wisest, Richest* man could not have that *one lifetime ago*! We even have the time, the opportunity to become whole happy people, who can grow in knowledge and wisdom, **true wealth**!

Recognize that a scant few generations ago, you might have been grubbing in the mud trying to grow your own food, hunting in a frigid winter, **just surviving**, a frontier sod buster here, a no chance serf in Europe, a *current* citizen of the third world, where they still haven't got the act together! **WHY**?

HOW did our fantastic modern world finally happen, and all in just a few hundred years? Do you realize that only a scant few hundred years ago, in the 1600's, Paris had but 40,000 people! That's only a lesser part of Kalamazoo, and with no sanitation they faced plagues with no known cause or cure, **died** helpless!

The History teachers completely missed the point. They teach us about battles, dates and treaties, the results of human ignorance! They should illuminate what really happened, a never ending succession of Genghis Khan types, power seekers, manipulative politicians, all with an excess of greed, and male hormones, and *a lack of real merit*, "true intelligence, wisdom".

One wonderful Professor at Ohio State, Sydney Fisher, miles ahead of his peers, understood, showed me how the human race kept fouling up for eons, the same flawed personalities going *nowhere*, that one should look to the negative and *positive "human movers"*, ***who*** and ***what kind*** of people were moving the human race ***backward, or ahead***. Humanity slowly managed three steps forward for every two backward, simply because there are the "*real movers*" who don't care about power or manipulation because they KNOW and CAN and DO! **LOOK**, recognize the ***genuine articles*** are ***always advancing***!!

People have been smart for a long, long time. The Greek Philosophers thought and wrote and created Euclidean Geometry. The Arabs had a better number system, still in use. The Egyptians built the Pyramids, better, the great Library at Alexandria. The Romans built great Aqueducts, roads and buildings. Their IQ's were just as high as ours! ***What was missing***? Why didn't the modern world happen sooner, why was progress so slow for eons, how come our progress and growth is suddenly so explosive? There is a ***specific reason!!!***

In the 1600's "men of intellect" came together in the great, if still prototype cities of Europe in "Philosophical Societies" to try to **define the basic "Natural Laws"**. They were really the **Prototype Scientists, Engineers**, and they ultimately found a number of very simple, but very *specific* little mathematical like formulas that ***exactly defined how the physical world works***. They "uncovered" the basic sciences of Physics and Chemistry, built on Mathematics, and the world was changed forever!!!

The great names, Galileo, Newton, Boyle, Charles, Bernoulli, Dalton, Kelvin, Avogadro, Priestley, Lavoisier, Thompson, Carnot, Watt, Cavendish, Coulomb, Ohm, Oersted, Joule, Gauss, Maxwell, Kirchhoff, Wheatstone, Faraday, d'Arsonval, carried the movement from the 1600's up through the 1800's and defined the physical world in a succession of fundamental, but strikingly SIMPLE relationships. There was not complexity, chaos, but **magnificent simplicity**, **precise logic**, in **nature**!

The Industrial Revolution was born, growing imperceptibly at first as the KNOWLEDGE GREW AND SPREAD. The ability to *manufacture products*, create usable engineering *materials* was born of opportunity and need. We've had vast advances in life, health, Pasteur 1857, products for living: The early mills, better cloth, more food, transportation, the "Iron Horses" of the 1800's (because there was not yet steel). Only as late as the end of the 1800's were there real steel mills, the first gas engines, early autos, *production* machines with the ability to manufacture *identical* parts in quantity -- and **progress just exploded!!**

The fundamental difference, the **pivital difference**, was now we could ***understand exactly how*** the physical world, the natural laws worked. We could **learn**, **solve** biological health problems, calculate precisely and design any product our need and imagination could **envision**. Sociologically it was the birth of **THE SCIENTIST, ENGINEER**, where the man that creates a nations destiny, is the creative man of intellect, not the obsolete Saddam Hussein type, the warrior, or the shallow power seeker.

We still have to corral the "power seekers", who still operate on hormones and manipulation from the bygone days, and learn to "move them out of the way" so society can move on to the next level! The people are catching on. The Swiss have a "citizen government"! We are into the age of **knowledge** and it **works**! It's **the Scientist Engineer** who has **created modern life**, our "**horn of plenty**", and there **need be no negatives**, because the ***thinking man knows*** *how to clean, recycle, preserve, as well*!

Recognize in this book how the wonderful, **simple laws of nature** go to work to make "duck soup" of the complexities of Aerodynamics, **let <u>you</u> see right through it, own it - watch!**

THE SCIENCE OF FLIGHT, SUCCESSFULLY TAUGHT

The barrier between what I know about The Science Of Flight and the great fun and satisfaction that you too can have in grasping the core of Aerodynamics, as it applies to intelligently flying an airplane - is - can we successfully **TEACH**, **LEARN**?

If I were just the engineer who understood the Science, the Engineering, we'd both be in big trouble here. The task at hand is as much about successfully teaching, and learning what is otherwise a complex morass, as it is about the science of flight.

There are three reasons why we're going to succeed here.

1. I **really understand** how airplanes work!
2. I see a **simple, clear logic path** through Aerodynamic complexity, that is essentially never completely grasped.
3. I am going to **teach it in a unique way** -- so you can't possibly fail to "get it", even if you normally shy away from things technical or mathematical, **<u>and</u> you'll get plenty of help**!

I've had a wonderful and unique education and career, a unique "forced education", that at various times required me to learn how to be: 1. A very good deep digging engineer who learned how to get to the very essence of any technology. 2. A leader who could "create", "lead", "teach", "sell" as the occasion required. 3. A consultant, who could accurately assess what a situation lacked, needed, and like a chameleon, fit, or teach, or provide, what the situation required. I <u>HAD</u> TO LEARN **a lot**!

With over 100 of my Spacecraft Control Products on the Spacecraft in the Milestone Of Flight Gallery, the central hall of the National Air and Space Museum, millions of dollars of leading edge products all over the world, there's a bunch I can teach you, and I know how to make it CLEARER, EASIER!

BONANNO PISANO'S CONST.
1174AD
YOU NEED A BETTER "FOUNDATION"
I WISH YOU'D LET ME HELP!
GET A BETTER FOUNDATION? WHO'LL EVER KNOW?
CENO'S TECH SERV
YEA RIGHT!

Our brains work quite differently than a computer. They work "**by association**", by "relating knowledge" in layers. Thus, if we have something to learn, **IF**, it is presented to us in "**a logical way**", if, we can see a "logic path" that "makes sense to us", if, we can see "an integrated story building on a central logic tree", we can learn quickly, easily, and cope with far more than we could ever sort out on our own! **I'll make it all clear to you!**

You will be amazed here to find that in just a few chapters you will already be seeing through the Aerodynamics of intelligent flight. It won't be an accident. We're going to be doing it differently than it's normally done. We'll be using quite a different writing style, quite a bit more emphasis, CLARITY, than the terse technical journal writer, the editor practiced in "good writing". The objective here is quite different than in a novel, a journal. It is to "show the clear path", "to lead", "TO TEACH", To HELP, a <u>broad</u> spectrum of "learners".

It would be wonderful, if I could write this individually for the many levels of interested intelligent people who will want to read and learn here. Of course you realize we must make it work for the beginner, right up through the pro who's looking to see if there's really something new here. (there are, things not quite grasped and said clearly yet!) To the potential critics, cut me a bit of slack, my objective is worthy, bigger, and different than what's been accomplished yet. I'm after a big, worthy goal, AN UNDERSTANDING FOR EVERYONE, OF THE THINKING MAN'S, LOGIC OF INTELLIGENT FLIGHT.

I'll introduce myself to you a bit as we go, tell you an anecdote here and there, along with the central story, that will give you extra insights you would not otherwise have. You may think, that I digress, but then quickly find that I'm teaching you HOW we'll be teaching you the subject at hand. HAVE FAITH. You'll even find it's a surprisingly EASY READ, even **fun**!

DEFINITION OF BASIC TERMS

A ready reference to clarify, stop confusion on abbreviations.

"q" -- The "Ram Dynamic Pressure" in psf, pounds per square ft., (#/ft.^{2} in engineering terms), sensed by the "pitot tube", after subtracting the "static pressure", sensed by the "static port". It operates the Airspeed Indicator. It turns out that airplanes fly on "q", Lift and Drag equally affected, "energized" by it. Constant "q" flight, at a given weight, becomes constant L/D flight, a fundamental of HUGE Importance, that permits A SIMPLE, LOGIC OF INTELLIGENT FLIGHT!

IAS -- Indicated Airspeed: The readout of the Airspeed Indicator in Knots, nautical miles per hour, or MPH, statute miles per hour, from sensing "q". IAS reads erroneously low at altitude as air density decreases, which turns out to be **good**, not bad because "it's sensing what the airplane is flying on"!

CAS -- Calibrated Airspeed: True IAS, corrected for any error in the instrument, or the static pressure pickup, or the Pitot location. Pilots tend to not use this term, though they certainly should. To keep the explanation simple, especially in the first part of the book, we'll just use IAS as pilots do, but what we'll mean is CAS, correct Calibrated IAS!

TAS -- True Airspeed: Literally the true speed after IAS, CAS, are corrected for the drop off at altitude due to the air density, ρ, decreasing at altitude. TAS = True IAS $(\rho_{SL}/\rho_{Alt})^{1/2}$

Lift -- L: The *pounds* of lifting force generated *perpendicular to the flight path angle*. Generally taken as equal to the weight in level and near level flight, though thrust or drag help hold up the weight as climb or glide steepens.

C_L -- Coefficient of lift: A simple dimensionless coefficient, or "multiplying factor", that shows the relative ability of an "airfoil" to lift, as its relationship with the angle of attack of the airstream changes. Used in the basic Lift and Drag Formulas.

α^o -- Alpha, Angle of attack: The angle between the flight path angle, and the mean chord line of the airfoil, a simple line drawn from the very nose of the airfoil to its trailing edge.

Drag -- D: The *pounds* of all types of drag generated *parallel to the flight path angle*. There are two basic kinds of Drag, **Parasite Drag** resulting from surfaces being pulled through the Airstream, and **Induced Drag**, the drag due to generating lift. (A third type, Profile Drag is simply the parasite drag of the wing area.) Two (or three) basic drag curves can be drawn, and/or be combined into a classic **leaning "J" shaped total drag curve**.

C_D -- C sub D: The overall total Coefficient of Drag, at any given speed, which is made up of two parts, C_{Do}, a constant, representing the Parasite Drag, and C_{Di}, the highly variable Induced Coefficient, actually dependent on $C_L{}^2$ and *aspect ratio*, used to create the **two basic drag curves**, and/or the classic *composite*, leaning "J", **total drag curve**.

Aspect Ratio -- A_R: The wing span divided by the chord or the average chord of a tapered, or shaped wing. That is also mathematically equivalent to Span^2/Area. Used in calculating theoretical Induced Drag.

e -- Oswald Span Efficiency Factor: A high class "fudge factor", nominally about .8 used to increase theoretical Induced Drag up to its actual value. Actual = Theoretical / e

FPA -- Flat Plate Area: The equivalent flat plate Drag Area, of the Parasite Drag of the entire plane. Numerically equivalent to C_{Do} x S, the wing area. Thus, Parasite Drag equals FPA x q.

S -- Wing Area: Simply the wing area in square feet.

ρ -- Rho (row): The "mass density" of air at a given altitude. Equal to the density of air at any given altitude in pounds/ ft.^{3}, divided by g, the acceleration of gravity, 32.174 ft./ sec^2.

L/D -- Lift over Drag Ratio: The ratio of Lift to Drag that separates "clean" planes from "draggy" designs, but also in the most fundamental way defines the "flight condition" that, if held constant, permits a simple, essentially exact, LOGIC OF FLIGHT! L/D is also equivalent to C_L/C_D numerically.

MPG -- Miles Per Gallon: Every bit as applicable to your plane as it is to your car, except that in your plane you have much more intellectual control over what it actually is, and it in turn "tells" whether you are a really smart pilot, or just another driver, who doesn't understand yet how the airplane works!

N_R -- Reynolds Number: A rather esoteric engineering term that actually relates the "size and speed" of the plane to the "density and viscosity" of the air. You might realize that would be significantly different for a "bee" than a 747, with model airplanes, sailplanes, personal planes in between, seeing different effects that vary from quite significant to a part of a percent. Generally larger faster planes have a 30% *max* C_L and a smaller C_D advantage, less so at high altitudes. HOWEVER, smaller chord, slower sailplanes more easily maintain low drag LAMINAR FLOW. Interestingly, models, too slow, too small, can build stagnant laminar wakes, sometimes use "turbulators", turbulent flow, to energize, shrink the wake.

GW -- Gross Weight: Simply the "total weight" of the plane at takeoff, or any given time under consideration. (See CG)

CG -- Center of Gravity: Simply the balance point of the airplane with its load. Actually, the plane can fly heavier than it actually is, if the CG, the "center of gravity" is far forward of "the Center of Lift", requiring the tail to exert a large downward balancing force, that acts like "extra weight", and furthermore causes extra "trim drag". However, that extra downward force is usually ignored in basic discussions! Clearly, a rearward CG is better for efficient flight, but too far aft is unstable, dangerous, prone to "flat spins", not smart, at all!

Center Of Lift -- C_p, Called the Center Of Pressure: Moves forward at higher angels of attack, *particularity on highly cambered airfoils*, which, as above requires excess downward tail balancing loads, which in turn favors more symmetrical, well designed airfoil sections.

η_p -- Propulsive Efficiency. "Eta" Sub p, the Greek Letter "h": Simply the "Power Required" for the *real* Drag of the airframe (multiplied by the true speed) divided by the *"actual* power" supplied by the engine. The subject was never adequately understood, because there was never a real test for full scale propeller airplane drag, other than free glide testing of a propellerless plane. Now with Zero Thrust Glide Testing, for REAL drag, and level flight Speed Power Testing, many old planes show poor η_p, that degrades with increasing power, whereas current tests of carefully designed leading edge homebuilts show good η_p, a valuable insight and design criteria.

Leaning -- Changing the Air:Fuel Ratio from the nominal13:1, or less, Max Power mixture, to the nominal 16:1 Max Economy mixture, that is usually done manually, to give the pilot *control*!

TABLE OF CONTENTS

NONSTOP
QUICKER
THIS BRUTE IS
REALLY FAST!
I POUR THE
COAL TO IT!
$ I LOVE IT $
THE RAPID RABBIT
BRUTE FORCE
AND AWKWARDNESS
ooo
HALF WAY

CHAPTER 1: INTRODUCTION

The Simple Insight

There is a BEAUTIFUL LOGIC to flight, to the way that airplanes work. It is beautiful because it is so logical, cuts right through the core of complex Aerodynamics, is so easy to grasp and use, has final conclusions that are GENUINELY SIMPLE!

Indeed, the great good feeling, and the funny insight into THE THINKING MAN'S WAY TO FLY, is the real intellectual satisfaction in actually seeing through the complexities of Aerodynamics, that perhaps you thought you could never really grasp, and find that all the important final conclusions on how to fly a plane efficiently, intelligently, are so **laughably simple**.

We can teach you the major bottom line conclusions in five minutes, and we will, just a few paragraphs down, so you *know at the beginning where you're going!* But, the true beauty and value is that in going through the whole logic, and making it your own, you can gain a wonderfully incisive grasp of how **Aerodynamics, Engines,** and **Flight** really work, -- and easily!

The problem has been that there has been no one to explain it to us in the clear, logical, comprehensive way that is possible! One might think that the real pros in Aerodynamics could do it easily. The problem is, a pro forgets to say and tie together the fundamentals that the learner needs, doesn't grasp what a pilot needs to know, gets lost in minutiae, seldom grasps the logic that integrates efficient operation of the plane and the engine.

Though the intent is "SIMPLE", don't underestimate what we're after here. This will be a book for the learner, but it will be much more. It will teach professionals *incisive* logic that they had not thought to tie together. The OBJECTIVE here is nothing less than **the definitive book** on THE BASIC LOGIC OF FLIGHT, that will work **for a broad range of readers**!!!

How does one set out to make a complex technical morass like Aerodynamics simple, logical, easy to follow and grasp, with very simple conclusions? Well, there is a *very specific way*. It only takes a lifetime of experience and knowledge to grasp the simple way to see through Aerodynamics complexity, and see how to teach it simply! We began to explain in the "Forward" that the modern age began when the first Philosopher, prototype Scientists and Engineers, first started to pin down the fundamental laws of nature, of Physics, in the 1600's.

They found *very simple*, *logical*, mathematical relationships, the Natural Laws of Physics and Chemistry, easy, but very specific little formulas underlying *everything*! You'll see that even though the basic formulas are very simple, the fact that they are very specific, allows us to draw multiple, very incisive insights, from each part of what is there! It's quite amazing really!

The genius, that we hope to pull off in this book, is to: 1. let the laws themselves show us the logical way to proceed, to teach you. 2. very carefully follow that simple, *incisive path* so that the relationships *stay simple*, *logical* and essentially *exact*. You may not even be aware that we'll be following a very specific disciplined path, based on a lifetime of experience and insight, you'll just be amazed by the clarity. We do warn you that you can learn a whole lot, so keep your receiver turned on high!!

Here's that early insight we promised you. There are <u>three</u> ideal flight modes, but for most flying, the only one that makes sense is the one that minimizes drag vs. speed, **maximizes speed vs. drag**, or **fuel** required, or fuel **cost**. Drag, gallons, cost, are all equivalent, decrease MPG!! For *a given gross weight*, there is an **ideal indicated airspeed, IAS** that **stays ideal** *at* **any altitude**, because constant IAS *means constant drag*!!! Now all you have to do to *optimize engine efficiency*, and *optimally match the engine and plane*, is to **climb** to the altitude where your engine is *just wide open*, ***unthrottled***, and ***lean** it properly*, holding that *magic IAS*!!! Naturally your True Airspeed will have greatly increased, you will have good speed, you will have optimized speed vs. trip cost, and optimally matched your engine and plane! IT'S THIS EASY! 1. **AN OPTIMUM IAS** (vs. GW). 2. **CLIMB, GET UNTHROTTLED**. 3. **LEAN**!

Believe it or not, there is an even easier way, that's even more all encompassing! Constant, ideal, IAS, Drag, vs. weight, lift, found IDEAL TRAVELING L/D, C_L, *angle of attack*, fuselage deck angle!! That means you can mount a small "level" on the cabin wall, at the *magic angle*, call it the "Super Science Flight Optimizer", and use it for any altitude, any gross weight!!! You climb up to where the engine is ***wide open***, ***(leaned),*** at the **MAGIC ANGLE OF ATTACK**, bubble in the middle. That's it, everything is optimum!! Do **not lean** in climb, **low**, at high power, where the engineers intended a **cool richer** mix!

Do you see that I wasn't kidding you? We're, going to have fun here, learning! You're going to get **as smart as you wish**, and the way we're going to do the explanation, it will work for first time learners as well as professionals looking for more insight!

Yes there are some subtleties that have minor effects, like Reynolds number. We'll cover those too, but not let them obscure the central truths!! There is indeed a BEAUTIFUL, and SIMPLE logic to the way that airplanes work, and it's all going to be laid out so everyone can grasp it!!! ENJOY!!!

It's always easier to learn something if, before you start, you get a feel for what the final conclusion is, where you're going, how you're going to get there, so we're going to build that into our presentation! That's why we already told you a concise version of the major conclusions of the book! Quickly now, we're going to start showing you the first look at how we're going to get there, an **early outline of how** we'll make a complex subject like the aerodynamics of intelligent flight, logical, easy, fun and satisfying! But **first** there's something about **how we learn**!

Looking back, I was fortunate to have some great teachers. Sylda Smith, red hair, school marm bun and all, was one of those wonderful high school teachers who cared, who was not about to let any of us dumb kids fail math, before we were mature enough and organized enough to know what was good for us! If you weren't up to her standards, you would, by command, be in for her special 7 AM class for "laggards". If it happened to be a frigid Cleveland morning, tough, that was just another part of the necessary growing up process, and she told us so! Abby Rush, signed Ab^2y, the other Algebra teacher, was a close second, won honorable mention.

Ohio State, one of the "Land Grant" Universities, at $47 a quarter, did a fantastic job, targeted "a broad engineer" in a five year program with the (3) quarter system, 15 sets of subjects, rather than 8 with a four year semester system. Somehow they all succeeded, because for a kid who started as a pretty unpromising student, I certainly had a fantastic career. It included, jet landing gears, complex servo flight control packages, spacecraft rocket maneuvering system controls, nuclear power plant earthquake protection, analytical instrumentation, and much more. In all, a whole fistful of technology, and not just as engineer, but as company founder, general manager, salesman, program manager, the whole game.

What I've really always been about is creating successful, leading edge, high technology products. That very specifically includes learning, solving the problem, teaching and leading the organized solution. Our task here is very much about complex technology, and specifically, **clarifying, simplifying, teaching**, and **learning**, **complex technology**. Those are pages right out of the book that I've been dealing with for a lifetime.

I've had the great fun and challenges of the major multi billion dollar complex technical programs, Aircraft, SPACE et al! That can be as important, challenging, interesting as any profession there is. You have to bring everything together, the entire human, technical, business spectrum, and prevail over any and all competition. There is no second place prize. What that takes, is really having the right answers, proving *absolute credibility*, and surprise, **teaching**, that is, leading intellectually! What you will see herein is the result of **a lifetime of experience**, not only with complex technology, but in making complexity clear, understandable, agreeable to deal with, *user friendly*!!

To learn complex subjects, you **need a better way**, and if you're **lucky**, someone to figure it out and **simplify** it **for you**. You'll see *both* here. Somewhere back there some enlightened soul taught me the classic old, and well proven **SQ3R** method of learning: **Scan - Question - READ - Recite - Review.**

SCAN what's coming. QUESTION. Get in mind what's apt to be involved, so you "start anticipating a pattern", a frame of logic and facts to learn on. READ it, RECITE it, the key facts, as you go, then REVIEW it -- and you got it. The central core of this classic method is SEEING EARLY -- WHAT THE PATTERN IS. **We adults both learn and remember by seeing the pattern, how everything in a subject fits together, interrelates. IF, it makes sense to us. we learn it quickly, easily -- we remember it. and we USE IT**!!!

That's the game we both want to play here, so I'll do my part, if you'll do yours. So, here goes, **here comes the pattern**.

As introduced in the Foreword to the book, the basic natural laws are very simple, and all tie together in a neat orderly way once you have enough experience to see the pattern. Thus, the way to see through the complex mix of science here, is to let the **natural laws show us** the **logical**, **incisive**, **exact**, **easy** way!

● ENERGY, is what makes airplanes fly! Looking at energy gives us a very enlightening fundamental grasp of how DRAG, INEFFICIENCY, HEIGHT, and DISTANCE eat energy, and launches us on the logical understanding of the whole subject of flight! DRAG and INEFFICIENCY become the twin enemies!!

● LIFT and DRAG. One simple little formula, gives us the first picture of how *both* lift and drag work, is the key to grasping several fundamentals of the logic of flight, including WHY AIRPLANES FLY ON IAS, so it will be the first and only formula we'll take the first time novice through!!

● DRAG is the chief culprit that eats energy! There are **TWO** basic types of drag. They are quite **opposite** in the way they change with speed, and they **add together** in a way that causes all airplanes to have the **same shaped drag curve**, but with important *rotation* differences. LEARN ONE, LEARN ALL! You begin to see through all planes as this picture emerges!

● THREE OPTIMUM FLIGHT MODES. Minimum power gives maximum endurance, minimum drag yields max range, but MAX SPEED / $ becomes THE THINKING MAN'S WAY TO FLY, where we learn how to purchase speed wisely!!!

● THE IMPLICATIONS. OPTIMUM CONSTANT IAS for a given gross weight, at any altitude, better still,

OPTIMUM, CONSTANT ANGLE OF ATTACK at *any gross weight*, any altitude, both yield IDEAL CONSTANT L/D, (LIFT/DRAG), flight. The **effect of weight** on drag and required IAS, of **lower air density** increasing True Airspeed at **high altitude**, **free speed** with ***higher* TAS** that doesn't hurt MPG, are all central to understanding! You really begin to grasp the whole subject!

- THE ENGINE. Understanding why your engine must be too big for your plane, and why that causes it to be even more inefficient, teaches you WHY you need to understand HOW to maximize engine efficiency and optimally match it to your plane. The key, **HIGH**, **UNTHROTTLED, LEAN.** Q. HOW HIGH?

- THE THINKING MAN'S WAY TO FLY. The amazingly simple conclusions on how you can fly intelligently, efficiently easily, and why it all works out so neatly! OPTIMIZING ENERGY USE vs. SPEED, sets up the CORRECT LOGIC OF FLIGHT, where BASIC DRAG CONCEPTS set up simple specific logic that obeys the controlling natural laws!!!

- THE GRADUATE COURSE. It's possible to explore lots of subtleties, and we'll do a good deal of that for those who want to have it all precisely correct. Of even greater value however, is to recognize the simple orderly relationships between everything, *intelligently ignoring minutiae*. A few extra Chapters and Appendices with worthwhile insights for those who choose, and we'll have a flexible presentation for all!

Now just how do we propose to teach the whole logic of flight, deal with the complexities of aerodynamics, make it easy for the novice and satisfying for someone much more knowledgeable? FIRST, we're going to attempt to write it in understandable English and understandable logic, in big enough type that it's an EASY READ, but that's not all.

SECOND, we are going to carry along a **deeper**, more probing, narrative, with more detail, the complete formulas, in slightly **smaller print**, for the advanced people who want it all together. That way the novice can simply skip what is too much for him, at least on the first pass, and the more technically inclined can go as deep as he wishes the first time, and be satisfied! Either way the novice will see and understand.

With due apologies to the more technical reader, we are going to try to write the finer print version in English as well, not "Greek Tech.", to encourage the novice to come back for a second pass after he grasps the basic picture, get an even better, deeper grasp, hopefully answer what questions he may develop.

THIRD. Purposely sacrificing what is properly considered "good normal writing style", we will use every available method to EMPHASIZE, TO CLARIFY, TO HELP the learner. If we succeed in our objective of making the complexities of Aerodynamics, "duck soup" for a broad audience, a task never accomplished, that's a miniscule sacrifice. We'll even have some sentences that are too long when we need to tie a whole logic string together, TO HELP the learner! I've taken bigger risks than that, for lesser goals, so be forgiving, if you're a pro!

You are going to end up clearly understanding that: 1. AIRPLANES FLY ON "q", IAS, INDICATED AIRSPEED! *but* that 2. ENGINES FLY ON TAS, TRUE AIRSPEED! and WHY those two fundamental facts are true. They are *both* fundamental to both THE LOGIC OF FLIGHT, and THE THINKING MAN'S WAY TO FLY, but you won't find that grasp, that concept in any other book!

Our fond hope is that even a pro will find some thoughts outside his usual thinking that will prove interesting, give him an insight, here and there, that he hadn't thought of.... On the Voyager World Flight, we found very competent people missing some key fundamental points!! Many years in the high technology business taught me the REALLY smart guys learn from everyone, even the sweeper, because everyone sees a different facet of the problem at hand!

I once saw my ace machinist solve an obscure metallurgical problem that had stumped me, my chief engineer, and every pro I could reach on the phone! Denby Karaharawa, our experimental machinist saved an Apollo contract that really had the best metallurgical experts in the country stumped! I long ago learned to respect the other guy. I found he can be very smart in ways that I and you are not, in areas, where we may not even recognize that there is something to be understood!

It is a huge misfortune that many highly intelligent people in the technical fields just love to make things obscure, complex, arcane, to write in technical Swahili, unfortunately like guys in a private clique, looking for elite stature! The technical journals can be awash in it! NEWTON'S TRUE GENIUS WAS CLARIFYING THE OBSCURE, FOR ALL TO USE!

WHAT IS TRULY VALUABLE IS TO DO PROFESSIONAL WORK TO SEE THROUGH COMPLEXITY WITH X-RAY VISION, THEN MAKE THE COMPLEX CLEAR AND EASY, SO IT CAN BE GRASPED BY ALL, PUT TO BROAD INTELLIGENT USE! THAT'S HOW THE REAL ORIGINALS DID IT. THAT'S HOW PROGRESS HAPPENS. **That's what we're after here!**

SUMMARY INSIGHT

Since the basic natural laws of Physics are so remarkably simple and logical, if we are only wise enough to follow the logic trail that the simple laws and formulas set up, we have found the ***magic way to decipher*** a complex subject like Aerodynamics, find remarkably simple concepts and conclusions! (Indeed, the engineering formulas, with those scary Greek letters that so quickly put off everyone not trained in the sciences, are the very mechanism that, in the right hands, make the complexities simple and understandable. ***Watch it happen --- for you!!!***)

In this book you will come to understand that an airplane is a "flying machine" that converts *energy*, fuel energy, into flight!

If **energy** is the fundamental commodity used to create flight, the lead we should look to is what are the basic laws governing energy. That quickly teaches us about big INEFFICIENCY, and that we need to come to UNDERSTAND DRAG, how to minimize it, how to get the least drag for the most speed, to intelligently fly any given trip!

That leads to **a remarkably easy and interesting logic path, a long list of equally remarkable insights and conclusions**!!

We'll learn WHY it's proper to **FLY CONSTANT IDEAL IAS** for a given GROSS WEIGHT, at **ANY ALTITUDE**, A **CONSTANT IDEAL ANGLE OF ATTACK**, CONSTANT IDEAL L/D at ***any* altitude, *any* gross weight** -- WHY you CLIMB TO MAX ALTITUDE, MAX TAS, WITH THE **ENGINE WIDE OPEN**, ***LEANED*** **for** ***MAX ECONOMY!*** YOU HAVE the ***LOW IAS DRAG,*** at the ***HIGH TAS SPEED!***

IT"S A CINCH TO LEARN THE THINKING MAN'S WAY TO FLY, **BUT**, THE REAL REWARD IS TO UNDERSTAND **WHY**, AND GRASP THE HEART OF AERODYNAMICS AND ENGINES IN THE PROCESS!!

The straightforward objective of this book is to take the complex subject's of Aerodynamics and Engines, make them fun and interesting, to understand and use advantageously in flight! We didn't initially intend to, but props are explained too, the ultimate morass challenge that took 138 years to explain here!

The subjects are: Understanding The Logic Of Intelligent Flight, and Grasping The Central Core Of Aerodynamics, and Engines, -- EASILY. The challenge is to do a better job of **TEACHING,** and **LEARNING**. We'll use every method to keep, you clear on WHERE we're GOING, HOW we're going to get there, every method to emphasize, make CLEAR. Since, you already have your first grasp of the **simple, final conclusions, you really can't get lost, because we'll keep coming back to the same central theme, building, refining your grasp.** *If you do get lost, or snowed, temporarily, don't panic*, because we'll be right back to the theme you already know. **1. Ideal IAS vs. GW, any altitude, (ideal α, any GW, any altitude!). 2. Climb, until wide open, 3. Leaned for Max Economy! --- EASY, in fact DEAD SIMPLE!!**

It's all set up to accommodate learners at any level. You need **only go as deep as you want to, and are able to**, but you have the option to get smarter than experts on how it all works and ties together. **The experts never state it clearly, completely! A learner needs that help to see it best, and easiest!**

My bet is that you'll be intrigued and challenged, get it in your own way, at your own speed, and end up smarter on how airplanes work, than you ever thought possible!!

YOU MEAN MY ENERGY EFFICIENCY IS WORSE THAN 18%, CAN'T BE BETTER THAN 24%!
YEH!
THAT'S THE TRUTH OF IT! WORTH GETTING SMARTER ABOUT IT!
USING MORE THAN 4 OR 5 TIMES THE REAL, BASIC ENERGY REQUIREMENT!

CHAPTER 2

ENERGY

The objective of this chapter is to show you, in a way that you will never forget, that the laws of thermodynamics are very unfriendly, that there is a gross inefficiency in the way that engines use fuel. That directly points a Thinking Pilot to get smarter on how to minimize drag vs. speed and optimize engine operation, to use fuel optimally, indeed to the entire CORRECT LOGIC OF FLIGHT -----

An airplane is a flying machine that converts energy, fuel energy, into flight, and thus it is bound by the simple laws of energy! Energy calculations can be SIMPLE ARITHMETIC, yet they PERMIT GREAT INSIGHT on matters that can be vastly more complicated analyzed by other means!

If you *push with one pound of force over one foot of distance*, or *lift a one pound weight one foot*, either way you have expended *one foot pound of energy*. There are 778.26 foot pounds of mechanical energy per BTU, British Thermal Unit of heat energy, (A BTU simply heats 1 pound of water 1° F)

Did you catch the clue there? Can you see where this simple logic is going, how easily we'll meaningfully interrelate drag, distance, fuel energy required, and even how much fuel it takes to hoist your plane up to 10,000 feet --- an answer you'll not easily or accurately get, by far more complicated means!

There is a BIG SURPRISE however, a very important insight that I'll bet you'll never forget, so we'll take you through a simple little example, and you can gain the insight just as I did!

Way back at Christmas 1950, driving home from college in my senior year, and looking for an airplane, knowing I was going to be called into Air Force Engineering at Wright Field, immediately upon graduation in June, I came across a beautiful 1947 Luscombe 8E for **$1225, *one third* of its new price!** As a mere student, it took nerve to buy an airplane with my model airplane winnings. With no income yet, it was a leap of faith. It became a real education flying back an forth, essentially every weekend, summer and winter, safely, not becoming a statistic!!!

As things turned out it became a lifetime, essentially free airplane, now a valued Classic, less than **22 dollars per year, 57 years later.** I got way too busy to have an airplane, but was not about to let go. The simple, reliable Luscombe, was practical to keep, always ready to go. It's been a lifetime escape vehicle that's been everywhere, every possible flying condition. It was the cover plane on the February, 1993 AOPA PILOT.

$1225 / 57 Years = $21.49 per year Capital cost!

By 1982, getting closer to retirement, and after years of just using the trusty, classic, Luscombe for R & R fun to get away from the hassles and stress of high tech. engineering work, my curiosity finally got the better of me, and I dove in to find out how THE LOGIC OF FLIGHT really worked!!!

I tried the very calculations we were getting into above and **found such a wild mismatch in the numbers** that it started an investigation that has yielded **a whole series of amazing results**, just one of which was my backing into being the **Technical Director in Mission Control on the world flight of the Voyager**. By 1986, I had a lot of insight others just didn't have, and it became amazingly easy to see through it all!!

Here's the simple enlightening calculation! To hoist a 1400 pound Luscombe up to 10,000 feet would require 14,000,000 ft. lbs. of mechanical energy. DIVIDING by the 778.26 ratio above (ft.lb./BTU), it would look like only about 17,989 BTU would be required, only about *one pound* of Av gas, at nominally ~19,600 BTU per pound. Any pilot with any feel for his plane would know that was not close to being enough!

Here is where we start getting smarter! The hooker is that there is a **huge efficiency problem!** The overall efficiency of a plane and engine combination which could fall in the range of 20% to 25% is actually most apt to fall well below 20%, *heavily throttled*, or in max power climb where the engine designer may dump in *more* than a max power mixture to cool the valves!

Thus, rather than requiring (17,989/19,000), .95 pounds of fuel to climb a small plane to 10,000 feet, it will take more like a 5.85 pound gallon of Av gas, which is lighter and has *less* energy than a heavier, 6+ pound broad cut auto gas, or Jet Fuel!

The main problem is that internal combustion engines are a disaster from an efficiency standpoint, 27% to 33%! To be efficient thermodynamically, you want the engine to burn hot, maybe ~4500°F, then dump its exhaust at absolute zero, -459°F below zero, after extracting *all* the available energy. In fact, the exhaust gets dumped *red hot*, maybe ~1500°F, for a number of reasons, so as an efficient heat engine it's a bigtime failure! Since the Aluminum/oil can't run above 460*/220°F, there's a big cooling loss as well, and in addition that adds a big cooling drag to the plane, which amounts to nominally 10% of the available horsepower, after it's generated at that terrible efficiency! It's a very unsatisfactory situation, not readily changed.

* Cylinders have a ~460 F° Red Line. Stay below 400, avoid cracked cylinders!

A really good engine requires about .4 pounds of fuel to develop one horsepower for one hour. If you run that through a calculator as above, and compare it to 33,000 ft. lbs /min. per HP you get only 33.5 % efficiency. A more common engine at a "specific fuel consumption" of .45#/ HP hr. would be only 29.7%, and a .5 engine, where **you** are operating it poorly, is but 26.8%. Engineers learn in college that Thermodynamics is a very inefficient process, and learn to accept it as inevitable, but if you look at those terrible numbers, there's a lot of argument for being a lot less accepting! There will be change, refinement.

I can get about 28.9% out of my 73 Octane Continental C 85!

So that's a quick look at the engine part of the efficiency problem, so you can start to get a little feel for what's really going on, and begin to recognize that if you can do something intelligent about how you fly your plane, to cut the fuel bill, mitigate the big fuel waste, you're making a right move. We'll look deeper into engines in a later chapter, so you can understand them even better. **You can make a big difference!**

Now to the airplane part of the game! Back in 1982 when I got curious about all this, I wanted to do a test and find out what the real drag and propulsion efficiency of my classic Luscombe was, just to really understand the *whole* LOGIC OF FLIGHT. I was told there was **no test**, I could *not* get a real answer! But pros told me props were about 80% efficient, so I could assume 80% propulsive efficiency, and calculate a drag considering .8 of my HP! Eight decades, into the age of flight and two decades into the space age, I saw the **inability** to get a REAL test answer as half **unacceptable** and half **ridiculous**.

The sailplane people, notably Dick Johnson, several times National Soaring Champion, get magnificent drag data on their craft by doing very carefully controlled, more than professional gliding sink rate tests.

Drag x TAS = Sink Rate x Gross Wt.

Drag # = Gross Wt.# x (Sink Rate / TAS)

To make that nifty, simple little formula and test work, and give correct answers, you keep drag and GW in pounds, and sink rate and TAS *both* in ft./sec. The very simple principle working there, is that the weight sinking at a rate of speed **releases vertical power**, to match the **flight path power required**. Power is force times speed, rate of use of energy, (ft. lbs./sec.)

It was never possible to do that glide testing on a propeller airplane because the propeller/windmill was always in the way, creating thrust or drag, providing or absorbing power! As we'll explain later in the book, we worked out a way, that, by sensing zero thrust, using the axial slop in the crankshaft thrust bearing, permitted glide tests, just like a propellerless sailplane! That solved the longest running, most fundamental gap in Aeronautical Engineering, 87 years after the first flight!

The results didn't agree with the party line at all, on my unsophisticated Classic plane. The **drag was much lower** than might have been predicted, but the **propulsive efficiency was far below a propeller's nominal 80%**. It had seemed completely illogical to me that a poor little propeller sitting in front of a less than perfect, classic airframe, blowing back on it, would still deliver a full 80% thrust efficiency. It did **not**! Testing showed it **got worse** as you increased thrust. **The prop was ~75%**, **another ~90% factor** in there at reasonable power, for **nominally 67% when both were multiplied together**. That would have been even lower if we would have installed closable cowl flaps, eliminated cooling drag, lowered the "real airframe drag" even more. Comparing low airframe drag and required power, with real engine power, would be **below 60%!**

In the process of working it all out, we learned of August Raspet's great work in 1954 at Mississippi State, where he did free glide testing of a *propellerless* Bellanca Cruisair, towed to 12,000 feet. Testing it with the engine cooling ducts sealed, he got such low drag, that, comparing it with the horsepower required in powered level flight, he found that propulsive efficiency, **airframe required power**, divided by **engine input power**, was only **58%**, **essentially, exactly what we found!**

1 /.58 = 172% more Power required, vs Gliding, WOW!!!

So, we have something like 30% engine efficiency, multiplied by roughly 60% propulsive efficiency, a *ridiculous* overall **<18%.** That means we may use more than FIVE TIMES the basic energy requirement!!! That number can float around between FOUR to SIX, depending on how smart your engine and airplane designer are, **and you**! **You can have a BIG effect!!**

HOMEBUILT planes have become A NEW LEADING EDGE OF FLIGHT! The slick, efficient, new homebuilt designs, with low drag and good propulsive efficiency, that are going faster, farther, on less fuel, are an IMPORTANT and necessary development for personal flight in this 21st Century. **YOU** knowing how to fly your plane intelligently, efficiently, is equally IMPORTANT. Challenging the laws of thermodynamics to yield more efficient engines is something you can bet on, as we better grasp the huge need, grasp that fuels may well be manufactured, not mined, in the new Century, that is here now! When the oil runs out we *better* have Fusion, liquid **H** fuel !!!

Before we finish, you'll understand that automobiles, which have been making great efficiency gains lately, are severely wounded by driving around **at low power, heavily throttled**, which requires them to **PUMP their air in**, which compromises their potential! The same problem occurs with your airplane engine, BUT, **if you grasp** the lessons in this book, you will understand that **you can fix that** problem by simply climbing up, flying at BOTH **low power** and **wide open**, in the thinner air!!

The engine is such an important item, learning how to very easily optimize engine performance, and optimally match it to your plane by simply flying it unthrottled, **wide open**, but at **low power**, **lean**, at the optimum altitude, IAS and α° for your plane and engine combination, *is* **the major objective of this book!**

The way the logic develops, however, it's proper to learn the airplane logic first, **how you intelligently fly a Drag Curve**. As we already showed you, for a given weight, all you have to do is learn how to fly your plane smart, at optimum IAS vs. drag, fuel, and fuel cost, then simply climb up until your engine is wide open, or close enough that your efficiency is not wounded by throttling. Climbing **high** at **optimum IAS**, constant drag, constant angle of attack and L/D, you increase speed, TAS, *without* hurting drag or MPG, then *improve* MPG by *increasing* engine efficiency by unthrottling it, leaning! That is the core, of the entire CORRECT LOGIC OF FLIGHT!

YOU'LL COME TO SEE THAT THE MOST BASIC PART OF THE GAME IS COMING TO UNDERSTAND A DRAG CURVE AND WHERE A THINKING PILOT SHOULD FLY ON IT, SIMPLY BECAUSE THAT DRAG IS DRAGGING OVER YOUR ENTIRE TRIP LENGTH, EATING FUEL ENERGY, JUST LIKE YOUR ENGINE INEFFICIENCY IS! Climbing eats energy too, but it's a **good investment** with a big payoff in free speed, and increased engine efficiency. You can get some of it back in an efficient letdown, but the startling insight there is that you already wasted nominally 80% of the energy and can only recover the 20% left in your plane's height!

I Fly 2 Miles High. At 12: 1 L/D, 24 Miles Range, at ~ 70 MPH, ~ 20 Minutes!

You can actually do better than that makes it sound, because you can glide down at zero thrust, as I have, at ~Max L/D cruise speed for over 20 minutes, at as low as 5 HP and not burn the fuel you would have been burning, with its built-in big loss. It's an intriguing little fun puzzle you can see with a fuel flowmeter!

Hopefully you're seeing, agreeing, that, true to our promise, we're going to "teach you a bunch". ALL the conclusions are going to be SIMPLE, one step at a time, and LOGICAL, *but very incisive*, and build to a very complete understanding of the whole LOGIC OF INTELLIGENT FLIGHT, an understanding you'll USE and ENJOY for the rest of your life!

A huge accolade is due the late August Raspet, head of the Aerophysics Lab at Mississippi State U, in the 50's, and mentor of many successful Aeronautical Engineering Graduates such as Richard Johnson, of Sailplane Fame. An activist, maverick, he **challenged his students to get out, creatively use their brain!**

Raspet began free glide testing propellerless personal airplanes in the early 50's, to get the first factual drag and propulsive efficiency data to evaluate and refine the breed. His data stood as the best and only REAL data on propeller airplanes. He clearly showed the real drag, and inferior propulsive efficiency of several planes of that time. Oddly that insight did not seem to be picked up by the industry when it was clearly, important, meriting further work and product improvement.

Richard Johnson became known worldwide for his refinement of his R J 5 sailplane, working with Raspet, his many soaring accomplishments, and finally, his ultra sophisticated glide testing of the ultra sophisticated **world class 60 : 1 Sailplanes**.

His work raised sailplane testing to an art form, and inspired our work which led to the development of **Zero Thrust Glide Testing.** Most significantly, whereas ZTGT would seem to be less pure and sophisticated, the availability of the engine makes it **possible to go find "perfect air"** for near-perfect test results. Such is the way that progress is made, the Science advanced. **This book is possible because we have real drag curves !!**

CHAPTER 2, ENERGY, (IN A NUTSHELL)

If a Flying Machine converts energy into flight, it is logical that we look to the basic laws of energy to find the start of the logic trail to decipher the basic and correct LOGIC OF FLIGHT!

We find a pound of Gross Weight lifted 1 foot is 1 ft. lb. of vertical ALTITUDE ENERGY, and a pound of DRAG flown one foot is 1 ft. lb. of FLIGHT PATH ENERGY!!

This can be related directly to fuel required, if we know the efficiency of the engine and propulsion system, since there are 778.26 ft. lbs per BTU of heat (fuel) energy, and nominally 19,600 BTU per pound of 5.85#/Gal. Av Gas. (Jet = 6.5#/Gal.)

Heavier Auto Gas, ~ 6#/gal., is generally above 20,000 Btu/pound.

As pilots we would never try an engineering calculation. BUT, doing that simple arithmetic calculation immediately provided our first major reward, the shocking insight that engines may only be 27% to 33% efficient, our propulsion efficiency perhaps as low as 58%, that we use FOUR to SIX times the basic energy requirement. We can get HUGELY SMARTER, EASILY, by just following this logic trail of the natural laws!!!

Since the distance to our destination is fixed, at 5280 ft./ mile, (6076.11549 ft./n.mi.), (1.15078 ratio), we realize that if we follow the logic trail to UNDERSTANDING DRAG, *flight path energy* --- more huge insight, knowledge, *intelligent flight rewards*, await us!!!

The simple calculation that we did to illuminate the huge efficiency problem, gave us a super easy way to find extra fuel required to climb to *any* altitude at *any* gross weight! It also gives us the insight to recognize that the game of an efficient let down, is to RECOVER THE (roughly 20)%, ALTITUDE ENERGY, THAT WASN'T LOST IN THE CLIMB!!

GET THIS
AIRPLANES FLY ON "q"
(DYNAMIC PRESSURE)
IAS ≅ q
CONSTANT IAS = CONSTANT q
CONSTANT DRAG!
ANY ALTITUDE!
ONE BASIC FORMULA
TEACHES YOU EVERYTHING
YOU HAVE TO KNOW
AERO 101 IN
ONE FORMULA
q
$L = C_L \cdot AREA \left(\frac{\rho V^2}{2}\right)$
$D = C_D\ AREA \left(\frac{\rho V^2}{2}\right)$
IT'S BASIC
MAN!

CHAPTER 3

LIFT AND DRAG

Chapter 3 and 4 are the ones with the most Formulas, and Math. Don't be faked out, read on. If you do get snowed don't stop, read on, get all you can get. We'll be back, end with Pictures!

This is where we start teaching you the basics of LIFT and DRAG, so you can completely understand THE LOGIC OF FLIGHT. There is ONE simple little formula that defines them BOTH, so it's even easier to deal with, well worth looking into. You'll see that we will get a simply huge amount of information and insight out of it by twisting it around, looking at it in different ways! A few chapters ahead you'll see that the whole game is simply flying at the SMART, OPTIMUM POINT on the DRAG CURVE, which will in fact be at the OPTIMUM L/D for SPEED vs. DRAG! Just hang in and we'll take you through everything you need to understand, and we won't even strain your brain! There is even one of those scary looking Greek letters in the formula, but you'll laugh when we show you that little part of the formula just gets the pressure in front of your pitot tube*, actually equivalent to your IAS!!

* We simply use rho, ρ, the air's Mass Density, to get q, Pitot Tube Pressure

You realize, of course, that Aerodynamics can be an extremely complex subject, that can require supercomputers to get answers that are only approximate, and here we are telling you that we can keep it simple, essentially exact, for you. Well, we can, but the key, the intelligent choice that we are making for you, is to choose the special logic path, the special cases where the relationships stay simple, essentially exact, easy to understand!! It turns out that there are very valid, very fundamental performance reasons for flying an optimum constant IAS for a given gross weight, a constant optimum L/D, because, *when we do that,* ***drag stays constant****, all the basic relationships stay simple and essentially exact*, so that's where we are heading. Doing that, we can start right at the basic lift and drag formulas, take you right up through the whole LOGIC OF INTELLIGENT FLIGHT and never strain your brain. Once you're there you'll have such great insight, that you'll be able to see through even difficult things. SO, HERE GOES!!

The most basic formula in Aerodynamics is the ONE that defines BOTH LIFT and DRAG by simply switching a few symbols!

$$\text{Lift} = L = C_L \times S \underbrace{(\rho V^2 / 2)}_{\text{("q")}} \text{ pounds} = \text{Weight}$$

$$\text{Drag} = D = C_D \times S (\rho V^2 / 2) \text{ pounds}$$

Do you see what we mean? Both those formulas are really the same one! We just switch C_D, the Coefficient of Drag for C_L, the Coefficient of Lift, and we convert the basic Lift formula to the basic Drag formula! We'll get back to C_L and C_D shortly. There is a lot of insight to be gained from C_L and C_D.

To jump ahead to see important things and see that a formula does not have to scare you away, the S is just WING AREA, in square feet, and that stuff in the parentheses equals "q", the ram dynamic pressure, in pounds per square foot, the pressure you feel *if you stick your hand out the window*, the *ram pressure in front of your pitot tube*. Once you see that it's that simple, you'll never forget "q" again!! **AIRPLANES FLY ON "q" !.**

Your Static Port Pressure is subtracted to get the Real q Ram Pressure!

That "Greek" term is rho, pronounced just like "row your boat". "ρ" is just the symbol for air density. V is just Velocity, speed, in fact *real* speed, TRUE AIRSPEED, TAS, (but in ft.per sec.)!

It turns out we can get a Huge amount of insight by understanding "q" better, looking at it from different aspects, and it's very basic to understanding THE LOGIC OF FLIGHT, so we'll do that now, but with just a few words to put the other items in the formulas on "relaxed hold".

Really, Static Port Pressure is subtracted to get the Real q Ram Pressure!

Obviously the wing area, S, how big it is, is pretty basic to how an airplane flies, but once the designer establishes it, it's not going to change. *How many square feet are there,* is not hard to understand. We'll discuss S further in the next Chapter, when we get more deeply into drag, because the aspect ratio, span/chord is *very* important to Induced or "drag due to lift". Aero's use *wing area* for drag, when obviously other surfaces affect drag. That needs a good discussion, intelligent answers.

BOTH C_L and C_D vary vs. speed and resulting angle of attack, (α^o). We'll discuss them shortly, C_D continuing into the next chapter on Drag. C_D has two components, "an almost" constant C_{Do} for Parasite Drag, and one that is *highly* variable C_{Di} for Induced Drag, based on $C_L{}^2$, and Aspect Ratio actually!

Induced Loss is just the Energy Cost, Loss of Throwing Air Down, to Make Lift!

It makes sense that **AIRPLANES FLY ON "q", RAM DYNAMIC PRESSURE,** that it's basic to BOTH LIFT and DRAG, each *equally* affected by it! What could be more basic to an airplane than **how *hard* the wind is approaching**?!!

It makes equally good sense, that "q" depends on ρ, how dense the air is, and the speed, V, in fact the V^2 you see in the basic equation. We can skip the math and Physics derivation of why it's V^2. It's this simple! Twice as fast you hit twice the air particles, each second, and the harder it is to shove them out of the way, (faster and faster), so it's a *double whammy*, V^2. It's that simple! Accept it, it's an absolutely fundamental fact.

YOU WANT A "q" SENSOR, so you can sense exactly what THE PLANE is sensing, what it is reacting to, **SO YOU CAN FLY THE PLANE BY ITS LOGIC OF FLIGHT**!!!

SURPRISE, YOU HAVE A "q" SENSOR SITTING RIGHT IN FRONT OF YOU!! IT"S a very special gauge that does an amazing mechanical and Aerodynamic trick to give you the data in a form that is more useful to you, easier to interpret for the average pilot. It's called the AIRSPEED INDICATOR, and it reads out as **IAS**. To read out in IAS it has to do the amazing mechanical trick of *extracting the square root of "q"!* (That's a pretty nifty trick mechanically!!)

In case that confuses you, don't let it. The key is real simple. **JUST RECOGNIZE THAT THE AIRSPEED INDICATOR IS SENSING THE 'q' AT THE FRONT OF THE PITOT TUBE**. That's what is important!!

Really, Static Port Pressure is subtracted to get the Real q Ram Pressure!

You don't really care if it reads Knots IAS, (or MPH in an old bird), not pounds per square foot. Pounds/ft.^{2} would not be useful to you. You probably always thought that it was bad that it didn't read out as TAS, but you see its *wonderful* that it reads IAS, drops off as the air thins, because by just sensing the ram dynamic pressure, **it ties what you're sensing to what the airplane is sensing, and flying on!!! BIG important point!!!**

YOU WANT TO FLY A CONSTANT IAS AT A GIVEN GW, BECAUSE YOUR DRAG WILL BE THE SAME AT ANY ALTITUDE , YOU WILL BE HOLDING 'q' AND L/D CONSTANT, ALL OF THE AERODYNAMIC RELATIONSHIPS WILL STAY SIMPLE AND EXACT.

You'll see when we get to drag curves in the next Chapter that we always **PLOT DRAG vs. IAS**! It stays the same at any altitude, super simple, the basis of the Simple Logic Of Flight!!!

GRASP HERE EIGHT ITEMS OF HUGE IMPORTANCE

1. **IAS IS EQUIVALENT TO "q"**, because "q" is what the Airspeed Indicator senses, ram dynamic pressure, **with static (-static)**, and **AIRPLANES FLY ON "q"**, the "wind" that **equally effects, energizes both Lift and Drag. THAT'S THE KEY!**

Really, Static Port Pressure is subtracted to get the Real q Ram Pressure!

2. **AIRPLANES FLY ON IAS** If IAS, "q" stays the same, **both lift and drag stay exactly the same**, and the **same ratio** to each other, at any given gross weight! (and that means **Constant L/D**!!) **YOU CONTROL IAS**. After direction, IAS and ALTITUDE are what you control about the airframe, (including angle of attack, α°), so you need to be a lot smarter about BOTH IAS and ALTITUDE!

3. **CONSTANT IAS FLIGHT** at a given weight, and later constant L/D, constant angle of attack flight at *any* weight, is what makes THE LOGIC OF FLIGHT easy to understand and use because the relationship of all the technical factors stay **simple** and essentially **exact**!

4. As you **climb** and the **air density decreases**, your **IAS**, Indicated Airspeed, **will tend to decrease**!

5. If you put on a little more power, hold IAS CONSTANT, fly constant "q", your DRAG WILL STAY EXACTLY THE SAME, but your **TAS MUST INCREASE** to make up for the fall off in air density, rho, to hold the same "q", if $q = (\rho V^2/2)$!

6. That sets up the wonderful condition that your "friction of flight", your drag, does *not* increase, but your real speed, TAS *does* increase!! Now you'll understand better as we go along, that friction, drag, controls MPG, miles per gallon, so you don't hurt MPG, but you go faster, Voila, **FREE SPEED**!!

7. As you'll learn in the Chapter on Engines, and as we already told you on a preliminary basis, you'll actually IMPROVE MPG as you climb higher, because you'll IMPROVE ENGINE EFFICIENCY by getting it unthrottled, then leaned!! That's a double win, **FREE TAS** increase, **MUCH BETTER MPG**!!

8. In the finer print version at the end of the Chapter, we'll actually derive the equation for TAS increase with altitude, based on the fact that q stays the same at constant IAS, and find that it changes as the square root of the ratio of rho at sea level over rho at altitude. $TAS = IAS\ (\rho_{sl} / \rho_{alt})^{1/2}$ **We'll give you a handy table**, (p38) so grasp or skip the math, as you wish.

Now I fully understand that we can be giving you so much, so fast here, that we can get your head reeling, if you're new, get you feeling that this *is* complicated. Not to worry. The items above are "a major part of the logic", as we build you toward a total grasp. We put it all together so you can see that it all fits together. You'll see that everything will come together in a very easy, logical way, to those *extremely simple* final conclusions that you've already seen! Don't sweat!

Engineers get correct answers out of these little formulas by keeping the units consistent. That simply means that in the English system, you use foot, pounds, seconds, for *everything*, That means you don't mix up MPH, miles per hour, with feet and seconds. You calculate speed in Ft./Sec., (Ft. per Sec.) and you calculate pressure in pounds per square foot, psf, not psi.

That will seem like a big bother to non engineers, but there are two huge advantages. First, you get the right answer! Second, if you have some big complicated equation, you can check the units through and if the answer comes out with the wrong units, you know you have a mistake somewhere. It's actually a neat trick and a big help when you're slogging through complexity!

Now to show you a good example, in the equations above, naturally you want lift and drag to come out in pounds. Naturally the wing area is in square feet, ft^2, and thus you want q, the ram dynamic pressure, to be in pounds per square foot, psf. That multiplies out to pounds, just what we want!

With that, we must get into the lift and drag coefficients. They are simply dimensionless *coefficients*, multipliers, that tell you "a multiplying factor". A plane has a lot of drag, or a little, that the wing is flying slow at a high C_L, or fast, at a low C_L. The C_L is interesting in that with a little help from you, it *adjusts to whatever it has to be* to do whatever you want to do! It has to be high enough to hold up the weight, pull "G's", considering the wing area and "q" available. It forces you to pull back on the stick, retrim, make it what it needs to be to make the *lift* equation *equal the weight,* or the "G" load.

Airfoils generally stall at a C_L of 1.2 to 1.6, depending on their characteristics, roughness, and the Reynolds number (a function of density, speed and length, divided by the absolute viscosity of the air). That gives a good, max C_L and C_D advantages to big, fast planes, less so at high altitudes where the air is much less dense, . Airfoils have the interesting characteristic of increasing their C_L almost exactly .1 for every degree of angle of attack, until they approach stall and bend over, (App.C), "losing it", an interesting and easy characteristic to remember.

The high lift devices, flaps, slats, can add a substantial increase to the max C_L capability and wing area and are one of the basic capabilities that make Jets practical. They have to lift huge loads at reasonable speeds. The Boeing 727 has the first of the modern wings. I sold the Aileron Control for it, conceived the fail safe mechanism to succeed. It sprouted multiple leading and trailing edge flaps, inboard and outboard ailerons, multiple spoilers, sequentially linked to the ailerons, based on q and control authority needed. The wing seemed to double its area in landing configuration, and looked like an aluminum porcupine hung on two naked spars at touchdown, spoilers up. Look close the next time you land in a 727, or equivalent plane.

At 200 MPH an RV C_L is ~ .125, thus at 50 mph it's ~ 2.0

Here's the interesting insight you should grasp on the relationship of "q" and C_L. For a small homebuilt plane, like the RV6 I'm building, that can go 200 MPH and land at 50 or less, the **speed ratio of 4** produces a **"q" change of 16** ("q" depends on V^2) so the $\mathbf{C_L}$ **must change by** a factor of **16**, to support the weight. C_L *increases* rapidly as you slow down to land.

Flaps are needed to get a factor of 16!

That's an interesting insight, but the even more interesting insight that you'll see working in the next Chapter and next page, in a real drag curve, is that the Induced or "Lift drag" coefficient $\mathbf{C_{Di}}$, **depends on $\mathbf{C_L^2}$, so it really goes up**, (256!) as you slow down! That makes practical sense because a "mushing wing" slogging through the air, logically would have a much higher drag coefficient than a fast level wing. The effect is so big, C_{Di} goes up so violently, slow, a **$\mathbf{1/V^4}$ effect,** that it **overwhelms the $\mathbf{V^2}$** in the drag formula and Induced Drag becomes a **$\mathbf{1/V^2}$ curve**, that DECREASES rapidly as you go faster, but INCREASES rapidly as you slow down!! You'll see Induced Drag plotted on the next page. **LOOK**, at the induced drag rapidly decreasing as speed increases, and visa versa!

It all ties back to that very high angle of attack wing plowing through the air, like a high drag, "nose high", motor boat. The math and the engineering logic *are always* in tune with good "horse sense" once you "get the right grip on it". Incidentally, that same "motor boat effect" is how pilots get on the "back side of the drag curve" in extreme cases, "hanging on the prop"!

Remember - Induced Loss is the Energy Cost, Loss of Throwing Air Down, to Make Lift!

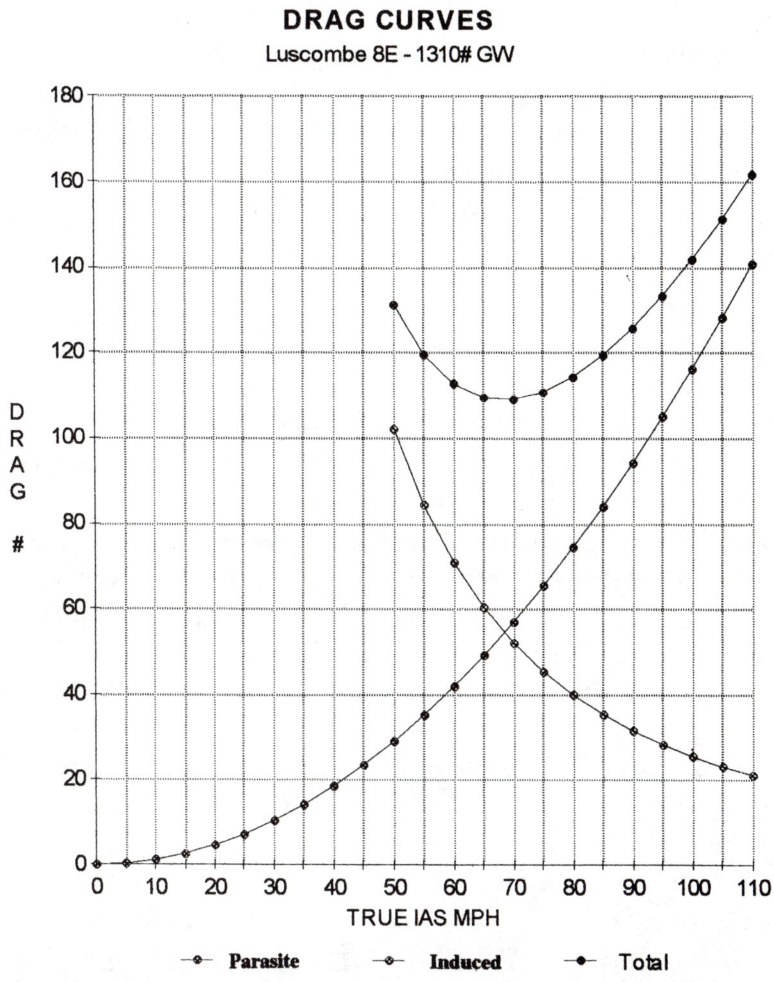

Coeficients ---- C_{Do} C_{Di} C_D

That transitioned us nicely from C_L to C_D and to the next paragraph which continues right into the fact that we have **two** Drag Curves, **two** kinds of Drag Coefficients, Parasite, C_{Do}, for all the surfaces hanging out into the airstream, that is treated as a constant, and C_{Di}, Induced, that we found is *highly* variable!! They add together, the **two parts** of **the total C_D**.

Most airplane people have heard that Drag (and Lift) are V^2 phenomena, that is, twice the speed, four times the drag, 3, 9, etc. You can look at the basic formula, and there it is, the V^2, just as big as life, sitting right in the middle of the "q" equation! **HOWEVER**, we've now seen there are TWO kinds of basic drag, Parasite Drag, which is a pure V^2 Drag, that **swoops up rapidly** as speed increases, with a constant C_{Do}, and Induced Drag, the Drag due to Lift which is just exactly *opposite*, a $1/V^2$ phenomena, that **swoops down** at higher speed! (It's a FACT!)

Since one of the components is DECREASING vs. speed, total drag is **LESS** than a V^2 phenomenon, even at high speeds! When the **TWO DRAG CURVES ARE ADDED TOGETHER**, you can see they produce the **standard, leaning "LAZY J" DRAG CURVE** characteristic of ALL airplanes!! (More proportional to mid speeds!!!) You'll get very familiar with that shortly, because understanding that curve and where you want to fly on it is *the most basic objective of this book*!

Now, that just about transitions us to the next chapter where we seriously look at REAL DRAG CURVES. We must soundly close out this introduction to the basic Lift and Drag Formula. We want to get to the Drag Curves, THE PICTURE we've been talking about. Words and math confuse learners until you see it, get it, so don't sweat it. We'll finally deal with Drag in PICTURES, to *see* THE LOGIC OF INTELLIGENT FLIGHT.

For that we have gained a lot of important insight, from just dissecting the basic formula and "q" We have dealt with each of the elements of the lift and drag formula, and we hope we have kept our promise to not strain your brain!. There is much more theory here than in the rest of the book. We'll proceed to DRAG and the drag curve, a PICTURE, easier. But now, a bit of **fine print**, additional insight for those who are interested.

Earlier we told you that we'd add a "finer print version" for those who are farther along on understanding this subject, more comfortable with math and formulas, and want a deeper look on the first pass. For those who want it, we'll give you a little "bonus insight" early.

Up in item 5, some of the more perceptive readers may have thought I made a BIG mistake. I said that if you "add a little power" -- to hold IAS constant as you "rise" in altitude, (up into thinner air) ------. I know from experience that seems absolutely backwards to many people. It seems that it should take *less* power to fly in thinner air! We'll cover that a few Chapters ahead under Power, because it's so important, but for those who are ready, here is an important early insight.

Power is drag times true speed, TAS, lb. ft./sec., in Engineering units. Now you already saw that we're going to be looking into constant IAS flight, at any altitude, for a given GW, because that means drag stays *constant*, and THE LOGIC OF FLIGHT STAYS SIMPLE, and because it ties right in with optimum, constant L/D flight! Now if IAS and DRAG STAYS CONSTANT, and TAS INCREASES at altitude, POWER MUST INCREASE by *exactly* THE SAME RATIO, as IAS to TAS , and it does!!!

You'll see that's very fortuitous, because as you climb more power is required, the engine is able to put out less power as the air density decreases --- and it makes it that much *easier* to get the engine ***wide open, at as low an altitude as possible***, so it can be as efficient as possible!! Higher altitude, less air density and power capability, *hurts* engine efficiency, **but not as much as throttling it**. So the game becomes getting it **wide open, as low as possible**, at optimum IAS, (EXCEPT*) and the need for *increasing* power at altitude, *flying high,* HELPS!!

***We'll learn later, that the gain in TAS, beats the loss of Engine Efficiency!** ✓

Before we go farther with the basic explanation, we have to deal with a housekeeping problem, for the advanced guys. Once we get a V^2 phenomenon, we have a "dynamics" problem that does an odd thing to the "consistent units" subject we explained earlier. It's not really important to the new guys (and gals), on their first pass, so we'll switch over to the finer print again, to keep the new folks concentrating on the important basics!

There is good news and bad news. The bad news is that we have a messy, esoteric little "units" problem to deal with here, that is so basic that we dare not skip it, even though I'd like to, since the objective is to keep this book on the easy way to grasp the meat of Aerodynamics. The good news is that at least we have the fine print method available to deal with it. I just hate to make this an early example, but you will quickly understand a basic about air density, ρ, that you simply must to have the whole story.

Once the Velocity is squared, and the airflow becomes a Dynamics problem, the units of V^2 becomes ft^2 / sec^2. That multiplied by an air density in pounds per cubic foot would not come out to be psf, the pounds per square foot for q, that we want, to keep the "units" correct.. It turns out that the mathematics of a dynamics problem requires that the air density be divided by the acceleration of gravity, ft./sec.^{2}, which puts the density into units that engineers call "slugs per cubic foot", (lb.sec^2/ ft.^{4}). (That looks weird even to "non Aero" Engineers.) When that is multiplied by the V^2 units above, the units do, in fact, come out to "psf" for "q"!

Recognize that although we have an odd new "unit" "slugs / ft^3", it's really just the result of using the acceleration of gravity, with everything still in normal foot, pound, second units. The math and "units" *always* work !

Sorry to have to subject you to such an esoteric engineering detail early, but it is SO BASIC to the subject of Aeronautical Engineering that it's an absolute necessity to cover it one time to be complete, for the advanced learners. ALL of the DENSITY TABLES are in these units because they *must* be!! **Standard sea level air density is .002377 slugs per cubic foot**.

For the more advanced guys who want it all on the first pass, here's the derivation of the IAS, TAS relationship. If you're going from sea level, where IAS and TAS are the same, to some altitude where rho falls off, and you want the exact TAS/ IAS relationship, recognize that q must stay the same in each case for an equivalent comparison.

For Sea Level, case 1, $(\rho_1 V_1^2 / 2)$ = Altitude, case 2, $(\rho_2 V_2^2 / 2)$

Canceling the / 2, substituting, $(\rho_{SL} \times IAS^2) = (\rho_{ALT} \times TAS^2)$

Rearranging, taking the Sq. Root $(\rho_{SL} / \rho_{ALT})^{1/2} = TAS / IAS$

$$TAS = IAS\,(\rho_{sl} / \rho_{alt})^{1/2}$$

That will be a cinch, for the people who are sharp on Algebra, harder for those who are rusty, and a stretch for everyone, who by now forgot whatever Algebra they had. The conclusion is as above and is written out in item 8, earlier in the chapter, for anyone who wants to understand exactly how the TAS / IAS relationship works, without going through the Algebra. Our objective, of course is to make it appropriate for any level of reader. Following is a chart showing the Standard Atmosphere, (p. 38) which includes the TAS / IAS relationship for standard altitudes. We've also put a more extensive Atmosphere Chart in Chapter 4

Notice that the TAS multiplication factor is 116.37% at a standard 10,000 ft., 126% at 15,000, 137% at 20,000, 201% at 40,000 ft. high commercial Jet altitude, and a huge 325% at 60,000 where the top fighters, U2's and SR 71's start coming into their own! Notice at 60,000 ft. you're trying to fly on 9.41% of standard air density, reaching out for those molecules!!

Before you start considering that as all "Free Speed", you must recognize that as you take reciprocating engines to very high altitudes, as we'll see later, compliments of Voyager engine tests, there is a significant drop off in efficiency, especially with high RPM friction, with the power capability dropping fast, friction an ever larger part of the total. Superchargers reinvigorate power capability, but increase back pressure, remove some work capability, and can hurt efficiency, except for special designs!

However, did you realize that for decades, the piston engine altitude record of **56,046** ft. was established October 22, **1938**! Mario Pezzi established it in a Caproni 161 biplane **unsupercharged**! Seeing that it was done at Montecelio, Italy (Sky Mountain), I thought that it may have been a wave flight. I called the NAA to check, but was told that it was real, used a very specially developed 700 HP engine and plane! I keep thinking that it was supercharged, but amazingly, the write-up in the NAA/FAI record book specifically say's, "despite the absence of a supercharger"! Nobody said it had to be efficient, reaching out for those air molecules! (I'd also bet that there's a good chance that he did find some wave lift too.)

Advanced folks may have noticed that we didn't derive, prove, the $1/V^2$ relationship of Induced Drag vs. speed, the $1/V^4$ relationship of C_{di}. We'll do that in the next chapter where we're plotting curves, making pictures of the actual drag. We'll do it in the fine print of course, but give learners help to see the **great insight, "Span Loading"**, that shows up!

A FEEL FOR "q"

Since this Chapter is very much about "q", constant IAS, constant "q" flight, it's worthwhile to get "a feel" for "q". If Airplanes fly on "q", HOW STRONG IS THAT RAM DYNAMIC PRESSURE --- HAND OUT THE WINDOW??? I once purposely, and *carefully*, put my hand out the flight engineer's vent window at 300 MPH in a B26 Invader. It did give me a good "practical feel" that I still remember, many years later. Below you can figure it was about 230 psf, 32 lbs. at 1.6 psi for a 20 in^2 hand! I also remember thinking that it was good that a 300 MPH bug didn't hit. I really would have felt that!

Back in the fine print, for the advanced guys, we established that ρ at sea level is .002377 "slugs"/ft^3. It's useful to remember that 60 MPH is 88 ft./sec., a 15/22 ratio. So 100 MPH is 100 x 22/15 = 146.6666 ft./sec. So if $q = (\rho V^2/2)$, **q at 100 MPH IAS** equals 25.5659, **25.566, or 25.57 psf**.

So at 100 MPH, Parasite Drag is **25.57** pounds per square foot of FPA, equivalent Flat Plate Drag Area. By just remembering that number, 25.57, and using a calculator that can "ratio" the V^2 factor, you can quickly do any IAS at any altitude!! Example: At 200 MPH "q" is **4** times as much, 102.264 psf, at 110 MPH, 1.1^2, 1.21 times as much. Get it? With a simple calculator, you can quickly ratio "q" for **any** speed, and using IAS (remember that means "q"), it works for **any** altitude! If you can't remember the 25.57, there's a great **"Patriotic Mnemonic"**, **.1776** psi x 144 in.^{2}/ft.^{2} equals **25.57 psf**.!!

A 1400 lb. Classic Luscombe with a 140 ft.^{2} wing , 10 psf wing loading, flies at only .07 psi. wing differential pressure, a C_L of .391at 100 MPH. At **85** IAS, q = 18.47 psf, C_L.541, **100 TAS** at altitude, economy range is **800 miles** at only **3.75 GPH!!!!**

CHAPTER 3, THE LIFT AND DRAG FORMULA

It is absolutely amazing how much insight can be gained on how one flies an airplane logically, intelligently, efficiently, from the **one basic formula** that describes **both lift and drag**! From a pilot's standpoint, it is *almost a total education in one package.* If one mentally walks around it, prods it, exposes **all** the many hidden secrets and incisive concepts, it offers up **huge insight!!**

We are by no means done with it because it's logic will underlie this entire book, and we will find that it quickly yields the insight to understand the effect on drag and necessary speed as weight changes in Chapter 6, an absolute fundamental in the LOGIC OF FLIGHT. **Really grasp, that basic formula!!!**

The **central fundamental that we wish you to grasp in the, formula, is the CONCEPT OF CONSTANT OPTIMUM "q", CONSTANT OPTIMUM IAS FLIGHT, at a given gross weight. That leads directly to the CONCEPT OF CONSTANT OPTIMUM L/D, CONSTANT OPTIMUM ANGLE OF ATTACK FLIGHT, AT ANY WEIGHT!** Both concepts make the logic so easy and exact. By starting you on constant IAS, constant "q" , teaching you **where** you want to fly on a **Drag Curve**, bringing in **weight effect** on **drag** and **speed**, you'll **"get it all"**, laugh, in hindsight, at how easy it all is. It's ridiculous, really, that there was never anyone to teach it.

There is so much basic insight in this fundamental Chapter, that, rather than try summarizing it for you here in insufficient space, might we suggest that you go back and skim the "highlighted" items in the text, the Review, in **SQ3R**, (p5)! For sure, **GRASP** that **TAS is free, does not cost fuel, hurt MPG, BECAUSE Drag does not increase, if you "nail" an optimum Lift vs. Drag, by simply holding "q", IAS --- for a given weight!**

ICAO INTERNATIONAL STANDARD ATMOSPHERE

Alt.	TEMP. (-3.56Δ°/k)	PRESSURE	RATIO	DENSITY	RATIO	TAS/IAS RATIO
h	t°	p	p/p_o	$\rho\ 10^4$	ρ/ρ_o	$(\rho_o/\rho)^{1/2}$
ft.	F°	Hg"	δ	slugs/ft³	σ	$1/\sigma^{1/2}$
0	59.00	29.92	1.0000	23.77	1.0000	1.00000
1000	55.43	28.86	.9644	23.08	.9711	1.01482
2000	51.87	27.82	.9298	22.41	.9428	1.02986
3000	48.30	26.82	.8962	21.75	.9151	1.04537
4000	44.74	25.84	.8637	21.11	.8881	1.06112
5000	41.17	24.90	.8320	20.48	.8617	1.07735
6000	37.61	23.98	.8014	19.87	.8359	1.09385
7000	34.05	23.09	.7716	19.27	.8106	1.11074
8000	30.48	22.22	.7428	18.68	.7860	1.12790
9000	26.92	21.39	.7148	18.11	.7620	1.14561
10K	23.36	20.58	.6877	17.55	.7385	1.16374
11	19.79	19.79	.6614	17.01	.7155	1.18217
12	16.23	19.03	.6360	16.48	.6932	1.20106
13	12.67	18.29	.6113	15.96	.6713	1.22055
14	9.11	17.58	.5875	15.45	.6500	1.24023
15	5.55	16.89	.5643	14.96	.6292	1.26056
16	1.99	16.22	.5420	14.47	.6090	1.28156
17	-1.58	15.57	.5203	14.01	.5892	1.30276
18	-5.14	14.94	.4994	13.55	.5699	1.32468
19	-8.70	14.34	.4791	13.10	.5511	1.34698
20K	-12.26	13.75	.4595	12.66	.5328	1.37005
25	-30.2	11.10	.3711	10.65	.4481	1.49387
30	-48.0	8.885	.2970	8.893	.3741	1.63479
35	-65.77	7.041	.2353	7.365	.3099	1.79630
40	"	5.538	.1851	5.851	.2462	2.01532
50	"	3.425	.1145	3.618	.1522	2.56279
60K	"	2.118	.0708	2.238	.0941	3.25945

CHAPTER 4

DRAG CURVES

This is where we will start constructing Real Drag Curves, "drag in picture form", so you folks who had trouble with the math in the last Chapter will find that, true to our promises, like Sylda Smith, my old Algebra teacher, "we won't let you fail". We will still have math, theory, in this Chapter, but since we'll end up with pictures, that will make very good sense, and fit with the final conclusions that you already have a pretty good feel for, you can't possibly get snowed or lost. All the math will show is that the Parasite Drag Curve "swoops up" as you go faster, and the Induced Drag Curve, due to the Lift that is generated, "swoops down" as you go faster. There's nothing scary about that! We'll add those two curves together to form the classic, leaning, "lazy J", composite drag curve, that all airplanes have as a Basic Drag Characteristic --- You'll end up seeing soon, that there are "three optimum places" to fly on a drag curve, but only one where a "Thinking Man Buys Speed Wisely, Frugally"! You'll see that it's so easy to grasp where speed is available "cheaply" for little drag, fuel and cost, that it's a real "laugher", that you'll never forget!

Remember, the conclusions to the whole LOGIC OF INTELLIGENT FLIGHT, are so easy that we started to show them to you in the first Chapter. We're taking you through the core of the Aerodynamics of Flight to show you that you can "grasp the whole subject" so if you're new to all this and not strong in math, don't worry, because if one part "snows you" we'll keep on coming back to things you can see, so there is no way that we'll let you get lost. We intend that this be fun, and that you end up proud of yourself, confident in your new knowledge, and that you remember and use what you learn here for the rest of your life, in SMARTER FLIGHT!!!

We'll cover the Parasite Drag Curve first, because it's simple, has a constant Drag Coefficient, C_{Do}, and is a simple V^2 Curve that simply "**swoops up**" as you go faster, twice as fast, four times the drag, three times as fast, nine times the drag! Get it? Look at the Parasite curve on the facing page. **That's what a simple V^2 curve looks like**. Remember, like in the last Chapter

$$\text{Parasite Drag,} \quad \mathbf{D_o} = \mathbf{C_{Do}}\ S\ \underbrace{(\rho V^2/2)}_{\text{"q"}}, \text{ pounds}$$

A real number for a C_{Do} for a classic Luscombe is .0325 from Zero Thrust Glide Testing, the first time in history that a real drag coefficient was obtained from a flyable propeller driven airplane! In 1952 August Raspet got .033, free gliding a propellerless Cessna 120, with the cooling ducts open, a good comparison of these two classic planes, which are very evenly matched.. In actual fact our Luscombe tests probably gave it a small advantage, because we found essentially perfect test conditions three miles out to sea, at dawn, in perfect dead air, a condition not normally attainable, the real advantage of ZTGT!

That .0325 **Parasite Drag Coefficient**, is also known as the **Zero Lift Drag Coefficient**, a logical name and concept to grasp! If that .0325 coefficient is multiplied by the Luscombe wing area, 140 ft.^{2}, we get 4.55 ft.^{2} another concept, **Equivalent Flat Plate Drag Area**, that is a particularly good, undistorted, way to describe Parasite Drag, as you'll now see.

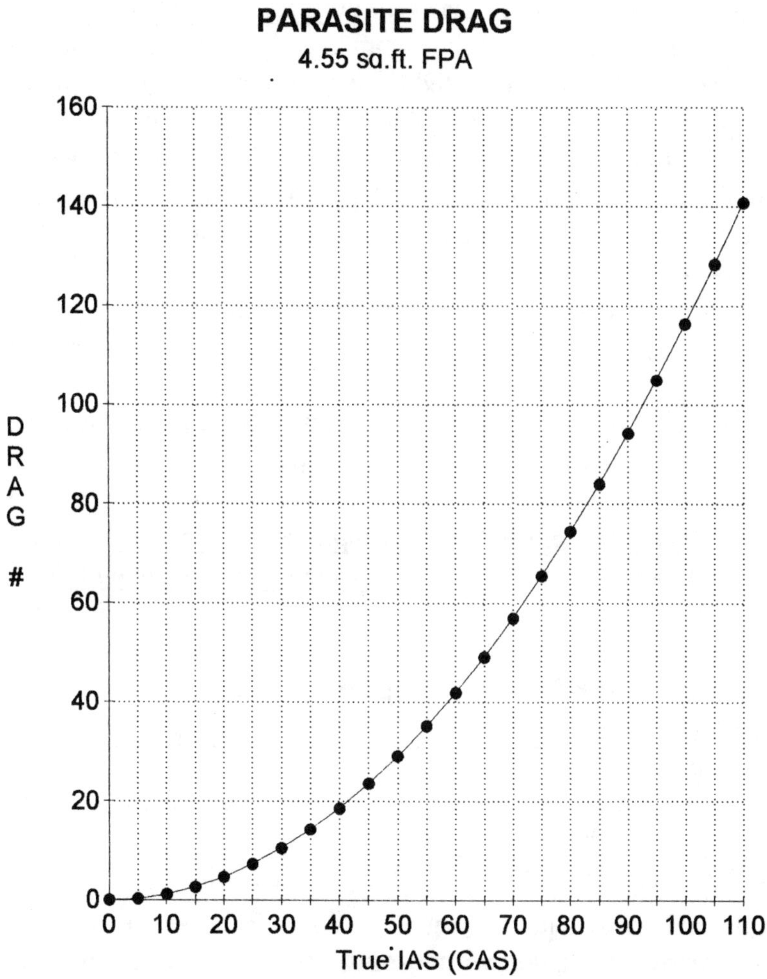

It is the convention, in Aeronautical Engineering to use S, the wing area in both the Lift and Drag Formula. It provides a convenient and consistent basis, which however is not completely logical, if one looks close, and it never seems to be specifically explained in the basic texts. So, we'll give it a good airing here to eliminate any possible confusion for you.

If one looks into the basic texts on drag for parts of planes, like wheel pants, struts, etc., they immediately start talking about "cross section area" and related drag coefficients. -- They would clearly be *different* than a coefficient related to a wing area!! In fact, experiences teaches that parasite drag is most closely related to "total wetted surface area", with due allowance for some "pressure drag", if there is some "separation" occurring!

Clearly then the use of wing area is consistent, and convenient, but a tad less than precise, especially when one recognizes that two planes with abnormally big and small wings would have somewhat distorted "drag coefficients" that should **not** be directly compared!!! Recognize *the* ***flat plate area*** *concept nicely steps around this distortion, eliminates wing area, and* ***permits accurate, undistorted comparisons between planes***!! Parasite Drag is simply, FPA x q, pounds, at any given CAS.

The FPA concept seems to imply a drag coefficient of "1", which is a tad low for a true flat plate (vs. cross section), unless it's two in tandem. Ignore that, just use FPA x q, pounds!

The good news, however, is that S, wing area is used consistently in both the Drag and Lift calculations, and in calculating C_L and both C_{Do} and C_{Di}, so all calculations remain simple and consistent, valid justification. Even when the wing area is greatly expanded with leading and trailing edge flaps, the basic wing area, S, is still used. A normal Flap C_L of 2, 2.5, can act bigger, maybe even 3, or more on Complex Jet Wings.

We'll cover the Induced, or "Lift Drag" Curve next. Remember it has that trickier, "highly variable" C_{Di} that increases as C_L^2, so that C_{Di} violently increases as you "slow down", overwhelms the V^2 effect by producing a $1/V^4$ C_{Di} effect, a **$1/V^2$ Induced drag curve** --- that **"swoops down" as you go faster!** ***Tricky!***

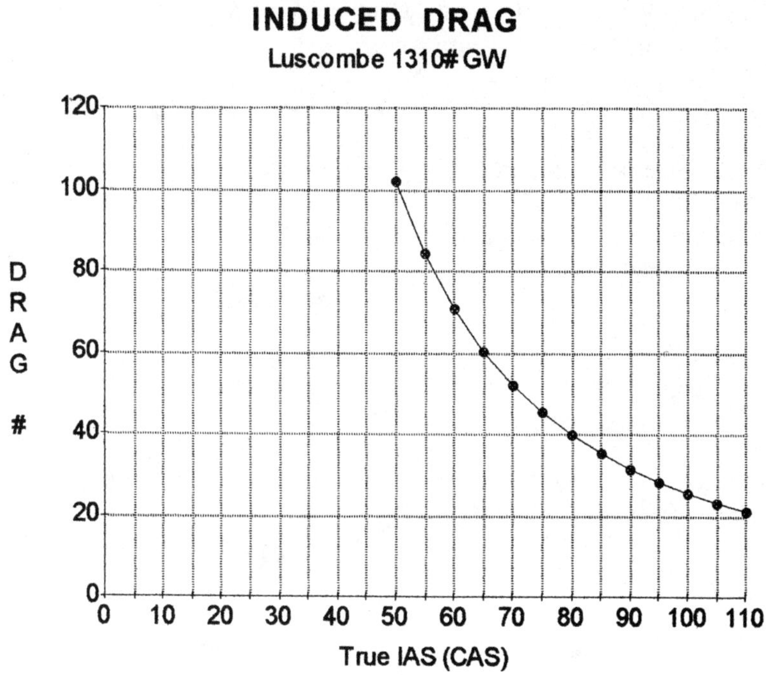

So **here's what a $1/V^2$ curve looks like**, as we saw at the end of the last Chapter. It would go to *infinity* at zero MPH, 1, or actually the number that makes the curve plot to the correct values, divided by zero IAS. **As speed picks up, it drops like a sinker**. To be practical, we just start D_i at stall speed, to keep it in a realistic range. Recognize though, it **never goes to zero!**

NOVICES: Do Not Try the next serious Algebra - for advanced guys! C_{Di} depends on $1/q^2$, $1/V^4$ -- thus D_i depends on $1/q$, $1/V^2$, but we'll learn, of great fundamental importance, **Induced Drag depends on SPAN LOADING²**, a simply marvelous insight to grasp!! It also includes the classic Prandtl/ Oswald formula $C_{Di} = C_L^2 / \pi A_R e$, that shows that the induced drag coefficient depends on C_L^2, the aspect ratio, and is "fudged up" to a correct value by the Oswald factor, e, "basic stuff" in Aero! That sounds terribly complex --- but it's just a curve that swoops down.

Dividing by, say .8 Oswald, adds 1.25 Induced - for Tip Vortex Loss!!!

DERIVATION OF THE BASIC DRAG FORMULA

Novices - Skip the Math - See the Bottom Line Answer in the Box!

PER THE BASIC LIFT FORMULA, IN LEVEL FLIGHT

Lift = Weight = Lift Coef. C_L x Wing Area S x q

$W = C_L \times S \times q$ Solve for C_L $C_L = \left(\frac{W}{Sq}\right)$

INDUCED DRAG COEFFICIENT, PER PRANDTL AND OSWALD

$$C_{Di} = \frac{(C_L)^2}{\pi e A_R}$$ (Below, we'll substitute formula for C_L above)

Where Aspect Ratio A_R is $\frac{\text{Span}}{\text{Chord}}$ or $\frac{\text{Span}^2}{\text{Area}}$ or $\frac{b^2}{S}$

Where e is the Oswald Span Efficiency Factor

$$C_{Di} = \left(\frac{W}{Sq}\right)^2 \frac{S}{\pi e b^2}$$ Substituting for C_L A_R Now rearrange

$$C_{DI} = \left(\frac{W}{b}\right)^2 \frac{1}{S\pi e q^2}$$ Notice $\frac{W}{b}$ is Span Loading

INDUCED DRAG, PER THE BASIC DRAG FORMULA

$D_i = (C_{Di})\ S\ q$ Now Substitute for C_{Di} from above

$$D_i = \left[\left(\frac{W}{b}\right)^2 \cdot \frac{1}{S\pi e q^2}\right] \cdot S \cdot q$$ Next, simplify

$$D_i = \left(\frac{W}{b}\right)^2 \cdot \frac{1}{\pi e q}$$ Recognize, $\frac{W}{b}$, is SPAN LOADING

$$D_i = \frac{(\text{SPAN} - \text{LOADING})^2}{\pi e q}$$

THUS, THE COMPOSITE DRAG CURVE FORMULA IS:

$$D = D_o + D_i$$
$$D = (\text{FPA} \times q) + \frac{\text{SPAN\_LOADING}^2}{\pi e q}$$

Where $q = \frac{\rho V^2}{2}$

A Little Insight on How Science Develops Smarts!

Ok now, this is a good time for *all* of us to **have a good laugh** ***together*!** To any normal non engineer, **that page over there looks ridiculous, funny!** Most regular folks would run for the trees, either laughing, or muttering about ridiculous engineers.

In fact, however, some individuals are quite Smart, Good at Math, love it, **can develop nifty New Scientific Insights** on how Nature, Science, Physics works. **This is a Good Example!**

Here we find Induced Loss, the **Energy Loss of Throwing Air** Down, to make Lift, **depends on Span Loading**, Less Loss if Span is Big. Loading Goes Down, we Don't Have To Throw the Air as Hard, to Make Enough Lift, and We Get Completely New Insight. **This is a Good Insight on Math Creativity.**

Would you believe that way back in **1918**, **Prandtl**, in Germany figured out that the Induced Drag Coefficient, was just C_L^2 **divided by π and the "aspect ratio"**, which is, span/average chord, or span²/area. It's an amazing fact in engineering, that there is always a smart guy somewhere, that has the basic theory worked out, while everyone else is still "hacking", in the dark. **Oswald** came along a decade later and **added his "fudge factor", (e), .8* ± , to account for Tip Vortex Loss**, to boost the **calculation of Induced Loss up to its actual value. Voila!**

Dividing by nominally .8 increases the induced loss by 1.25, for Tip Vortex Loss!

So we get a BIG SURPRISE INSIGHT. **SPAN LOADING²**, controls the Induced Loss! q is on the bottom, so as we've seen, the **Induced Loss** decreases fast, as you go faster, but **goes through the roof if you slow down, mushing!!!!**

*** On Long Span Wings, Less Tip Vortex Loss, Oswald's e is Bigger!**

That Math is a lot more complicated than I intended to put in this book, but it covers an important basic subject on how Induced Loss Works, can be understood. The book would have a hole without it, **so we just put it in for the Advanced Folks**.

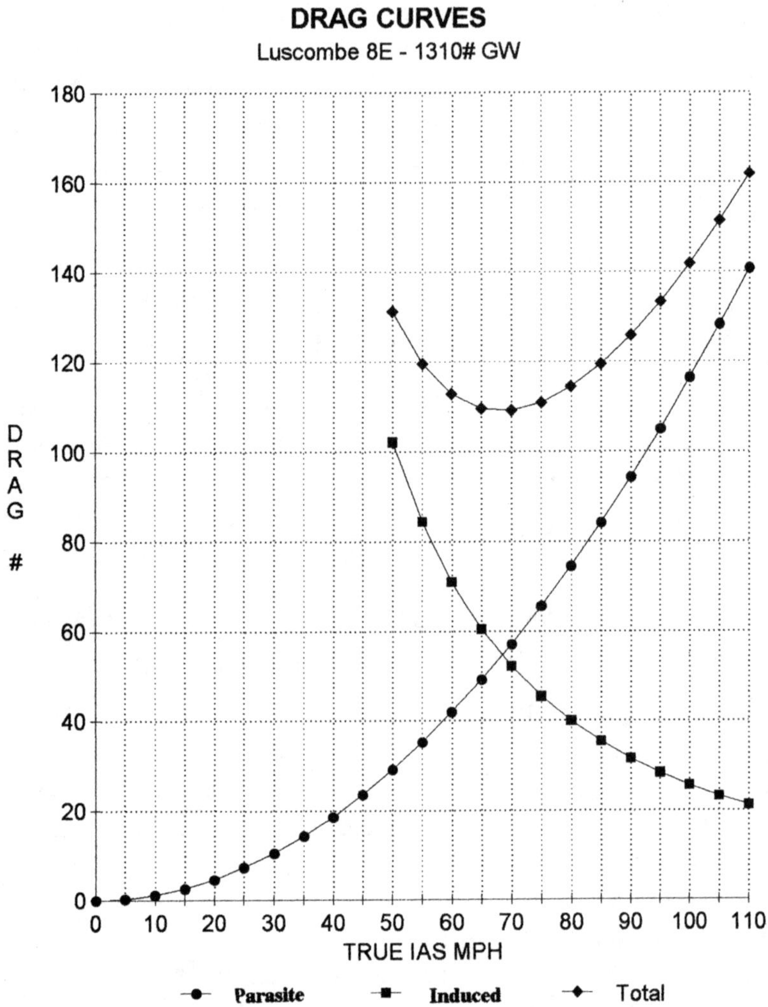

C_{Do} C_{Di} C_D

Now, if we just plug that Drag Coefficient into the Basic Drag Formula, (and that means we just multiply it by S q), we get a pretty simple conclusion and expression.--- **INDUCED DRAG equals SPAN LOADING^2 /π e q.** Not only does that give an airplane designer, *and* pilot, an easy way to understand "LIFT DRAG", there is the second great insight. The q^2 in the bottom of the drag coefficient formula, overwhelmed the q in the top of the drag formula, we end up with q on the bottom , **Induced, (Lift Drag) works exactly opposite to Parasite Drag, Induced swoops down vs. speed** (q on the bottom)**, while Parasite swoops up,** a $1/V^2$ curve and a simple V^2 curve!!!!

Now Look, of course that page can look scary, if you're new, but **hell**, **anybody can understand** that there are **two drag curves**, **one** that **swoops up**, and **one** that **swoops down**. **IF YOU GOT THAT, YOU GOT THE CONCEPT, THE GRASP YOU NEED , AND THE WHOLE REST OF THE GAME, THE BOOK, IS GOING TO BE A CINCH!!!**

SO HERE IS WHERE WE START GETTING REALLY SMART! LOOK AT HOW SIMPLE THE TWO PARTS OF THAT COMPOSITE DRAG CURVE ARE. PARASITE Is Simply FPA x q, it swoops up! INDUCED Is Simply SPAN LOADING² / πeq it swoops down vs. speed. $q = (\rho V^2/2)$ **(Learn as deep as you choose.)**

YOU CAN PLOT EACH OF THOSE SEPARATELY AND GET THE TWO, AS WE JUST DID, --- AND/OR --- YOU CAN ADD THEM TOGETHER AND GET THE STANDARD COMPOSITE LEANING "LAZY J" DRAG CURVE --- ON THE FACING PAGE.

NOW THE WHOLE GAME IS UNDERSTANDING where THE THINKING MAN FLIES ON THAT CURVE. Look.

Having (conquered??) the theory, the math, and converted it to PICTURE FORM that EVERYBODY CAN UNDERSTAND for sure, we'll jump to the next Chapter to grasp the three optimum places to fly, in picture form. We'll see there are SMART places to fly, where we **BUY SPEED FRUGALLY**, and places where you get **BIG DRAG**, a poor speed payoff!

CHAPTER 4, THE BASIC LIFT AND DRAG CURVES

Anyone can grasp the simple final logic of intelligently flying an airplane. Our objective in this book is to help any Pilot, anyone interested in THE LOGIC OF INTELLIGENT FLIGHT, to "actually grasp the central foundation of Aerodynamics", to see what the airplane senses, what it's logic is, and **why**!

Of course you'll recognize that we face the very real problems of helping the full spectrum of people: the guy who's never been good with math or science, who's afraid of this, even if he doesn't need to be, who needs lots of encouragement and help, to soon realize that with help, he CAN get it,-- the funnier case of the fiercely free spirit pilot who'll actively fight learning, unless we bowl him over with "insight", the high IQ critic who'll want to be critical that we adopted a "super emphasis" writing style that aims at making it impossible to finally fail!

These were the two toughest Chapters, where we dove right into the basic formulas, the math, the curves. Ahhh, the curves; now it's in "picture form" that will make it impossible to misunderstand, impossible to not get it. **My gamble, my bet, is that almost everyone will see that there is so much to be learned here, that they'll get themselves through it, which ever way works for them, get it, and we'll go from thousands of pilots who don't have a clue, to a bright well informed throng. Help me, help yourself win on that bet!!!**

Now that we have the drag curves in picture form, it will be fun and easy to see, fully understand, "the three optimum ways to fly", the concept of "MAX SPEED/$", the logic of maximizing speed vs. drag, fuel burn, cost; the way the thinking man does **not** go for speed on the drag curve, but gets it free at altitude, with the engine far more efficient. **Be amazed at your insight!**

CHAPTER 5

THREE OPTIMUM FLIGHT MODES

The objective of the whole game is MAXIMIZING SPEED, MINIMIZING DRAG, and this is the Chapter where we show you the first and most fundamental step, WHERE A THINKING MAN FLIES ON A DRAG CURVE!! Earlier in introducing you to the logic of flight, we showed you HOW to HOLD an optimum point for a given weight, by holding IAS at any altitude. But this is the crucial missing information, where that optimum point, optimum IAS is on a drag curve! We purposely are not taking you first to the most simple, constant optimum L/D, constant optimum angle of attack method, --- because, you really need to understand this most important fundamental first, then understand the effect of weight, on drag and speed, IAS required, (next chapter) and then later, headwind or tailwind, all of which are best seen on a drag curve!!!

Now, we will show you the THREE optimum flight conditions, and why the third, the Thinking Man's Way, that *maximizes speed vs. drag,* is clearly the way to go!!! We will use the composite drag curves, that we just developed, but we will also "borrow the power curve", that we develop a few chapters hence, -- along with a bonus "L/D Curve", so you can see how it fits in, and the "magic tangent line", That Makes The Logic Of Flight CLEAR!

The Minimum Drag Point -- The Voyager Flight

The **classic, optimum flight mode** taught in all the basic text's, is the **minimum drag, max L/D** point, (IAS). It nominally occurs at the IAS where the Parasite and Induced Drag Curves cross and are equal. LOOK. Quite logically this is the point where the least energy is consumed for any given flight range, the least fuel is required. **Range will be maximized**, along with **MPG**, miles per gallon of fuel. Look, it's at 68 MPH IAS.

Recognize it's equally logical that this minimum drag point is also the max L/D point, since in level flight, lift *exactly* equals the unchanging weight, that the drag curve was drawn for. You'll see L/D plotted on the *next* curve, so the one on the facing page won't be cluttered. L/D is like a *mirror image of drag*, but do notice the scale change, because it's a ratio ≅ 12:1.

Obviously the minimum drag point is one optimum, it's cheapest, MAX L/D, max range, but UGH, look at the speed, less than 70! You can go a long, long way, but you may die of boredom first. No THINKING MAN is going to fly there --- unless you're trying to **double** the distance record, **fly all the way around the world, unrefueled**, get every foot of range out of every drop of fuel. That's why the world flight took **nine, 9 DAYS**, 3 minutes, and 44 seconds. It flew 26,358.6 miles! I was the Technical Director, Mission Control on the **Voyager**. There are a few books left, all the facts, if you're interested..

Wildly, we predicted the Tail Wind to .007 MPH, and it was Science, genuine!!!

Jets start getting some compressibility wave drag as they approach Mach 1, which causes the drag curve to go even more vertical, faster, but the fact of HUGE SIGNIFICANCE is that other than that, and some favorable, ***drag dropping rotation*** that we'll show you, all drag curves have the **same characteristic shape.** So what you're going to learn in this Chapter, works for **any** airplane and is **the most important thing you'll ever learn about airplanes,** Cub, Classic, Bonanza, Homebuilt, 777!

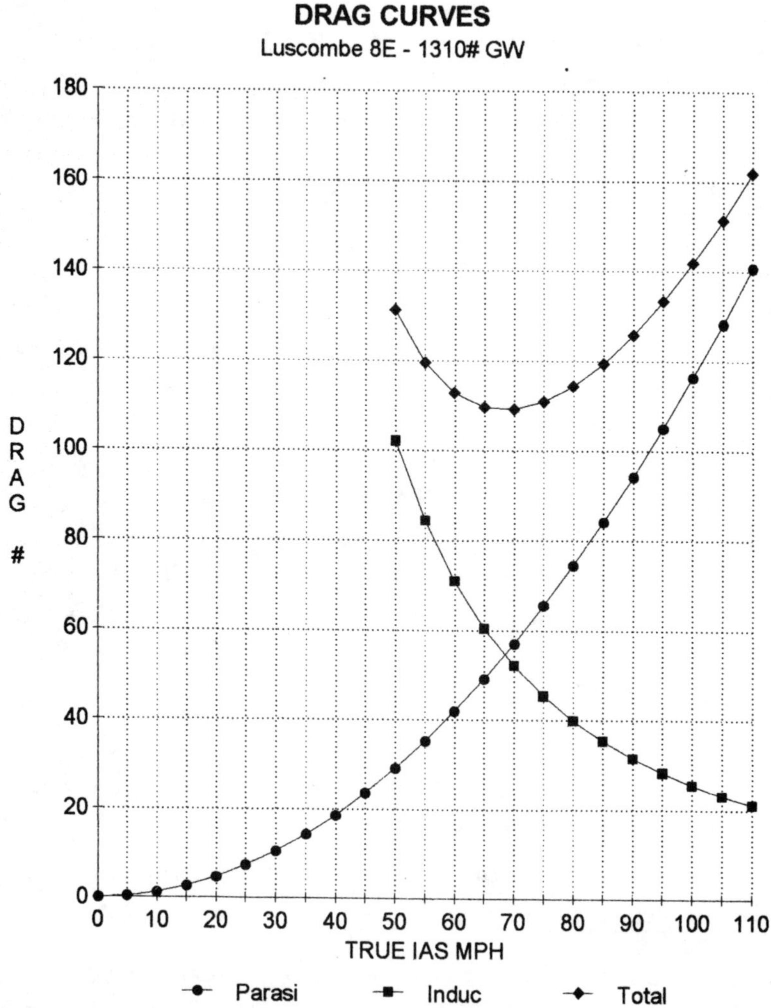

We're going to look at **Minimum Power, Max Endurance, the 2d Optimum** next, but first the thing to start noticing above is **how flat the bottom of the drag curve is**---but how quick it starts to go vertical! That's where the smart insight is going to come from! At first, **as you leave the min. drag IAS**, going faster, **extra speed is almost free, almost no extra drag!** **THERE'S AN OPTIMUM POINT** to fly, before "drag goes near vertical"!!!! First, let's dispatch that Min. Power Optimum.

Once you have a drag curve for an airplane, it's a straightforward job to construct a **"POWER REQUIRED Curve"**. Remember, Power is simply Drag, lbs. x true speed, TAS, ft./sec, therefore, ft. lbs./sec. It's **rate of use of energy!!!** You divide by 550 ft.. lb./sec (per HP) to get the answer in HP. That's the lower curve, on the facing page.

Recognize the Drag Curve (vs. IAS) is valid at any altitude, **but power depends on TAS, which increases with altitude, so that Power Curve is only valid at Sea Level where IAS equals TAS**. There's a "family of curves" as altitude increases!

As you'll see in the Chapter on Power shortly, you get the **"ACTUAL ENGINE POWER CURVE"** by running a low level "SPEED POWER TEST", correcting for any pressure, altitude, temperature, power loss. Comparing the two tests, actual/calculated, yields **real overall propulsion efficiency, η_p**. What you can find on a Classic, like the Luscombe, is markedly **decreasing** η_p as more engine power is applied, even a rough curve indicating poor flow, separation, which we've smoothed here a bit to keep the picture simple for you. LOOK, you can still clearly see the "actual HP heading skyward" as η_p degrades --- you'll see that's **another powerful argument for not trying to brute force a plane faster than it was designed to go!!!**

Ah, but the **central point we want you to learn here is the POWER KEEPS DECREASING, BELOW THE MAX L/D IAS!!!** The simple thing that's happening is that the speed is decreasing more than the Drag is increasing, so the product decreases. Theoretically, mathematically, the min. power speed is .7598 x the Min. Drag IAS. Actually, LOOK, the curve is so flat, you can hardly discern the low point, it's *pointless*!!! Also, the min. power is barely lower than at the min. drag IAS!

LUSCOMBE 8E CURVES

Drag Power L/D 1310#

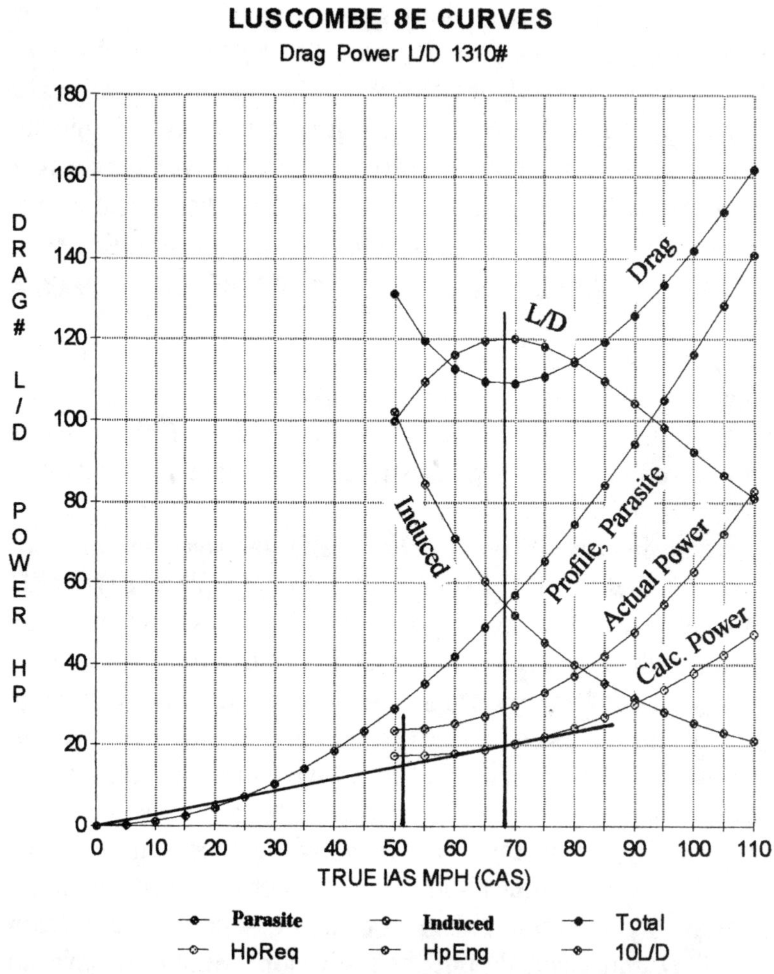

What practical use is this optimum? Well, if you're trying to win a (duration) model airplane contest, it's the "**max duration glide**". Ditto if everyone is scrambling to get life jackets on before you "ditch". If you want to go for a "duration record" (at a tad below 52 MPH in a Luscombe) maybe dangerously close to a stall, heavily throttled, you might consider it. The MAIN POINT TO RECOGNIZE is, **<u>slow</u>, at min. power, the most power is available for climbing, to hoist your GW skyward, so you get theoretical <u>MAX CLIMB ANGLE</u>**, for

The Fourth Root of 3 is $\sqrt[4]{3}$ = 1.31607, Inverted, <u>.7598</u>

clearing trees. The **max climb rate,** is at **Min Drag IAS, if** you have a controllable prop for max H.P. RPM. **Fixed pitch, faster gets more RPM, more HP**, so the **real max climb rate IAS, by test is really some MPH faster**, especially recognizing the curve is very flat, **BUT** if you're really going **for max angle** to clear the trees, going faster is **not** the right theoretical move. Soaring at "min. power", which is also "**minimum sink rate**", you'll climb the fastest, **the most**, "crossing" a lift area, "**slow**"!

Frankly, I think that's pretty incisive insight, where you hardly ever get good answers on subjects like that, so for the folks who are ready for that level of X-Ray vision, I think we're coming through on our promises to get you a quality education.

The really important point, the really incisive lesson here however, is that there is **a BIG FAT SAFETY ITEM HERE!!** Two local guys were recently killed, doing a max performance climb out, when they really didn't have to, **nose way up, weight partially supported by thrust --- the engine died, instant stall, it's more than a matter of getting the nose down.** They just fell like a brick, died in an ugly fire. Terrible, worse because it was pointless. The airplane always does *exactly and unemotionally*, what the Physics book decrees. It was not guilty. Nice guys, good pilots, I'm told, not technical, their sin was probably no bigger than not having a clue of the death corner they unnecessarily put themselves in. Simply, I have no critic in me for such tragedies. I'd rather take the time to be **sure you do understand.** If you "gotta flat hat", don't do it in a "coffin corner"! **THAT INSTANT STALL IS REAL, specifically because in a powerful plane, nose way up, that engine is supporting weight, and if it dies, you will, if low!!!**

Compliments of Zero Thrust Glide Testing, and the EAA sponsored CAFE Organization test programs we did, that we'll tell you about later, we have some really incisive, never before

THERE'S OLD ONES
AND BOLD ONES
THIS WILL IMPRESS
THE HELL OUT OF EM!
CLANG
BANG
WHY WOULD
A GUY DO THAT TO
HIMSELF?
I DON'T WANT
TO WATCH

available, *specific numbers.* The RV series of homebuilts, designed by Dick Van Grunsven, are marvelous airplanes, one of a new leading edge of flight actually. They "go fast", "land slow, thus safe", "are strong as a bridge", aerobatic, agile, light to the touch, "a joy to fly". Clean, efficient, they step out to higher speeds readily, 170 to over 200, at impressive MPG, flown intelligently. At just over gross, an RV6 has a min. drag, max L/D speed of 106 MPH. That calculates to 80.5 MPH, min power, max climb angle. That's a fat margin above the nominal 50 MPH ballpark stall speed, with flaps, depending on GW.

The "lead sled" max performance, planes that do not have that "speed ratio of four", will not survive a slow, nose up, engine failure. How do I know that? Because the guys that were killed had it happen to them in an RV4, with all that generous margin, great aileron control and a gentle normal stall, because of wide wingtips, that results in low required C_L tips, that "hang on well". A designer just can't do more for you than that.

The pertinent hidden detail to know here, is that the RV also has an exceptional airfoil, the 23013.5, the same class as the DC-3, the Bonanza, many other famous and excellent planes. That airfoil "hangs on" so well, that when it lets go, it's all done and stalls totally. So what you can read into that accident, is that in all probability, the nose was way up, they could have been even slower than "min. power speed", they didn't get the nose down, possibly going for the fuel valve, which some people thought they bumped, and the "Physics book got them"

I had no intention of covering that subject here, but as I started to think it through, to nail it completely for you, it became clear just how dangerous that "slow, nose up" case can be, even in a plane that gives you more "margin" than most designs would. If you're going to fly high performance planes, you better understand where the coffin corners are, be properly careful.

The Rocketdyne P.A. on my Gemini Controls, Dom Novelli who became a valued friend, after I saw him solve some serious internal mistakes, like a real pro, was killed years ago, when that American DC 10 crashed at Chicago. An engine blew, took out the left wing leading edge flaps hydraulics, but not the right. The pilot was "good", but couldn't see the wing condition, knew the book, "slowed" to best 2 engine climb speed, *and lost it,* stalled, no longer had enough q and control authority to maintain the asymmetrical flaps, he had been handling!. There's a terrible photo of a **low** DC 10, **wings vertical**. My conclusion, I'll just forget, the max climb case, just not get myself into cases where I need max climb angle!! The ***only*** time to be low and slow, high α^o, is flaring for touchdown.

Now, before we go after the "third optimum" MAX SPEED/$, the real objective of this whole book, let's go back and finish completely understanding the relationship of the "Power Curve" and the "min. drag, Max L/D" case. Notice that we've drawn a line from the 0,0 origin, "tangent" to the airframe power required curve. Notice that it's tangent, (just touches), right at the min. drag, max L/D IAS! That's no accident. In this mathematical, natural law lash up, everything relates to everything else in such an orderly way that it's quite extraordinary really, and it's why we can show you a LOGIC OF FLIGHT that is so orderly, simple and exact. Science is so vastly more extensive, so all encompassing, that if you grasp its wonderous extent, you finally decide that the first Chief Engineer was so vastly smarter than the best of us, that we're a real pack of neophyte hackers! Science's logic is everywhere!

The main thing we want you to notice is that **everything lines up at the same IAS: min. drag, max L/D, tangent to the power curve, the two drag curves cross, equal, for this classic optimum case. The case has everything going for it, except it's just too slow to be used, except for max economy!**

Mathematically, the Power Curve has a tangent at min. drag IAS because: 1. "the rate of change of the drag curve is max at its bottom minimum", 2. the power curve was calculated from the drag curve, so it just "kisses" the tangent -- The reason for that will seem obscure, -- but recognize the tangent line can be thought of as a slope line *that relates power to IAS,* -- now notice that *every point on the power curve lies above the tangent line, except the tangent point, the min. drag point*-- because the "upsloping" drag curve calculated the power curve!

Now that's far more subtle and obscure than we'd normally want to deal with *except*: 1. that nicely and totally puts both the "min. drag" and also the "min. power" cases to bed and behind us and, 2. **the most important point in the whole book is just about to show up with a more simple version of a tangent line!**

LOOK, on the facing page, we've brought back our simple three basic drag curves and added a tangent line from the origin. **THE TANGENT POINT ON THE COMPOSITE DRAG CURVE IS THE POINT WITH THE MOST SPEED IN PROPORTION TO DRAG ---- ON EVERY AIRPLANE!!!**

This point is simple, not subtle and obscure. **NOTICE HOW EVERY OTHER POINT ON THE DRAG CURVE IS ABOVE THAT TANGENT LINE! THAT MEANS EVERY OTHER POINT HAS MORE DRAG IN PROPORTION TO SPEED -- THE TANGENT POINT HAS THE MOST SPEED IN PROPORTION TO DRAG!!** Did you get that for sure? Again, all points on the tangent line have the same proportion of drag vs. IAS. All other points on the drag curve are above it (extra drag). So the one point of tangency is the **optimum speed vs. drag, traveling IAS**, to fly **on every reciprocating plane in the world**, (and many Jets.)

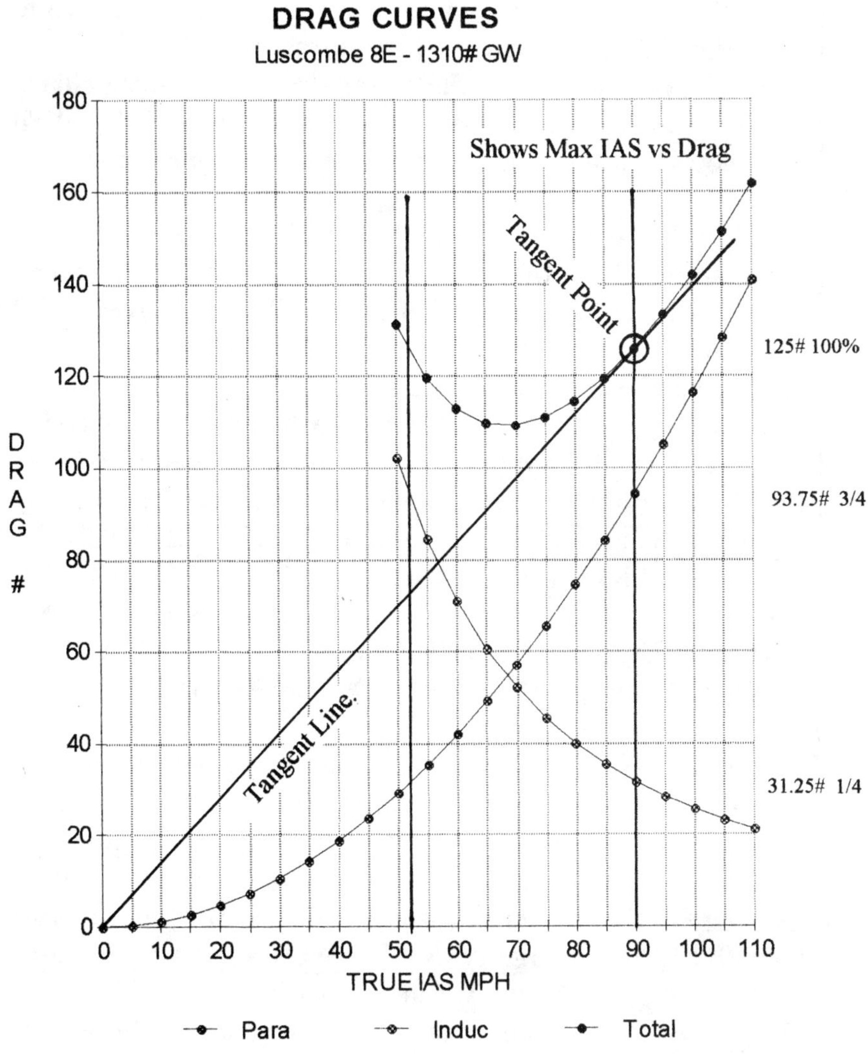

Now <u>if Drag, Fuel Used, and $ Cost are all equivalent</u>, you can see that tangent point is <u>MAX SPEED PER $</u> --- except that we're going to <u>go much faster by climbing high, (TAS)</u>, then we're going to maximize engine efficiency, minimize fuel burn, <u>spend much less</u> on fuel, <u>high</u>, by getting the <u>engine wide open and leaned</u>. <u>SPEED / $ IS MAX HIGH</u>!

Notice Drag is 3/4 Profile, 1/4 Induced, at Max. IAS / $ -- or Visa Versa at Min Power IAS,

Now do you see how wildly easy it is to be a really sophisticated pilot. There's just no reason to be afraid, it's so easy! You just go for that smart IAS on the drag curve, climb up until you're "wide open", and leaned!!!! We'll show you Jets just do their own version of the same game!

That just has be the biggest and best hidden joke in all of flight. There are hundreds of thousands of pilots, and almost no one knows that there is such a SIMPLE LOGIC TO FLIGHT, TO HOW AIRPLANES "WANT" TO BE FLOWN. Many would actively fight learning it, scarred stiff that it's some big technical drill that they could not possibly cope with, when it's something you can explain to anyone, once you grasp it. **Remember the PICTURE, the drag curve, where you want to be on it, fly at that IAS, go high, wide open, and lean. RECOGNIZE WE'RE MAKING IT MUCH MORE COMPLEX, BY USING IT AS A TEACHING MECHANISM, TAKING YOU THROUGH ALL THE STEPS, TEACHING YOU A GOOD BIT OF BASIC AERO ENGINEERING!!**

That term MAX SPEED per $ tends to jar people at first, they're not quite sure what it means. That's almost justification for the term, because it forces people to stop, and question, and think. You can see now, that we ***maximize speed vs. drag, fuel, cost***. That boiled down is where you **buy speed most frugally!!! Now recognize that it is NOT some wimpy, slow, time wasting, too frugal speed ---- because the ENGINE IS WIDE OPEN, the plane is at V MAX, only where it is ALSO at its MAX POSSIBLE EFFICIENCY!!!** The plane can go faster at sea level, but down there it's like flying through JELLO! **The atmosphere is only 2/3 as dense at just over 13,000 ft. --- and that means it's 1 1/2 times as dense, down there vs. 13,000!!! Recognize to a Jet at 40k, where the density ratio is .2462, the air down there is FOUR HUNDRED PERCENT THICKER!!** Ref. your ICAO Table.

The ANSWER is on the BOOK COVER

It's this EASY!!!

Fly 1.316 Times the Handbook *Min. Drag, Max. Climb IAS*! ✓
Just Climb up until the plane quits climbing, finally, fully leaned! ✓
(Now, look at the ICAO Atmosphere Table p. 38, 152, marked up for you!)
Your TAS will be up 16% at 10,000 ft., above IAS Speed!
20% at 12,000 ft.
24% at 14,000 ft.
32% at 18,000 ft.
Jets Twice as Fast - 200% at 40,000 ft.

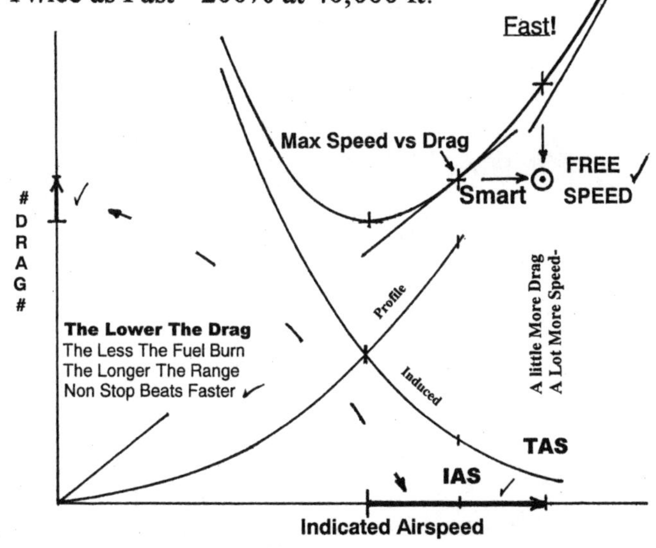

The IAS Drag Curve -- vs. Altitude TAS Speed
The Basis of Intelligent Flight!

It's all in Picture Form on the Book Cover!
The Max IAS vs. Drag is 1.316 times Min Drag, Max. Climb V!
(The drag is up only 15%, 1.15, a bargain over the absolute minimum!)
The TAS / IAS Speed is up a lot, as above, and that's FREE!
1.2 x 1.316 at 12,000, is 1.58, 158% for 15% - a Great Bargain!
1.316 is $\sqrt[4]{3}$, the fourth root of 3, $\sqrt{3} = 1.732$, $\sqrt{1.732} = 1.316$ ref p.68.

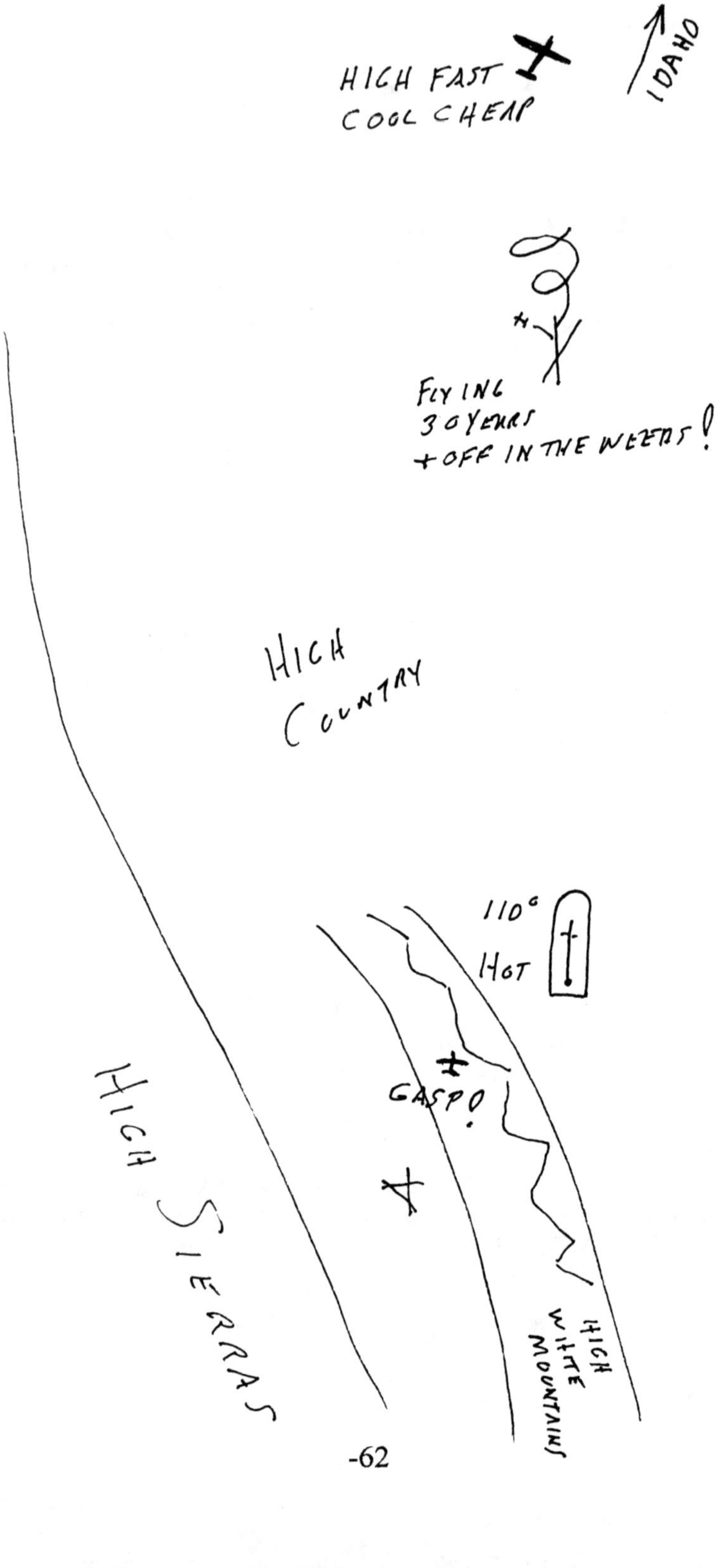
HIGH FAST
COOL CHEAP
IDAHO
FLYING
30 YEARS
+ OFF IN THE WEEDS!
HIGH
COUNTRY
110°
HOT
GASP!
HIGH SIERRAS
HIGH
WHITE
MOUNTAINS

Do you suppose you'd get those bargain tickets if the Jets had to fly through that? Now remember you're burning something like **4 to 6** times more fuel, than the basic energy requirement, so you better be able to "do something at least half way smart"!!

Now **at this point you have the book licked**. The hard part with the basic theory is behind you. You just got the most basic and important part of the whole book, **where** you really want to fly on your drag curve. There is a huge amount of valuable insight, yet to come, that we can pass on, but it's all going to be easier and interesting, not subjects you'll have to sweat over.

Now in conformance with our policy of always keeping you clear on where we are, -- and where we're going:

1. We just learned **where** to fly "an optimum" on a drag curve.
2. Earlier, we learned **how** to **hold IAS**, **to hold an optimum.**
3. Next, we'll learn the **effect of weight, on drag, and IAS.**
4. Then we'll learn **constant L/D, optimum α^0 flying.**
5. Then we'll be ready to get into Power, the Engine, Leaning, the Propeller, the rest of the book. There is a lot of good insight that can be passed on, but it won't be a sweat.

Let's take a paragraph detour, and I'll give you a snapshot insight into what understanding this fundamental logic can do for you: In June 1994, a gang of us took off, "for a fun break", for "a Classic fly in" in the mountains of Idaho, 804 miles from Camarillo, California, most of the United States, S. to N. I left two hours after the other guys, got there an hour before them, went non stop, burned 70% of their fuel, had **half** their cost using auto gas, listened to them struggle over high ridges out of Bishop in *110F° sea level air*, avoided the wild dust devils at God forsaken Winnamucca, that ran one of them off the runway into the weeds.. They wondered how I could do that. **It was easy, cool, comfortable, fun.** I had a No Numb Buns Seat!!! They landed TWICE, lost time, \$, sweat a lot; had *worries*!!!

Now, before we go on to wind up this Chapter with some valuable additional insights let's answer some obvious questions.

IF UNDERSTANDING THE DRAG CURVE IS THE GAME, CAN I GET ONE FOR MY PLANE? --- At this writing using Zero Thrust Glide Testing, we have the breakthrough complete Aerodynamic data on my Luscombe 8E*, published in the July-August 1993 AIAA "Journal of Aircraft", including the real drag curves, the calculated theoretical, and actual power curves, and **thus the real propulsive efficiency at various Indicated Air Speeds**. We also have **CAFE ZTGT** tests on an **RV6*** and a **Whitman Tailwind**, and base line testing of a **Cessna 152**. As time permits, we hope to do power testing on an RV 6, so we can see its Theoretical Power based on the Drag, and the Actual Power, and thus the Actual Propulsive Efficiency Curve, which we know will be much better than the low Luscombe η_P. Unfortunately, the new Lycoming engines on the additional planes that we wanted to test consistently had inadequate end play in their main bearings to permit easy, good ZTGT Tests.

***EAA Sport Aviation, Mar., Apr., 95, Lusc, RV 6. Sep. 93 CAFE RV 6.**

The simple enough answer for you is that the **Max. L/D IAS** for your plane is usually published, and available, or reported as the **Best Climb Speed**. That IAS times the Fourth root of 3, $\sqrt[4]{3}$, **1.316, is the Max.IAS vs. Drag Speed**, and as you've learned, its reciprocal, **.7598** is the Min. Power, Max Angle of Climb IAS Speed, so there is a way for you to find your needed IAS Speeds.

ISN'T IT KIND OF RIDICULOUS TO USE A LUSCOMBE TO EXPLAIN FLIGHT IN THE SPACE AGE, TO JET PILOTS? --- No, it's entirely appropriate for several reasons. **First**. It's the classic basic that every pilot can understand, from the student, to the 30,000 hour Captain who first learned in a Cub, T-craft, Cessna, or "the first modern all metal production plane, the Luscombe", "*no sticks, no nails, no glue*"! The guys who figured out how to do that in 1938, in the Depression, when not everyone was eating, deserve to be remembered!

Second. Somewhere in the explanation, everyone will realize the fact that I really mean it, **all** planes really do work the same, only the speeds change, so the simple basic Classic becomes an ideal teaching tool and a basis for comparison, which leads to: **Three**. On p.182, the Appendix, we superimposed a Luscombe and an RV6 drag curve. It's clear that the RV 6 "J curve **rotates**", much lower Parasite Drag vs IAS! It's a much lower slope vs IAS. A modern design can "squirt" out to a much higher speed at no increase in Drag!!! The "short wing" RV6 has an induced drag curve **twice** as high, but a parasite curve that is **only half** the Luscombe's, so you cruise at 170 TAS MPH, *smart, fast,* not 100, at the same or better MPG, with low drag, better propulsive efficiency. Amazingly, you have almost the same low induced drag, despite the 2x curve, because you're going FAST, where induced drops like a sinker, the reason you see a 20,000#+ F104 with a, maybe, 8 foot, 10 inch half wing stub!!!

At 1600# G.W., more Induced, RV's can Fly at 140 IAS, 170 TAS, 12,500', 27MPG!

Space Age. At the crucial all up Apollo 5 unmanned test launch, Astronaut Gordon Cooper was enthusiastically telling me all about his wife Trudy learning to fly in a Luscombe in Hawaii. Over 2000 are still flying, plenty with senior Captains having **carefree fun**. A break from designing, manufacturing spacecraft controls, twice, coming back from skiing at Mammoth, I had mine in the Sierra wave at 18,500, at idle, Mt. Whitney and the whole Sierras on a fast down elevator. (To do it safely, you go in and out under the *forward* ends of the cap cloud.) It's **spectacular**, no kid stuff, all the fun a guy needs!

OK, back to work. The great thing about the orderliness of the LOGIC OF FLIGHT is that while it is all very precise and mathematically definable, you can see it just as well, maybe even better, with just plain old **horse sense**, really understanding the drag curve, **the picture.** You'll also see soon, and in Chapter 6 that there is a lot of room for **judgment**, using grasp to suit the occasion. My job is to be sure you **see the horse sense of it all!**

Way back on page 51 I showed you that **the bottom of the drag curve is very flat**. It would be **actually dumb, to fly at the classic min. drag IAS, because it costs almost no drag to go significantly faster!** Pretty soon, however, the curve is heading up fast and you're buying speed at a greater and greater drag penalty!! The math optimum is at the **90 MPH IAS** tangent point. There is a *trap there however, and I want to be very sure you see it, understand it!!!* It LOOKS like the drag curve is very parallel to and close to the tangent line over a broad speed range, nearly the same ratio of drag to speed --- and that's **true, BUT that ratio is not the real bargain. The BARGAIN IS TO GET THE ALMOST FREE IAS BETWEEN 68 AND 80, GO 85 --- THEN GET TO 100+ TAS REALLY FREE, BY CLIMBING HIGH!!! LOOK, THE GAME IS TO GET THE BARGAIN IAS, *STAY LOW ON THE DRAG CURVE***
If you go up on the drag curve, your MPG goes DOWN -- *grasp that.*

Up your drag from 113# to 140#, at 100 IAS, drag is **up 23.8%** and your **MPG is DOWN to 80.7%. Much smarter to grab the bargain and get your speed FREE by climbing HIGH.** In the next Chapter that varies weight, I'll show you how I fly 85 IAS at all gross weights, go normal Luscombe speed of 100+ high, and get super range and economy. **DO YOU SEE THE JUDGMENT, FLEXIBILITY, HORSE SENSE --- ROOM FOR REALLY UNDERSTANDING WHAT YOU'RE DOING HERE. You can tailor what you do to fit the flight you *want* to do, *BUT* IT'S HOW HIGH YOU FLY ON THE IAS DRAG CURVE THAT CONTROLS REAL MPG!! I simply fly 85 IAS. at 113# Drag Light, for less drag, better MPG..** I burn only 3 3/4 GPH at 100+ MPH, 800 mile range!

In the next Chapter, 6, you'll see clearly that *weight costs you drag!!!* There you can see that *with judgment, really understanding* all the overall facts, you can be even more sophisticated than the mathematical optimum. You'll see that by flying a thoughtfully selected CONSTANT IAS

rather than Max IAS/Drag. **I fly a lower part of the drag curve especially "at high gross weight, heavy with fuel", stay at a lower IAS, for better range --- that allows me to go farther non stop, --- beat the socks off the guys who are trying to go fast, but are always down there on the hot desert floor buying $$ gas, tortoise vs. hare!** (This is all an early look at the advanced meat of Ch. 6). Realize, you won't be doing it at a wimpy speed --- **You'll be *wide open*, flying where the plane and the engine are at, or right up tight against their max efficiency, depending on your insightful choice!!!** ---- Are we having fun , are we getting smarter, painlessly???

Don't get misled by those slow sounding Luscombe speeds. Remember we're teaching principles here with the basic plane. We have the numbers on the Luscombe and the RV family. An RV6 at 140 IAS, is actually going 170, or more at 12,500'. Rather than burning 8 or 9+ GPH you can burn 6.1 GPH. That will give you over 1000 mile range on 37.5 gallons, LA to Seattle non stop, in 6 hours. With oxygen, a high H.P. RV does 185 SMART. The planes designed for speed alone do it proportionally faster. Jets do their own twice faster TAS/IAS.

Of course on days where you want more Speed, you just up the IAS to what you choose.

We're teaching SMART here, not slow, but slow on the drag curve, high, is how it's done. **We're teaching the Thinking Man's way to Fly, the Simple Logic Of Intelligent, Efficient Flight. More than that, we're teaching you how to really understand what you've done and loved all your life, or will do, if you're one of the young ones getting smart early.**

OK, hopefully we've shown that this is **not** about being a "Techie", but rather the joy and advantage of really understanding what you're doing, actually deserving to be seen as a real Thinking Man, with real *insight, able to make incisive judgments* on how you fly "this one". But let's switch back and see more of the incisive insight that an inside look can show us.

$\sqrt[4]{3}$, a good laugh. I'll bet in your wildest dream you never thought you'd be interested in, "the fourth root of three, 1.316". I'm sitting here laughing at this dumb computer, thinking of my great 33,000 hour QB friends, who have been known to hoist a few, saying "what the hell was that you just said there Jack" I'm particularly thinking of Starr Thompson, who's going to take a crack at drawing better cartoons for me, to help make it clear that we can have some fun here getting smarter. He has this wonderful, perfectly facile face, for a natural cartoonist, that would "instantly screw up to the perfect quizzical expression", upon hearing that one! He'd say something like "what could that possibly have to do with anything", and his face would match!

Pick up your home calculator, that probably does square roots, and take the square root of 3. That's 1.732, a number that shows up all over science. Then take the square root of that, and you get **1.316**, a number that shows up absolutely nowhere, except that a Math drill can prove that it and **its reciprocal .7598**, .76, relates the **MIN. DRAG**, max. L/D speed to **MAX IAS/$** speed, and the **MIN. POWER** speed, helpful, key insight. √

With Starr's talented help, we've made a joke of **"the funny fourth root of three"**, but of course we're sneaky, and I'll bet you remember that silly number, *or* **how to get it instantly** on almost any calculator. You see, if I tell you that the max L/D speed on an RV6 is 106 MPH at 1650# GW, in a flash, **you can instantly know the SMART IAS, 139.5, and the min. power IAS is 80.5!!!** As soon as you tumble to the fact, that, **that little trick also works on a 787, or any plane in the world, I'LL BET YOU DO REMEMBER THAT FUNNY NUMBER, "1.316", OR, HOW TO GET IT quick!!!** !

3/4 Profile, 1/4 Induced, at Max IAS/$, or Visa Versa at Min Power IAS, because--- $\sqrt[4]{3}$ multiplies V, but V is a V^2 function, so effect is $\sqrt{3}$, 1.732., and 1/1.732 = .57735, 1.732 exactly 3 times bigger,so induced is 1/4 of loss, profile 3/4 at Max IAS/# Drag, or Visa Versa at Min Power!

There are more orderly things about these three optimums, and their logic, that everyone can see and appreciate. At the **Min. Power IAS**, the **Induced Drag is 3/4 of the Total Drag, and the Parasite is 1/4**. **At the Max. Speed/Drag, or $,IAS, that exactly reverses, Parasite Drag is 3/4 of Total Drag, Induced has fallen to 1/4.** Of course, at min. drag they're *equal*

Now, that looks nice and neat and simple, but there are always additional subtle insights that can be seen once you walk around them a bit. On the modern, **low Parasite**, low flat plate area designs, that relegates the Induced Drag to 1/4 of a **smaller** number as well!!! As explained above that makes Aspect ratio much less important. Physically you scoot out to higher speeds, end up with **"low lift drag", even with a short wing**, high span loading, like the RV6 and the fighters. The old, graceful, long wing, high Aspect Ratio Luscombes, cruised well on little power. Surprise, the **low Parasite homebuilts** scoot out, go fast, get **great MPG *fast***, use the Physics book much smarter!

Of course there are always little subtleties that mess up the simple, precise, order of things, all pro's know that. The more important thing, however, is to not let the minutiae obscure the central facts and logic. Pilots never had a chance to see easy conclusions here, because the important central story never got out to them in an intelligible way, and that's bad and wrong.

So let's illuminate, but then ignore some of the minutiae, as we go. A good part of parasite drag is the wing profile drag, and a part of that is α sensitive. That can be used, ignored, or lumped with induced. We were able to get near perfect air in our Zero Thrust Glide Tests, at dawn miles out to sea, with our engine available for safety, so we were able to separate it out, have a 3 part C_d, below. It acts like extra induced and that steepens the apparent Induced curve -- which 1: causes a slightly faster min. drag speed, 2: causes the Parasite and Induced curve to not cross exactly under the min. drag IAS, with the α sensitive profile plotted separately. Our simplified curves are generated using the proper flat plate area, and the proper span loading. We used a poorer Oswald factor, .7, not a correct .74, kicked up the slow end of the drag curve, simulated the α sensitive profile.

Parasite + α sensitive Profile + Induced

$$C_d = .0325 + .009444(C_L - .4)^2 + C_L^2 / (\pi A_R e) \quad e = .74$$

Also, modern laminar flow airfoils used on leading edge homebuilts can have a drag bucket, lower drag in a "sweet spot" lift coefficient range, but that can readily be seen and accounted for if you have an actual drag curve. The key, really, is to ignore the minutia, learn the basic big picture, and then account for the refined cases, which is easily done once you understand. Recognize with laminar Drag Buckets, the designer needs to be sure to pick an airfoil with a drag bucket, where you really want to fly!!

Summing up MAX IAS/ DRAG vs. Min. Drag, Max L/D

- **We go 31.6% faster, that's .7598 time, save 24% on time.**
- **Drag, by calculation is up to 1.155, +15.5%** , costs MPG.
- **MPG,** (the reciprocal of drag), **drops to .866, loses 13.4%**
- **Power required, Drag x Speed goes up to 1.52, + 52%**

Demanding more Power is Good, Gets the Engine Wide Open at a Lower, More Efficient Altitude! ✓

CHAPTER 5, OPTIMUM FLIGHT MODES

The meat of this Chapter is the most important insight that you will ever learn about airplanes, Classics through Jets!

The Min. Power IAS, is max endurance, max climb angle,
Min Sink, Longest Duration Gliding - but dangerous, if near Stall!!!
The Min. Drag, Max L/D, classic optimum flight mode, is max range, best economy, best MPG, but really too slow. (max climb rate, if a controllable prop), Fixed Pitch, More IAS is Best!

The true optimum flight mode is the IAS (vs. GW) that maximizes IAS vs. DRAG, Fuel or $Cost, especially when it is flown at high altitude, for FREE TAS, with a properly leaned, wide open engine, maximizing engine efficiency, for this Plane, the TRUE MAX SPEED/DRAG, Fuel or $ Cost!

The fourth root of 3, (1.316, and its reciprocal .7598), relate all of the speed and time ratios of the 3 optimum modes!

Thus, flying max IAS/Drag flies you 1.316 times min. drag IAS, takes .7598 as long, saves you 24% on time. A drag calculation shows that drag is increased 1.155, which reduces MPG to .866, the reciprocal, costs 13.4% more fuel. Since Power is Drag x Speed, required Power goes up 52%.

That last point proves to be the clinching advantage, because, as you will learn in Chapter 9, your engine must be "too big for optimum efficiency", since it must be sized for a good climb rate at sea level, *which forces throttling, thus lowered efficiency, by turning the engine into an air pump*! THE WHOLE GAME BECOMES FLYING THE PLANE AT THE IAS AND/OR ANGLE OF ATTACK WHERE the IAS / DRAG IS MAXIMIZED, WITH THE ENGINE WIDE OPEN, LEANED, AT *MIN*. ALTITUDE!! Recognize I said *MIN*. Altitude, because you want your wide open altitude to come out as low as possible, because the higher it is, the less your max potential efficiency. (Ch.10)
BUT — We'll Learn Later, in CH 10, the TAS/IAS gain beats the Effciency Loss!

"JOE GROSS"
WOULD YOU SIT ON THE CEMENT? THE KIDS WILL FIT IN THE "BARROW."
GOOD GRIEF
SHOULD HAVE JOINED THE TEAMSTERS

CHAPTER 6

WEIGHT EFFECT ON DRAG, SPEED

We tend to think that weight costs us climb performance, without thinking further to its effect on drag, speed required, and fuel consumption. However, now that we're pursuing an orderly, logical approach to our flying we'll quickly see that the effect is substantial, but happily, easy to see and understand, because it's quite proportional!

It's easy enough to see and understand, that this Chapter can be short, (so we'll practice a few calcs). To make it easy, however, we'll think ahead to our next Chapter where we'll address constant L/D, constant angle of attack flight. We'll still focus on our "ideal IAS method", to see weight's effect on the ideal IAS, and Drag, and the Drag Curves. You'll find that you'll quickly "tumble" to what's really going on, start to grasp the whole big picture, if you haven't already.

Right from the very start, we made it clear that, although we'd pursue an Ideal IAS method, and go through all the steps to really teach you the "WHOLE LOGIC", we'd really be working on an "integrated concept", where ideal, constant L/D, and angle of attack, also worked, with both methods yielding the same optimum final result. Well, right here is where it all comes together, in an amazingly easy concept, you probably didn't see!

Amazingly, it's this simple: If you're really heading toward flying **constant, optimum, L/D**, if you **double your Weight**, thus your Lift, you **have to double your Drag.** ***SIMPLE!!!*** There's a good chance that you never thought of that before, even though it's dead simple. People seem to not tumble to that. So, if you did see it before, give yourself "special good marks".

The corollary is just as simple and true, and not obvious also. Double your weight and **drag** and you **double your fuel consumption, cut your MPG in half!!!** WOW --- but wait, that seems severe, what's really going on here? This is a point where your insight can go way up!!! The hooker is, if you're really trying to fly constant L/D, constant angle of attack, **you must speed up**, to "hold" that same angle of attack, not "mush out" to a steeper angle, under load. You don't have to be an Aerodynamicist to grasp that, your Pilots horse sense will tell you that's true. Sure, I realize that still probably seems "a little severe, unfair", but you see, if you don't do it that way you're not comparing "apples to apples", it's the only way that is fair!!!

We'll see real clearly, shortly, what's really happening, by plotting a **family of curves** for **three gross weights**. We can even accurately **see** what **relief** we get if we **purposely do not speed up**. We now have a wonderful capability to get **specific answers**, either by number calculation, or **just looking** at our drag picture, the Drag Curves, with our now very orderly way of flying, and **looking at specifics**. (You're becoming a pro!)

Before we plot curves however, we should first understand **what increasing weight does to** our **required speed, IAS**. We're going to quickly find that IAS is also proportional, but in this case, **proportional to the square root of the gross weight ratio**! That sounds like a mouthful, is apt to strike fear in the heart of the non math folks, but don't panic we'll lay it out, **so you can see "as much as you wish"**. **It's valuable**, because making "Drag and IAS corrections for gross weight" is basic in flight testing, isn't hard at all with "calculators", once you see it. If you're not a "calculator person", have courage, "dive in" here, we'll help you with examples, *You'll see it's quite easy! We'll risk making it look hard, with a derivation of the method, then calculations, to help you, learn what you wish!*

Now, what we're going to do here is just write our basic lift formula twice, for two gross weights, drop out everything that doesn't change, make a simple proportional statement that we will take the square root of, to find a new "heavy speed", V_2 ratio.

$$\text{Lift} = \text{Weight} = 1400\# = C_L \times S \times (\rho\, V_2^2/2)$$

$$\text{Lift} = \text{Weight} = 1310\# = C_L \times S \times (\rho\, V_1^2/2)$$

Recognize we want to see how much our V, our IAS changes, if we change our GW from 1310#, half fuel, up to MAX GW, 1400#, full fuel. We don't want anything else to change, not altitude, ρ, certainly not S, wing area, certainly not C_L since we want the L/D to stay the same. Now, since those things don't change, trust me, if you're new, it's mathematically legitimate to just drop them out, and create a "proportional statement" as follows, by just treating what remains of the formulas, as a big "ratio" --- to find the new "unknown" V_2, its "ratio" to V_1.

$$\sqrt{\frac{1400\#}{1310\#}} = \frac{V_2^2}{V_1^2}$$

Now, do you see that if we take the square root of the "weight ratio" we get the "speed ratio". (permit me to not use the square root sign, but its equivalent, a 1/2 power, since I'd have to switch to another computer program to create a sq. root sign.)

Be Consistant, stay with IAS, 1 and 2, or TAS, 1 an 2. Either OK, if Consistant!

$$(1400\# / 1310\#)^{1/2} = V_2 / V_1 \quad \text{or, rearranging}$$

$$V_2 = V_1 (1400/1310)^{1/2}$$

(Ignore the derivation, if it confused you, just use the result.)

So, go ahead, be brave, if you're new, punch in 1400/1310 in the calculator. You get 1.0687022, the weight (or Drag) ratio, then just hit the square root key and you get 1.0337805, which is the speed ratio! (If you can navigate an airplane, that won't strain your brain, so relax, if you're new, decide you CAN do it. (Would you believe that I just did that on a calculator that I bought at the local drug store check out counter, for $1.00. This modern life is truly amazing, if you stop to appreciate it!)

Now, do you really appreciate and realize what you've just done? You've calculated BOTH the DRAG and SPEED, ratio, going from 1310# to 1400#, specifically 1.0687, and 1.0338!!! **Exact and easy!!!** (We were working and thinking in IAS, and we left C_L and ρ unchanged, so our answer is in IAS, but if we had been working in TAS, the ratio would work for that also.)

(The inverse, or reciprocal of those two numbers .9357143, and .9673233, shows you how much you would decrease your drag and speed too, if you decrease your weight from 1400# to 1310# burning off half of a Deluxe Luscombe 8E's fuel, 180# total, 30 gallons. Don't be confused, the math works either way!)

Now we've got the subject "nailed" (!?) so on the opposite page I've plotted three curves, Full Fuel, 1400#, at Max GW, Half Fuel, 1310#, Zero Fuel, 1220#, so we can see what happens when we burn off a full load of fuel starting at 1400#, max GW! Now we can **see** the **effect**, the **picture**, it's **significant drag!!**

The three key Max IAS/ Drag, IAS MPH speeds are **94.877**, **91.777**, and **88.568**, all calculated as above, (from the original

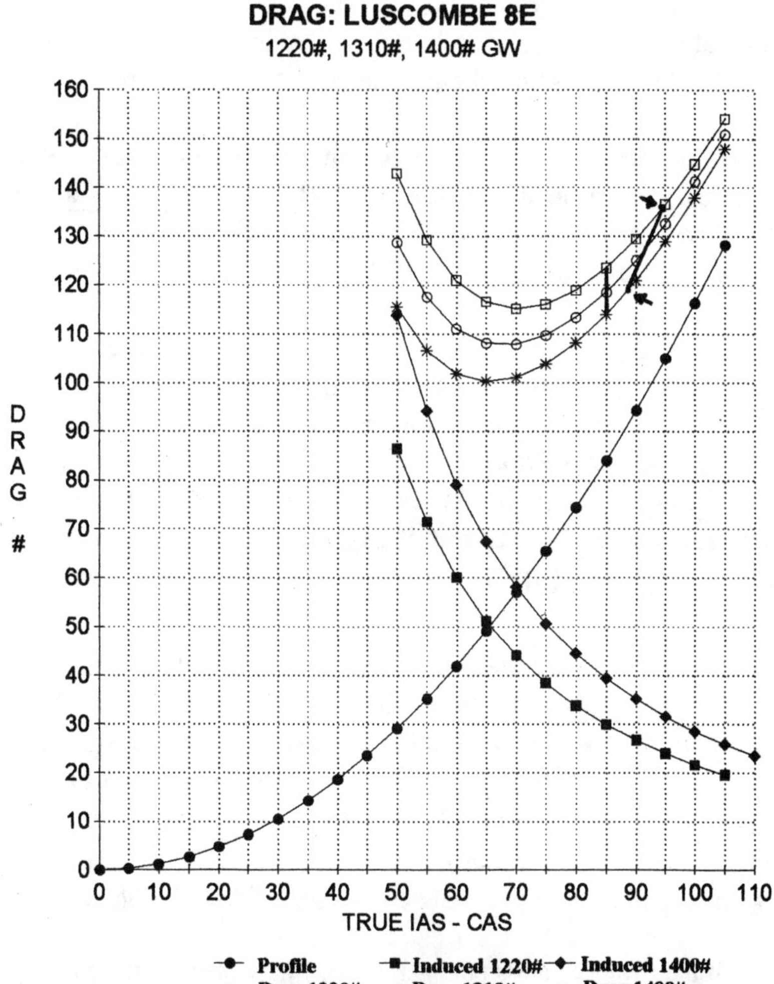

ZTGT test data, <u>Min. Drag IAS of **68.12** MPH at **1250#** GW</u>. I multiplied <u>68.12 by $3^{1/4}$</u>, as we learned in the last Chapter to get the <u>1250# Max IAS/Drag point, of **89.65096** MPH</u>). Practice on one or more, to prove to yourself that you <u>can</u> do it, **and you can forevermore <u>handle GW speed changes</u>, just using <u>the square root of the weight ratio</u>!!!** As above, you <u>get the **drag even easier**</u>, since it's a **<u>simple GW ratio</u>**, starting from **<u>136.34#</u>**, <u>at</u> **<u>1400#</u>** <u>GW</u>, <u>127.57</u># and <u>118.81</u>#. Go ahead, do that, see that you <u>can</u>!!

Of course I hate to "complicate" my nice clean pages, with busy work calculations, and there's risk in having "too much", but recognize "new guys" need encouragement, help, practice!

The big picture thing to see is that weight costs. **Look,** we're **well above the min. drag, light, of 101#,** or114 to123# at 85 IAS. ***Don't treat your plane like a truck!*** **The designers *"get livid"* when you pile on weight!** Before we go on, let's *think*.

Recognize, the calcs work out easy, because you've simply looked at **the same point on the drag curve, as you change GW,** straddle 3 Drag Curves for the 3 GW's. In this case we're looking at the *optimum tangent point*, so it works. The method works for any other point like the Min. Drag point, actually **any other point you choose, but it must be the same, point --- where the α^o, C_L, L/D, all stay the same, get it?!!**

(You can **not** use, say, the Min. Drag of 103.5#, at 68.12 MPH, at the test 1250# GW, to "ratio" to a "tangent point" drag. You would come out wrong, all confused, **because it's a completely different point. When you move around the curved curve, you can't use the simple ratio method** --- you've completely changed the α^o, the whole game!!! You need to use the $3^{1/4}$ trick, or calculate and plot new curves as I did -- as follows.)

We'll tell the advanced guy's (and gal's), that I generated the three curves simply using the basic formulas we developed back on page 44, using a flat plate area of 4.55 ft.^{2} for Parasite, and the (Span Loading)^2, using the three gross weights, a span of 35 ft. and an e of .7 to generate the Induced. Remember: Drag = Parasite + Induced = FPA x q + (GW/Span)^2 /πeq

Incidentally, EVERYBODY RECOGNIZE that generating the three curves **only required the one Parasite Curve**, but then **three induced curves for three Gross Weights!** Notice that the middle induced curve is missing -- only because "Works for

Windows" only copes with 6 printed curves. I could and did calculate all seven, but six seems to be the print limit. It's easy to eyeball where that middle curve would be, so, not a problem.

Now, let's look at the great insight that we get **easily**, from the ratios that we've done. **Starting at 1400 GW, burning 180# of fuel off, we slow from 94.877 MPH IAS to 88.568, 6.31 MPH IAS, drop our Drag from 136.34# down to only 118.81#, a significant 17.5# difference, to 87.14%, (Invert, MPG is 1.1475, +14.75%). That top Drag is a lot higher than 118.8#, in fact it's 14.75% higher, MPG to 87.14%!**

THE CLEAR MESSAGE IS THAT LOADING UP YOUR PLANE DOES COST YOU FUEL AND $$$!!! ---- BUT ----

Now here is a very significant insight, where you really start learning what's going on. Look at the picture, the curve. If you just fly the 88.5 IAS, your drag only goes up ABOUT HALF AS MUCH, about 9 pounds --- and you're automatically flying A LOWER SEGMENT OF THE HEAVY CURVE!!! --- At 85 IAS, drag is even less, ~123#!

Per the Cartoon at the beginning of the Chapter, Cessna 182 Pilots going out of the country, to build their "vacation cabins" long ago learned that "great truck" hauls whatever you can fit in (within some reasonable limits of course). Literally, cement, wheel barrows, wood, literally the "kitchen sink" in addition to the wife and kids, fly regularly. BIG LOADS COST FUEL --- BUT HOLDING, OR LOWERING THE IAS IS THE WAY TO FLY HEAVY ECONOMICALLY, OR FARTHER - BUT, THAT COSTS SPEED. **Understand, pick your choice!**

That Drag Curve shows that when I need a long range to go non stop, I just fly a constant 85 IAS, go high, 10,500', lean, 100 TAS, beat the socks off the guys trying fast, - who have to stop! !

CHAPTER 6, WEIGHT EFFECT ON DRAG, IAS

In this Chapter we risked making an easy subject look harder than it is, by encouraging those who aren't yet comfortable with calculations to do them, square roots and all. We showed the actual numerical calculation results,. Since this is a teaching book, for pilots at all levels -- a little risk, trying math, is OK -- especially if, by now everyone has caught on to the fact that: *you really can't get lost in this book because we always come back to the same theme, the same simple conclusions. HERE, WE'RE SIMPLY TEACHING HOW MUCH DRAG AND OPTIMUM IAS CHANGE, ---- IF YOU* ***CHANGE WEIGHT!!***

1. **Flying our optimum IAS, optimum angle of attack system, if your weight goes up some percentage, say 21%, to 1.21, your drag goes up 1.21, 21% -- dead simple, but $$**

2. **Ditto, if your weight goes up 21%, to 1.21 times the original weight, your IAS has to go up, the square root of 1.21 , to hold the same angle of attack, the same flight condition, --- which is 1.1, on a simple calculator --- +10%.**

For those of you who are afraid of math, please don't let yourself get faked out by any of this. You can skip it! Our objective is to give you help, not let you get lost, snowed or confused. Sure it may seem hard at first, but hang in. Practice, you'll see you can , you'll be proud.

ANY PART OF THIS BOOK CAN BE READ, IN A FEW MINUTES, A FEW HOURS. WE EXPECT PILOTS AT ALL LEVELS WILL READ, AND REREAD, GETTING SMARTER AND SMARTER AS THEY GO! We'll both succeed.

The central message in this Chapter is that treating your plane like a truck is not free. Don't build it, fly it too heavy!

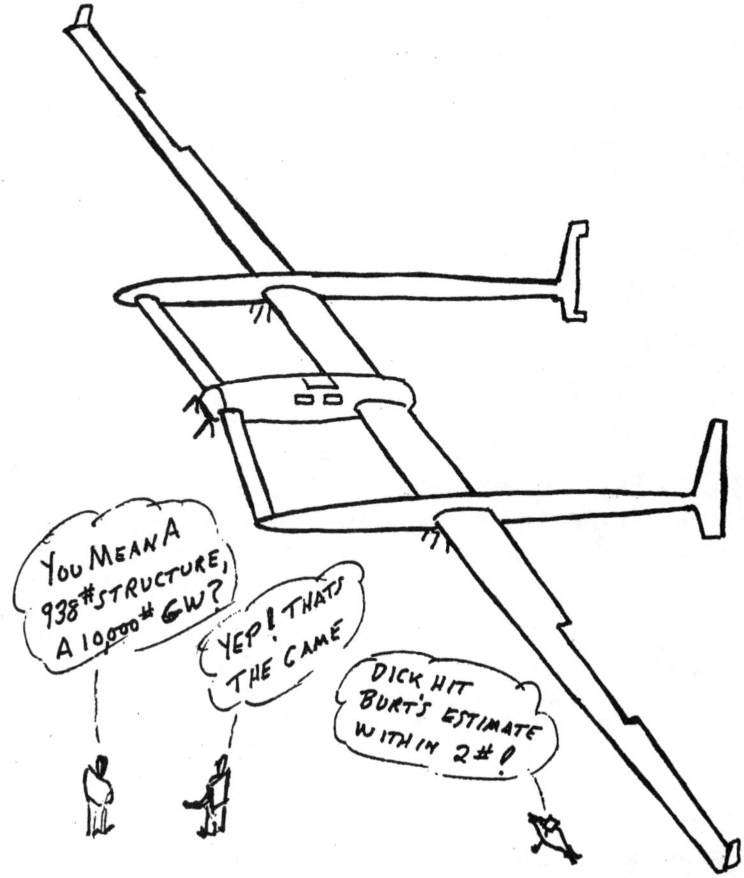

The basic design requirement on the Voyager, was to make a fantastically light *yet* strong structure, so it could take off FOUR times heavier than it lands, *75% FUEL*, the heaviest/ lightest plane, by far, in 1986, a weight ratio of 4, a speed ratio of 2, efficient at every weight!!! Actual numbers, 9694.5# T.O., 7011.5# fuel, 6796.4# used, 26,358.6 statute miles, equator 24,903.1+ 1455.5! If you're interested, there are still a few copies of the Voyager Official Log, Flight Analysis, and Narrative Explanation, identical to what we put in all the National Archives, a collectors jewel! 818 360 1105

Weight Effect, The Last Chapter, 6, A Little Extra Insight

Unlike a ground vehicle, a "flying machine" has to "hold itself up in the air". THAT'S NOT FREE! It's no big surprise, then, that the more a vehicle weighs, the higher it's drag, the more energy it consumes "levitating itself", flying, all in nice neat proportion. Just plain old horse sense says we should find that sort of relationship in the basic Physics of the situation, and of course , we do!

Worthy of notice, "we can slide down the induced drag curve", decrease "the drag due to lift", but we must do it "by sliding right up the Parasite curve", going faster, so there is no escaping either the Physics or the "horse sense" of it all!!! All of Science is like that. It's thoroughly logical, the math follows right along, because it was there all along. We just aren't smart enough to see it unless, or until, we work at getting our act together as we have been here. You are to be complimented, if you're "hanging in", learning how your world works, one of the small percentage of humans, using your intellect to "really understand".

Most good planes can fly at an L/D of 12, or 13 to 1, but drop to something like 10:1 at Max IAS/$, *equivalent to a coefficient of friction of .1, 10%,* just like sliding a block across a table, in high school Physics class, at .1 to .25. It's not super low, like the rolling steel wheel of a train, .004, but it's pretty good for a "self levitating" *speed vehicle* !!

The corrections you learned in this chapter, are a real basic in Aero Engineering, because you can use them for correcting all test flight data points to the same GW, in addition to getting the feel and understanding, you just did. Get as smart as you wish, skip, or practice the math. ✓

Chapter 7

Optimum L/D, α^o

(and insight on airfoils too)

Since we've been talking about constant α^0, optimum, L/D flight from the beginning, it will almost be an anti-climax to finally get to it, but there are many things to learn about it, and airfoils. It's an amazingly simple concept, but you'll be surprised to see how sensitive it is, how small the angle differences are at higher IAS, CAS speeds You'll see, almost with x-ray insight when we finish, that the basic "flight condition" is set by the relation of C_L to C_D, which controls L/D, both controlled by angle of attack, α^0 which is controlled by IAS, for a given weight. It's all tied together, and by this time you are primed to get your mind around the whole concept. In the final analysis, it is holding α^0 constant that is the most simple concept, that permits holding any selected "optimum flight condition" for any weight, any altitude, an amazingly simple and sophisticated "technically elegant" concept of flight!!! It would clearly be "the only way to fly", except as we have already hinted, and shown you, in Chapter 7, p. 44, 47, One Constant IAS has less cost for weight! Grasp it all, and You Can Become a True Master of the subject, to a degree that you never realized was possible -- a true "Thinking Man of Flight". The learning always takes some effort, but then "doing it" is laughably easy - either Constant Deck Angle, or One IAS!

We realize that we may have dragged you screaming through more formulas, Algebra, Graphs and Engineering like calculations, than you ever thought you'd tackle in your life --- but you realize that you can skip, forget what you wish, or dig for more, **as you choose.** If you're going to spend your whole life flying a plane, loving it, loving flying, it's almost like an "incomplete life" to not end up "savvy" on how "the birds really work", especially if you can find the whole basic grasp in one short book, written especially for you. You can understand the whole logic, the real meat, relatively easily, as deep as you wish!

This book is intended for Pilots, to let you have a sophisticated insiders grasp of flight, not be a limited, unknowing outsider, on your life's work, or avid hobby. Though you won't be an Engineer, when you finish, as a pilot, you can build insights the engineer will miss. As a side benefit, you will have gained a very good insight into what engineering is all about, how our wonderful modern life has been brought to fruition, by just getting to the very essence of each subject. It's not the warriors and "political hacks" that created a world where we can fly, it's the guys that "conceive, manufacture, produce". The game is not to do each other's laundry, be *a service nation,* but "create", produce, be a little smarter, leave more than we took.

The concept now of bringing all our newfound knowledge and insight together, in a grasp that is as simple as holding an "optimum angle of flight" **is truly technically elegant**, an almost perfect example of the wonderful simplicity, classic logic that one finds in the **natural laws,** even for such a complex field.

The driving force is that the basic coefficients of lift and drag change in the sophisticated, yet simple way that you have seen, the optimum found in a simple "discerning" angle of attack, that you will now see is *elegantly precise.*

YOU MEAN WE WENT THROUGH ALL THAT STUFF AND ALL WE HAVE TO DO IS HOLD THE ANGLE?
YEH! BUT NOW YOU KNOW HOW IT ALL WORKS!
YOU DO DON'T YOU!?

Lets be sure, before we go further, that you **understand specifically *WHY***, simply controlling angle of attack, α°, controls L/D, which yields a dead simple, precise logic of flight.

(Recognizing the unchanging Parasite Drag Coefficient), it is the angle of attack that controls the C_L, thus Lift, and all the variable C_D, thus Drag, and thus the Lift/Drag Ratio.

Lets look at the NACA 2412 airfoil data on the facing page, a typical section used on a typical Cessna. The sloping vertical line relates angle of attack, on the horizontal axis, to C_L, the coefficient of lift, on the vertical axis. Notice that if you were to mount the wing on the plane at a 2° angle of attack, it would develop about a .45 C_L at that angle, a reasonable ballpark, for a slower, efficient IAS. If you speed up enough to get down to zero angle of attack, your C_L will drop to about .25, a C_L more fit for a fast cruise, roughly spanning the useful speed range!

That creates several insights that are very much worth noting. Let's list them, walk around them, come to really recognize their existence and significance.

1. 2° got us a .2 C_L change! "**.1 C_L per degree α" is typical!**
2. **ONLY 2° α change covered a whole .45 -.25 cruise range!**
3. **C_L vs. α is a very straight line, predictably proportional over a wide speed range, right up until nearing stall!**
4. **"Standard roughness" wings, and low Reynolds numbers, small or slow, stall at a 1.2 C_L, not 1.6!**

Each of these insights are "good rules of thumb" that pro's know and use. It's interesting, easy to remember, that the C_L vs. α° is such a straight line, and almost exactly .1 C_L per degree α.

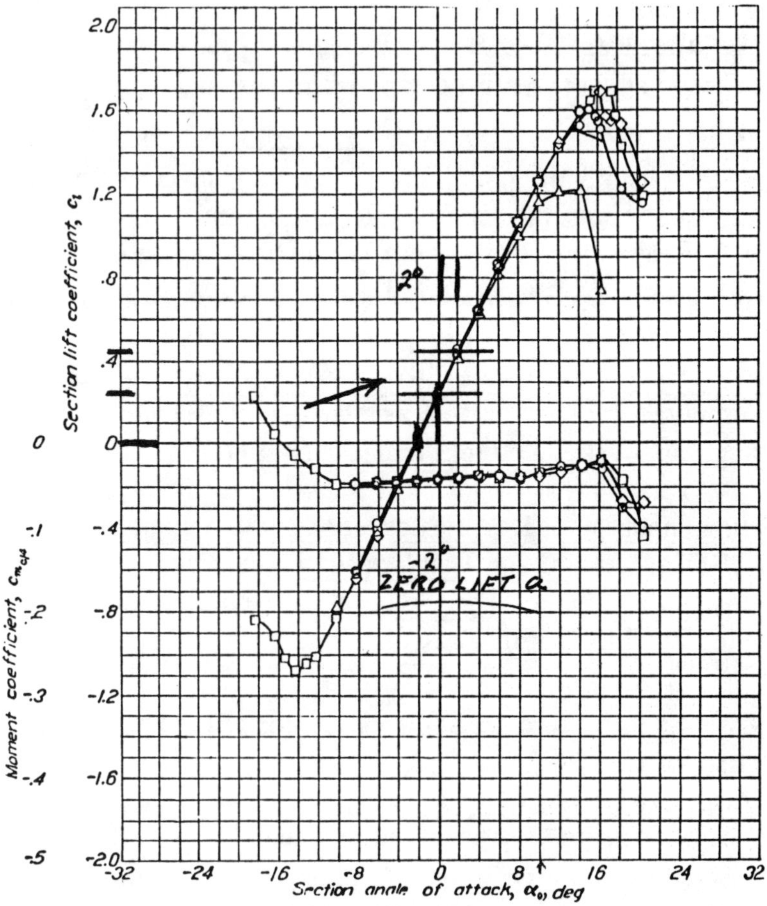

NACA 2412 Wing Section

It's interesting to recognize that we only use a very skinny part of the chart in normal flight, a few degrees α, a few tenths C_L. You'll see the angle is very sensitive shortly. We'll do a "fine print calculation", that shows that, and justifies the "roughly" 2° α nominal cruise range for you. The (-) α^o and (-) C_L only come into play in inverted flight, the high C_L's only come into play in slow flight, landing and takeoff. **Most flying is done in a very small and sensitive angle of attack vs. C_L band!!!**

It turns out the newer, faster planes, with a bigger speed range, fly an even narrower, lower C_L range, since all planes start at max C_L at stall, at takeoff, perhaps with flaps, and the fast ones fly at much higher q, thus tend to fly with even lower C_L's, unless they have inordinately small wings. Conversely, a high aspect ratio "cruising classic", like a Luscombe, flown at a lower IAS, as we're teaching here, will fly at a bigger C_L, vs its Vmax speed, especially heavy. We'll show a Luscombe and an RV6, light and heavy, at max gross weight, at cruise IAS and Vmax, shortly, and plot them on their airfoil charts, because that produces a lot of valuable insight --- **will give you a good basic grasp, of the pertinent insights on airfoils!!!**

But first, courtesy of Dover Publications, 180 Varick St., N.Y, N.Y. 10014, "Abbott and Von Doenhoff", their classic 1949/59 basic airfoil reference, "Theory Of Wing Sections", that all the pro's use, we'll look at the drag of a Cessna airfoil, C_D vs. C_L.

Now here again there are some classic insights to recognize!

1. Most foils have .01 min. "Standard Roughness" Drag. Coef.
2. Most airfoils have a .006 to .007 min. Smooth Drag. Coef.
3. Most airfoils are used near their minimum Drag Coefficients.
4. Again, the negative α and C_L range tends not to count.
5. The big swoop up in the drag coefficient, has a lot less effect than it appears because it happens at slower and slower V^2.
6. The big jets, with huge Reynolds numbers, can approximate "smooth coefficients", even with leading edge device gaps!
7. Notice C_D is plotted **vs. C_L not** α^o, but it could easily be α^o.

Nail these key points and you will have nailed the basic insights on airfoils. The small private plane, with "rough wings", suffers a penalty vs. the big guys, .01, more, **but**, those slick composite homebuilts, with near perfect wings can get .005 C_D's, even less

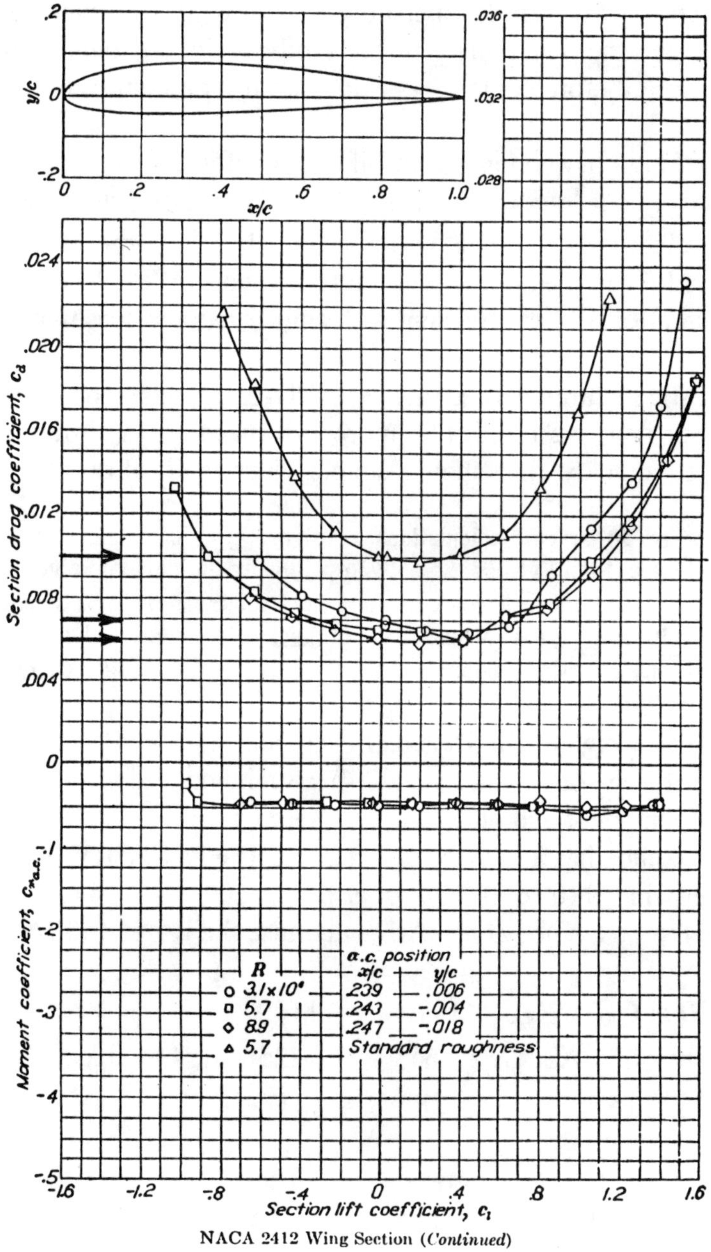

NACA 2412 Wing Section (*Continued*)

using "modern laminar flow airfoils with drag buckets" actually get the wing drag that Aerodynamicists could only dream about, before composites came into use on homebuilts!!!

Now that big upsweep in the airfoil drag coefficient, though not as effective as it looks, since it happens at lower speeds, where the V^2 effect falls off, must be reviewed. It was purposely not accounted for in the drag curves we taught you. We'll deal with that right after we finish dealing with angle of attack sensitivity.

To investigate **angle of attack sensitivity** on **fast** and **slow** airplanes, first we'll tell the more advanced guys, how to calculate it quickly, then show a simple, clear, chart of results.

If I have to do several calculations involving q, I calculate it very accurately once, for 100 MPH, 25.565956 psf, using the modern value for ρ, .002377, put it into calculator memory, and subsequently "just square the ratio to 100 MPH", for a whole string of quick calculations. Solving the basic lift formula for C_L, you get: $C_L = GW/ S\,q$ **Find C_L and α change.**

The empty weight of a Deluxe Luscombe, and an RV 6 are **both** apt to be something over 900#, so minimum flight weight with a skinny pilot and no fuel can be set at **1100 #**. The **1400#** GW Luscombe has a 140 sq. ft. wing, the **1600#** GW RV 6 has a smaller wing, 110 sq. ft., but goes faster, 200, vs. 116.4 max. I cruise the Luscombe at 85 IAS, will cruise the RV at 140 IAS.

C_L AND α CHANGE		1400#	1100#		
Luscombe	85 IAS	C_L = **.541**	**.425**	.541	max.
	116.4 "	" = **.289**	**.227**	- .227	- min.
				= **.314**	= **3.14°Δα** ✓
		1600#	1100#		
RV 6	140 IAS	C_L = **.290**	**.200**	.290	max.
	200 "	" = **.142**	**.100**	- .100	- min.
				= **.190**	= **1.90°Δα** ✓

LOOK at how low that RV lift coefficient gets, **only .100!** Over its whole usable speed range, light to heavy, the C_L only changes **.19**, the angle of attack, α, **1.9 degrees**, 10 times as much, and still very small, at 1 degree, per .1 C_L!

The RV's wing chord, about 4.78 feet, rotates almost exactly **1 inch per degree.** At 200 MPH, if you burned all the gas, threw out all the baggage, bailed out the passenger, and "lost weight" to get down to 1100#, the wing would only change .42 degrees, .42 inches!!!! Over its whole flight regime, light to heavy, slow to fast, it can only change **1.9°**, **1.9 inches!!**

The Luscombe, a slower old cruiser, changes more, because it's operating at much lower q, so it must change its lift coefficient more, especially slow and heavy, and you can see that in the numbers with the .541 max C_L, heavy at efficient cruise, 85 IAS 5 MPH below its mathematically optimum "max IAS/$" speed.

Now here is where we get to the objective of the whole Chapter, get smart on how practical it is to fly constant α, constant optimum L/D!!! Burning all 180# of gas out of a Luscombe, from 1400# to 1220#, the C_L drops from .5414 to .4718, an .0696 change, **.7 degrees!!!** Burning all 225# of fuel out of an RV 6 from 1600# to 1375#, the C_L changes from .2903, to .2495, a change of .0408, equivalent to **.4 degrees!!!**

My **Sears Craftsman bubble level**, bubble moves .125 inches. **1/8" for 1 degree** change, **.1** C_L. To hold **.1 degree, .01** C_L the bubble only moves **.0125"**, 12 1/2 thousandths, *very close to read!* You need to make a little adjustable mount. You can read a ***centered*** set point that accurately, and once set, it's good for life!

You'll see I set, and use, this surprisingly precise angle, but mostly I fly one IAS, all GW!

Changing IAS 5 MPH in the Luscombe changes C_L .052, .5 degrees, .1 degree per MPH. The RV's only .037 degrees per MPH, so you can hold 1 to 3 MPH, right at the limit!

Now on the premise that the way to really learn something is to get physically and mentally involved, go get a pencil and a "highlighter". In Appendix C look up the Luscombe's "old fashioned" NACA 4412 flat bottom airfoil, and the RV's famous 23012 airfoil, the first smart, fast airfoil used on the DC 3, Bonanza, and a bunch of other great planes. Van actually used a 23013.5, half way to a 23015, (13.5% thick), so you'll find the comparable 23015, compliments of Dover Publications.

Actually lightly, but accurately, mark up the C_L vs. α^o, and C_D vs. C_L limits, with the pencil, and then highlight them, when you're sure you have them right. That effort should really bring home to you what we've been talking about here. We really fly a very "skinny" part of the data plot, and you'll better appreciate that the numbers show, there is "very little angle change"!!!

Now you may think that I'm going too far, telling you to actually mark up the data, but you see there is a "method to my madness" a lot of experience in my recommendation. Many of you will be looking at this close for the first time in your life. We are going through the drill here, so you can see for the first time, how it all works, even the math, the formulas, and the curves. But sure, that's not the objective, that's the effort you have to get past, to earn your right to be one of the smart guys.

What we both know, is that like everything you ever learn, it seems hard at first, but then you start "tumbling to it all", pretty soon "you actually get it", and then pretty soon you start "getting a feel for it all", and it quits being a lot of "gibberish", funny engineering stuff. All this stuff ends up making good horse sense, and you can really "make it your own".

I'm giving you the straight line guided tour, the easiest way, through the center of a hugely complex subject,

Aerodynamics. I'm leading you by the hand "to all the key points". I'm hoping that you're reading it as deep as is correct for you, "not letting yourself get snowed". I'm putting in all the key points, making it possible for the broad, diverse audience, to pick up as much as is correct for each individual. I know that those that "highlight" the skinny part of the airfoil plot, are most apt to "grasp" the simple, but sensitive, elegant relationship of lift drag & α.

We purposely did not talk about "moment coefficients" in the airfoil data. That has to do with tail loads and stability, C.G. Perhaps we'll cover that later, less basic, advanced design..

Now, the last thing we have to do in this Chapter, is to dispatch the nasty little fact that there is a big "α sensitive" increase, slow, "in the parasite drag coefficient of all airfoils", that we "purposely did not include" in our drag curve construction. Though the effect looks huge on the airfoil drag plot, its effect is greatly reduced, since it comes into play at low speeds, where the V^2 effect is smaller and smaller. Since it's not induced drag, it must be part of parasite, **but** it swoops up at low speed, just like induced does, and therefore to not complicate the constant, fixed parasite drag coefficient, it's usually either ignored, or treated as an induced "add on".

We found "stable air" Zero Thrust Glide Testing conditions so good in our breakthrough original testing of my Luscombe, "out over the stable ocean at dawn", that we were able to "separate out that alpha sensitive parasite drag". **Look on page 109** in the next chapter, where we discuss "real power, and propulsive efficiency" to see the **real, complete test data curves**. You'll see the extra "alpha sensitive parasite drag curve" at the bottom. It's there, it's real, but it only exists slow and we can ignore it, on the first time through, for the simplified curves for teaching you, with only a small unimportant error, that we'll show you.

MINUTIAE -- SEE IT CLEAR, CORRECT - IGNORE IT

There is a lot of detail in engineering, and it's important for the engineers to dig in, understand it all, to get it all correct, so their work is correct -- "so wings don't fall off", *important*! The problem is, if you're locked onto detail, you can lose track of the big picture. Worse, the minutiae can confuse the learner, stop them from seeing the big picture, any picture!!! This "α sensitive parasite drag" is just such a detail. But we'd make a big mistake if we left it fuzzy, because that would leave some people confused. They'd go off unclear, unsure, not end up with a firm, sure grip on all this. We'll blow two pages to put this chapter clearly, firmly to bed, so all of you can go forward, clear, sure, **confident, with a clear grasp of the big picture!!!**

We very purposely taught you **basic, classical, aerodynamic theory,** the simple, clear version, with 1. a constant parasite drag coefficient, a definite "flat plate area" a simple V^2 parasite drag curve, 2. a normal $1/V^2$ induced curve, with the minimum drag, exactly where they cross and are equal! That's exactly like the great texts like "Aerodynamics For Naval Aviators", so you have the classical grasp, not some "non standard version".

The problem is that big "swoop up" of the airfoil's "profile drag coefficient" throws "a small, but real, spanner in the works" The trick is to deal with it as an "extra little detail", **not** let it confuse you, **not** let it mess up your simple, clear grip of the "big picture". We'll make it perfectly clear, then recommend that you put it back in the minutia category of your mind, at least while you're learning the big picture.

As you can see in the "real Luscombe curve set" that you just looked at, in picture form, that little extra "third basic drag curve" - "swoops up the slow end of the composite drag curve". It will do that little addition to every plane in the world -- **but** --

It's a detail, specifically because that's not where you fly!!!! The variable profile drag has the following characteristics:

1. It acts just "like a little extra induced drag", swoops up slow.
2. It simply "hooks the slow end of the J curve up a little more"
3. It's really a part of "Parasite Drag", an unwelcome extra and "non constant" part of the constant parasite drag coefficient.
4. It causes "the real min. drag speed to increase a couple of MPH", faster, than where our two basic curves cross - equal.

It's a subject that you can pick up, accurately remember, now, if you're at that level of advancement, or forget and roundly ignore if you're a first time learner.

MOST SIGNIFICANTLY -- HOLDING L/D CONSTANT, BY HOLDING α, ANGLE OF ATTACK, CONSTANT, STILL WORKS, ABSOLUTELY ACCURATELY. We said it absolutely accurately at the top of page 86 -- α, the angle of attack, controls all of the variable C_D including the basic "α sensitive profile coefficient"detail. OUR SIMPLE BASIC FLIGHT LOGIC STAYS PRECISELY CORRECT

Our original excellent ZTGT flight testing accurately sorted out this drag, confirmed it agreed with theory. At 1250# G.W. we found a 68.12 MPH min. drag speed. To simulate this accurate data, for the "teaching curves" used in this book, we used the 4.55 ft.^{2} Flat Plate Area we found, times q, for Parasite Drag, the proper Span Loading^2 /πeq for Induced Drag -- except we used .7 for e, rather than the correct e of .74, which kicked the slow end of the drag curve up, a bit, to simulate the missing "third curve drag" while increasing fast drag only a tad, a fair MATCH. The simplified curves still show the min. drag speed 1 to 2 MPH slow, a small error. **IGNORE MINUTIA - GRASP "THE SIMPLE BIG PICTURE", CONFIDENT!!**

Once an "optimum, (or any desired) flight condition" is selected on a drag curve, as represented by a "constant L/D", the most simple, direct way to hold it is to, **simply "hold α^{o}, constant". That works for any altitude, any Gross Weight, SIMPLE!!!**

The specific reason **why α^{o}, angle of attack, holds L/D is because, it is angle of attack, that directly controls C_L thus Lift, and all the variable C_D, thus Drag, and thus L/D ratio.**

Looking into required C_L, shows that a very narrow C_L range is used in the "normal cruise and weight range", especially on fast, high q planes, about .2 C_L, a little more on slow, low q, planes.

Looking at basic airfoil data shows C_L varies essentially .1 for each 1 degree of angle of attack change, until relatively close to stall, so a 2 to 3 1/2°, change will handle the entire IAS, weight range of a fast or slow plane, but **only .4 to .7 degrees change occurs as a complete fuel load burns off typical light planes!**

With the bubble of a good carpenters level moving a scant .0125" for a .1° change, a good level is right at the edge of handling a fuel burn off, or .1° for 1 MPH change of a slow plane, ~3 MPH, (~2.7 calc), for a fast plane, at .037° per MPH. I actually use an adjustable, centered bubble, actually readable that close, .0125" for .1°, set once, usable for life, but I actually use one constant 85 IAS, all GW's, less weight penalty.

Airfoil data shows an increasing profile drag coefficient, and drag (part of Parasite), that acts like "extra Induced", as the plane slows down, below the normal cruise range, that raises the slow end only, of the normal J shaped drag curve, typically raising the minimum drag IAS, a few MPH. This detail can be ignored for basic flight theory and logic, (or optionally used).

A LITTLE INSIGHT ON FLAPS

Flaps are a whole other subject, but we'll include a bit of insight for those who wish it.

Flaps, especially the huge extending Fowler flaps, that are used on jets, greatly change the "camber" of wing airfoils, their center of lift, and their moment coefficient, and greatly increase their lift coefficient. In the process of all this, the original airfoil "pitches down" considerably, lowering the nose, because the original wing becomes more the leading edge of the new bigger, much more cambered airfoil.

This puts the tail at a much more negative angle of attack, greatly increasing its down load, but the change in the center of lift, and the moment coefficient of the now much more cambered airfoil, is great enough, that retrimming to get even more down load is typically required. That's all pretty routine, something pilots are well aware of, but there is enough happening there that it's all worth pointing out in passing.

Thinking that through, a bit, the designer has to be sure that the tail is big enough that it can never stall in an extreme case. The pilot must be sure he never gets into an extreme forward C.G., a fluke case like ice on the tail leading edge, some unusual combination, the designer didn't account for. Very highly unlikely, but a dive into the ground, would be a bad result.

That all, of course, is well removed from efficient flight logic, but this book will be short enough that we have enough space to improve the pilots feel and insight, wherever the opportunity arises. Understanding how planes really work, that which is usually not understood, is what this book is all about.

Power = Drag × Speed (TAS)

$\# \times \frac{Ft}{Sec}$

550 = 1 HP

or $33{,}000 \ \frac{Ft\#}{Minute} = 1 \ HP$

Chapter 8

Power (and its hidden insights)

(and propulsive efficiency too, η_p)

Airplanes fly on IAS, (CAS), but ENGINES FLY ON TAS!! A simple, basic insight, but no other source that we've found seems to offer it --- and you'll really see it in this Chapter on Power where all the engine logic shows up tied to TAS! We'll again make it absolutely clear that, flying at constant CAS, "constant drag", required power increases as air thins, at altitude, because "TAS increases". Amazingly, you'll learn that the RPM increase at altitude is amazingly SIMPLE, fixed pitch. It increases, essentially EXACTLY, as TAS increases, q, all angles constant, Simple, Easy. As complex as propeller math based logic is, that's fantastic!

Remember, Engines have less Power at Altitude, We'll match H.P. to Plane Power!

You'll see that although it takes more power to fly at altitude, that does not hurt MPG, because power increases only as much as TAS, so you use energy faster, but not more energy because your true speed increases just as much as your energy use rate, (power) does! In fact, your MPG will actually improve, because as your throttle opens, the engine breathes easier, pumping losses decrease, engine efficiency increases, and MPG goes up!!! Additionally, with the throttle plate less angled, acting less like a "liquid separator", fuel mixture distribution can greatly improve, permitting much better leaning, and MPG can improve very substantially, much more than you'd expect!!! However, if you try forcing a higher IAS, CAS, higher q, you can wound propulsion efficiency, on many old planes. That's very Important! (Note: CAS is just correct IAS).

This is an interesting, and insightful chapter, because you'll come to understand that it's a ***transitional*** chapter from **Aerodynamics to Engines!** And a very interesting, and incisive part of that transition, is to recognize the obvious change from CAS Aerodynamics to TAS Aerodynamics, something that we've found to **not** be generally recognized, in the aeronautical literature!! **But that's a very important fact!!**

One of the significant contributions of this book is to equally see and integrate "flight logic" and "engine logic". It becomes obvious that those who specialize in Aero are not thinking about engines, and quite logically, those thinking about engines, quite a different field, never grasped the heart of the Aerodynamic logic. This is the book that found the opportunity to put both subjects together! That's VERY SIGNIFICANT, because the way you want to fly the airplane is specifically dependent on how you also get the engine to function efficiently!!!

The engine must have excess power for safety, speed, and to make it a "fast climbing, high performance craft" --- but that relegates it to being throttled at low altitude cruise --- and that relegates it to a life of even greater inefficiency --- unless the insight in this book is put to work. You certainly do not want to fly it at a wimpy, heavily throttled, min. drag, max L/D IAS, which makes the engine efficiency worse, despite what all the basic Aero literature says. Climbing high, at "max IAS/Drag", getting the engine "wide open and optimally leaned", is "where efficient, frugal Vmax. is found"- The intelligent "Logic Of Flight"!!

THAT'S THE CONDITION WHERE BOTH THE PLANE AND THE ENGINE CAN BE EFFICIENT, MATCH EACH OTHER IN A --- TRULY INTELLIGENT LOGIC! AT A FASTER TAS, BETTER MPG -- TRUE MAX V / $!!

NOW AT THE HEART OF THAT LOGIC IS THE FACT THAT THE ENGINE CAN PUT OUT LESS POWER AT ALTITUDE, WHILE THE AIRPLANE REQUIRES MORE, WHICH MAKES IT POSSIBLE TO GET THE ENGINE and PLANE MATCHED, --- *COMBINED* AS EFFICIENT AS POSSIBLE. IT'S TRUE, THE LOWER THE WIDE OPEN ALTITUDE, THE MORE EFFICIENT THE ENGINE CAN BE, BECAUSE, LOWER THE POWER IS GREATER, THE ENGINE LOSSES LESS IN PROPORTION, THE EFFICIENCY GREATER!!! YOU DO NOT WANT *TOO BIG* AN ENGINE IN YOUR PLANE, BECAUSE THAT RAISES THE WIDE OPEN ALTITUDE, LESS EFFICIENCY -- But it can also get too high to Breathe! Go for sleek low drag, not brute force HP!

In CH 10 — the TAS/IAS Gain -- Beats Efficiency Loss Breathing Decides!

The wonderful part of this logic --- and a main point that most people miss for a long time --- is that you are NOT flying at a WIMPY SPEED, but WIDE OPEN, V MAX!!! You just happen to be flying at a low drag IAS, where you are buying speed intelligently, frugally, where the engine is perfectly matched to the plane, OPERATING AT THE COMBINED MAX POSSIBLE EFFICIENCY, MAX SPEED / Drag, Gal., $ That's the winning combination!!!

The good joke is that it's so laughably easy!!! What could possibly be easier than 1. Flying a smart IAS, or smart α^o, 2. Climbing until your engine is wide open: 3. Leaned!!!

That's the wonderful situation here. You take people right down the center, through a comprehensive and incisive course in basic Aerodynamics, and you can make all the conclusions come out so easy, it's honestly funny --- but also a wonderful statement on the incisive eloquence of Science, the simple natural laws! **Our fond hope is that everyone will come learn, appreciate, attracted, charmed by the beautiful final simplicity of it all.**

So, we've used a lot of very black ink for the last few pages. Specialists in proper writing form would probably go into shock, over such excesses. But as we told you in the introduction, our central objective here is to really communicate, **teach**, make it possible for you to **learn**, as **easily** as possible. **If** you really got the meat of the last few pages, you already grasp the central core of the book, everything else you learn is just valuable additional insight. Those few pages are so meaty, so much to the point, so clear, it would be a big mistake if we let them look bland, so a "book skimmer", anyone, could miss them. **We told you this book is as much about teaching, learning, as it is about Aerodynamics. We have to get past the BIG teaching, learning barriers, past the considerable complexities of Aerodynamics, so you can actually grasp the meat of Aerodynamics, see that you really can grasp the meat, accept that there are conclusions that are so easy they are actually funny.** I'll willing to risk, if you'll end up **smarter!**

We're using another mechanism here that's worth recognizing. We tell you what we're going to tell you, the conclusions. Then we tell you. Then we tell you what we told you, summing up in "an ever bigger, more comprehensive picture". In effect we're telling you what the course will be, you "see the pattern early", then we say it, then we walk you all around the subject, so you "really get it", really learn it. That's a form of the SQ3R method we introduced you to back on page 5. Some may think that excessive, hard over opposite to the usual terse technical journal, but that's very much on purpose. Our objective is to make it possible for many thousands of pilots to actually see through, and learn the relatively complex Aerodynamics, and grasp the eloquence of the disarmingly simple conclusions.

If we succeed at that here, you've won big, and I've won big, and as many of the pilot population as we can bring to the table, can win. Any minor excesses, or misses, are of little importance!

Those first few pages are so dense, so much to the point, that we really have this chapter *licked*, so per our format, let's be positive we have the fundamentals nailed.

Power is "Drag" times "True Speed", TAS, pounds times feet per second, # ft./sec. in engineering units. Physics defines a horse power as 33,000 ft.# / min., or 550 ft.# / sec., so you just divide by 550 to get H.P. required. See, engineering doesn't have to be scarry, it's just very orderly and logical. Good that it is, so I can cope with the really obscure, complicated stuff!

It's perfectly straightforward then, that if you fly constant IAS, constant drag, (at constant G.W.), at any altitude, **power increases, exactly** as **TAS** increases at altitude, dead simple. Just keep the logic straight.

Now here is a far deeper, but truly elegant insight that we can build from that, and what we've now learned - At constant IAS, constant GW, **RPM** also **increases exactly as TAS** increases at altitude. Grasp this, this is beautiful! ---- **If your plane continues to fly at constant IAS, constant drag, constant L/D, constant angle of attack, it just speeds** up as **TAS increases, as altitude increases --- and your prop does too!!! Fixed Pitch, the same thrust, against the same drag, it "just sits at exactly the same angle of attack, the same L/D"- it just "winds up" to a proportionally higher RPM!!**

That's a truly elegant insight, one of the best I've ever found in all of Aerodynamics, truly eloquent, in its incisive simplicity. It really makes it easy for the pilot to grasp and use --- and for "The Simple, Thinking Man's, Logic Of Intelligent Flight".

There are some great guys in engineering, smart, funny, with a great sense of humor, not at all the "green eyeshade types" that the public too often misunderstands. But then, of course, there are some "pills and oafs", like any group suffers. The first guy that I ever pointed that out to, was one of those "full of himself, know it all" guys, that had it in his mind that he was smarter than everybody else. He stormed out of the room, blustering "it couldn't possibly be that simple". I had to laugh at him, but I felt sorry for him too, wondered if he ever did catch on. Life can be so much fun, and discovery. The poor guys that get too full of themselves, or just never learn how to simply enjoy good friends, suffer a terrible penalty, in a life where people just like a good egg, someone who's simply genuine. That alone almost guarantees a "full happy passage". I cherish "Auntie Mame's admonition, "Life is a banquet, don't bring a brown bag"!

Ah, but now, just look at the wonderful insight we've created. **A plane doesn't know, or care, how fast it's going, it can only sense IAS, q. -- But the engine sure cares, cares a lot -- Fixed pitch, constant q, it "linearly senses TAS", puts out proportionally More RPM --- More Power!!!! That proves Airplanes fly on IAS, Engines fly on TAS! I'LL BET YOU NEVER FORGET THAT, SUCH A GREAT INSIGHT!!!!**

Regarding the next basics, I'll bet that by now you've got it clearly and correctly in mind that: **DRAG, thus energy required per mile, controls MPG. Higher TAS, more power required, simply using energy proportionally faster, as you go faster, does not hurt MPG. MAKES SENSE. Believe it!**

We'll be covering engine efficiency, pumping losses, altitude effects, leaning, not letting your throttle plate be a liquid fuel separator, thereby ruining fuel distribution and leaning, those things that we introduced in the opening, -- in the next chapters.

So now we're ready to look at a really **new** subject, one that's right on the leading edge of flight, "less than wonderful propulsion efficiency", η_p, "eta sub p", the Greek h, even if it looks like a fancy n, --- and a "lumpy engine power supplied curve", fundamentals that have not been well understood. Before Zero Thrust Glide Testing, for Drag and η_p, fixed Pitch there was not, for ninety years, a good way to test for Prop Planes, *flight test* answers, η_p. **It was the most fundamental, longest unsolved problem in Aerodynamics - with Props too!**

Gus Raspet, Miss. State, did the great, breakthrough, towed up, Propless Glide Tests.

So, let's discuss it all a bit, "then look at the insightful curves, *a picture you can see* ". Jets took over the research during WW II, before propellers and things like propulsion efficiency were completely figured out. With sparse tests, the facts were sparse. To a degree the technical end of the industry charged off in a new direction, before it was completely done with prop planes. And it never went back to finish the job, because with jets, the post war world, it was a whole new challenge, a new era.

There were some early insights. NACA Report 680, (1939) showed a 10% efficiency loss if the cowl was half the prop diameter, none at 1/3, but the party line was pretty much, if you have an 80% efficient prop you have 80% propulsive efficiency, (overall), no loss for blowing back in your own face! That just never made sense, never rang true to me, because there were some pretty ugly airplanes out there, and most were a lot less than perfect. It hardly seemed possible that a poor little propeller sitting on the front of a normal airplane, would work with the same efficiency as a prop in a free stream. Not likely!

The first ballpark testing that I did in 1982, with a preliminary method, clearly said "no way". Even if the first answers were more inaccurate than I thought they could be, the test said there were big losses there, for sure. Dr. Andy Bauer, a lifelong friend and a fully experienced, serious Aerodynamicist, from Douglas Long Beach, and I went to work on it testing, analyzing, looking for a valid test method, recognizing we were working on the longest unanswered challenge in Aerodynamics.

A few years into a serious, if intermittent, part time, hobby effort , we knew we were right, there were big losses. Then presenting our findings to a local EAA group, one Friday night, **Bob Archer**, a local Bellanca Cruisair owner and antenna expert introduced me to the lore of the famous, ***Dr. August Raspet***.

"Gus" Raspet, was head of the Aerophysics Department at Mississippi State, teacher and mentor to many well known and respected Aeronautical Engineers, among them Sailplane and glide test expert "par excellence", **Dick Johnson**, and **Bruce Carmichael**, a friend, author, Personal Aircraft Drag Reduction, a jewel, and "laminar flow expert". Raspet, genuinely creative, had done propless, towed, free glide testing, in the 50's, with a series of light planes, a Cub, a Cessna 120, an early short wing Whitman Tailwind, then a Bellanca Cruisair, Bob Archer's craft.

Gus was great, challenged students to practical tasks, George Lambros's Bellanca tested

Voila, he found the well respected Bellanca had terrible "propulsive efficiency", only 58%, comparing level powered flight, with "sealed cooling duct" glide testing. That gave the true gliding drag of the airframe, unpenalized for cooling losses. He found cooling losses could be 10%. The nominal η_p, 58%, started better at low power, sagged at cruise speeds, and inexplicably got a little better at max speed. That was just wonderful, because that was what we had found, η_p down in the lower 60's. Cooling drag would drop it into the 50's. **Science demands independent corroboration, before work is accepted as truth, and we now both had that, confirmation!**

Significantly, Raspet seemed to "introduce propulsion efficiency" in one of his papers, which pretty well confirmed to me that it was not a well defined, understood subject earlier. Unfortunately, the field never seemed to either appreciate or catch on to his very fundamental breakthrough insight, and with dead stick landings required, no one ever took up his method. Dead stick landings involved more risk, but he found the **truth**!

The way Science works, the next guy starts, standing on the previous guy's shoulders. Though I never knew Raspet, didn't know he existed until half way through our work, he surely placed us on his shoulders. His student, Dick Johnson became world famous, working his milestone R.J. 5 sailplane up to a 40:1 glide ratio, then World Championships, then his "more than professional" sailplane glide testing. Those even showed subtle knees on the drag and sink rate, power curves, as laminar flow broke to turbulent. It was reading his work in "Soaring" magazine that set me on course, challenged me to find the final answer, for the propeller airplane, that had been so long denied! What August Raspet had begun had found its way to Andy and I through Dick Johnson, before I even knew of Raspet.

When our work, Andy's and mine, prepared me to be able to see and invent the "thrust sensor switch" in 1989, a simple, "stiff, vibration proof, feeler" that lit a bulb, when the prop and its crankshaft face moved in the engine's thrust bearing clearance, from thrust to drag, we finally had the key to a valid, "bias free, glide test method". Zero Thrust Glide Testing, ZTGT, was practical, safe, accurate, in fact more accurate because we had the engine available, could go find truly dead air, "stable air at dawn, out over the ocean", for near perfect test results. Raspet had been tied to the airport, and dead stick landings, unsatisfactory test conditions, that could distort desired precise test data accuracy!

Fixed Pitch ZTGT just needs practical Engine Bearing End Play, say ~.015"!

The following is from ZTGT! Drag is measured by "sink rate", accurately knowing "Gross Weight and TAS". The real key is finding true stable air to "eliminate any biasing lift or sink, while gliding". Working out how you also get truly accurate data from your instrumentation, is always the heart of any good engineering test work. The bottom line however, is that we are about to see "***flying* propeller plane facts**", **unavailable for 87 years! The CAFE group in Santa Rosa, CA did fantastic instrumentation and test work on an RV6, Whitman Tailwind!**

LUSCOMBE 8E Aerodynamic Data

We'll list the many basic insights that a complete Aerodynamic data set shows, but the breakthrough insights are Real Drag and, readily obtained Propulsion Efficiency!!

1. **A classic V^2 Parasite Drag Curve, 4.55 ft.^{2} F. P. A.**

2. **A core $1/V^2$ Induced Drag Curve, from $(W/b)^2 \cdot (1/\pi eq)$**

3. **The actual Induced Curve, increased by a .74 Oswald, e.**

4. **The additional "α^o sensitive profile Drag Curve", Ch. 7.**

5. **The Total Drag Curve, combining the 3 basic curves.**

6. **The Gliding H.P. Required Curve, (Drag# x V ft/sec. /550).**

7. **The *Actual* Engine Power Curve, from speed-power tests**

8. **The real Propulsive Efficiency %, (Reqd. H.P./Actual).**

All of the test data and curves are exactly what one would expect from theory, *until* the Actual Engine Power Input, from near sea level "speed- power tests "were plotted, and the resulting propulsion efficiency percentage plotted!!!!

The Luscombe propulsive efficiency started off reasonably good at low power, considering that the propeller efficiency is nominally 75%, **but** η_p fell substantially, and proved to be quite irregular as power and speed were increased!! **The obvious favorable place to fly is at 85 MPH CAS, right where I do fly! You can now see why I fly there, a favorable η_p knee!!** ✓

FLYING THE THINKING PILOT'S WAY

LUSCOMBE 8E
Drag - Power 1250# GW

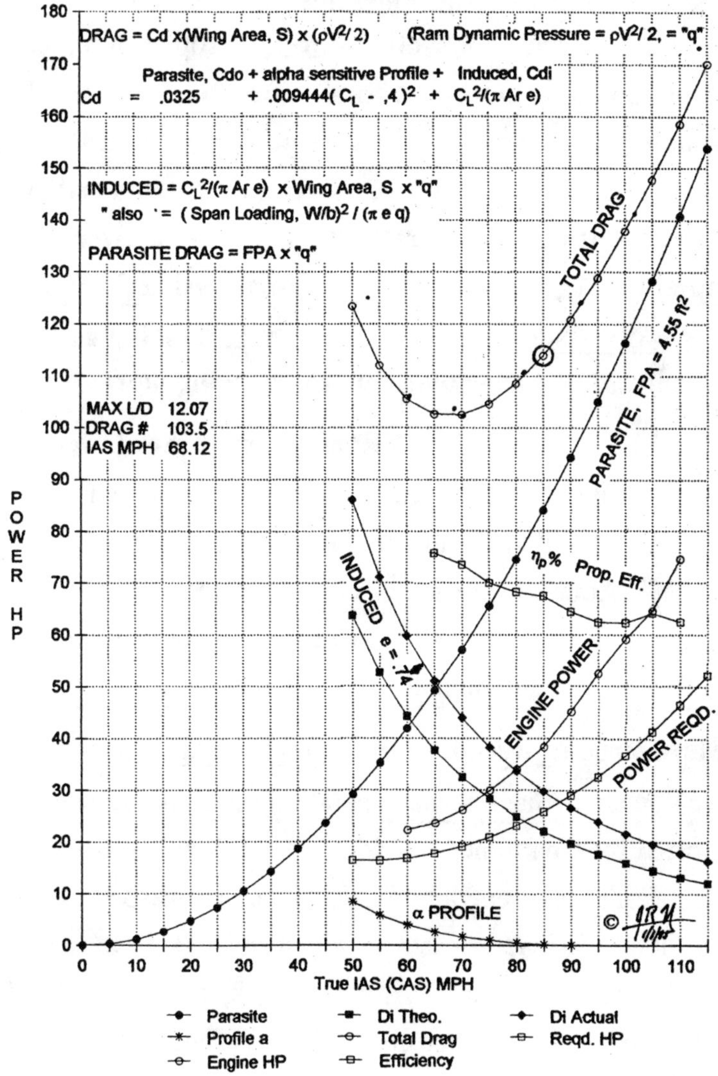

There is considerably more valuable, but semi hidden insight available there. The numbers do some tricky things. If one takes Raspet's 10% factor for cooling drag, (if we had rigged tightly closing cowl flaps), thus multiply the apparent .67 η_p times .9, the result is only 60.3% overall propulsion efficiency.

The brutal facts apparent there are that you have lost nominally 40% of your available power, no small loss, *off the top*, and if you have a ~30% engine thermal efficiency you are at the 18% overall efficiency we showed you in Chapter 2. Again that says that you are using 5.5 times the true energy requirement, at your best, smart point! You can easily go well over that, far worse, being more sloppy!!!

Further 11,100' Camarillo CA Runway length, Sea Level, Speed Power Testing of Propulsion Efficiency is intended, using CAFE RV6A, ZTGT glide test Drag Data, as time becomes available. Approximate conclusions from accurate Theodorsen Prop design results show the modern sleek low drag planes will be far better, enlightening, interesting to designers, and pilots.

The Luscombe η_p **irregularity** suggests variation in the airflow, under power, perhaps some separation, and that seemed to be supported by a patterned variation in the drag coefficients that the data analysis generated. Looking at Raspet's data on propulsion efficiency, each plane seems to have it's own particular characteristics! With the sleek, far cleaner modern designs we'll find far better propulsion efficiency. A P 51 with a **big, geared, slow** turning prop, a relatively small fuselage, has far better potential. A good RV 6, 6A η_p test will be valuable.

<u>Facts:</u> <u>Being able to nail Real Propulsion Efficiency</u>, with ZTGT is the first time we've been able to accurately accomplish that on functioning Propeller driven planes. **We got good data on a Classic Luscombe, an RV 6A, and a Whitman Tailwind**. Excessively tight new Lycoming Engines, some even zero slop, hot, bad, prevented more testing. If we can get good planes with loose engines, in the future, we can do a lot more!

Readily Testing, Knowing Real Propulsive Efficiency, That's a Breakthrough!!!

<u>Minutiae</u>: Aero engineering pros will see minute variations from the clear simple rules pointed out in this chapter, but they truly are minutia, that should be avoided, for a clear understanding of the big picture. A small Reynolds number change, as prop RPM goes up with TAS, would naturally occur, but its effect would truly be minutia, as would second order effects, that could be pointed out as Power, RPM and TAS changed. We purposely use terms like "essentially exactly", throughout the book. Other places we purposely avoid the ill advised complication, when the central task is to inform and teach. The picture presented here <u>is</u> an accurate, clear concept of what happens physically in logical flight, and scant apology for minute simplifications, is needed. There are important concepts here that have never been pointed out. Pro's are not normally thinking of engines, or the pilot's lack of insight, so this presentation can be a good mind jogger for the pro to look from a new perspective, as we teach the pilot who has never had the chance to grasp what a pro can see! If you're a pro, see further insights, refinements, corrections, call us. We're after the ultimate truths.

Max L/D IAS. The three element drag curve causes a subtle change. The α sensitive parasite curve increases the slope of the induced curve, increases the "min. drag speed" an "extra amount" due to the extra slope, <u>faster</u> than "where the combined curves cross", <u>because</u> of the extra slope!!!

CHAPTER 8

POWER, Its Hidden Insights, and Propulsion Efficiency, η_p.

Airplanes fly on CAS, (correct IAS), **ENGINES FLY ON TAS**, **more power**, and fixed pitch **RPM**, required as **TAS** increases!

POWER is **rate of use of Energy**, **(ft.# / sec.)**, DRAG #, times true V, ft./sec., (or 2π RPM / 60 x Torque, ft #). One H.P. is 550 ft.#/ sec.

Flying at constant CAS, constant q, both **Power** and fixed pitch **RPM** increase essentially **exactly** as **TAS** increases at altitude!

Realize Prop q, and all Flow Angles stay essentially Exact!!!

MPG, which is dependent on **Drag**, thus "**energy** use per mile", is not hurt as altitude increases TAS and Power, up, same ratio!

Actually,

MPG is a lot better at altitude. Throttle open, pumping losses decrease, fuel distribution can be more even, *low HP*, leaning safer!

August Raspet, and **ZTGT**, have both shown **major η_p losses as power is increased**, on old design planes. With ZTGT, true facts are available for new homebuilt designs, any plane. All we need are engines with normally loose end play, ~.016" hot!

Realize ZTGT uses Fixed Pitch Props, No oil Pressure Bias!

Engine **efficiency strongly favors flying wide open, at altitude, leaned at lower power, where it can be properly, safely done, with the plane at a low drag IAS, but gaining free, faster TAS at optimum IAS/Drag. That yields "frugal speed, max efficient Vmax.", wide open, engine optimally matched to the plane, MAX TAS/$, "best *fast* MPG"!**

AN IMPORTANT SUMMATION

Believe it or not, we have completed the basic course in Aeronautical Engineering! YOU MADE IT! This chapter began the transition to "engines, an equal fundamental" in this book, and all the side issues will be easy to pass on to you, now that the big, hard, central issue, "Aero", has been swallowed!!??

You can see that your view and grasp is becoming broader and deeper, but **still centers around the beautifully simple conclusions we started with**. Your view is getting far more knowledgeable and sophisticated at this point, however!

Hopefully, you can see that it's unnecessary, undesirable to have too big an engine, to try brute forcing a fast speed, when the plane and engine really want to fly at Vmax., but at a not excessive IAS, at moderate drag, at an IAS where speed is maximized vs drag, fuel used, and cost, at altitude, where TAS is *essentially free*, (except for climb fuel, a good investment, even at some loss of unusable engine efficiency, that we'll soon show you.) The engine which "must be too big for the plane", for climb performance, "**can't fly efficiently at low altitude**, but would love to fly wide open, properly leaned, with good potentially undistorted fuel distribution, at altitude, where it can be safely and properly leaned, at lower power, actually at a point where it, and the plane are at **Max Possible Combined Efficiency.** There is quite a hunk of favorable logic, and you're getting closer and closer to making it your own.

The real way to go fast in an Airplane is with low Parasite Drag, low flat plate area, where you can effortlessly scoot out to higher IAS at low Induced Drag, low total drag, go fast, but we also *want to understand how to fly it economically*.

We're going to look at propellers, engines, all usable for modern homebuilt designs, that give us such great characteristics, **but NOW, let's really nail, the sophisticated core of this book!**

1. It is necessary, but not sufficient, to know how to fly the plane efficiently. You must know how to also fly the engine efficiently, furthermore, you must know how to get both, at the same time, know what flight condition does BOTH!!!

2. You must grasp, that a plane is properly not really an economy vehicle, (and you can't get the engine efficient at max L/D anyway), but rather a "speed vehicle", and the real proper objective, is maximizing speed vs. cost -- efficiently!!

3. You will learn an engine is an "efficiency disaster", its best "efficiency" is supposedly wide open at sea level -- but oversize, to fly the plane, AND climb it fast at the same time, throttled for cruise, low, it's an unmitigated disaster! So - How do we maximize Combined efficiency?

4. With the plane at max IAS / $, high, wide open, but at lower power, leaned, with good even fuel distribution, the engine is at max possible efficiency. Combined with your Plane, max. TAS /Drag, Gallons. $, and *fast*, at High MPG!

RECOGNIZE

You're going to see in a Voyager engine chart in the 10th. chapter, **why** you really don't want too big an engine, **why** you really don't want to fly a slow Max L/D speed. BOTH force you to a **"too high wide open altitude", which starts killing engine efficiency, killing BSFC. Excessively high altitude causes a significant reduction in the max. engine efficiency.**

BSFC, Brake Specific Fuel Consumption, #/ HP Hr, shows Engine Efficieny!

That MAX IAS/DRAG plane is at an efficiency optimum for a "speed vehicle", a valid technical justification, but it also is a "practical balanced objective" where you can get a properly selected engine "wide open, lean", at a reasonable altitude - thus "as efficient as it can be, used in that plane", and at a true, technically valid, MAX TAS /Drag, $ point!

We hear the engine is supposedly at its max efficiency wide open at sea level, but it can't be leaned there, will be set "over rich to cool, protect the valves", can't be run too long at excess power, certainly doesn't match any efficient "Aerodynamic IAS" point for the plane!!!!! Conversely, "too high", power has dropped , losses too big, a fraction of the power, resulting efficiency low, and excess RPM hurts!!

The Voyager Engine Chart gives Incisive Overall Insight!

If your engine is a reasonable size, you'll be going V max, at a place where your engine is as efficient as you can get it in your plane, you'll find good MPG, Max TAS/Drag, Fuel, $.

NOW, TO AN INCISIVE LOOK AT, PROPS, ENGINES!

Are you a Tech Pro, *one of the Smart Guys*, or a Novice?

Even if you're a pro, it's probably best if you go back to the **Book II Introduction**, and **Primer** to start to learn Prop Logic, because it's designed to **Blast Everyone into Huge Awareness,** ***Quickly***. Though one wag described the **key 2 pages** as like *drinking from a fire hose*, (I laughed at that funny insight), it has very good practical purpose, because **everyone gets his first look at the big picture, specific Logic ASAP**. I don't expect anyone, especially Novices, to really fully grasp the basic logic immediately, but it's then much easier to proceed, look back, come forward more slowly, for the Novices, show everyone the *simple enough* horse sense of Propeller Logic. Everyone gets it *Quicker, Better*, if they **quickly see the bottom line insights!**

I get Everyone aboard in an hour, or less, then 2 pages of X Ray Insight!

In the early Primers, and Chapters we start from the same basis, but soon go ever broader, deeper, start tying in new concepts, **superemphasizing everything** so novices see it super clear, and pros who thought things were different don't miss it either. **If pros didn't understand the system they'd quickly think I was being too repetitive,** *probably not like the super emphasis.*

Pros, here's what I expect you to clearly understand! With your training and experience **you can read this much faster, get it much quicker** **than** **the Novices I intend to help get it!!** I expect you to see that everyone from salty old pros to brand new greenies, Pilots who never tried anything like this before represent *the full bell shaped normal distribution curve*. The ***Book is written to help the New Guys. It should be a snap for you!***

Accept it, do your part. Help the new guys!

We went through all the Math
to make it easier for all ---
Help, Don't complain.
Grab your chance
to get it all!

Simply: **Understand Aero 101, don't let your prop load inside out,** DEAD WRONG, **like this!**

Preparing You to Understand Propellers

Exactly How - A Prop Pulls In and Throws Back Air, to Make Thrust

Look at the Prop Airflow Picture on the Book Cover*. **Betz** teaches us to create a **Constant Pitch, Pure Helical Air Inflow** to each propeller blade, a **slightly stretched, slightly higher Pitch, still perfect helical screw shaped outflow, downwash, backwash**. It's just a perfect helical **Archimedes Screw inflow, stretched outflow** *from each blade*, ***two*** **perfect screw surfaces, constant stretched pitch out, all radii!** (Inflow and Rotation Combine.)

Archimedes died 212 B.C. His Screw still pumps irrigation water in Egypt.

Now, what that means precisely, is that **the inner radii are at a progressively steeper pitch angle**, the smaller radii are screwing, **flowing in exactly the same distance each revolution, equal pitch, like a perfect screw at all radii**, then **flows back**, still a perfect screw surface, **at a slightly stretched pitch**. If we think of **the prop** with a constant angle of attack, thus constant C_L, it would have exactly the same pitch as the air **inflow**, but **with the constant angle of attack *added*, slightly steeper, not quite a perfect helical screw surface***. Now here are **3 secret, hidden insights**.

***Of course the prop is *basically* pitched for Plane Speed vs. RPM, inflow extra.**

1. The prop is pulling in air, a slightly reduced pressure in front, Δ**V/2**. **Inflow adds to the prop pitch**, not outflow. **2. The air speeding up a similar amount flowing out**, Δ**V/2**, a slightly higher pressure behind the blade. **3. The Prop Slips a little**, overpitched a little to pull in the air, thus **actually Advances a little less per rev than the pitch! Yes props are tricky, the challenge!**

ΔV, Delta V, is just how much airflow speedup is needed to make the thrust.

Now that's great, but **how do we make that magic happen??** The **secret there** is that **we load the prop exactly correctly**, and we do that **by precisely Shaping it, with exactly the proper twist,** simply because **chord represents area, thus load capability**, as we integrate out along the blade. Now if you jump ahead to **page 147, or 11-II** you'll see that perfect, **constant α^o, constant C_L Betz props** have **very characteristic loading and Blade Shape**, that **varies vs. Advance Ratio**, (most easily understood as the actual advance angle of the prop tip), very much tied to Low Pitch or High Pitch, Slip being the difference). We'll teach you the 2D, 3D **Airflow Geometry**, in **Ch. 3 - II**.

 Of course that's complex, dense, when you first see it. Think, Go slow, Get it.

Better understanding of Superemphasis, and the wording.

Now, as a pro, naturally I started out to write this in the **"good professional writing style"** that I'm used to and capable of. The smarter I got on props, the more I found out that I was **NOT in charge of the wording, or the writing style!** Getting smarter, I found out that there were enough things happening at the same time, that had to be tied together, qualified, that the **Logic**, the **Physics**, the **Math** (unseen, operating behind the scenes), **were telling me very explicitly what wording, what phrasing, what qualifying phrases *must* be said to most incisively nail the Logic**, to make it **clearest, easiest for you!**

Swell, here I am with a lifetime of professional experience, not in control, being a typist, a slave to what the complex 3D flow logic dictated, as it did its task perfectly. The **good news** was that I could see through the wild technical complexity, never explained before, got to where **I could explain it** as the ***technical horse sense*** found in **1. Newton, 2. A Rotating wing, 3. An Airscrew, 4. Betz's Logic - exactly solved by Goldstein Theodorsen Math** -- then many more great finishing insights.

Many of the **great insights**, and **the phrases that best state them, and qualify them** were ***learned over a 10 year period***. Naturally, I realized that the little gems could be missed by brand new technical Novices, Pilots, in many cases struggling to get aboard on all this new technical logic they never saw before. Many might need to have things repeated a few times before the little gems really register in their brain. Pros sometimes don't catch on too quick to completely new thoughts, I've seen, so don't feel too alone and lonely out there. ***Everyone needs time with this!***

Props are just not for speed reading, a challenge for us all.

So naturally we highlight, superemphasize the key phrases, and the little qualifying, limiting phrases, that the Logic, the Physics, the Math demands, to **help the novice see, grasp the crucial key insights. Often it gets busier than we'd like, but the game is to nail the key insights for you, make it as easy as possible for a Novice to grasp the keys!!! Notice, it is possible to speed skim in review!**

Preparing you to Understand Propellers

The **Simple Enough way to Basically Understand Prop Logic** is to just be able to grasp the **Horse Sense of the basic 4 steps** that the **11 page prop Primer** and this **Chapter 9 are based on**:

Here's a compact way of seeing the basic Overview of Basic Prop Logic.

1. Learning that There is *a tricky* basic Logic to Props, their Induced Efficiency loss, most easily learned through **Newton's Laws, Minimizing the ΔV,** Delta V, **throwing of air to make Thrust.**

Big Ṁ, (Big D", or just go Fast), for small ΔV, Small ΔV/2V, axial loss.

2. Counteracting Excess Tip q with **Narrow Tip Chords**. Especially on **slow planes**, the **dynamic pressure**, q, **of rapidly rotating prop tips is so much greater** than the **slow inner radii** that the prop is **trying to work Inside Out, Dead Wrong** vs. the **horse sense of a proper lift distribution** of a proper wing, high at the center, zero at the tip. ***Excess* Lift or Thrust** at a wing or **prop tip *must* fall to zero**, the **air escaping, swirling around the tip into excess** *costly* **vortex loss. So here's the dumb prop trying to operate inside out,** *excess Tip Load!* **Our job is to be smart enough to counteract that** with **tapered tips.** ***Simply*, chord represents area**, and **low area lowers the loading. Ideal Props are Properly Loaded by Proper Shaping,** and twist, **tapered tips!!**

Get It --- Narrow Chords Counteract excess Tip q, Dynamic Pressure!!!

3. Only Hi Pitch Propellers can have max efficiency, a **short spiral path to the destination, less profile drag energy cost, min induced loss, a minimum of ΔV wind thrown, vs. the V_1** wind speed coming at the prop, **helped by a big Ṁ, mass flow rate, for a min ΔV, a big diameter prop disk size, or twice helped by just going fast**, a small **ΔV/2V_1 axial, (and in addition total) induced loss.**

4. Betz Constant Pitch, constant Slip, Pure Helical Inflow, stretched pure helical outflow, Ideal Min Induced Loss by **Ideal Blade Shape, thus Ideal Radial Loading,** (if Zero Drag), proper Twist and Pitch, **constant dT/dQ, a Constant Ratio of Thrust to Required Torque, equal efficiency at every radius!!!** ***WOW!***

And Goldstein-Theodorsen Math ~ perfectly solves the Ideal Betz Prop!!!

Watch how we build this same Basic Logic into answering the 10 basic questions a prop designer wants to know!!! If you just work your way through the explanation, we lead you to all the secret insights!!!

The Wise Old Bird

Props!!! Here's the Big challenge ---
Props can be the ultimate technical Swamp,
if not explained Incisively. We've found a logical way to
fairly quickly get you a great, basic, incisive grasp, 20 pages.
***Then* - It's much easier for you to learn the whole story,**
***if* you choose to go for *the whole Magilla* in Book II ---**

Props can be Pretty Efficient, nominally **85% to 90% +,** for **slow to fast planes**, ***Good!!! But***, **Gus Raspet** showed us **Overall Propulsion Efficiency**, eta sub P, η_P, can be **terrible! 58%** on a **Bellanca Cruisair**, propellerless **glide tests, (cooling sealed)**, **vs. Engine Power Required**. ***(1/.58***, is ***172% More H.P.!!!)*** **Ducts open,** (charging cooling drag to the airframe), it's **maybe .67 to ~.85** η_P. An **Embedded Body** can have, **extra scrubbing drag, pressure drag, variable separation**, worse faster, **power on!!!**

We invented a way to Glide at Zero Thrust, to get real drag, real η_P, *facts*.

There has never been a comprehensive, understandable, *incisive* explanation, *Zilch*, if you're looking for the genuine bottom line insights, the clear Truths. I've looked for that as a kid, an industry pro, it never happened, not even close!

We did all the Math, but we teach it all in Words, Logic, Pictures.

Two of us tackled it, two lifetimes of professional insight, the **biggest, most fundamental gap in Aeronautical Engineering, a total technical morass, as you first wade into it, *a pro killer!***

For Pilots, *this is for pros too*, to Learn the Logic - before the Math.

But, **Nature, Science, is always Orderly, Logical, Explainable. Marvelous Technical work was done by 7 historic Masters, over 83 years, 1865 to 1948, *for us to use, explain*. Once we show you the logic of it all in 20 basic pages, the complexity starts to melt away --- so you can go as far as you choose ---**

This primer, a 10 & 20 page core, incisive, is extended, can get you a great base.

Props try making, max thrust and drag, at the tip, at high q, where thrust must fall to zero, zero inboard, at low q --- *inside out, dead wrong! We teach you* - just tapering the tips fixes that error, and you're on your way to a *marvelous*, optimum solution.

CHAPTER 9

THE PROPELLER

PROPELLERS FOR A PILOT'S *Insightful* GRASP

Fixed Pitch Propellers are harder to design than a "Constant Speed" that adjusts Pitch. First, we'll teach you the Ungeared, Normal Disk Loading, Fixed Pitch Prop.

The Norris - Bauer Law: It is the ***Ideal Shape*** **of a Prop, with the** ***correct Twist and Pitch*** **that determines if you have a** ***Triple Ideal Prop*****, or not. The Game is to actually achieve** ***a constant Ideal Angle of Attack, a Constant Ideal C_L, with pure Helical Inflow*****, by** ***Ideally Loading the Prop vs. Radius, by setting Ideal SHAPE vs. Twist that*** **actually achieves** ***Ideal Minimum Induced, Profile Loss, and Torque*** **- *min. area optimally placed!***

AT A LIGHTER WEIGHT, NEAR IDEAL, .55 C_L, DRAG IS NOMINALLY 1/3, INDUCED 2/3 OF THE LOSS.

An explanation like that has **never been available. Let's list the proper questions, right up front, then explain the core logic, then answer these key, core questions**, get you some **great early insight in an hour or two, only 20 basic pages!**

1. Is there an **Order** and **Logic** to Propellers? If so, What is it?
2. **Why** do **Speed** and **RPM** set the **max efficiency limits**?
3. Exactly what **Shape** should a prop blade be?
4. Exactly what **Twist** is correct?
5. How do I understand how to get **Pitch** exactly correct?
6. Do I want a big area and diameter prop at a low C_L, or a high angle of attack, high C_L smaller prop? **What C_L** is Ideal?
7. Do I want a big **Diameter** with narrow blades, or a smaller diameter with greater **Blade Width**? In other words, what **Aspect Ratio** is best, stubby blades, or narrow blades?
8. Can I get super performance, with a super **Airfoil**?
9. Are **Mach 1, Sonic Tips**, a real private plane problem or not?
10. **What Thrust is required**? **What must we design for**?

DIAMETER? THE PLANE SPECIFICS AND PROP SPECIFICS -- SCALE "D" UP OR DOWN.

Jack Norris's Basic Horse Sense Insight into Propeller Logic. Just about everyone sees a prop as ***"kind of a rotating wing"***, but right there answers stop. ***No one ever tells you what's next.***

There is a **very fundamental**, *very important difference between a wing and a prop*, and **grasping the simple horse sense** *is the first thing to learn*, ***perhaps*** --- ***the easiest, most important insight that you will ever learn about propellers!!!***

A prop, in spinning as it moves forward, has **much higher velocity at the tips** vs. the root, and since **dynamic pressure, q**, the **basis of lift and drag**, is a strong V^2 function, **q** is *dramatically greater at the tips!* They are **trying to load up, work much harder, produce the most lift, Thrust and Drag way out at the tips --- and that is *Bad*, quite undesirable**!!!

A ***Wing center*** **can maintain high lift** --- **but it must, and does fall off to zero at the tip**, swirling off each tip in high energy, wasteful vortices --- nothing to dam the high pressure bottom from the lower pressure top. So here's the prop with low q, low thrust at the root, ***trying for max thrust, max drag at the tip, Inside Out!!!*** Do you see the horse sense of the problem? It makes: **1. max drag at the tip**, and **2. at a max lever arm, bogging down the available engine torque, limiting RPM, and thus Power, 3. extra tip vortex loss, 4. loses the *paid for* Lift-Thrust, which must Fall to Zero --- *losing four ways!!!***

We'd like max lift at mid span, zero at the tip --- and q is opposite to that!!!

The optimum configuration is narrower chord tips to set up the mathematically optimum Thrust vs. Radius Loading - to limit the tip overloading, get optimum Thrust vs. Torque!!! We do have an **optimum theory**, optimum, indeed **marvelous math**, but do you see, **once you understand**, this can be **just good technical horse sense?** You can **see the optimum T vs. r loading and the blade shape** that gives it on **page 11-II**, for a Luscombe and an RV 6 -- **see it as a picture -- GRASP WHY!**

Betz will teach us how to do it right - optimally.

The **problem is worst on Slow, high RPM** planes, **tip q 25 x the root** on a slower Luscombe, **7 to 10 x** on a faster RV, **2 1/4** on a fast, geared down Reno racer. Those post WWII square tip paddles - absorb power - ***but inefficiently*** - avoiding .9 Mach.

 Creating extra Induced and Profile Drag *At* the Tip, losing T#, is just Not Smart!

The Basic Actuator Disk - Logic and Insights - NEWTON

It's much too oversimplified, inaccurate, to be used for final design of a propeller, but we can gain a **huge amount of very basic, very important insight** on how propellers work by **seeing that propellers**, in essence, **pull in and throw back air**.

Newton teaches us **T = $\dot{M}$ ΔV**, Thrust equals **$\dot{M}$, M dot, Mass Flow Rate,** times the ΔV, delta V **speedup**, a **very simple** basic of Physics, a basic insight of **huge importance, dead simple**.

What it teaches is Tricky, but the formula is Dead Simple, just 2 items multiplied. (*Later*, for those who go on, we'll learn that mass, M, *slugs*, is like Weight, W, but actually **pounds weight divided by the accelleration of gravity, (32.174 feet per second, per second), used because we're speeding up, accellerating**. We use **$\dot{M}$, a flow rate**, because we're not throwing a hunk of air, but reather **dealing with a steady flow rate**. ***Mass, slugs*, is just a smaller number of bigger units, to get Thrust in Pounds***, the answer we want.

***g gets involved in Dynamics problems - gets answers in pounds!!!**

Skip the fine print for now, realize **T = $\dot{M}$ ΔV *is easy***, **helpful insight!!**

You'll learn that the mass flow rate of **air through a prop disk is much heavier than you would ever guess**, even pros are surprised, and thus **a surprisingly small ΔV, delta V, speedup** is all that is required to make even huge thrusts on big, fast planes. You'll learn an RV 6 at 200 MPH, Vmax on the deck, can get by with **only a 10 MPH ΔV**, and make **280 # of thrust**, because it's throwing back **615 # of air per second!*** To fully grasp the significance of that number, realize that's a **1 1/4 cubic foot block of steel**, that three guys would strain to lift, and only if they had a good way to grasp it, and that's **every second**. Would you believe a Reno Racer at 480 MPH, with a Skyraider prop is pulling in and throwing back **4 1/4 TONS per second, 8500 # /sec?!** How's that for a wild insight you'd never guess?!

*615 # per sec./ 32.174 = 19.11 slugs per sec. x 14.66 ft / sec. ΔV = **280 # T.**

Notice the **RV's 10 MPH ΔV is only 5% of 200 MPH**. You'll learn that ~ **half the speedup of the air is pulled in as inflow, a slightly lower pressure created in front of the prop**, and **half the speedup happens as outflow, downwash, backwash**, the second 5 MPH, a slightly higher exit pressure caused by the prop. An RV at economy altitude cruise of 140 IAS, 170 TAS, less thrust, ~ 163 # needs just about 9 MPH ΔV, in thinner air. A

 60 MPH is 88 ft./sec., so 22/15, or 15/22 converts. (for ft., lb., sec., system)

slower 100 MPH Luscombe at ~136 # T needs more, maybe 13 MPH, **(a much bigger13%)**, depending on the case. A **Reno** racer **with that Huge Mass Flow Rate** needs **less ΔV_{Avg} than an RV**.

The **Stream Tube concept** is a basic of the **Actuator Disk** concept, pulled in and thrown back, **assumed to be uniform velocity***, **(which is quite incorrect, inadequate for detail design,** *example* **ΔV varies a lot vs. radius)**, but can **teach us a huge amount**, relatively **easily**.

***Also* Backwash at the prop has a much higher ΔV than the average ΔV.**

It turns out that the **ΔV/2**, **the average of the speedup**, is the most basic driver of the **Induced Drag Loss** of the Prop, raising the **power loss cost** of **throwing the air, making the thrust**. **$\Delta V/2V_1$** is the **basic measure of axial efficiency loss** for Induced Drag, half the **ΔV, ΔV/2,** the average, vs. the plane speed, or the speed of the air coming toward the prop, **V_1**, ***before*** the prop effect. (We say it that way, because later, in the advanced explanation, we deal with a variable velocity at the nose, *slowdown*, where we *adjust the ideal prop for velocity profile*.

That $\Delta V/2V_1$ formula is a key teaching insight! We'll use it again!!! .

So, **(ungeared)** ***realize*** **the basic game of propeller efficiency** is to **go fast**, **get both a big mass flow rate $\dot{M}$**, **thus a smaller ΔV**, *and* **also divided by a bigger V_1**, --- **a two way lower loss ratio** ***winning twice going fast***. It turns out that **Going Fast** is the **most important factor in getting an efficient prop**, **more important than perfect design**! ***When you grasp that, you have grasped the core of prop logic, of huge importance!!!!!!***

(We'll cover *geared props, heavy disk loading, Advance Ratio* soon.)

But the **marvelous, quick insight goes on**. It turns out that for ***two reasons***, you **want a High Pitch, High Advance Ratio prop for max efficiency**. First, quite naturally, a **High Pitch, High Advance Ratio prop goes with a fast plane, a big $\dot{M}$, a low ΔV, high efficiency**. Second, an equally marvelous insight that no one ever tells you, **a high pitch prop* has a short steep spiral path to the destination** --- which **saves profile drag energy**, less skin friction energy loss, **marvelous basic insight**, a **low pitch prop spinning an excess distance** to get home. The ***High Pitch Prop Wins Twice, Least Induced, Least Profile too.***

 We'll learn a **~ 45° Blade Angle is Best**, Air Inflow, φ, a tad Less!

Now, of course, **you can only use the pitch you really need**. You just can't arbitrarily slap high pitch on a slow plane. The **game finally becomes using the genius level math that is available to pros, to precisely predict the air *inflow angles into each radius* of the prop, to actually get the ideal angle of attack, the ideal C_L *to get an ideal prop at its design point*.** To know how it works at its other test cases, Vmax, Climb at S.L. and Altitude, Takeoff, we do **an incisive study in Ch. 3-II**

Remember, a prop is *pitched for inflow only* half ΔV, ΔV/2, NOT ΔV .

Now, with all that insight, you can see that **you must finally Pitch your prop to accurately account for Speed, RPM, Inflow, (ΔV/2 + rotation) and desired angle of attack, alpha** and finally to account for the highly variable speed profile vs. radius one finds at the nose of a plane --- in the plane of the prop.

You Must leave the **Design Task** to a Pro, a Computer with **Theodorsen's Math.**

Now this brings up the point that **RPM is very important**, much more important than you probably thought, because, **with Speed, it is fundamental to Setting Pitch. Excess RPM, will limit the Max Efficiency Cap that your prop can attain.** In **wanting Low RPM for High Pitch, High Advance Ratio, High Efficiency**, a fixed pitch prop is in **direct conflict with the engine, which wants high RPM for High H.P. at low Weight**. **RPM** ***is second only to*** **Speed in setting max Efficiency**!

You've been wondering what Advance Ratio is? .

Now, we've brought up the subject of **Advance Ratio**. Look at the graph and explanation **on page 58-II**. Finally done with all up Math, it's the professional's **graph of max potential Efficiency for *any* Prop**. In its most simple *modified* form it is a graph of the **Actual Advance Angle of the prop tip**, vs. **Max Potential Efficiency: Poor efficiency at low Advance Ratio, Low Pitch, it swoops up as Advance, and the closely allied Pitch increase.** Have you figured out yet that **since a Prop must have extra Pitch to account for inflow and angle of attack, the *prop seems to Slip*, Advance a little less than Pitch, but each closely related, *the Slip + α° the difference*!**

Pitch over Diameter, P/D Ratio, is the **even easier great insight** into **Max Potential Prop Efficiency**. The game is simply to **hope you need a P/D Greater than 1 --- because** (at a nominal Aspect Ratio of 14:1, based on the outer 90% of the Diameter), a **P/D of 1** gets you a **max efficiency Cap of ~88%**, only 85% at a Luscombe's 51/71, .718 -- **a very good 89% Cap on an RV 6, 79/70, 1.126. About 1.27 P/D, RV 8 gets ~ 90%**, but in that range the prop is **high pitched enough that you might want a variable pitch constant speed prop**, because if fixed pitch, the inboard blade angles are so steep they're **stalled at runup, slow, at takeoff**. But, most interestingly, **we'll learn that the major inboard blade angle *correction for Slowdown* gets an RV Prop fully unstalled at liftoff, 50MPH!**

You'll learn Excess Disk Loading, HP/ ft², hurts efficiency, *but not normally*.

Aspect Ratio Since the job of a Prop is to make Thrust - like a sailplane, **High Aspect Ratio minimizes Induced Drag**, the **higher the A. R., the more efficient a prop can be. A Big Diameter**, low span loading, **narrow blades**, **tapered tips** are the **real keys to efficiency**. The problem is that a **metal prop is just as vibration prone as a tuning fork**, and **too long and skinny is more likely to get into trouble with ever lower natural frequency, multiple modes of vibration, vulnerable to all the forcing frequencies that can come out of an engine, a total can of worms**, technically, that takes a lot of experience.

Aspect Ratio, Definition? See Below. **See Full Definitions after Ch. 1-II**

Indeed, the primary job of a metal prop designer is not to give you an ideal prop, but rather, **one that can be made from one of the long paid for forging dies --- that won't kill you**. We base the Aspect Ratio on 90%* of the Diameter, divided by the Average Chord, **[or (.9D)² / Area]**, and **use a nominal max A.R. of 14:1, that of a Luscombe Prop, as a working start, here. It took 2 years to develop a set of safe props for the RV's ---**

***90% -- of course the prop hub makes no thrust.**

We don't try to cover all of the vibration subject here, because it takes a graduate course understanding of vibrations, and a **ton of special experience. Composite props can be less vulnerable**, more damping. ***But always be watchful, Caution!***

 ***With an 18% D, RV Spinner, later we analize 19% to 99%, ten 8% steps.**

A 14:1 A.R. over 90% R, equals a 12.6 A.R. over 81% R., .81/.9, a .9 ratio.

Constant Slip: The Core Logic of the Ideal, Minimum Induced Drag Prop

A Prop, like a wing, *would* like to have max thrust at its mid span, like an Elliptically Shaped and Loaded Prandtl Wing, with its Constant q, C_L, and Downwash, controlled loading dropping to zero at the tip for min induced drag --- BUT a prop has highly exaggerated q at the tip, low at the root, ***inside out***, --- and also acts as **two separate rotating wings, 180 degrees apart** - with **a root vortex too**. How about that surprise, deeper insight?

We've learned that ***we counteract the excess q based tip loading by more severely tapering the blades*** to **pull the max loading back away from the tips, toward mid span, the inner blade q loading weak, a weaker root vortex, *teardrop loading!***

Notice that ***High Pitch*** **Betz Props Taper Inboard Too**.

Thinking of the *Actuator Disk* it might be **nice if we really did have a *uniform axial velocity across the disk****, but we'll soon find the **axial and rotational inflow vary hugely**, but there is an ***amazing trick in the Betz Pitch that makes it act uniform!***

*We actually don't want uniform load, rather **Betz's Pure Helical Flow**.

The **Plot Thickens**, ***or so it seems***. The tips at small blade angles pull more straight forward, the steep root more sideways, actually pulling against engine torque, not pulling forward well, and at low q, weak! An angular trig mess to sort out? No! Look.

***As a Screw, Drag free, Betz gets a constant ratio of Thrust vs. Torque at all radii!**

Now, **here comes the Central Secret of Propeller Design!!!** Next, (and Ch 3-II), we show you how we set up and understand the air flow geometry into, (and finally out of), a prop blade, the **airplane forward velocity**, the **prop's rotational velocity** *at any radius, and* the **axial and rotational components of the airfoil inflow**, (and ultimately outflow). It turns out that **it is possible to set up *constant helical pitch of the air inflow*, a perfect screw surface, stretched helical outflow**, which *looks like* **constant velocity at all radii, a constant slip, vs. the constant airplane speed. The constant slip isn't, but acts like constant axial velocity!!! *That Constant Slip minimizes radial flow, tip induced loss!!!***

Here, in Two Pictures, is What we're Trying to Do!!!

Pure Helical Airflow *Into* a Propeller, Stretched Helical *Out*

Picture shows Outflow

Note: Pitch looks like half of what it is -- ***a dual screw!*** Follow it 360 degrees ---

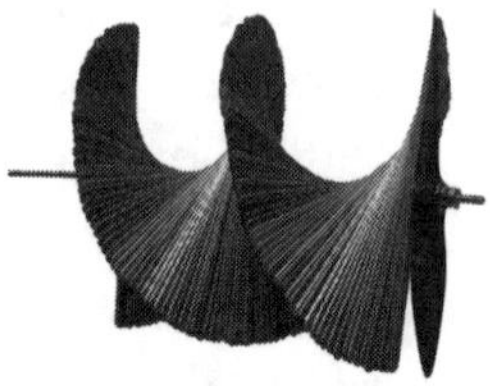

Each adding ~ half ΔV, $\Delta V/2$

Betz's Rigid Vortex Sheet

***Outflow*, Two Perfect, Archimedes Screws, Blended, at the center, *one from each blade, 180° apart, two* perfect screw surfaces. Next, *Constant Pure Helical Pitch of the Air Inflow*, as in the sketch below, the Wind line W at all radii is pitched exactly the same to point D - But *with inflow, a and a'*, it *Actually Advances Less, only to point C* --- Giving the Constant Slip in the sketch below, at all radii. The prop** is at a **small angle of attack to the W wind, which sets Prop Pitch**. Then, the *equal outflow* downwash, backwash, **$+\Delta V/2$ (not shown)**, not involved in prop pitch.

The Following is How we get Pure Helical Inflow.

A First Look at the Constant Pitch, Constant Slip Geometry

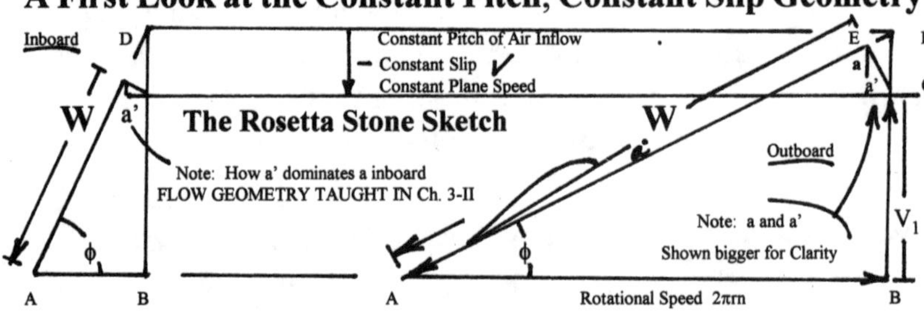

Notice that the **a and a' representing the air inflow and rotation,** (as a decimal ratio to plane speed V_1, and rotational speed, at each individual radius), are ***not*** constant at each radius, but are **highly variable, their resultant perpendicular to the wind line W. The tricky thing that happens here is that, quite non uniform, a and a' work together to raise the *Pitch* of the air inflow to be constant,** to what we call the tippy top, (corny, but people get it,) **pt. D, *constant pitch, constant slip* vs. plane speed, pt. C, at every radius. It's the constant slip that creates the most uniform, least loss airflow!!!!!**

 Realize the 3D Potential Flow Math also **handles 3D Radial Flow ⊥ to the page!**

Albert Betz was a contemporary of **Ludwig Prandtl** (who conceived the Ideal Elliptically Loaded, Elliptically Shaped, Minimum Induced Loss Wing Theory). **Betz**, amazingly, conceived the **Ideal Minimum Induced Loss Prop Concept in 1919, post WWI, in Göttingen**.

- **It is First Done with Zero Profile Drag, drag added later.**
- **It Yields an Ideal Minimum Induced Loss Prop by**
- **Creating the exactly Ideal Thrust vs. Radius Loading by**
- **A Perfectly Shaped and Twisted Prop from Great Math** *also*
- **Setting up Perfect Helical Air Inflow, stretched outflow.**
- **Which is Constant Pitch Inflow, stretched outflow which**
- **You'll learn sets up a "Constant Apparent Slip Prop".**
- **All yielding a Constant dT/dQ, Thrust / Torque Ratio.**

REMEMBER, THIS IS ALL DONE PROFILE DRAG FREE -- AT FIRST.

The **Ideal Helical air Inflow, Stretched still Helical outflow**, a **perfect Screw Surface** downwash, **Backwash, the Ideal Thrust vs. Radius Loading**, the **Ideal Shape vs. Twist** can all be **seen, understood with pictures, No Math needed to Understand!!**

The Constant Slip produces the most uniform, least loss airflow.

The ***constant dT/dQ***, simply a **constant ratio of Thrust to Torque or HP, equal efficiency at every radius**, (drag free), becomes the **Magic Bullet that *makes the complex flow and forces* SIMPLE**

If you wish to go all the way through propellers, in Chapt. 3-II, we teach you how to set up the geometry, little sketches of the airflow geometry, that does all this. It's easy enough in picture form, but most interestingly, it does all these subtle, tricky things, that all work together to make all this work out. The **constant apparent slip**, above, happening with **constant plane speed**, **constant helical pitch of the wind inflow**, (which accounts for the prop also pulling in air, and needing an angle of attack) --- is **the nifty, tricky key, because it achieves constant slip** even though the axial inflow is not equal at all radii, but the **geometry makes it act like it is**. And this, it turns out, is ***the real hidden key to Betz theory***, because **it's the constant apparent slip that minimizes the radial flow, basic to minimizing the tip & root vortex, the induced loss, the little secret, obscure point that is central to everything!**

Betz logic demands highly tapered outboard blades to prevent excess tip loss, max chord set vs. J/π, Advance, with weak inboard thrust due to low q, at constant C_L twist, the blade shape and loading you can see on p. 147.

 This is the clearest statement of Betz Logic you'll ever find!!!

Despite wild 3D flow complexities, **the key fact is, we have an essentially exact math solution for Betz Logic, that allows a pro with a computer and a program to totally analyze an ideal prop in seconds after entering all the spec. details.**

Betz's technical insight was marvelous, but it finally **took 83 years** from **Rankine's** first Newton based water prop analysis in **1865 to finally solve the math. Goldstein** did a genius level basic solution of the **3D Potential Flow** Differential Equations in **1929**, which he thought was only valid for multi-blade, light loading. --- Then **Theodorsen** recognized that it was good for **heavy loading**, Betz's helix moving back through the Stream Tube *faster than the Stream Tube average*, all if you *simply considered* **a Higher Advance Ratio,** and ***a slightly reduced effective diameter far back***, usually ~ 99%, (low advance, maybe 97%). His 1948 book has all the design formulas, extra multi-blade and counter rotation solutions using very accurate analog Voltage Field simulation, **a rare genius level of work too!!**

You'll see Ribner and Foster have checked it all by Modern Computer.

The beauty of the method is that it uses a **simple chart of prop circulation** equivalent to **Thrust** --- **vs. Advance Ratio and Radius**. You put the chart in the **computer**, which **accurately interpolates the correct blade loading from the chart - designs and gives you full details, lightning fast!**

The key Kx Circulation Chart is held to be accurate to 1% or better.*

Now, that gets you an Ideal Prop design quicker, easier than you would ever believe. There is actually much more to know and understand, of course, and you need the computer program with Theodorsen's math to actually do that nifty trick, but we have the math and program available, conquered, actually in the most simple **"*Basic* Language"**, (and we expect to make it possible to buy a copy). **We'll stop the *core* explanation here, because our first objective is to help you quickly enough get the basic picture of prop logic so you can grasp how to fundamentally understand --- a 138 year mystery!**

To understand fully, the final sophisticated game is: You 1. Add Low Reynolds Number Drag, and its Early Stall. 2. Adapt the Ideal Twist and Chords, for the Nose Velocity Profile, "Slowdown". 3. Use a special Program to analyze Performance at different Speed, RPM, Altitude. 4. Learn how to pick a "Design Thrust" for a *poor* Interference Efficiency! The last steps are Andy Bauer's *final completing contributions* to the field.

*** In 1963 Tibery and Wrench at Divid Taylor Model Basin got Higher Kx Blade Loading for very High Pitch, High Advance Props above a Lambda, of .5**

 Now, we'll just refine and expand on what you've learned in 10 pages.

How Betz's Narrow Tip Chords and Load Distribution fixes the Problem of Excess Dynamic Pressure at Prop Tips.

See in Pictures <u>How to Load</u> an <u>*Ideal*</u> <u>Wing or Prop by Chords</u>!

- **Lift or Thrust --- <u>MUST</u> --- <u>DOES</u>, fall to Zero at the Tips!**
- **Shaping Chords, <u>*thus*</u> Loading vs. Span, creates Prandtl's Minimum Induced Drag Loss Ideal <u>Wing</u>, chords representing area.**
- **Tapering Props <u>more severely</u> <u>does the same thing for a prop</u>, <u>*counteracts*</u> greatly excess Dynamic Pressure, q, at tips.** √
- **Lift or T# = Dynamic Pressure, q, x (Area-Chord) x C_L**

WING | **PROP**

1. q, Dynamic Pressure, ($\rho V^2/2$)

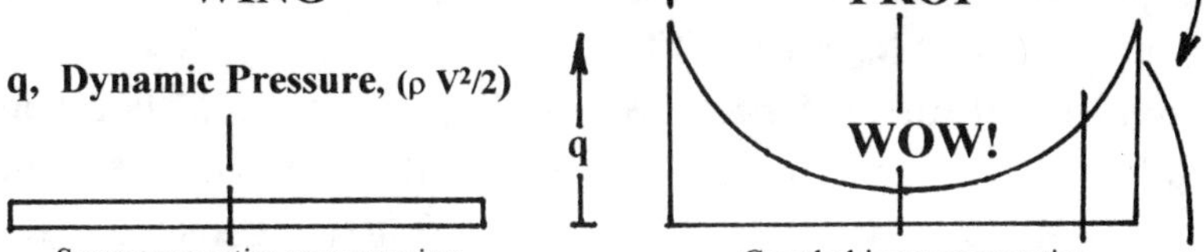

<u>Same</u> over <u>entire span</u> on <u>wing</u> | <u>Greatly bigger</u> at <u>prop tips</u> x

----

2. Area, (represented by chords), as you integrate across entire span.

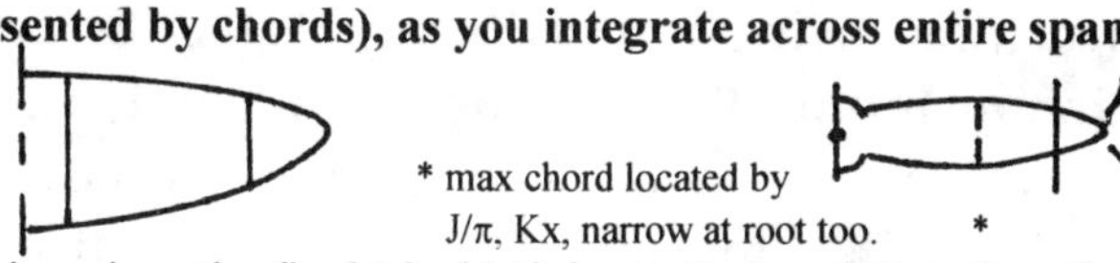

* max chord located by J/π, Kx, narrow at root too. *

Exact (K_x) Shaping gives the final Ideal Minimum Induced Drag Loading =

----

3. Radial, Spanwise Loading --- <u>The Result of: Chords</u> (Area) x q x C_L

T# per ft. of Span

Elliptical Loading | Teardrop Loading

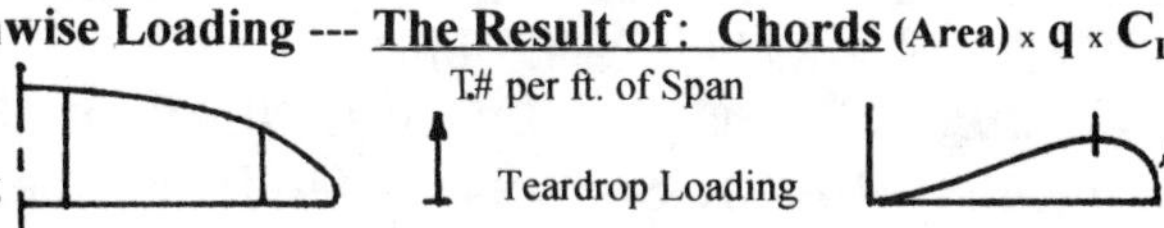

Ideal Loading for a Wing | vs. a Prop

<u>The wing is Ideally Loaded</u>, <u>high at the center</u>, far from tips and their Vortex Loss --- <u>shape sets the Loading to be Ideal</u>.

*Theodorsen's **Kx (Circulation)** **Factor** shapes **exact loading vs. Advance & Radius.**

<u>The Prop when Ideally Loaded</u> --- <u>can't be high at the root</u>, <u>at low q</u>, <u>is heavier outboard</u>, but <u>*not excessively*</u>, <u>*max load pulled back from tips*</u>, to 75 to 80% r. ideal, <u>actually constant dT/dQ, if drag free</u>.

The Max Chord on a Betz Prop is **set by J/π, Advance, Kx,** losely tied to ~45° β, you'll see. *

Now, let's give you the 10 Basic Answers to Propeller Logic:

1. The Basic Understandable Logic of an Efficient Prop is: **Go Fast** to have a **BIG $\dot{M}$,** Mass Flow Rate through the **Prop Disk**, thus need **only a SMALL ΔV**, a **Doubly Low $\Delta V/2V_1$**, for **Low Axial Induced Eff. Loss* - LOW enough RPM** for **HIGH PITCH**, **HIGH ADVANCE RATIO**, which naturally goes with that **Low Induced Drag**, and with a **Short Spiral Path** to the destination, for a **Low Drag Energy Loss *also***, ---- Then, with that foundation, use a **Betz Detail Design**, which produces the **Ideal Minimum Induced Drag Prop**, initially **done Drag Free,** (which requires **correct radial loading, Ideal *Shape*,** vs. Twist).

***Eta, η, the Greek h, is the symbol for Prop Efficiency**.

2. (***If not geared***)^, **High Speed**, **Low RPM** are the **two most Important Fundamentals** to achieve **High Propeller Efficiency**. They produce **High Mass Flow Rate,** **thus** **low ΔV, low $\Delta V/2V_1$, thus Low Induced Drag Loss --- High Pitch,** **thus** a short spiral path to the destination, **minimizing** **profile drag energy loss too.**

Gearing^ can *also* give Max η, Lo RPM, Hi Torque, Hi Pitch, Bigger Diameter.

3. It is the **Characteristic Tapered Narrow Tip Chord Shape** (and accompanying proper Twist) that **gives an Ideal Betz Minimum Induced Loss Prop**, the **Ideal Thrust vs. Radius Distribution**, best done with an **Ideal Constant Angle of Attack, Ideal C_L**, for any given case, because that produces **the ideal min. area, properly placed, the least profile drag *too!!!*** (Chord vs. α° and C_L can be swapped, but that is less Ideal.) It is **the loading of the narrow tip chord Betz Prop** that fundamentally, **Ideally corrects the tendency of excess tip q to overload prop tips, create Extra Induced and Profile Drag!!!**

GET IT --- IDEAL PROP SHAPE vs. twist is the ***Key BASIC FIX!!!*** (p.147)

4. Ideal Prop Twist (and Shape) is **automatically produced by Betz-Goldstein-Theodorsen Math, for Ideal Thrust vs. Radius Loading.** Since the fundamental objective of a Betz Prop **is to produce a Pure Helical Air Inflow to the Blade Airfoils**, and subsequently a stretched, still pure helical outflow, downwash, backwash, **and the best chord distribution is**

produced by an Ideal Constant Angle of Attack, Ideal C_L, the **ideal twist is a pure helix air inflow, plus a small constant angle of attack, perhaps only 1. 1°, for a .5 C_L, very close to pure helical blades, but just a tad different.** This is the easy way to understand ideal twist --- (but do understand that in advanced chapters we correct twist and chords for slowdown).

5. To understand **Precise Pitch***, realize that a prop must be very accurately **overpitched to account for the inflow of air, ~ half** the ΔV **plus a small angle of attack,** (the **second ΔV/2 thrown back by pressure is NOT involved in prop pitch).**

***Theodorsen *Does* account for Rotation and 3D Radial Flow too**,and Heavy Loading.

The whole concept of a **Betz Minimum Induced Loss propeller** is simply to accurately - **Shape - Twist - and Pitch** - the Blade to **precisely pull in a perfect helix, a perfect screw shape of air at all radii, *all at a constant inflow pitch*, big enough pitch to account for the air pulled in, plus the necessary small angle of attack.** When the (*Constant*) **Plane Speed is subtracted from the excess pitch airflow, you get a CONSTANT SLIP**, ***each radius of the prop working equally hard***, for the **most uniform airflow, no radial unbalance causing extra radial flow, the perfect minimum induced loss design, drag free.**

The **constant dT/dQ** that also results, the **Thrust perfectly proportioned to the Torque at every radius, every radius equally productive, equally efficient** --- is not only a magnificently ideal outcome --- but it **beautifully cuts through all the flow, math, and conceptual complexities** of **propellers**, and teaches us **the bottom line is Dead Simple drag free when you do it right, do an ideal design. It's actually easier to do it right, ideally, than to flail around guessing!!!**

Constant dT/dQ Shows there is NO FUNDAMENTAL PENALTY TO PROP TWIST, (drag free).

Now, all this is true **only at the Design Point**, and **only on a zero profile drag basis**, but that's ***not*** **bad**. You can easily **run the computer with and without drag, easily see where it costs and how much**, *also* **evaluate what happens when you fly off design point. A computer can give marvelous insight.**

 ***Heavy Loading** also has **significantly more than average ΔV and Pitch at the prop.**

6. **<u>Ideal C_L .5 to .55</u>** Propeller design, which can be extremely demanding, requires great precision to get computer answers <u>that match actual results</u>, (but with Theodorsen's math it works fine), but can also show the propeller can be quite forgiving, flexible, (in some ways). Nothing shows this better than the final **C_L** study we do on page 131-II through 134-II. It turns out that if you design with a **.6 C_L** an **RV 6 prop** only has to have a **69.1"** diameter, whereas a **.3 C_L** requires **83.7"**, and there is **surprisingly little difference in efficiency**. **C_L vs. D** can float around with only minimal effect, in some ways, *(some ways not)*.

<u>Induced</u> Drag, and <u>Profile</u> Drag simply <u>swap</u>, <u>surprisingly evenly</u>.

When you leave the **Design Point, Altitude Cruise**, look at Altitude and Sea Level **Climb, Takeoff, Vmax** on the deck, **there is a winner, .5 C_L,** (72.68" D) with enough differences, that it's **worth picking the Ideal, the winner, but a .55 C_L is quite close, and lighter, -7.6%,** a <u>*Diameter Cubed*</u> volume function, diameter x chord x thickness. (The 83.7", .3 C_L costs **+53% more weight!!!**)

But now the catch ---

Understand, **here is where props are not at all flexible!** Design a **big prop** for a **low C_L**, miss your calculation, get a **high C_L**, <u>overpitched</u>, and your engine won't turn up enough, a real dog from a performance standpoint. Slow down and it's <u>worse</u>. A **small prop** with a **low C_L**, <u>underpitched</u>, will overrev, unacceptable, though slow it will correct the loading at high α^o, but you're slow, missed your design point. **The flexibility only works if you very <u>accurately hit your design point C_L</u>.**

You need math that gives <u>*accurate, Pitch, Twist, Size answers*</u>.

How accurate do you need to be? McCauley's shop tolerance is +/- **.1 degree**, yes one tenth degree, to ship a +/- 2 % consistent product. **Theodorsen can beat that accuracy**, but amateur computer programs just won't correctly calculate the ***<u>real</u>*** **flow**.

A better version of a much used misguided fix ---

On **fast**, **ground loving planes with a <u>High Fixed Pitch Prop</u>, <u>Stalled α^o and C_L inboard at Takeoff</u>**, *bad*, an **unwound broader outer prop, less α^o slow**, are tried to shorten the T.O. run, the **lower angle, broader chord tip, seemingly better slow. Better, to do that at the 3/4 radius, <u>not create extra tip loss</u>!**

Going for <u>more tip thrust is a bad idea</u>, creates Greater tip Loss, Loses the Thrust!.

7. **What Aspect Ratio do I Want?** The **job of a Wing** is to **make lift**, hold the plane up, the parallel **job of a Prop** is to make **Thrust**, **both cause induced drag loss**. You **Reduce Induced Loss by going Fast, *or* having High Aspect Ratio!!! The Plane Controls Speed here! High Aspect Ratio is GOOD!** That's how Sailplanes go slow with very low total drag, both low Induced, and low Profile, Slow *hurting* Induced, helping Profile. High A.R. is an absolute necessity for Slow Soaring.

Remember, it's really lower span loading that's key.

The **higher the Aspect Ratio** ***the better it is for Prop Induced!*** **EXCEPT long skinny things vibrate at lower frequencies,** potentially get the prop down to reacting to more engine frequencies, failing, tearing the engine out, maybe killing you. That's nothing you want to risk. ***Don't tinker with props***. **Prop engineers lose sleep, serious sweat getting you a safe prop!!!**

Definitions, A.R., Rn, etc listed after Chapter 1 Book II.

We use a favorable 14:1 Aspect Ratio, that of a Luscombe, which works OK for vibration on a Luscombe, but may not in another case --- simply to be consistent in our examples. It's **actually two 7:1 blades, two separate blades 180 degrees apart, with root vortices**, but props look like high A.R., so to not confuse people we do it that way - but do our calcs correctly.

Blades generate *separate* Archimedes Screws *affecting each other*.

8. **Airfoils The truth is that private plane props have the Speed of Jets, but the *small chords* of Model Planes, and thus regularly have Reynolds Numbers, *Rn, down as low as* 200,000, up to 1,000,000 +, bad news, *significantly higher drag coefficients*, also bad, *early stall*. Flat bottom airfoils**, unsophisticated, do help a lot for **easy accurate measurement**, an important basic. The **Low Drag Coefficients** that the **airfoil guys go for are just not real here!!!** Of course, you can always do better tailoring a perfect airfoil, **but there is *no super performance to be found here*. Thin airfoils, especially at the tips, maximize L/D, a key routine plus on metal props, composite props a little thicker tips - inboard always thicker.**

9. Mach Limit The easy way to understand the Speed of Sound **Drag Rise Limit** on Props is **simply stay below .9 Mach tips, compared to the vector combination of the rotational and forward speed.** Do you remember back in School how the square root of the sum of the squares of the sides - finds the hypotenuse of a right triangle? So a good calculator can find that vector speed - or by trig. Sure, I know most never did serious math since High School --- just understand.

The **Standard 59 F° day Speed of Sound is 1116.46 ft./sec., 761.243 MPH**, and decreases as the **square root of the Absolute Temperature,** (**-459.7 F°** --- thus **F°** + **459.7**, ie. **59 F°** = **518.7 R°**), thus **968.1 ft./sec., 660 MPH at the -69.7 F°, 36,089' constant temp. Stratosphere. For Info**: 60 MPH = 88ft./sec., so **15/22, or 22/15 converts. R° is Rankine Absolute.**

So, **.9 Mach is 1004.8 ft./sec., 685.12 MPH sea level std. day.** Just keep your prop tips below that and you're OK, a specific answer, easy enough, and it's faster, hotter in Summer. But, it's **slower, colder at altitude!!!** You get a speed-up going over an airfoil, but the reason **it stays simple, .9**, is that there is a **tip relief, overpressure escaping out the side door, the tip.**

Remember 1000 Ft./sec., the easy to remember limit!!!!!

On most private planes, Mach is just not a problem, but faster planes, a bigger diameter prop, high RPM can get you into it. A fast Harmon Rocket going 255 MPH with a 7', 84" prop does ***not*** go faster when you go from 2500 RPM to 2550.

Next, **Gus Raspet tried to open our eyes in the 50's ---**

10. Thrust Required, a basic question with an intriguing insight. If you knew an **accurate gliding drag** for your plane, or had an accurate drag calculation that, big surprise, happened to be correct, that may well ***not*** be the design thrust you need. **Thrust may *not* equal real Airframe Drag!!! When you turn on the fan it can cause an extra drag, a significant interference inefficiency**, and actual power required may be 10% or more, than drag **and prop eta**, η, would predict! That has been a very big hole in Aerodynamic knowledge, and ***a central objective in our work***. There is Scrubbing Drag, Pressure Drag, separation. A ***Luscombe flown Ideally has ~.893 interference efficiency, a ~ 75% prop η, *only 67% overall.*** **More to come!**

 ***.67 inverted demands 150% more H.P. vs. Plane Drag!!!**

Actually Designing a Prop --- 10 Specifics define the task!!!

1. The **3 Basic items** of a **Prop Specification** are: **Speed**, **RPM**, and **Altitude Density rho**, ρ -- set by the application ***objectives***.

2. The **3 Chosen items** are: **Aspect Ratio**, **Coefficient of Lift** C_L, **Coefficient of Drag**, C_D, **Physics sets the** C_D -- ***but poorly defined for low Rn***.

3. The **3 Design Steps**: **Sizing**, **(Diameter)**, **Pitch**, **Twist vs. Shape**

   Use H.P. (or find the correct Design Thrust is # 10 --- covered later).

a. **SIZE the Prop Right** to **deliver the required Thrust** at the **3 Spec Conditions**, the **3 Choices** above. It turns out that with the **highly variable Dynamic Pressure q**, **Diameter**, **Area**, **Shape** ***are interdependent***, **can't be separated**, but ***simply setting*** a **desirable**, **high aspect ratio**, puts them **in strict proportion** **so we just scale them all up or down** to give the **H.P.** or **(T# Reqd.)**

b. **PITCH** You'll learn that one of **the important capabilities of Theodorsen's ~Exact Math is to set actual Advance and Pitch to *account for inflow and rotation with heavy loading* - where the *ΔV at the prop* - is faster than the stream tube average --- everything to better than .1° accuracy, < 1%.**

   Geting the Precisely Correct Thrust vs. Radius Loading is the Game.

c. **Shape vs. Twist** Getting these **Precisely Correct** for the spec is the ***core objective of Betz-Goldstein- Theodorsen Math*** and happens automatically, *2 printout pages nails everything.*

One computer run solves everything, at every radius.

So, you can see that **the bottom line** is that **using the Aspect Ratio you set**, the **math and the computer simply scales the prop up, or down to meet all the spec conditions, the H.P.**, or **T # reqd, 2 or 3 pages of data covering everything** you would ever want to know about the prop **at every radial division set.**

We now use 10, 8% Divisions from 19%, outside the 18% Spinner, to 99%.

Andy has actually set up his program to be more in his control. He **guesses at Theodorsen's $\overline{w}$ factor, essentially the full ΔV of the pure helical downwash sheet, maybe 2, 3, 8 times *bigger than the stream tube average ΔV*** (!), and iterates **D"** vs. **A.R.**, **T#**, **H.P.** to just what he wants, **a modern computerized miracle.**

 If you don't know the Thrust Required -- Go for Tmax, at design V, HP available. ✓

Gearing for Ideal Efficiency --- Disk Loading H.P./ ft²

We've just learned that the **Plane** ***spec specifics*****, sets the prop design**, the **math** simply carries out those given Specifications for an **Ideal minimum loss design**. We also saw that once you've **selected Aspect Ratio, and all** the other spec specifics, the **math simply scales the prop diameter, the blade area and shape up or down to fit the demands of the case**, pretty **simple, logical** really, **something you can easily understand**.

Designing a 4 Engine Plane? Simply divide the Prop Thrust by 4!!!

But the case may, or may not have an **ideal efficiency, ~ 90%, maybe 91%, depending on Aspect Ratio possible, pitch, etc**. The **Advance Ratio Graph on p.58-II shows us that**. If you can just get an **optimum Advance Ratio**, closely tied to the **actual advance angle of the prop tip**, thus the **pitch**, or P/D ratio, which always has a little extra pitch to account for inflow and angle of attack, **you can get such great prop efficiency!!!**

Gearing down RPM increases Pitch, Torque, and Diameter. ✓

So theoretically, ***actually***, if you could **gear the RPM vs. Speed** for the **Steep Advance and Pitch angles**, (***~ 45° Blades best**), you can *(always)* get those amazingly good propeller efficiencies. Of course we've already shown you that there is an interference efficiency loss that can really hurt the overall results in some cases, and gearing is often not worth the weight, and complexity. Man powered planes, the Wright Flyer, Ultralights, **all slow**, are **all geared**, chained, belted, and do great - the maybe 5000 RPM ultralights, undergeared to not have excessive diameters, poorer.

*We'll soon analyze how we get ~ 45° optimum blades from ϕ - drag angle/2.

Disk Loading: The Luscombe, 85 H.P./ 27.5 ft², **3.09 H.P./ ft²** max, the RV 6, 160 H.P./ ~ 27.5, ~ **5.8 H.P./ ft²** max with ~ 6 ft. props, man powered, ~.25 H.P./ 100 ft², **.0025 H.P./ ft²** all work out with prop efficiencies right in line vs. Advance Ratio, Pitch. **Reno Racers, NASA Advanced, very high H.P., multi-blade props, 30 to 37.5 H.P./ ft² are hurt, below 80%, what's going on?** Simple, the ***Diameter is Artificially Restricted by Mach***. **If you can let math and a ~.5 C_L set D - Advance** ***&AR*** **sets efficiency!** ✓

Jack Norris's Incisive, Bottom Line Understanding of Props

Propeller Logic starts with a **terribly exaggerated "q"**, **Dynamic Pressure at the Tips**, trying to **grossly overload the Thrust and Drag at the Tips, where *Thrust must fall to Zero*, Paying for extra Profile and Induced Drag Created -- Losing the Thrust.**

Not Counteracting that Flaw is The Basic Mistake in Propeller Design!!!!

Since the formula for *Lift*, **Thrust** *and ~Drag* is **T# = q (Area) C_L** we can **counteract the excess q** by **simply tapering the Tips,** (cutting the Chords which represent the area per inch of Span) ***thereby pulling back excess loading from the tips*, cutting profile drag at a max radius - and stopping the generation of excess Induced loss!**

In fact, you can *taper them back ideally*, for *minimum induced drag* by simply using Betz-Goldstein-Theodorsen Math which will **exactly Shape, Twist and Pitch the prop** for any *chosen* **Speed, H.P., RPM, Thrust, Aspect Ratio, C_L, Altitude** ρ, **density**, (on a drag free basis). By choosing and designing for a **constant, optimum C_L** (and adding proper Low Reynolds Number Drag C_D's) **you can also achieve min. Profile Drag, (min. Area precisely placed).**

Analysis of the case shows that there is ***no basic inherent loss because of prop twist*, a constant dT/dQ** results, **a perfectly constant ratio of Thrust vs.** ***required* Torque** at every radius, *if drag free*. Also **Constant Pitch, perfectly Helical Inflow,** ***stretched* helical outflow, Constant Slip vs. Constant Plane Speed, which minimizes radial flow, tip vortex induced loss, *no extra radial imbalance* to the inflow and rotation of the air, though highly variable radially, work together to produce the Constant Pitch, Constant Slip, acting like the axial inflow were constant --- constant dT/dQ, *if* drag free.**

Newton - Actuator Disk Logic - teaches **Fast Planes produce Big $\dot{M}$, min ΔV, min ΔV/2** x T# = ft.#/sec., the axial induced drag energy loss/sec **=Min Power** loss, **min $\Delta V/2V_1$ axial efficiency loss**, also tends to produce **High Pitch, High Advance Ratio**, thus with sufficiently **Low RPM, High Pitch**, ***a min steep helix path, min profile drag loss also!***

Adding Andy Bauer's: 1. Low Reynolds Number Drag, Early Stall. 2. Retwist, rechord, for nose velocity profile, (slowdown). 3. Analysis Off Design Point. 4. Design Thrust accounting for Prop-Body Interference Loss, (Gus Raspet). --- Finishes the Job that took 138 years to complete, 1865 to 2003!!!

There Are Practical Confirmations that Betz is Correct

The **Curtis-Reed,** (and the Fairey-Reed props from the UK), had blade shape pretty much right in the 30's, with highly tapered blades, and the smart guys remember those were the best props of that time. Reed, actually a smart Dentist, got it reasonably correct early. In WWII the big engines forced the pros to wide tips to absorb gross power and stay below Mach, but less efficiently. Some believe *unwound wide tips* are better for takeoff, fast, high pitch props, excess α^o slow. The ***falacy is you lose the tip thrust you try to make***, the proper insight on loading vs. radius, shape lost.

And every time you Compromise with a Wide Tip it Will Cost you all day.

But in the 70's **Paul MacCready**, an indoor model builder in the 40's, conquered man powered flight by **creatively making an outdoor-indoor Gossamer Condor, substituting Mylar Film for a modeler's bathtub Microfilm! He won the long unfulfilled *Kramer Prize* for *Man Powered Flight.* But then Bryan Allen just couldn't pedal an incorrect broad tip prop on the Gossamer Albatross for 45 minutes to win Kramer's next challenge of crossing the English Channel by Muscle.**

Paul did have a Geared, Big Diameter, High Pitch Prop, smart like the Wrights.

To the rescue came Eugene Larrabee of MIT and his Star Student, Mark Drela with their own code for narrow tip ~ correct props, that had everything except Theodorsen's perfection, and MacCready made history again! Simply, Bryan Allen *could* pedal a ~ properly configured narrow tip Prop, the kind of Practical Proof that everyone sees is real!!

The final proof is B-G-T Props are *Quiet*, the Excess tip Loss Gone!!!

Confirmation by Ribner and Foster, University of Toronto.

Ribner and Foster, using modern computers in the 90's, .did a great service to the engineering profession and all people interested in propellers by **checking and confirming that the key Goldstein-Theodorsen work was correct to nominally 1%, modern confirmation**, a little more error on some of Theodorsen's pre computer, less important peripheral numbers. No one need think the work was old, unreliable, ***it was genius work.***

Everyone missed that in 1963 Tibery and Wrench at Divid Taylor Model Basin got Higher Kx Blade Loading for very High Pitch, High Advance Props above a Lambda .5.

 This completes the basic 20 page core --- but there's more, then Book II.

An Overview and a Look Ahead. I started looking for that explanation in the mid 40's, a model designer competing and winning at the National Championship level, (46 and 48 in the Senior Age Division) where I needed real, factual answers to optimize my props. Working as an Aero Industry pro, I looked and watched for a half century, a professional lifetime. The explanation never happened. **It remained the one central unexplained major subject in Aeronautical Engineering**, which had refocused on Jets, **left this job of bringing together and explaining undone!**

Yes, there had been decades of testing, a lot of reports written, **unfortunately most before Theodorsen got it right**. There was marvelous, genius level work done, that made our work possible. **Rankine, Froud, Betz, Prandtl, Goldstein, Glauert, Theodorsen are the 7 Historic Giants who made all our work possible**, each standing on their predecessor's shoulders, the ***building block milestones*** that created **the Ideal Propeller in 1948, 83 years after Rankine's 1865 first *water* prop analysis.**

What remained undone was reading and truly understanding, ***using*** **Betz, Goldstein - Theodorsen's marvelous book, *adding* Low Rn Drag and Stall, *figuring out what it all really meant*, then accurately adapting the theoretical ideal prop to a *real* variable velocity nose, what happens "off Design Point", *realize* Gus Raspet** showed us **propulsion efficiency** could be **as bad as 58%** on a Bellanca Cruisair, **172% more power required than the plane's drag said. WOW! Raspet deserves to be Historic Person # 8. Doctor Andy Bauer, my partner, brought it all together, finished the engineering job, deserves to be # 9, a decade long final red ribbon on the propeller!!!**

I invented ZTGT, Zero Thrust Glide Testing to get Real Drag, Real η_P.

My job in the collaboration has been how the entire order, logic could be understood, made logical, explained. It's now been 138 years since Rankine's 1865, maybe the longest challenge in Engineering. **If you grasped those 20 pages you are well launched, the comprehensive explanation is Book 2.**

A Look Ahead to Book II, Propellers

In hindsight, you'll see that this **Primer on Prop Logic** can get you **a giant step toward a complete understanding**. Just give it your best shot, **maybe reread it, think about it, soak it up**.

I've found that no matter who you are or how smart you are, it **takes at least a little time to soak it all up, truly get it**, be comfortable with it. None of us, including Engineers, usually come across a subject where so many things happen, all interdependent. Figuring out just how to get this basic meat across, frankly, is almost as tough as figuring it all out in the first place. **The challenge is to not drown the new guy, help him!**

Superemphasis, always tieing together, finally helps Novices see it all.

In Book II, Propellers, we simply, flesh it out, for those who want to go all the way, to dive in, learn it in more detail, **slowly enough that a novice has a shot** at what engineers never quite got their head around well enough to write it out in understandable English ------ but also with enough disciplined detail, that your final grasp ends up genuinely correct, so Engineers, everyone, can get the final correct grasp.

---- I send you to the final conclusions early, which helps a lot too ----

Looking to Book II, *a few examples*, $\dot{m}$ to $\dot{M}$ to start. In pulling in and throwing back air, **the prop pulls air in from a slightly bigger diameter**, and that, naturally enough, ***raises*** **the natural streamtube mass flow rate $\dot{m}$**, (the plane Speed -- times the prop disk area -- times the density ρ) --- to **the final correct $\dot{M}$**.

Happily, that $\Delta V/2V_1$ factor, we learned and used for efficiency loss, is also the correct increase here. $\dot{M} = \dot{m}\,(1 + \Delta V/2V_1)$. **We teach you the $\dot{M}$ does NOT increase behind the prop as the second $\Delta V/2$ is added** - understandable enough when you think about it, because the prop can't create air, only shove along what it pulled in. We found that when we laid that extra complexity on Novices some drowned, so we left it to Book II.

But, here it is if you can grasp it now.

There are many finer points like that -- you learn in Book II.

The concept of Δ**V**, and Δ**V/2V**$_1$, is really useful, helpful, because it helps you understand the basic logic of the **Actuator Disk - Stream Tube**. You'll learn that it most directly ties into Glauert's **a factor**, ***the axial inflow***, **(a decimal ratio of Plane Speed, or V$_1$)**, 2a equal to the full, ΔV, ***(but different at every radius)*** ------- his **a'** and **2a'** likewise handling the rotation. You saw that first on the ***airflow sketch*** on ***p. 124***. Now in the real world and Theodorsen's math, the ***"Vortex Sheet" speeds up much more than the Stream Tube Average*** Δ**V**, maybe **2x, 3x, 8x** so you'll ultimately learn Theodorsen uses his $\overline{\overline{w}}$ **factor**, bigger than 2a.* His **kappa** factor is used to knock down to the average effect, but the **prop correctly pitched for the bigger** $\overline{\overline{w}}$ **factor, actually** $\overline{\overline{w}}$**/2!**

You can see that it **would be a big mistake to drown everyone with all the detail at first, better to help them get the basic concept first**, then become aware of the precise **details** in a **longer, slower more comprehensive Book II.**

I expect many professionals to hate the Super Emphasis, the constant repeating, summing up that I do, but I've found that's the only way the Novices can get it, and frankly I find the pros need more help than they realize. **Everyone needs time and all the help they can get**, because it takes everyone, amateur and pro alike, time to soak it all up, **absolutely needing the summing up, because no one, not even the pros can sort this swamp out while they're learning**. I think, I hope you'll end up appreciating the Super Emphasis, and always summing up, because **all you have to do is read through it, and I keep leading you to the correct conclusions --- that took a long time to sort out, check out**. *No vanity here, it took long hard work!*

If you can get a good first grasp in an hour or so read, a professional grasp of what's never been explained, understood ------ in a day or so ---- we've both won Big!

We'll look at Engines, then Leaning, then those who wish can nail props in Book II, *all laid out on a platter for you.* **Now a little Extra.**

* **With Theodorsen's Heavy Loading his $\overline{\overline{w}}$ is bigger than Glauert's 2a, which is NOT normally in a BGT Analysis. We synthetically calculate one for insight, less than $\overline{\overline{w}}$.**

Prop Vibrations

Prop Vibrations are such **a complete snakepit**, that frankly, I hesitate to get into it very much, since sorting out props is enough complexity and potential success for one book, but there are several insights that are worth grasping, and ***one great big, pretty simple fundamental that I want you to grasp clearly.***

We probably all think that the marvelous Emmmuuugghhhh **Doppler Frequency Drop Sound of a V 12 Merlin Engine passing overhead and away** is the **most beautifully smooth sound in flight**, and it is, at least it's my vote. But to fully appreciate the fatigue failure challenge the prop engineer faces, look at the vibration output on the facing page of that Merlin, a **hidden nightmare for the prop Vibrations Engineer**, all that rotating, reciprocating, twisting, flexing machinery, a **myriad of forcing frequencies, and notice Torque Vibration changes as Blade Angle**, β°, **thus Torque Load changes, a *tech horror.***

It's chance matching of forcing and response freq.'s that's trouble.

Now, realize **the prop has more than the one main natural frequency (the bending vibration, generally perpendicular to the blade face),** that it will respond to. It can twist, lag-lead in torque, bend with nodes, the center of the prop moving forward, the tips and hub moving back, for example, respond to **harmonics**, multiples of the forcing frequency or natural frequencies, getting hit in tune every other cycle, or 1 out of 3, for example. Should I use the word **nightmare again? Sharp Guys like Brian Meyer at Hartzell, with huge experience and feel for the real possibilities hiding in that swamp of chance matches deserve our profound respect and thanks for a safe prop. Could you cope? Marvelously they develop a feel for where trouble can be!**

Don't start reshaping, messing with your prop. You could get dead ---

Now here's the **one simple key insight** I want you to **grasp clearly.** If you go for a great, but **too high Aspect Ratio, a long skinny prop, you lower the natural frequencies of the prop, *maybe* down into matching more engine frequencies!** ***Caution!!***

Torsional Vibration of Rolls Royce V-12 Merlin Engine

ENGINE VIBRATION

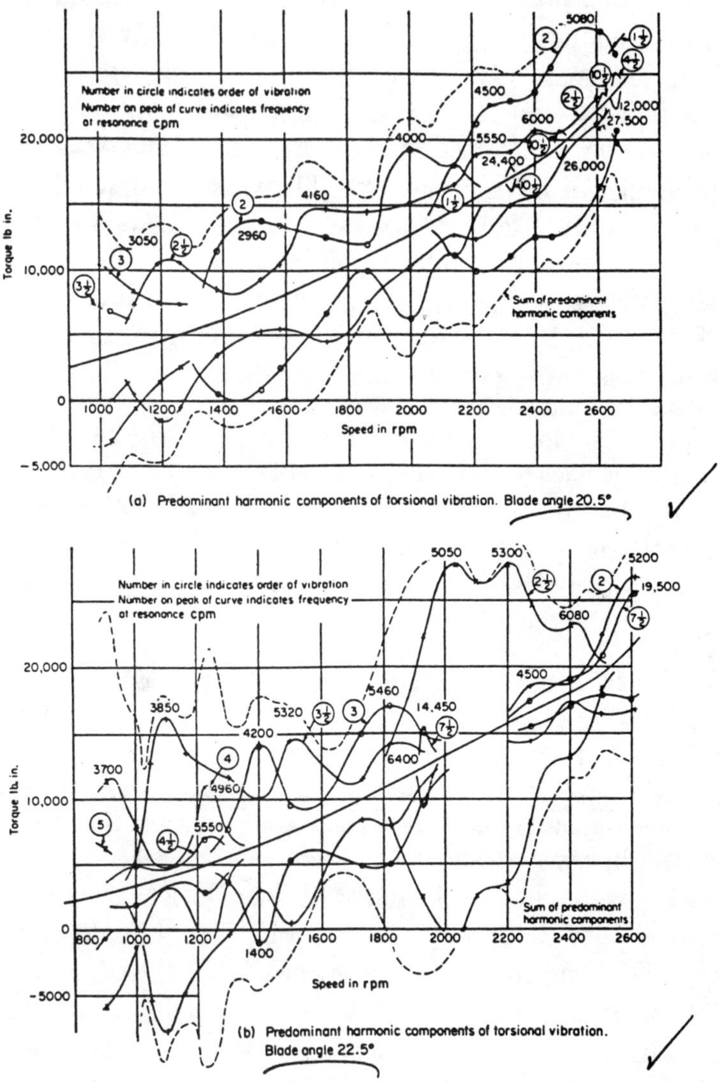

Fig. 8-29. Torque at drive end of crankshaft of Rolls-Royce *Merlin* II aircraft engine with variable-pitch propeller, 1-to-2 reduction gear (Carter and Forshaw, 8.492).

Data compliments of Will W. Mathews, Director of Manufacturing Engineering. Engine and Foundry Division, Navistar International Transportation Corporation.

The Ideal 45 Degree Blade Angle, and **where it's located**

When I was a late teenager competing and winning Nationally, I had very good plane designs, but was really frustrated trying to find correct prop design specifics. What was the ***really correct shape, twist, pitch?*** **It seemed that those high angle prop root radii, pulling sideways against engine torque, more than forward, could not possibly be good.** ***But***, we've learned here, that **drag free, a Betz prop can have a constant Thrust vs. Torque Ratio, every radius equally efficient, productive,** ***(if first considered drag free).*** We'll look more at high inboard blade angles shortly, see that they're perfectly OK, **working as a screw** on very high pitch props. We'll also see a **Blade Angle of ~ 45° is optimum**, thus it's insightful to see ***where*** the **45° blade angles are vs. Advance.**

Also --- Where is the Max Chord vs. 45°?

The study below looked at **J/π**, (Lambda's, λ's), of **.2 to .88**, **(of low to very high pitch props)**. **(J/π, λ Lambda, is the advance ratio of Prop Tips.)** The chart below and the chart and Shape Plot vs. load distribution on the next page give great insights into the characteristics of ideal propellers and are worthy of your study.

	Reasonable Advance		**Extremely High Advance**				
J/π, λ	**.2**	**.3**	**.4**	**.5**	**.6**	**.7**	**.8**
Overall η	.8607	.8940	.9068	.9125	**.9149**	.9144	.9112
45° β,% r	23.25%	33.60%	44.15%	54.82%	**65.74%**	76.70%	87.82
Local η	.8781	.9018	.9104	.9141	**.9159**	.9158	.9135
Max Chord.	25%r	38%r	43%r	47%r	**49%**r	53%r	56%r

Amazingly, a Reno Racer can have a .88 λ, almost a 45 degree Tip!!!

So we see the most efficient **45°** blade can fall way inboard to way outboard - but is **most efficient at the ~2/3⁺Radius at .63 λ**. Now if the ~**45°** Blade Angle is most efficient, it might be that the splendid Math might put the **Max Chord** there. Well, if you look close at the data above, the interesting insight is that's trying to happen, at the lower **J/π**, λ's where fixed pitch props operate --- but are pulled outward a little at the low **J/π**, λ's away from the inner root vortex --- **much stronger inward at high J/π, λ's away from the stronger tip vortex!!! Now, isn't that *smart*! It fits right into the logic, not overloading tips!!**

 Interestingly, the max chord is at ~ 50% r. on the max η ~ .63 λ Prop!!!

The Change in Blade Shape and Twist --- Low Pitch vs. Steep Pitch.

We know it's the **Slow planes, Low Pitch, <u>Low Advance</u>**, that have **the *huge magnification of Dynamic Pressure, q,* <u>tip vs. root</u>**, thus **need the <u>most highly tapered blades</u>**. ***<u>Conversely</u>***, the **<u>Steep Advance</u> props will be <u>less severely tapered</u>**. Below, see the blade outlines for this blade study, λ from **.2**, low, to a steep **.88λ**, (a Reno Racer class prop, though this study of all 6' props for comparative consistency is a miniature for Reno). Note how the **<u>steep λ props</u> *<u>taper in greatly inboard</u>***. (**<u>Slowdown</u>** ***lowers blade angles, especially inboard, widens blades.***) ✓

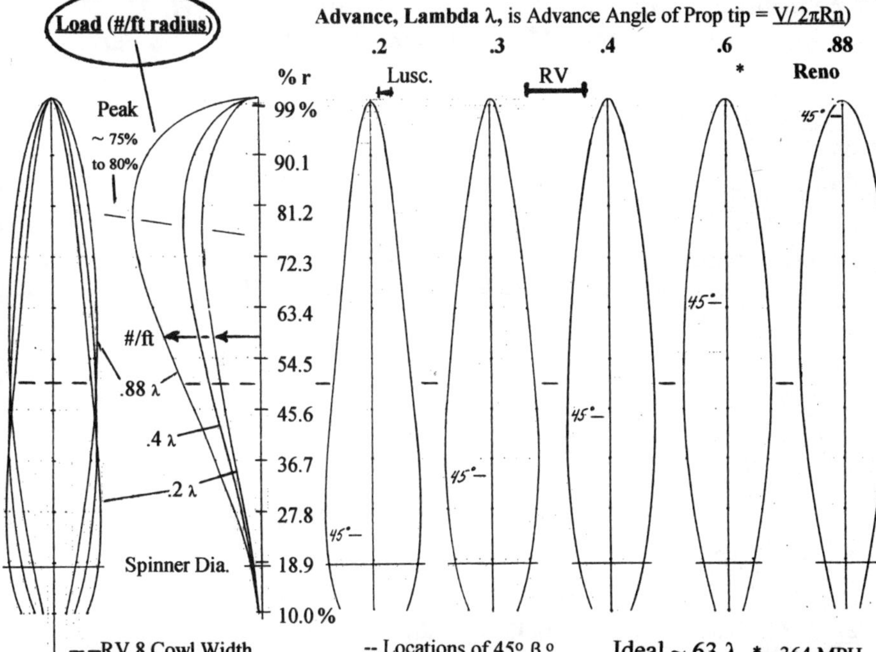

--RV 8 Cowl Width -- Locations of 45° β °. Ideal ~.63 λ * ~364 MPH.

To keep all props on the same apples vs. apples comparison, starting with a 6 foot prop we end up with low thrusts and H.P.'s for the fast planes, like a miniature, slick Reno Racer finally, but this shows shapes and twists well, the intention here, so just imagine and understand.

***<u>Prop Blade Study</u> <u>basis</u>*: 6'Dia, α° 1.1°, .5 C_L, AR 15, <u>2700 RPM</u>, 8000', Area 1.944 ft²**

Ref. Data pages 99.56 -1 to 99.56 - 8.1 BLADE ANGLES ARE *<u>BEFORE</u>* SLOWDOWN CORRECTION ✓

λ	Vmph	HP8000'	Thrust#	Eff. η	β°, 18.9% rad.	β°, 99% rad.	ΔTwist°
.2	115.6	56.5	155.3	.8607	50.365°	13.600°	36.765°
.3	173.5	94.4	179.7	.8940	60.583	19.046	41.536
.4	231.3	140.5	203.6	.9068	67.053	24.263	**<u>42.790</u>** ✓
.5	289.2	195.9	228.5	.9125	71.382	29.142	42.240
.6	347.0	260.5	253.8	**<u>.9149</u>** ✓	74.440	33.634	40.806
.7	404.8	334.5	279.1	.9144	76.715	37.760	38.955
.8	462.7	418.4	304.3	.9112	78.482	41.556	36.926
.88	508.9	494.2	324.6	.9058	**<u>79.653</u>** ✓	**<u>44.411</u>** ✓	35.242

Multiple Blades (If You'd Like a Little More Insight)

Generally, it's Most Efficient to use 2 blades, a Max Diameter Stream Tube -- **use *multiple blade* props if you have *BIG Power* to absorb,** particularly if you're **diameter limited, up against the .9 Mach limit. It will cost you a little to a lot in efficiency** vs. what it might have been with 2 Blades.

Staying with equal Aspect Ratio, 3 Blades gets a smaller Diameter.

People like **3 Bladed Props**, often feel they're **smoother**, but it **costs them a ~1/2% efficiency penalty, *smaller Diameter!!***

3 Blades, it can come out narrower, High AR, balance Diameter Loss!!!

The **WWII big recips**, had to go to **3, 4, 5, 6 bladed props to absorb the gross power of bigger and bigger engines, as do Turboprops,** right up against **Diameter and the .9 Mach limit. As you go faster vs. tip speed, less q ratio**, you can use **less severely tapered, broader tips and still have ideal shaping.**

But Physics works and you still DO NOT go to Excessively Broad Tips.

Unfortunately our guys seem to have lost their way, going for square, or nearly square, broad tips, not really an OK move, pumping big power into Excess Tip Vortices. **A High Activity Factor**, good Power Absorption with broad tips, must have mesmerized our people into Thinking Broad is OK. If you look at a **Mark 22 Spitfire**, you see a **5 bladed, (probably close to Ideal Shaped) Rotol** prop used to absorb the 2375 H.P. of the Rolls Royce Griffon Engine. **The Brits were maybe a bit smarter than our guys there, *credit to where credit is due.***

BUT, *if Smart*, A *Thin*, *Unwound*, lower C_L broader tip can work better near Mach!

An extremely high pitch, geared down, low RPM, Reno Racer, **37.5 foot advance** per rev, can have **a Theodorsen's $\overline{w}$ factor, a local ΔV, ~ 8 times bigger than the average ΔV of the stream tube -- hardly utilizing the stream tube! More blades helps there** logically - but you can have **a Very High Disk Loading vs. Power**, in **a limited Diameter, which costs efficiency, (in a way the opposite of a Big $\dot{M}$ for a small ΔV),** though the **$\dot{M}$** is huge. The many bladed swept Turbo Props NASA tried to replace Fan Jets were **79%** efficient at best, better than the **Fan Jets, 65% or less, with duct loss.***

 * Later BIG Fans are getting better.

A Pro's More Subtle, Incisive, X-Ray Grasp

When you first dig into Prop Analysis History, Betz's Ideal Minimum Induced Drag Prop Theory, as we did, you first realize that the game is to simply produce **a constant pitch, pure helical inflow, stretched pure helical outflow, constant 2 π r tan ϕ** mathematically, ***which is constant air inflow pitch.*** It takes a lot of time for all the full significance of all the other characteristics that are created to become completely clear to *us mortals*, and I expect it will take time for it all to become completely clear to you. ***I can help you by superemphasizing insights***, **hammering home the many key basics!**

Only in time do you **fully realize** the **subtlety** that the **constant slip** that results, despite highly variable axial and rotational inflow, **acts like constant inflow at all radii, balances the flow across the full radius, *minimizes* the radial flow, *minimizing* the tip vortex, thus minimizing induced drag - right at the Heart of Betz Minimum Induced Drag Logic!**

Theodorsen Minimizes Root Vortex too -- Tapered Inboard Blades!!!

Only in time do you pick up and **realize the full significance that a constant dT/dQ results,** and that implies that there is **no fundamental penalty to a twisted prop,** that on a drag free basis it means that **every radius is equally productive, equally efficient, drag free, that it slices like a razor** through the **twisted trigonometry and complex 3D heavily loaded flow, genius level math, to a magnificently dead simple ideal outcome** --- that you **can run the computer with and without the low Reynolds Number drag and precisely see the drag effect at every station & overall.**

You only fully understand, when you grasp props from *all* vantage points!!!

Only in time does one fully realize the **Really Simple Point** that the **Dumb Prop has a Bad Basic Problem, *trying to get max thrust at the tip* --- *where it Must Fall to Zero Thrust* --- making excess profile drag at max lever arm, a stronger tip vortex, *thus* unnecessary excess induced drag, losing the paid for thrust** - that we **play right into that fundamental flaw by making wide tip props, UGH! - But Great Fundamental core Insight!**

We purposely start you out with the Norris - Bauer Law, emphasizing SHAPE!!!

So we purposely start out showing you that narrow outboard chords can physically fix that Dumb basic problem of overloaded tips, that enough taper can actually ideally solve it all in such a magnificently logical way!

It's funny to watch some busy Critics -- Just not get it!!!

I expect plenty of grammarians, and technical purists will be all offended, and in blind wrath criticize that I've broken the usual style rules, used *super emphasis* - **never realize** that this has always been the subject that was so technically complex that it was never comprehensively, incisively, understandably explained, but that there is magnificent, simple enough order and logic, ***if one can only figure out how it can be explained - understood. My core objective has been to figure out how such a complex subject can be explained and understood by everyone who wants to understand props!***

Reorienting Your Thinking --- to a 45° Helix Blade Angle, β.

Thinking of your propeller as an Actuator Disk, whose job is to produce Thrust, a **Disk, perpendicular to the crankshaft Pulling Straight Forward - Limits Your Thinking, *vs. the final truth*!**

A little bit of math will show you that **nominally a ~45° Blade Angle, Beta, β, is most efficient, screwing you ahead!!** A little classic math derivation shows *the* **optimum airflow angle phi, ϕ, is 45° - γ/2,** 45° minus half the drag angle, gamma, that is **half the C_D / C_L.** If the **L/D** is **nominally 50:1**, $.5C_L/.01C_D$, the inverse is .02, a **2 % D/L**, an interesting insight. Trig will show us that's a ~ **1.1457°** drag angle, half that ~**.573°**, so the **optimum air inflow angle ϕ is ~ 44.43°**. But, you see that if we want a .5 C_L, that needs a ~ 1.1° angle of attack, **45.53°**, so the bottom line is: forget the detail, the complexities, ***about* a 45° β Blade Angle is optimum** - **Easy**!

Now that you've been through the build up of the basic core Logic of Propellers --- It's now proper to **fundamentally reorient your thinking** and ***realize the proper bottom line*** way to think about an ***optimum*** **propeller is as a 45° Helical Screw** *(average)*, **Most Efficiently Screwing to the Destination!** ***Voila!***

Let's Understand Flight Physically - Get a Proper Feel for it.

It's this simple. The **Atmosphere** is far more **massive** than we all realized, and what we learn here is **the efficiency game** is to **use its Mass, disturb it as little as possible, the least ΔV.**

Think of a **long wing 22 Meter Sailplane skimming along,** ***only*** a **whisper of downwash**, with a **60:1 glide ratio**, able to **land** ***anywhere*, out past the horizon, *120 miles*** from 10,500', 2 miles! **Wow!** A **747-400**, 940,000# GW, **nearly a million pounds**, also able to skim along in/on the atmosphere, **at any given C_L** only the **same angles of downwash as a 1400# *put put*, *if equal AR*, Wow!**

The **Job of a Prop** is to **Screw through its Stream Tube**, at a **High Pitch Angle, use the considerable Mass of the Stream Tube, *but* disturb it as little as possible -- only a small $\Delta V/2V_1$.**

 Betz teaches us to go for a Pure Helical Downwash Sheet, *not* a constant angle!!!

Physical Examples Really Help cut through the Complexity.

An **RV at 200 MPH** x 22/15, **293.33 ft/ sec,** at **2700 RPM** /60, **45 RPS** divides out to **6.5 feet per revolution Real Actual Advance**, thus **needs more than a 78 inch pitch or it can't physically do it**. **There's a really interesting insight there:** We know now that a prop must be **overpitched to allow for Slip**, the necessary **inflow**, plus a small angle of attack, yet a typical wood, fixed pitch prop is apt to be **70"D x 79"P --- no allowance for slip, what's going on???** The actual velocity profile at the cowl nose, a highly variable **slowdown**, that requires **depitching** the prop to be correct, **almost exactly equals the Slip**, (in this case)**!!!**

The Dreadnaught, or Furias, British Sea Furys, **Reno Racers** are the ultimate example for **teaching us about Props Screwing Ahead**. Pratt & Whitney 4360's with a Boeing C - 97 **3/8 ratio gearbox**, driving a ~13'5" Douglas Skyraider prop, 1125 RPM at 3000 RPM engine Speed, **18.75 Rev's /second**, at **480 MPH** x 22 /15, **704 ft/sec** divides out to **37.5 feet / rev Advance, WOW!!!** ----- **That's real easy to do on a calculator.**

Do the numbers yourself, for the RV and the Furias and you'll **really get it, believe it, never forget it, really grasp props correctly. Props really are an Airscrew - and that's really the correct way to understand them.**

Let me show you that there is a **big side benefit to correctly understanding props as a Steep Airscrew**. As a kid, looking for the good prop explanation and finding absolutely nothing incisive - I was really bugged that I could see that **steep inboard blade *angles hardly pulled forward***, ridiculously pulled against engine torque -- very undesirable *it seemed, hurting, hardly helping*. **Wrong!**

Visualize that prop blade screwing up a perfect steel screw.

Well, the prop doesn't have a steel screw to push against, its **Airfoils have to fly,** (lift) **it up that imaginary screw, min. slip.** That's why steep pitch angles are OK, (**Maybe a lot more than 60°**) **The prop is *not* trying to pull forward*!!!* - *rather* *screw forward steeply!* *The blade angle is correct when it's correctly alligned with its helix angle!***

 Simply Taper the Tips, Go for Ideal Betz Shape, Hi Speed, Lo RPM, Hi Pitch!!!

Seeing Through the Morass of Prop Physics - Our Hero Dr. Andy Bauer

The whole History of Science, of Invention, of man's progress toward a better life has been one of years of Perspiration, followed by an instant of Inspiration, of seeing the final answer. Seen more clearly, *only the prepared mind* can finally *see through all the complexity*, to the simple final, *invention*, solution. Well this task was **a long string of inventions**, insights to cut through the prop logic swamp **to finally see the total logic of props**.

My partner, and lifelong friend, **Doctor Andy Bauer**, had the raw **intelligence, advanced education, and lifelong experience**, specifically in **Aeronautical Engineering**, to incisively go through this grossly complex, multifaceted problem of propeller logic, **that no one else ever mastered well enough to explain**, see the logic, make it explainable in words, not impenetrable math, that prevented real understanding --- **the final insights**.

Over a ten year retirement intellectual challenge, we sorted out all the obscure hidden insights that are seldom, if ever, seen. Andy saw that Glauert's 1934 great work, was based on Prandtl's F factor at best an approximation. Goldstein's brilliant 3D 1929 math, and Theodorsen had the essentially exact math!!! Others failed to understand. Andy, smart enough, read, understood, and correctly used Theodorsen's masterful, final, exact heavy loading solution with Goldstein's 3D math. With Thrust and Pitch Errors in Glauert-Prandtl, **Theodorsen-Goldstein was clearly the way to go.**

Others who tried B-G-T math simply didn't understand all it implied!!!

Most significantly, Andy was smart enough to see that the bottom line result of Goldstein-Theodorsen was a **Kx blade loading chart vs. radius for the various Advance Ratio's** that a computer able guy could use, a **practical world usable tool**, for **quick, easy ideal prop design**. Most significantly we sorted out all the separate integrated great, but very obscure, basic points of **Betz's brilliant 1919 Historic Insight**, many never appreciated before. **Constant dT/dQ**, drag free, showed that there is ***no penalty to a twisted prop***, drag free, cutting through all the complex Trig., 3D flow and loading complexity to **a dead simple, magnificently desirable final outcome, every radius equally efficient, valuable and usable**. **Constant Slip** is the obscure balanced flow vs. radius core of **Minimum Induced Drag**, that is based on **Ideal Loading, through Ideal Shaping** of the blade, with matching **Twist**, that gives **constant Helical Pitch** of the **air inflow, constant Slip**, a one page explanation that nails what was **just never correctly understood**.

Andy, very early was able to see that there is **a highly variable velocity profile in the prop plane**, that acts like a **slowdown** in inflow, the body pushing a bubble of air forward. **Unlike others, Andy knew how to correctly calculate that profile** and **accurately correct the prop for it!!!!**

Slowdown - Adapting a Perfect Prop *for* Nose Velocity Profile.

Amazingly, the Slowdown was **nominally 50%** at the 19% radius next to the Spinner -- 67% at the next 27% station, 33% slower -- only 75% at the 12.6" r., 35% radius, on an RV 8, ***Big, 2 1/2% out at the tip!!!*** Obviously **we needed both a Blade Angle and Chord Change** to adapt a mathematically perfect prop **correctly** to a slower, lower q, variable Nose Velocity Profile.

What **Betz** did **in 1919** was **adapt a Rotating Wing, a prop**, for the **Major Rotational Slowdown**, Root vs. Tip, and he did that perfectly, logically, by demanding, **holding Constant Slip, Constant Pitch of the Air Inflow. We had to now deal with Axial, not Rotational, Slowdown.** At first, it seemed we should set up the modified blade angles and chords to **continue to hold, Demand Constant Slip**, still hold constant angle of attack, constant C_L in the **new Velocity Profile.** If we'd **widen the chords to counterbalance the now lower Dynamic Pressure, q,** we could potentially have **a still perfect Betz-Goldstein-Theodorsen prop, perfectly adapted for the Real Velocity Profile of a Plane's nose**, hopefully to the desired and Mathematically possible .1 degree, the proper prop tolerance for a consistent manufactured prop product out the door, McCauley's tolerance.

Well that doesn't work, and neither does holding the original thrust vs. radius load distribution. We started generating **ridiculous wider chords inboard, truly ugly props mathematically, 15", 18" root chords.**

More happens with the lower q field. Running in a slower airstream, you can generate the same thrust at lower H.P., but you develop a counterbalancing pressure drag on the fuselage that requires that you raise the gross thrust, now requiring the original H.P. to get the same original net thrust, potentially a new situation causing no new loss, cost free drag!!?!!!

What makes sense is to design a new ~perfect Theodorsen Prop to get the original required (gross and) net thrust, at the original H.P., now running in the slowed air to achieve **an essentially perfect new Theodorsen prop** that, of Prime importance, **holds the thrust distribution on the crucial outer blade.** What happens is that we **get wider blade chords on a still normal looking Theodorsen Prop with a little lower Aspect ratio**, to make the higher gross thrust in the slower air, at the original H.P. Actually we get a tad less thrust, at maybe **.3% lower efficiency, a little less thrust as we move inboard toward the already weak root, no big bad effect**, since we designed for the proper gross and net thrust. We do lose the constant Slip, but blowing less in the plane's face, and less root vortex, it's **the best fix.** Using **Source Sink Analysis**, we got **new insight on the whole flow field.**

 An Accurate Slowdown Correction makes the prop much closer to Ideal, SLOW!!

Maybe Explaining Terrible Overall Propulsive Efficiency - ***on Slow Planes.***

Gus Raspet, somewhat of a Maverick, head of the Aerophysics Department at Mississippi State in the 50's, was a creative force in Aerodynamics, very much deserving of important Milestone Recognition. Among many other accomplishments, he did towed propellerless glide tests of several private airplanes. He found normal, expected drags, but **often terrible overall propulsive efficiencies, a startling 58% on a Bellanca Cruisair**, when a taped, sealed, clean plane's drag was compared with actual power required!! Other light planes, a Cub, a Cessna 120/140, etc. showed similar results -----

Still almost unaddressed a half Century later, this is perhaps the largest data and basic technical hole in all of Aerodynamics, matching the prior lack of a valid propeller explanation, a key part of our overall propeller quest. By inventing Zero Thrust Glide Testing, a simple, insulated feeler wire that lit a bulb when the prop quit thrusting and slid rearward in its, maybe .016" axial crankshaft endplay slop, I found a 14.94 RPM per MPH relationship *near cruise* on my 71"D - 51"P Luscombe Prop. I found **Real Drags**, for the first time in a normal flight configuration, a nifty little breakthrough in itself, and like Gus, found terrible overall propulsive efficiencies, (reported at the Reno, Dec 1990 AIAA Conference, and our J of A. Vol. .30, No 4, July - Aug. 1993 paper).

Leaving the cooling open, *of course*, for otherwise normal flight, we found only a 67 % overall propulsion efficiency, charging cooling to the airframe, quite comparable to the 58 % Bellanca, since cooling drag is often expected to be ~ 10 % of overall loss. Interestingly, glide testing 5 miles out to sea at dawn, (out of heavy local sea-shore lift, sink, and circulation), in dead stable air, I got excellent, amazingly reproducible data, as good as we might ever expect to get, we found ragged, but speed reproducible overall propulsive efficiency data, η_P showing, I thought, variable flow conditions over the embedded body, tail etc. Interestingly, with subsequent CAFE team testing, we found little extra propulsion efficiency loss on a faster, clean, RV 6 and the excellent Whitman Tailwind --- that, perhaps most significantly, would have faster, more efficient props, smaller ΔV's and $\Delta V/2V$ axial loss, the ΔV extra propulsion wind, a much lower percentage of the flight speed, much less extra propulsion wind effect over the ~ 1/2 of the wetted area exposed to the slipstream Now, **What was the Explanation of all this???**

Early on, in our work, we presumed that my McCauley Luscombe prop, which could have an 85 % B-G-T efficiency, might be down to only 82.5 % η. However in Van's RV Prop Test Program we found a European RV 8 prop had double the B-G-T loss, an 80% η, not the 90 % η of a proper prop! With low pitch props logically bigger losers a 75 % η was quite possible.

To explain: Low pitch props like the Luscombe, a P/D less than 1, 51/71, at a slow 100 MPH, there is a huge q ratio, tip vs. a slow root, airplane speed, a ~ 500 MPH Tip, and the need for an almost triangular blade shape to pull excess tip lift inward ***to better place Thrust, kill excess tip losses***. The very non triangular, square klip tip McCauley prop could easily be worse than the European RV 8 prop, easily lose 10 % efficiency vs. an 85 % max. potential BGT η, we realized. Most Significantly, at 100 MPH TAS Altitude Cruise, at 75 % η, the 49 H.P. Altitude Cruise would only produce 137 # Thrust. Then, at .67 overall propulsion efficiency, η_P, a .75 prop efficiency, η, that would mean a .89333 interference efficiency, η_i. (Realize .75 x .8933 = .67.)

So, bottom line, the 137# Thrust developed at .75 η, at 49 H.P. at 100 MPH, TAS, at altitude, (not 182.6 # T, at 100% η), the resulting 137# Thrust, working at only .98333 η_i, would only produce **122.38 #** of resulting Thrust, a long way from 182.6 # T!!! (Notice 186.6 x .67 = **122.38 #**). Now that's **just beautiful**, because the Luscombe Zero Thrust Glide Tests at 85 MPH, IAS, 100 MPH TAS at 10,500' Cruise Altitude shows ~ **123 # Drag**, ***Viola!***)

If you go through the drill of determining the extra ΔV wind to make 137 # Thrust at the $\dot{\mathbf{M}}$, mass flow rate at Altitude, at 100 MPH TAS, you get ~ 12 MPH. Now 100 MPH + 12 MPH, $(1.12)^2$ = 1.2544. Now, the 123 # gliding drag divides up into ~ 83 # Profile Drag, and 30 # Induced at the 1250 # Gliding test weight. <u>Nominally half the profile drag skin friction, and messy Aerodynamic intersections would see the stream tube ΔV</u>, and 41.5 # exposed profile drag x the 1.2544 scrubbing factor would be 52 #, 10.5 # extra drag, a major part of the 123 to 137 # difference, 14 #, leaving 3.5 # to be explained by variable separation drag, just the kind of variation we saw on the bumpy, slightly irregular Propulsive Efficiency Plot that went from ~ 76 % at extremely low Power and Thrust to only ~ 62 % at high power, hi Speeds, all seemingly making sense with some variable separation drag, just what we'd expect on a less than pristine first Sheet Metal airplane.

All the other gliding data plots were smooth, gliding in pure air, at dawn, out to sea, all looking like the best data we could ever possibly hope to get. So the slightly bumpy Overall Propulsive Efficiency looked absolutely genuine, just exactly what would seem reasonable, certainly making sense with better efficiency slow, degrading as more and more ΔV and power was applied

Subsequent CAFE Testing on the RV 6 and Whitman Tailwind, far better Aerodynamic Planes were harder to do because new Lycomming Engines, with hardly any axial slop, after the hot aluminum crankcases swallowed the axial slop vs. the steel Crankshaft. Very significantly, a ~ 9 MPH ΔV vs. ~ 180 MPH on a fast plane, is only <u>a 5 % Stream Tube Speed Increase</u>, <u>a ~ 10% increase on maybe half the profile drag</u>, <u>much less crucial</u>. **More work on this subject, to further nail it, is fertile research ground!!** ✓

The UnThrottled Thermodynamic Cycle

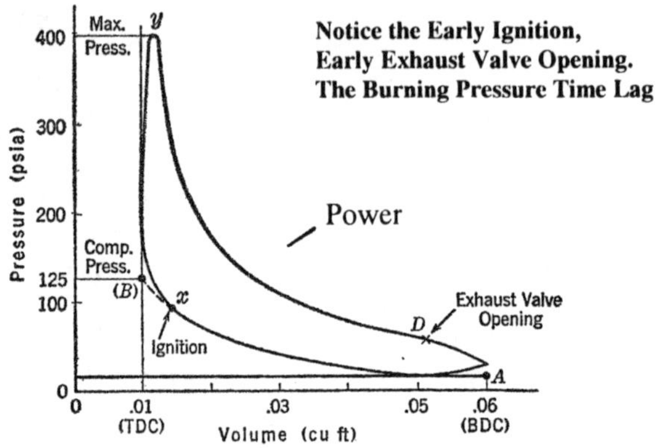

pV Diagram for SI Engine at Wide-Open Throttle.

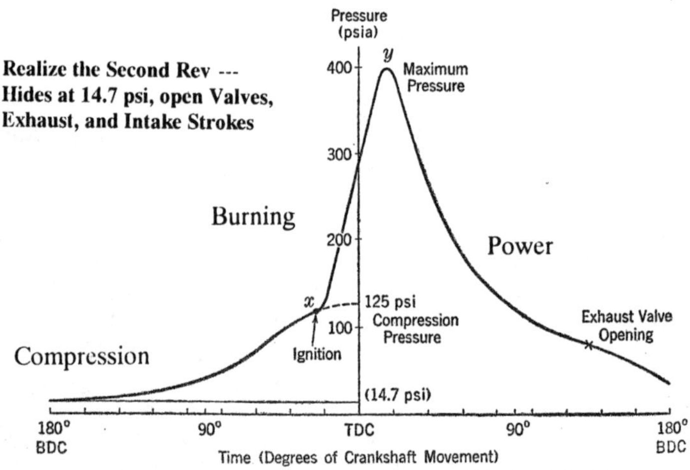

Actual pT Diagram for SI Engine at Wide-Open Throttle.

The PV, **Pressure, Volume**, and the **PT**, **Pressure, Time** Diagrams for a Spark Ignition Engine. A **Four Stroke Engine** takes **Two Revolutions** -- **Intake**, **Compression**, **Power**, **Exhaust**. This page emphasizes the Actual **Compression**, (burning), **Power** Revolution, the **Exhaust**, **Intake** on the Second Revolution, nominally at the 14.7 psi Atmospheric Pressure line. The next Throttled drawings makes it more clear.

BDC is Bottom Dead Center TDC is Top Dead Center -- Crankshaft Position

CHAPTER 10

THE ENGINE, Leaning

Where do we want to run our Engine? Wide Open, Unthrottled, Max RPM, less RPM, where vs. Friction, Pumping Losses, Efficiency, Sea Level, Altitude, best Fuel Burn, Economy, best matching our Plane?* It will be **Leaned for Max Economy!** Of course, we can't do that at High Power, low, or we'd burn the Exhaust Valves. **You can't lean at max. power!** You may well realize that a typical carburetor will kick in an **extra rich, wet cooling mixture at WOT**, Wide Open Throttle! ***Full story*:** **Rich delays the pressure rise, more margin vs. detonation!**

***In a few pages, a Voyager Engine Graph will help sort it all out!!!**

We **need an extra Powerful engine** to be able to **climb, hoisting the plane's weight**, at a fast enough rate, **just like Elevator Power! We also need to hold up the plane**, the **Induced losses from throwing air down** to get a **Newton's equal and opposite reaction force to hold up the Weight, of course, more power to overcome Profile Air Friction Drag**.

So, the **normal mode** of an engine is to **run well throttled**, at **Lower Power** for moderate cruise Speed, either **Plane or Auto!** We might be running our car at ~25% Power, our plane at ~65%.

Look at the Labeled **4 Stroke, PV**, Pressure Volume **Thermodynamic Graph** of a **Spark Ignition Engine** on the facing page. The **available Work, Energy** is represented by **the area inside the 4 Strokes**. The **Efficiency is that enclosed area, vs. all the area below the power stroke to absolute zero temperature**, and Pressure, 0°R, Rankine, **- 459.67 F^{o}!!! Throttled that area is greatly reduced**, **worse**, the **lower throttled loop is subtracted from the reduced work area, next!**

The Throttled Thermodynamic Cycle

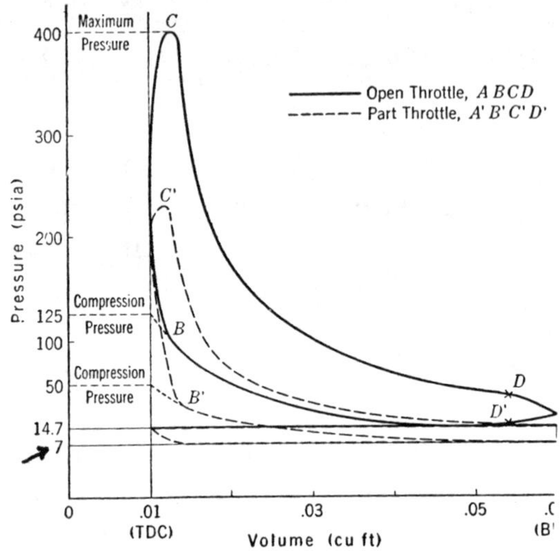

Effect of Throttling on pV Diagram of SI Engine.

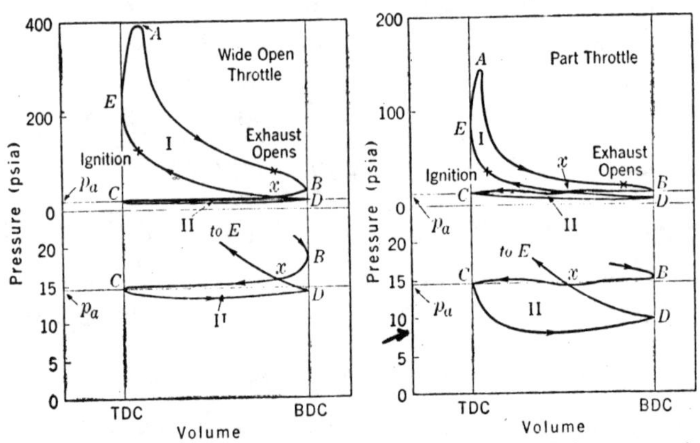

pV Diagram of SI Engine (Induction and exhaust to magnified pressure scale).

The quite instructive **Throttled PV Diagram, shows why we don't want to throttle an engine, rather, simply cruise it at Altitude, where less power is available, matching the plane's real cruise need, where we want best efficiency.** The **Throttled, engine is Required to Pump In it's Working Air --- against the Vacuum created by a throttling plate --- thus work, energy demanded - subtracted from the working Area of the PV Diagram!! It's much better to simply cruise unthrottled at Altitude!**

The Thinking Man's Way to Fly -- Low Drag IAS, Max TAS Altitude Cruise -- Engine, as Good as Possible!

We can do what the Automobile can't **We can climb up, fly at Low IAS, Low Drag**, but needing **ever higher required Power as our True Airspeed, TAS Increases.** As the air **Density Thins,** our **Available Power Falls,** so maybe **by 8 to 14,000 feet we don't have to Throttle, we can eliminate** that little **pumping loop, no longer having to subtract it from the now weaker work available area inside the PV Diagram**!!!

That is where the Logic of Flight Maximizes!! **Your Engine loses Power as you Climb, so at some Altitude your unthrottled Engine Power exactly matches required power. Altitude decreases the Thermodynamic Efficiency**, arguing against too big an engine. **But**, we'll learn shortly, **the Altitude TAS/IAS Speed increase -- beats the Fuel Ecconomy loss!**

In a nutshell, that's how we can **optimally, efficiently match your Engine to your Plane, for Max Efficiency Cruise!!! Pretty Simple, Nifty, no Brain Strain there, just Fly High!!!**

When I fly the Luscombe at 85 IAS at all Gross Weights, at 10'500' Density Altitude, Lean of Peak, I burn only 3 3/4 GPH at 100 MPH, 26.66 MPG!!! Van Flying his original **160 HP RV 6 at 140 IAS, at 12,500'DA, 170 MPH, burned only 6.1 GPH, 27.87 MPG,** even better, Faster!!! **That's 6 hours Range,** at **170 MPH,** a **1000 Mile LA to Seattle*,** no wind, no reserve **Range, so on any tail wind day, we could do that*!**

I've Flown LA to McCall ID, 800 Miles*

Realize what the Game is Here: You'll learn next how terribly inefficient your Thermodynamic Heat Engine is. Worse, it must be considerably oversize to give you Climb Performance Heavy, is normally run throttled, significant extra loss. You'll soon see, next, how really bad it can be, **using 5, Five times, 500% extra fuel. SO, THE WAY WE GET THE BEST USABLE ENGINE EFFICIENCY IN OUR PLANES, IS TO CRUISE THEM INTELLIGENTLY, WIDE OPEN, LEANED, AT ALTITUDE, UNTHROTTLED!!!**

A Little Insight on Thermodynamics: Real Core Insights!!!

We don't want to get all complex here, and turn this into a graduate course in Thermodynamics, *rather* there are some **quite simple, quite Important, Big Picture things to realize here!**

Thermodynamic Efficiency is Terrible, for a Simple Reason! Quite simply, in a heat engine, to get the max., full energy from burning fuel, **we'd like to compress the mixture to very high Pressure, burn it as hot as possible, expand it to Absolute Zero Temperature, -459.67°F, get all the energy out!! BUT**, in the real world, we **can only have a limited compression, or the fuel detonates**, and we **dump the gasses while they are still red hot***, no where near room temperature, Absolute Zero, not even a dream, **can't have a longer expansion ratio mechanically than the compression ratio**, have to get the heat out, to not overheat the engine. Steam Engines can have a secondary expansion, energy recovery, but with heavy machinery

*** They're really Yellow, Blue, White Hot, DC 6 or 7 Exhaust!**

A moderate Compression Ratio **Otto Cycle** gasoline engine may only have **a ballpark 32% Thermal Efficiency**, maybe 25% heavily throttled, worse when mechanical friction is added. With a 67% to 85% overall propulsive efficiency, η_p, worse, if cooling drag is charged to the engine, not the Airframe, like Raspet's eye opening 58% η_p 1950's benchmark glide testing of a Bellanca Cruisair, and you easily get to the **18% overall ballpark** we speak of in the prop book, which **uses more than FIVE, 5 times the theoretical fuel need!!!**

For those who are interested, want to look at the Pressure- Volume Thermodynamic Graphs a little deeper, you can see that **we light the spark a little before Top Dead Center**, TDC, going over the top of the crank throw, where it doesn't really hurt, **let the burn pressure lag, build, max. ~16° after TDC, open the Exhaust Valve before BDC**, get the hot gasses out, for less cooling need. We even open the intake valve, before the exhaust closes, optimum **valve overlap!** There could be additional Chapters on Thermodynamics, Engine Design, Lightweight Engine design, Vibration. Rather, we want you to **grasp how to optimize your engine** - ***in your plane***

Realize, Understand all that's really happening here. A four stroke engine, Intake, **Compression, Power**, Exhaust, is only really **working, firing, making power, every other revolution**. The second revolution, it is dumping the exhaust, intake valve closed, exhaust valve open, then opening the intake valve, closing the exhaust, starting the next intake cycle, pulling air in.

You'd like to get a good full, strong, Power Stroke, but most of the time throttled, you have a weaker, throttled power stroke, as the picture shows, but worse, that little throttled, pumping intake stroke, must pull the air in against a throttled vacuum, requiring work, energy, that must be subtracted from the weakened power stroke. **WOT, at altitude, engine power now exactly matching the plane demand, helps a poor situation!**

Now, of course, we'd like to get every bit of Power out of the Power Stroke, **expanding the gas out to zero temperature, and pressure**, but instead **we have to dump it as very hot gas**, still well above atmospheric pressure, sonic flow barking out the exhaust, **a major part of the heat energy lost, wasted**, the reason we have terrible thermal efficiency, maybe like the 32% efficiency stated. Then we lose the engine cooling drag loss, usually blamed on the plane, maybe 10% of Plane drag, and the overall propulsion efficiency loss, often a lot worse than the propeller efficiency loss, especially on old draggy slow planes, that require an extra big Delta V, ΔV wind thrown back, increasing the body and intersection drag, maybe some flow separation, a **pulsing, higher ΔV prop flow - overall terrible!!!**

In the propeller book you can learn that the fast planes have a big $\dot{M}$, Mass Flow Rate through the prop, only need a small ΔV to make the necessary thrust, but the slow planes, low $\dot{M}$, Mass flow rate, need a big ΔV*, Big vs. a slow plane speed which creates a bigger prop efficiency loss. **A clean, fast, low drag plane, efficient prop, an efficient engine are all important!!!**

***ΔV, Delta V is is simply the Speed Up of Air, Thrown by the Prop, to make Thrust.**

Marvelous, Rare Insight --- BSFC vs. H.P., RPM, Altitude

The Great Continental, Water Cooled Voyager Engine, WOT, Wide Open!
Extra Efficient, But, you can see How Your Engine Works

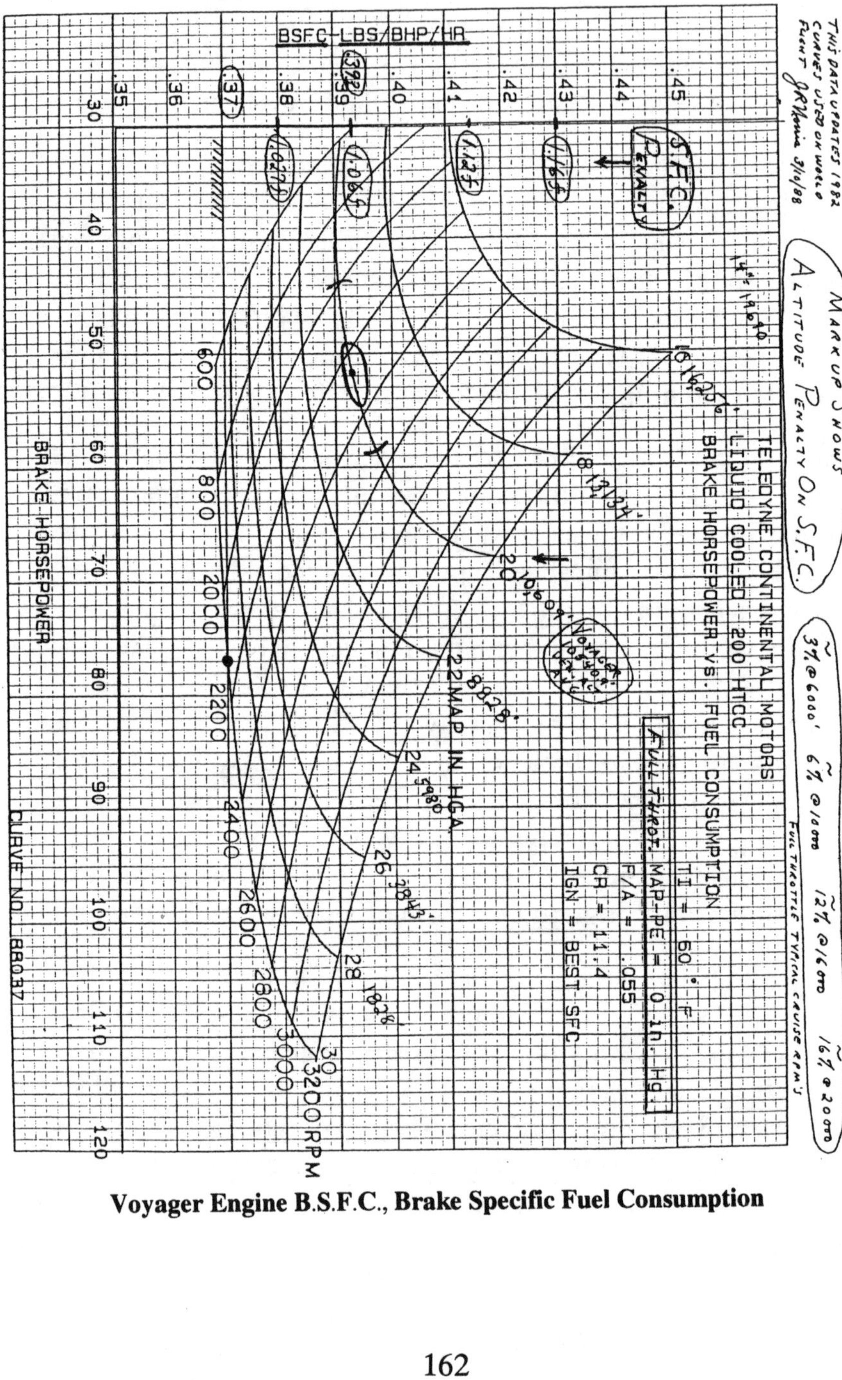

Voyager Engine B.S.F.C., Brake Specific Fuel Consumption

Altitude Loss of Engine Fuel Efficiency, BSFC

You're about to learn how much Higher Altitude Flight, Hurts Engine Brake Specific Fuel Consumption, BSFC --- --- But then find out that the TAS/IAS Speed Increase at Altitude gives you a net gain in MPG, Big Important Insight!

BSFC, Brake Specific Fuel Consumption, is in lbs / H.P. Hr.

On the Facing Page, is the best Engine Insight ever, the actual **BSFC Graph** vs. **Altitude**, **H.P.**, **RPM**, for the **Wide Open**, very Special **Water Cooled, Rear Cruise Engine that made the Voyager World Flight possible. With a high 11.5 : 1 Compression Ratio, a Special Combustion Chamber to tame detonation, the BSFC of the 110 H.P. Continental** is much better than you'll get on your engine*, but **Great Eagle Insight!**

***Though this Engine is Superior, The RPM, HP, Alt. Insights all Work!**

Now remember, this Graph is all WOT, Wide Open Throttle, no Throttling Losses!!! Notice this Engine can have .38 Pounds of Fuel per H.P. Hour, where .4 is great! Notice that at 10,500', the Voyager Average Altitude, the BSFC was hurt ~6%, the 1.06 increase marked on the Graph! On the next page the Square Root of the Air Density Ratio that sets the TAS/IAS speedup is ~1.16 - Viola, 10% Gain!!!

Of great interest to me, 10.5 Density Altitude, is where I regularly fly my Classic Luscombe, a great old Classic, used as an accurate test bed to nail all the Insights you see herein!

There's a lot more Great Insight on this Graph. Notice how the very High RPM, high H.P. hurts BSFC, especially as you go higher. Notice how WOT Manifold Pressure is a telltale of Altitude! LOW RPM, HP, ALTITUDE are BEST!

The Atmospheric Chart on the Next Page, shows that at 16,000', the TAS/IAS factor is 1.2815, still well ahead of the 1.16 BSFC increase that handles 16,000' and high RPM, HP. Study this Graph, and Altitude Chart, and be Smart for life!

Altitude Atmospheric Chart

ICAO STANDARD ATMOSPHERE (NACA TN 3182)

Geopotential ft	t °F	T °R	P lb/sq ft	P in Hg	$\rho \times 10^3$ slugs/cu ft	$\frac{T}{T_0} \times 10$	$\frac{P}{P_0} \times 10$	$\sigma \times 10$ $\frac{\rho}{\rho_0}$	$\sigma^{-1/2}$	$\rho g_s \times 10^2$ lb/cu ft	$\mu \times 10^7$ lb sec/sq ft	c_s ft/sec	$\frac{c_s}{c_{s0}} \times 10$	Geopotential ft
−2 000	66.132	525.820	2273.71	32.1481	2.5192	10.1375	10.7442	10.598	.97136	8.1052	3.7860	1124.54	10.0685	−2 000
−1 000	62.566	522.254	2193.82	31.0185	2.4473	10.0688	10.3667	10.296	.98552	7.8738	3.7656	1120.73	10.0343	−1 000
0	59.000	518.688	2116.22	29.9213	2.3769	10.0000	10.0000	10.000	1.0000	7.6475	3.7452	1116.89	10.0000	0
1 000	55.434	515.122	2040.85	28.8557	2.3081	9.93125	9.64387	9.7106	1.0148	7.4263	3.7248	1113.05	9.96556	1 000
2 000	51.868	511.556	1967.68	27.8210	2.2409	9.86249	9.29809	9.4277	1.0299	7.2099	3.7042	1109.19	9.93101	2 000
3 000	48.302	507.990	1896.64	26.8167	2.1752	9.79374	8.96241	9.1512	1.0454	6.9984	3.6836	1105.31	9.89633	3 000
4 000	44.735	504.423	1827.69	25.8418	2.1109	9.72499	8.63661	8.8808	1.0611	6.7917	3.6629	1101.43	9.86153	4 000
5 000	41.169	500.857	1760.79	24.8959	2.0481	9.65623	8.32047	8.6167	1.0773	6.5896	3.6422	1097.53	9.82661	5 000
6 000	37.603	497.291	1695.89	23.9782	1.9868	9.58748	8.01377	8.3586	1.0938	6.3923	3.6214	1093.61	9.79157	6 000
7 000	34.037	493.725	1632.93	23.0881	1.9268	9.51873	7.71628	8.1064	1.1107	6.1994	3.6005	1089.68	9.75640	7 000
8 000	30.471	490.159	1571.88	22.2249	1.8683	9.44997	7.42780	7.8601	1.1279	6.0111	3.5795	1085.74	9.72110	8 000
9 000	26.905	486.593	1512.70	21.3881	1.8111	9.38122	7.14812	7.6196	1.1456	5.8271	3.5585	1081.78	9.68567	9 000
10 000	23.338	483.026	1455.33	20.5769	1.7553	9.31247	6.87703	7.3848	1.1637	5.6475	3.5374	1077.81	9.65011	10 000
11 000	19.772	479.460	1399.73	19.7909	1.7008	9.24371	6.61432	7.1555	1.1822	5.4722	3.5162	1073.83	9.61442	11 000
12 000	16.206	475.894	1345.87	19.0293	1.6476	9.17496	6.35980	6.9317	1.2011	5.3010	3.4950	1069.83	9.57860	12 000
13 000	12.640	472.328	1293.70	18.2917	1.5957	9.10620	6.11327	6.7133	1.2205	5.1340	3.4737	1065.81	9.54264	13 000
14 000	9.074	468.862	1243.18	17.5773	1.5451	9.03745	5.87453	6.5002	1.2403	4.9711	3.4523	1061.78	9.50655	14 000
15 000	5.508	465.196	1194.27	16.8858	1.4956	8.96870	5.64340	6.2923	1.2606	4.8121	3.4309	1057.73	9.47032	15 000
16 000	1.941	461.629	1146.92	16.2164	1.4474	8.89994	5.41969	6.0896	1.2815	4.6570	3.4093	1053.67	9.43395	16 000
17 000	−1.625	458.063	1101.11	15.5687	1.4005	8.83119	5.20322	5.8919	1.3028	4.5058	3.3877	1049.59	9.39744	17 000
18 000	−5.191	454.497	1056.80	14.9421	1.3546	8.76244	4.99380	5.6991	1.3246	4.3584	3.3661	1045.50	9.36079	18 000
19 000	−8.757	450.931	1013.93	14.3360	1.3100	8.69368	4.79126	5.5112	1.3470	4.2147	3.3443	1041.39	9.32399	19 000
20 000	−12.323	447.365	972.490	13.7501	1.2664	8.62493	4.59542	5.3281	1.3700	4.0747	3.3225	1037.26	9.28705	20 000
21 000	−15.889	443.799	932.429	13.1836	1.2240	8.55618	4.40611	5.1496	1.3935	3.9382	3.3006	1033.12	9.24996	21 000
22 000	−19.456	440.232	893.715	12.6363	1.1827	8.48742	4.22317	4.9758	1.4176	3.8053	3.2786	1028.96	9.21272	22 000
23 000	−23.022	436.666	856.313	12.1074	1.1425	8.41867	4.04643	4.8065	1.4424	3.6758	3.2565	1024.79	9.17533	23 000
24 000	−26.588	433.100	820.188	11.5967	1.1033	8.34992	3.87573	4.6416	1.4678	3.5497	3.2344	1020.59	9.13779	24 000
25 000	−30.154	429.534	785.308	11.1035	1.0651	8.28116	3.71091	4.4811	1.4938	3.4270	3.2122	1016.38	9.10009	25 000
26 000	−33.720	425.968	751.638	10.6274	1.0280	8.21241	3.55180	4.3249	1.5206	3.3075	3.1899	1012.15	9.06223	26 000
27 000	37.286	422.402	719.148	10.1681	.99187	8.14366	3.39827	4.1729	1.5480	3.1912	3.1675	1007.91	9.02422	27 000
28 000	−40.852	418.836	687.803	9.72488	.95672	8.07490	3.25016	4.0250	1.5762	3.0781	3.1450	1003.64	8.98605	28 000
29 000	−44.419	415.269	657.575	9.29748	.92252	8.00615	3.10731	3.8812	1.6052	2.9681	3.1225	999.362	8.94771	29 000
30 000	−47.985	411.703	628.431	8.88541	.88927	7.93740	2.96960	3.7413	1.6349	2.8612	3.0999	995.062	8.90921	30 000
31 000	−51.551	408.137	600.342	8.48826	.85695	7.86864	2.83687	3.6053	1.6654	2.7572	3.0772	990.743	8.87054	31 000
32 000	−55.117	404.571	573.279	8.10561	.82553	7.79989	2.70898	3.4731	1.6968	2.6561	3.0544	986.405	8.83170	32 000
33 000	−58.683	401.005	547.212	7.73705	.79500	7.73114	2.58580	3.3447	1.7291	2.5578	3.0315	982.048	8.79269	33 000
34 000	−62.249	397.439	522.113	7.38218	.76534	7.66238	2.46720	3.2199	1.7623	2.4624	3.0085	977.672	8.75350	34 000
35 000	−65.816	393.872	497.955	7.04060	.73654	7.59363	2.35304	3.0987	1.7964	2.3697	2.9855	973.276	8.71414	35 000
36 000	−69.382	390.306	474.710	6.71194	.70857	7.52487	2.24320	2.9810	1.8315	2.2798	2.9624	968.859	8.67460	36 000
36 089	−69.700	389.988	472.679	6.68322	.70612	7.51874	2.23360	2.9707	1.8347	2.2719	2.9603	968.465	8.67107	36 089
37 000	−69.700	389.988	452.434	6.39698	.67587	7.51874	2.13794	2.8435	1.8753	2.1746	2.9603	968.465	8.67107	37 000
38 000	−69.700	389.988	431.203	6.09679	.64416	7.51874	2.03761	2.7100	1.9209	2.0725	2.9603	968.465	8.67107	38 000
39 000	−69.700	389.988	410.968	5.8106	.61393	7.51874	1.94199	2.5829	1.9677	1.9753	2.9603	948.465	8.67107	39 000
40 000	−69.700	389.988	391.682	5.53801	.58512	7.51874	1.85086	2.4617	2.0058	1.8826	2.9603	968.465	8.67107	40 000
41 000	−69.700	389.988	373.302	5.27813	.55766	7.51874	1.76401	2.3461	2.0645	1.7942	2.9603	968.465	8.67107	41 000
42 000	−69.700	389.988	355.784	5.03045	.53149	7.51874	1.68123	2.2361	2.1148	1.7100	2.9603	968.465	8.67107	42 000
43 000	−69.700	389.988	339.088	4.79439	.50655	7.51874	1.60233	2.1311	2.1662	1.6298	2.9603	968.465	8.67107	43 000
44 000	−69.700	389.988	323.176	4.56940	.48278	7.51874	1.52714	2.0311	2.2189	1.5533	2.9603	968.465	8.67107	44 000
45 000	−69.700	389.988	308.011	4.35497	.46013	7.51874	1.45548	1.9358	2.2728	1.4804	2.9603	968.465	8.67107	45 000
46 000	−69.700	389.988	293.557	4.15061	.43853	7.51874	1.38718	1.8450	2.3281	1.4109	2.9603	968.465	8.67107	46 000
47 000	−69.700	389.988	279.781	3.95584	.41795	7.51874	1.32208	1.7584	2.3848	1.3447	2.9603	968.465	8.67107	47 000
48 000	−69.700	389.988	266.652	3.77020	.39834	7.51874	1.26004	1.6759	2.4428	1.2816	2.9603	968.465	8.67107	48 000

↑

The Square Root of the Altitude Density Ratio Shows the IAS-TAS Ratio

A Really Good, Incisive Overview

No big surprise, the **Aerodynamic Pros seldom understand the Engine, the Engine Pros even less likely to understand the Aerodynamic Implications**, especially the subtleties, how you can put it all together, in the **simple but sophisticated way we do it here in this book: Intelligent High Altitude Cruise, The Logic of Flight, The Thinking Man's Way to Fly.**

We're purposely avoiding going too deep on the engine subject, rather trying to give you the **key essentials** to **incisively optimize how you fly your plane, matching the engine and plane to each other**, and the **Prop Book** shows you how you can **understand Props correctly, precisely, essentially exact, Optimum Designs, preferably targeting Altitude Cruise!** ✓

Since, to a degree we're **just Skimming Engine Logic here,** purposely avoiding drowning you with yet another technical subject, **it may be best if you read this concise overview a few times, be sure you really grasp the charts and graphs, the really quite incisive grasp that is here, easily enough --- gain a really incisive pro grasp of the whole flight Logic!**

If we went a lot deeper, took a lot more time you might just **remember, and use what is here**, so maybe it's desirable to **keep this simple, but to the point**, of **how the Engine is used in High Altitude Cruise**, **Wide Open Throttle**, WOT, **leaned at low power**, where you can **avoid burning the Exhaust Valves, often exotic materials**, that **often requires cautious, intelligent insight, covered next! Leaning covered next!**

(Would you believe, on the GE 90, 100,000 Pound Thrust Jet Engine, used on the Boeing 777, a 40 : 1 Compression Ratio, far above our *old* engines, day in and day out, they are running the Gas Temperature above the Melting Point of the Turbine Exotic Alloy Material, Bleed Air Cooled Blades, WOW!!!!! I once asked a very smart engine designer of our engines what exotic Valve Alloys were used, and was shocked by his answer: TRW won't tell us what they're using!) ***Considering that*** **-- Next, Leaning!**

Leaning - Fuel Mixture Management

One might think that a chemically exact, ~**14.7 (stoichiometric) mixture*** of **air and gasoline** would be central in teaching fuel mixture management, but, looking back at my graduate course engine text books, that word is not even mentioned! Rather, they immediately make clear that the **chemically correct mixture does <u>not</u> produce either the max. power, or max. economy**. ***See the Graph below.*** Rather tests of <u>a one cylinder engine</u> teaches the **12.5/1 max. Power** mix, and the **16/1 Max. Economy Air / Fuel Mixtures, <u>by weight</u>. Simply, with less than full vaporization, all the fuel, especially the heavy ends of the distillate, do not evaporate, and do not find oxygen.** Also, as previously mentioned, at **WOT, an extra rich cooling mixture is purposely used to protect the exhaust valves. (The truth is more complex: <u>A richer mix burns slower, delays peak Pressure, thus extra margin vs. detonation, especially at slower RPM, fixed Pitch</u>.)** Engines are very much a subject of **testing, learning** to **find the real truths**, in a complex subject, **to find what is really true !**

*** Get it: <u>Mixtures by weight, Air the Heavy part</u>. O_2 weighs 23% of Air!**

First, we'll **learn what is fundamentally happening** in a **<u>perfect one cylinder test engine</u>**. But even before we start, it is necessary to say that **air-fuel distribution will almost certainly vary a lot**, and **only rarely will *<u>all</u>* cylinders act this perfect! But <u>learning in orderly steps here</u>, is <u>the way to the truths</u>!**

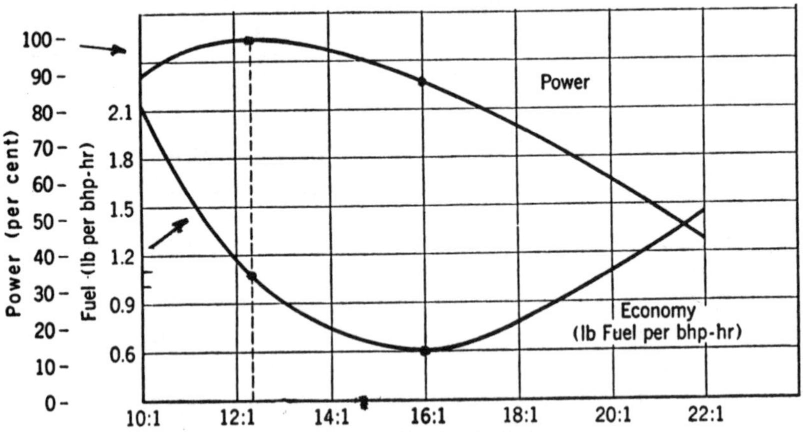

Fig. 123. Variations of Power and Economy of an SI Engine with Carburetor Adjustment (Constant speed and wide-open throttle).

That is a great old correct teaching graph, for 1940's auto engines, from my old text, but notice that it's minimum BSFC is .6 # Fuel / H.P. Hr., not the .38 we saw on the Voyager engine graph. But **it correctly shows max. Power is 12.4, to 12.5 Air Fuel Ratio, max. Economy is 16/1 A.F., much leaner!**

Now below we have a more modern Continental Graph for a **Fuel Injected TCM IO 550 Operation Manual** that brings in **EGT, Exhaust Gas Temperature**, crucial for Exhaust Valves, and **CHT, Cylinder Head Temperature, for Temp. critical Aluminum Cylinders. That gets us into the nitty gritty of the temperatures, the leads, and lags that we need to learn very well to protect our engines, and be proficient here!**

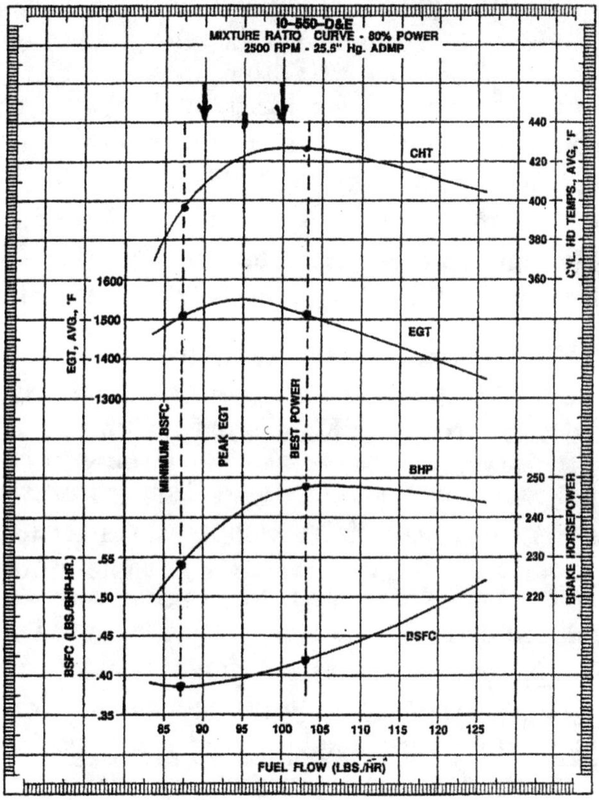

Figure 13-14. Mixture Ratio Curve 80% Power
IO-550-D&E

Leaning - Fuel Mixture Management

First, let me make very clear, that now we'll be dealing with a **modern, high performance, fuel injected engine, in the very best cases, even with Matched Injectors**, so we'll assume for the moment, that all cylinders are acting the same, (a very big assumption), also, **full EGT, and CHT sensors, and readouts**, so we'll **have a proper check of whether, or not, there is something close to mixture balanced cylinders, or not!** **Later**, we'll deal with our old 1940's, 50's carburetor engines.

Notice that at **Max Economy we're dealing with 227 H.P.**, say **~16:1 Max Economy Mixture**, ***not*** the **249 H.P. Max Power case, ~91%, ~9% less Power!** If we were low enough, we could open the throttle, get back up to 100%, but say we're at **high Altitude**, with **plane and engine power matched**, just as the book advises. **Our Engine reacts much like the curves!**

Should we try 50^0 F Lean of Peak?

Look, our **CHT is down from 428 F, to 395F, a lot better** for the **marginal Aluminum Cylinder Alloys, that crack when mistreated**, and the **EGT is at 1500 F°, not a 1550 F°+ max.**, the **same as 50 F° rich of Peak, often recommended**, **LOP is potentially significantly better for the Cylinder Material, the same temp. for the valves, and their critical material!!!**

Here are the key things to see ---

Max Cylinder Temp is near Max Power! - Don't Fly There!
Max EGT is between Max Econ. and Max Power! a No No!
50° Rich of Peak, often recommended, Delicate Cylinders Worse!!!
Max Econ. cools Key Cylinder Head Temperatures to OK!
Max Economy has EGT Valve Temperatures back Lower!
Max Economy: 12.5/16, .78125, ~22% less Fuel Burn?, No!
Here we see .383 / .423, BSFC, Max Power, ~10% saving, Real World!!!
Now, that's the core of the story, BSFC about .385 / .41, = .939, about 6.6% better BSFC 50° lean of peak, vs. 50° rich of Peak EGT! An oxidizing atmosphere lean, seems OK. The key is balanced mixture, good instruments, *or not!!!* We get **less Power at Max Economy**, but that **just lets you fly optimum lower, on Shorter Flights, more choice, flexibility!**

So WHY would an Engine Manufacturer recommend 50° F Rich of Peak when it is clearly worst, bad for Cylinder Head Temperatures, no better for critical exhaust valves? Cracked Cylinders is a real, and prevalent problem.

Well, first, they know they are shipping Engines with Bad to Terrible Fuel Distribution!!! Simply, they don't want a bunch of Amateurs, with "bad to none" Instrumentation and Knowledge, mucking around, burning valves, because some of your valves, *unchecked* could easily be at max. EGT!!!!

The big Airline Round Radials, very symmetrical, fine distribution, finally, in the 50's, had excellent distribution, real scientific, careful monitoring, professional flying -- and pro pilots could often see, the yellow, blue, or white exhaust, have an absolutely valid visual check, learned at night!!!

Our flat 4's and 6's have bad distribution, unless, on rare later top engines, the manufacturer got serious about mix! I'm told the carbureated Lycomings with the Plenum feed Chamber in the mixture ducts are worst, many expensive injected engines worse than you'd hope, the reason the most expert pilots, with the critical advanced engines, often go for swapping injectors, professional balancing -- pro work!

Obviously, on any advanced, serious engine, with a serious, knowledgeable Pilot, you want first class EGT, and CHT sensing, and readout, easy, quite available today!! Wishing isn't good enough, it's not likely to work, likely harmful!

If you do the right thing, get yourself good instrumentation, learn this, "easy enough insight", you can, with just the knowledge herein, know where you are, run your engine intelligently, safely, based on good facts about EGT, and CHT, both important to safe, optimum Engine Operation!!! I'm tough here, no excuses, because if I weren't it would be irresponsible! Higher Altitude insight, next, is important!

Flying at higher altitude, we have two countervailing factors! At Much Lower Power, we should, logically, have more leeway, margin, you might first think. *But*, we still try to adjust to the same mixtures, the same Temperatures, ***careful!***

Let me digress a little, a life of Flight. At Christmas 1950, the Korean War on, being in the Air National Guard, to play with B 26 Invaders, with the big kids, who fought WW II, knowing I'd be called into the Air Force at Wright Field, in Dayton, as an officer, looking for a Cessna 140 to fly home to see Milly on weekends, I found a near new, **2 1/2 year old $3400 Deluxe Luscombe 8E for $1225. A student**, with **no income, I bought it on faith!**

Life turned into a Whirl Wind, leading the Aircraft Laboratory Group on Mock up and Engineering Inspections, on all the new planes, with the top Industry, and Air Force People! I had a fantastic life, founded and ran companies, got Primary Flight Controls on Boeing Jets, my Spacecraft Controls in the Smithsonian, consulted, had a life most could only dream about. ***I sure didn't have time to have a plane* - *but I never let go***. Simple, reliable, rarely needed fixing, it was **my great escape when work got wild! I kept it as a Classic**, an Aerodynamic Test Vehicle, the Voyager, this book, Zero Thrust Glide Testing, a known 67% overall Propulsion Efficiency, η_p. **Lots more**, a 12:1 glide ratio, I do Zero Thrust or Idle powerless approaches, from 10,500', 2 miles, from 20 miles out, at ~70 MPH have 4 miles extra range, ***fly for 20 minutes with the engine essentially off***, **the Sierra wave, 18,000'**. The huge joke, last Christmas, 2006, 56 years, it cost **$21.88/ year!**

Money not a problem, safety ever more valued, only keeping it for fun, escape, **I hardly leaned, very conservative**, no need to push the envelope, flying just for fun, and to keep my hand in, **too busy to buy a Bonanza, and be more serious, my pro work great fun**, and **I always Overhauled the Engine Early. Guess what, it always had way too much Carbon Build up!!**

Flight was my hobby, since I was a kid, won the Model Senior National Championship Two of the 3 years I was there to compete. I decided I wanted to have a hand in the Engine Overhaul, know everything in there, Lean the Engine Properly, get intellectually serious, nail everything in these two books. A lifetime of pro insight, I should do no less! You can make that lifetime of Insight, fun, Work for You!

Over the years, I read everything good I could find, that seemed technically accurate, sensible, passed the horse sense test, a good rigorous technical look, vs. the **Graduate level Engine Design Course I took years ago! Truth, Facts, Physics don't change!**

Recently, boning up to check myself on all this, I found an excellent, concise article by **Peter Garrison**, a well respected, long experienced, Pilot, Designer, Builder, Melmoth I, and II, and respected, well proven author for AOPA Pilot, among others

Like many, Peter had bought all the instrumentation, studied, did the full Science Approach, over a long, long time. But, he **finally ended up more simply, using the old approach, lean to a bit of roughness**, that would be on the lean side of peak, **then richen a bit** to ***smooth enough*** operation, no misfires, no real roughness, (my words not his), and has no problems. I should immediately say that **Peter has GAMI Injectors, knows that his cylinders have well matched mixtures**, and the **full EGT, CHT instrumentation**. As he leans by sound, feel, he can glance over, **see exactly what his real EGT, CHT facts are, like a pro should. He has a 6 Cylinder Continental TSIO 360F, 200 H.P. Engine.** Just like the book, **he usually flys high, efficient, at 65% Power, or below**.

In the same vein, I fly high, typically at **10,500', wide open, 2300 RPM**, with **a higher pitch cruise prop**, **aggressively leaned**, but never rough, specifically at 49 HP, on a Sea Level 85 H.P. Engine, **57.65% Power**. With a pedestrian engine, nothing wonderful for valves, I've been doing that 1000 hours over 23 years, on my R&R Classic. When I look in with a borescope, there is ***no carbon buildup anymore***, everything looks just fine It's been to every corner of the country, many times over 56 years, transcontinental several times, first from Cleveland, from LA since 62. **It's right to be knowledgeable, careful, don't do dumb things, gamblers don't last, flying!**

It's proper to lean, not carbon up your engine, wrong not to!

Any aggressive Engine, I'd demand full instrumentation, exact insight, monitoring - *on the most aggressive engines, balanced injection.* A wise Pilot, cautious, learns, has great fun, a long full life!

Since we're trying to run at the same temperatures at Altitude, I'd be cautious to argue that flying high is easier on the specifics of Exhaust Valve Protection, but at Higher Altitude, at only 57% to 65% Power, one has the help of a much lower heat load, the chance for less challenged heat conduction through the metals. It's Hell in there, caution is good! Bottom line, I never lean really aggressively until I'm above 10,000', where I know I can't pull even 65% Power!

I fly High, WOT, low HP, Max IAS vs. Drag, then even faster TAS, at no additional Drag -- and Lean, just like the book says, at low HP, the Engine perfectly matching the plane's demand, without Throttling losses!

People misunderstand that I'm somehow flying slower, when, in fact, I'm always flying as fast as the plane will go, Wide Open at Altitude, where everything matches perfectly, loping along efficiently, cool, air conditioned where it may be 100F°+ down on the Western desert.

The day we all flew LA to McCall ID, for a fly in, I left two hours after everyone else, got there an hour before them, burned about 3/4 their gas, went non stop. The gang made two stops, sweat in 110 degree heat at Bishop, had big trouble climbing over the Mountains North of Bishop, hot. 2nd stop, one guy landed in a dust devil, ended up out in the weeds, shaken if OK.

A 747 takes off at ~170 MPH Heavy, only goes ~270 IAS, but at 40,000 feet thats 540 MPH. That's a good act to copy! The Game in Life is to just learn, Get Smart as You GO!!!

Picking up where he left off, it was clear that the **overall efficiency** was far worse than any propeller (in)efficiency could be, and that **there is a significant interference (in)efficiency**, η_i.

We found a **.67 overall propulsion efficiency,** η_P, a **.8933** η_i **interference efficiency** = **.67 / .75 prop efficiency, eta,** η, at **85 IAS** at sea level, ***or*** 85 IAS, **100 TAS** altitude cruise at **10,500'**. That **.8933**, you'll see, ***explains Gus Raspet's big mystery loss***!

It takes 49 H.P. at altitude, just what's available, lean, ~2280 RPM.

In the big picture, the **interference efficiency** logically seems to show up **on the high drag, slow classic light planes**, where the larger ΔV added to the Stream tube, is a much bigger ratio to the slower plane speed. Thus, it's sometimes quite small on **modern, fast homebuilts.** **Fast is twice better** with a more efficient prop ***and also*** better interference efficiency too, the **slow, high drag planes losing twice** --- all much **better when flown at very low power, even worse when flown at high power, Vmax.** ***It's quite Variable but logical, Understandable!***

The best thing we can do, as usual, is to **show you all the basic numbers, right up front**, so you can immediately ***see where it's all going***, so it makes easier sense as you learn all the details. We're ***purposely*** going to use too many decimal places so we can flip around from any number to any other and hit it ~exactly.

Books are printed in 64 page signatures. This book, already longer than intended, this longer Appendix, just would not fit.

If you are interested, and would like a copy, simply send a STAMPED, SELF ADDRESSED ENVELOPE, WITH $ 1 TO COVER COPY COSTS, and we'll forward a copy. Include your phone number, so we can solve any problem!

Jack Norris 11613 Seminole Circle, Northridge CA 91326
The pressure interaction with an Embedded Body, with slowed air, which requires more thrust, in slowed air, at no extra cost, no net loss, is most interesting!!! 818 360 1105